ADP-RIBOSYLATION IN ANIMAL TISSUES

Structure, Function, and Biology of
Mono (ADP-ribosyl) Transferases
and Related Enzymes

ADVANCES IN EXPERIMENTAL MEDICINE AND BIOLOGY

Recent Volumes in this Series

ADP-RIBOSYLATION IN ANIMAL TISSUES

Structure, Function, and Biology of
Mono (ADP-ribosyl) Transferases
and Related Enzymes

Edited by

Friedrich Haag
Friedrich Koch-Nolte
University Hospital Hamburg-Eppendorf
Hamburg, Germany

SPRINGER SCIENCE+BUSINESS MEDIA, LLC

Library of Congress Cataloging-in-Publication Data

ADP-ribosylation in animal tissues : structure, function, and biology
  of mono (ADP-ribosyl) transferases and related enzymes / edited by
  Friedrich Haag, Friedrich Koch-Nolte.
      p. cm. -- (Advances in experimental medicine and biology ; v.
  419)
    "Proceedings of an International Workshop on the Biological
  Significance of Mono ADP-Ribosylation in Animal Tissues, held May
  19-23, 1996, in Hamburg, Germany"--T.p. verso.
    Includes bibliographical references and index.
    ISBN 978-1-4613-4652-4   ISBN 978-1-4419-8632-0 (eBook)
    DOI 10.1007/978-1-4419-8632-0
    1. ADP-ribosylation--Congresses.   I. Haag, Friedrich. II. Koch
  -Nolte, Friedrich.   III. International Workshop on the Biological
  Significance of Mono ADP-Ribosylation in Animal Tissues (1996 :
  Hamburg, Germany) IV. Series.
    [DNLM: 1. Adenosine Diphosphate Ribose--metabolism--congresses.
  2. NAD+ ADP-Ribosyltransferase--metabolism--congresses.   W1 AD559
  v.419 1997 / QU 57 A2408 1997]
  QP625.A29A346 1997
  572'.7921--dc21
  DNLM/DLC
  for Library of Congress   97-7429
                                                               CIP

Proceedings of an International workshop on the Biological Significance of mono ADP-ribosylation in Animal Tissues, held May 19 – 23, 1996, in Hamburg, Germany

ISBN 978-1-4613-4652-4

http://www.plenum.com

10 9 8 7 6 5 4 3 2 1

PREFACE

Although ADP-ribosylation has been known as a post-translational modification of proteins for approximately thirty years, the study of endogenous mono-ADP-ribosylation in animal tissues has remained somewhat of an orphan field during this time. Until recently, interest in the field has concentrated on two types of phenomena: (1) poly-ADP-ribosylation of nuclear proteins in eukaryotes as a mechanism possibly involved in DNA excision repair and (2) mono-ADP-ribosylation by bacterial enzymes, either as a toxic mechanism in eukaryotic host cells or as a reversible regulatory mechanism for control of nitrogen fixation. The identification of diphtheria, cholera, and pertussis toxins as mono(ADP-ribosyl)transferases and their subsequent purification and crystallization have shaped our current knowledge of the biology of mono-ADP-ribosylation reactions and the structure–function relationships of the enzymes involved. In contrast, endogenous transferases of animal tissues escaped molecular cloning, and for a long time their biological relevance was merely postulated by analogy to their bacterial cousins.

The molecular characterization of the first mammalian mono(ADP-ribosyl)transferase from rabbit skeletal muscle, as well as the consequent realization that a well-studied surface membrane protein of T cells, RT6, is a mono(ADP-ribosyl)transferase, have changed this situation. With their genes at hand, the question of the significance of these enzymes in animal tissues can now be addressed more directly. The international workshop "Biological Significance of Mono-ADP-Ribosylation in Animal Tissues," held May 19–23, 1996, in Hamburg, was the first of its kind and was felt by many members of the community to be a necessary focal point whose time was now due.

The conference was held in the beautiful setting of the Elsa-Brandström-Haus, overlooking the shores of the Elbe river in Hamburg-Blankenese, and attended by 70 scientists from Europe, North America, and Japan. It is an interesting historical anecdote that the first scientific conference to be held at this site was one of the early international workshops on ADP-ribosylation, organized twenty years ago by Helmuth Hilz, Professor emeritus of the Department of Physiological Chemistry and a special guest at this meeting. Thirty-nine papers were presented by invited speakers, 22 as posters and short oral communications. Almost all of these are united in this volume, essentially arranged in the order they were presented at the conference.

The first section deals with the lessons to be learned from the studies of prokaryotic mono(ADP-ribosyl)transferases. In these enzymes, structure–function relationships have already been well established by crystallography and site-directed mutagenesis. These studies facilitate the prediction of catalytic residues of the eukaryotic transferases, generating hypotheses that can be tested experimentally by site-directed mutagenesis. The regulation of the dinitrogenase reductase system in photobacteria also provides an attractive model for reversible regulation of metabolic functions by ADP-ribosylation in animal tissues.

The second section focuses on the structural and biochemical characterization of the eukaryotic transferases. In the forefront are the reports on the elucidation of the primary structure of transferases from diverse vertebrate tissues and species. Transferase activity has been demonstrated also in fungi, although the enzyme(s) involved has/have not yet been cloned. Some of the cloned vertebrate enzymes have already been expressed in a recombinant form, and mutagenesis experiments have been conducted to test some of the structural hypotheses generated in the first section.

The third and fourth sections address the biological function of mono-ADP-ribosylation in the immune and other systems. Several reports sum up what is known to date about the biological significance of the RT6 T-cell mono(ADP-ribosyl)transferases. RT6 has been well studied over the years as a T-cell differentiation marker, especially in light of the observation that RT6 expression defects coincide with increased disease susceptibility in different animal models for autoimmune diseases. Thus, although its enzymatic activity has only recently been recognized, RT6 is the transferase whose gene structure has been the most extensively studied and the system in which hypotheses regarding possible disease association of defective transferases are most advanced. ADP-ribosylation reactions have been implicated in the modulation of activation of not only T cells, but also other cells of the immune system, including monocytes/macrophages and granulocytes. In skeletal and cardiac muscle, target proteins for ADP-ribosylation are being identified, leading to testable hypotheses as to the biological significance of these reactions. Regulatory effects of ADP-ribosylation have also been demonstrated in central and peripheral nervous tissue. In the latter case, interestingly, an association between a metabolic disorder (diabetes mellitus) and endogenous ADP-ribosylation could be found. Finally, extensive work has gone into the characterization of ADP-ribosylation reactions in the context of regulation of cellular transport processes.

The fifth section attempts to shed some light on a puzzle that still impedes a clear understanding of how ADP-ribosylation works in animal cells. Although NAD+ is a classic intracellular metabolite, all vertebrate mono(ADP-ribosyl)transferases (and also ADP-ribosyl cyclases) cloned to date are predicted to be localized extracellularly, the majority as GPI-anchored cell surface enzymes. Therefore, the cellular compartmentalization, dynamics, and interaction with signal transduction processes of GPI-anchored molecules have been addressed.

The sixth section examines the mechanisms and biological significance of other ADP-ribose transfer reactions, especially the production of cyclic ADPR and related calcium-mobilizing second messengers by ADP-ribosyl cyclases.

A special session of the conference was dedicated to Prof. Heinz-Günter Thiele, one of the pioneers of research on the T-cell mono(ADP-ribosyl)transferase RT6, at the occasion of his retirement. This session was highlighted by an eloquent plea by Prof. Jonathan Howard, another one of the "fathers" of RT6, to consider the implications of the extensive polymorphism within the RT6 system when assessing its function, as this may point toward a role of RT6 within the context of a defense system that needs to adapt to a changing environment.

We wish to express our sincere thanks to the Deutsche Forschungsgemeinschaft (DFG) and the many other groups whose generous financial support made this conference possible, as well as to our students and laboratory staff who worked hard to make the meeting a rewarding experience for all participants.

We believe that the present volume represents a condensation of present knowledge of mono(ADP-ribosyl)transferases and related enzymes and hope that it will fulfill its intended function to stimulate and facilitate further research into the significance of these intriguing enzymes.

Friedrich Haag
Friedrich Koch-Nolte
Hamburg, Germany

ACKNOWLEDGMENTS

Financial support for the conference was provided by the following sponsors:

THE DEUTSCHE FORSCHUNGSGEMEINSCHAFT
THE UNIVERSITY HOSPITAL HAMBURG-EPPENDORF
ELIAS
BIOTEST
BOEHRINGER MANNHEIM
DIANOVA
DYNAL
SANDOZ
SORVALL
WILKE & WITZEL
BECKMANN
BECTON DICKINSON
BYK SANGTEC
PHARMACIA
SIGMA
SCHLEICHER & SCHUELL
THE BINDING SITE

The organizers thank the department secretary, Ms. Eva-Maria Hobbje; their graduate students Rickmer Braren, Katrin Firner, Martina Matthes, Stefan Rothenburg, and Karsten Wursthorn; and their technicians Sandra Bauschus, Roman Girisch, Maren Kühl, and Cornelia Ritter for their support.

The organizers thank their significant others for their understanding.

CONTENTS

Poster Reports

MOLECULAR APPROACHES TO EUCARYOTIC MONO(ADP-RIBOSYL)TRANSFERASES

Plenary Lectures

Poster Reports

MONO(ADP-RIBOSYL)TRANSFERASES IN THE IMMUNE SYSTEM

Plenary Lectures

MONO-ADP-RIBOSYLATION IN OTHER ANIMAL TISSUES

Plenary Lectures

Poster Reports

PHYSIOLOGY OF GPI-ANCHORED MEMBRANE PROTEINS

Plenary Lectures

RELATIONSHIP OF ADP-RIBOSYLTRANSFERASES TO NAD+ GLYCOHYDROLASES AND ADP-RIBOSYL CYCLASES

Plenary Lectures

Poster Reports

SPECIAL LECTURE COMMEMORATING THE RETIREMENT OF PROFESSOR HEINZ-GÜNTER THIELE

APPENDIX

ADP-RIBOSYLATION IN ANIMAL TISSUES

Structure, Function, and Biology of
Mono (ADP-ribosyl) Transferases
and Related Enzymes

MONO(ADP-RIBOSYL)TRANSFERASES AND RELATED ENZYMES IN ANIMAL TISSUES

Emerging Gene Families

Friedrich Koch-Nolte and Friedrich Haag

Department of Immunology
University Hospital
D-20246 Hamburg, Germany

ABSTRACT

Mono-ADP-ribosylation, like phosphorylation, is an enzyme-catalyzed, reversible post-translational modification that modulates protein function. It was originally discovered as the pathogenic principle of diphtheria-, cholera-, and other potent bacterial toxins. By analogy, corresponding enyzmes were postulated to exist in animal tissues, and mounting biochemical evidence indicates that such enzymes, indeed, play important regulatory roles in cellular functions. The molecular cloning of the first mammalian mono(ADP-ribosyl)transferase from rabbit skeletal muscle, the finding of its homology to a well-studied T-cell marker, RT6, and the molecular cloning of additional gene family members from mammals and birds is providing fresh impetus to research in this field. Intriguingly, these vertebrate enzymes are predicted to be secretory or membrane proteins. They are expressed in lymphatic tissues, muscle, testis, bone marrow, and erythroblasts. Here we review the relationship between this novel family of eucaryotic mono(ADP-ribosyl)transferases (mADPRTs), ADP-ribosylating bacterial toxins, the poly(ADP-ribose)polymerase (PARP), the ADP-ribosyl cyclases, and the ADP-ribosylprotein hydrolase (ARH) in terms of their structure, enzymatic properties and possible biological functions.

REACTIONS CATALYZED BY MONO(ADP-RIBOSYL)TRANSFERASES AND RELATED ENZYMES

Mono-ADP-ribosylation involves the enzyme-catalyzed transfer of the ADP-ribose moiety from NAD^+ to protein while the nicotinamide moiety is released (Fig. 1) (see Chapters

ADP-Ribosylation in Animal Tissue, edited by Haag and Koch-Nolte
Plenum Press, New York, 1997

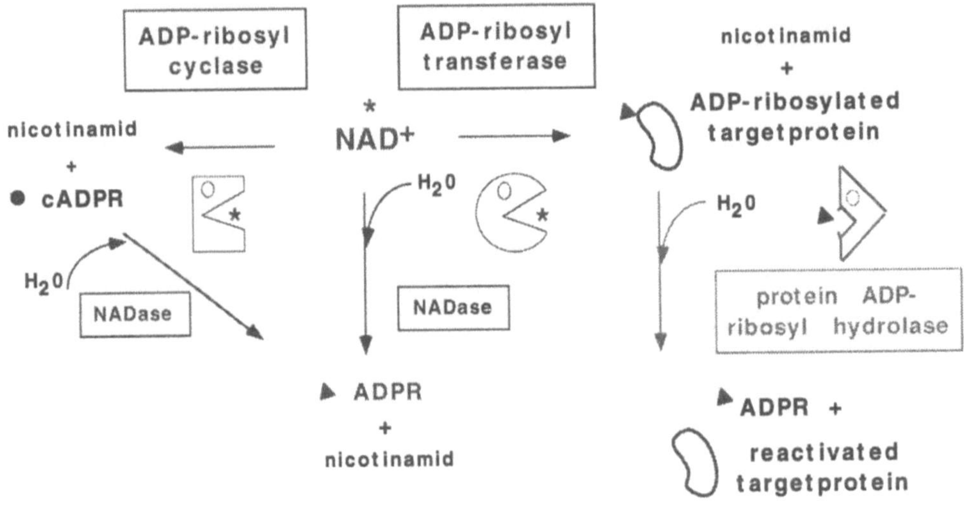

Figure 1. Schematic diagram of reactions catalyzed by ADP-ribosyltransferases, ADP-ribosylcyclases, and ADP-ribosylprotein hydrolases.

2 and 3). NAD$^+$ is presumed to be bound and oriented in the active site of *mono(ADP-ribosyl)transferases* so as to allow nucleophilic attack on the glycosidic linkage by an acceptor molecule. The physiological and most efficient acceptor usually is a specific amino acid residue in a specific target protein. ADP-ribosylation often inactivates the function of the target protein. Some of the transferases can also use simple amino acids or water instead of the amino acid in the target protein as an acceptor. When water serves as the acceptor, the net result is the glycohydrolosis of NAD$^+$ to nicotinamide and ADP-ribose (*NADase* activity). Some of the transferases also catalyze the covalent attachment of ADP-ribose to themselves (*automodification* activity). The known transferases differ markedly in their relative transferase, NADase, and automodification activities. A special subclass of ADP-ribosyltransferases, the *poly(ADP-ribose)polymerases*, catalyze the polymerization and branching of chains of ADP-ribose units onto target proteins. PARP also shows NADase and automodification activities. Posttranslational protein mono-ADP-ribosylation can be reversed by *ADP-ribosylprotein hydrolases* which catalyze removal of the ADP-ribose moiety. This can reactivate the target protein. *ADP-ribosyl cyclases* use the same substrate as ADP-ribosyltransferases, NAD$^+$, but after cleavage of the nicotinamide moiety, catalyze the formation of an intramolecular bond yielding cyclic ADP-ribose. Use of water as acceptor in this reaction or subsequent hydrolysis of cADPR to ADP-ribose also results in net NADase activity (see Chapter 52). Known cyclases differ markedly in their relative ADPR-cyclase and NADase activities. While mADPRTs and PARP appear to be related in evolution, the ADPR cyclases and possbily also the ARHs constitute distinct enzyme families (see below).

MONO(ADP-RIBOSYL)TRANSFERASES AND RELATED ENZYMES IN THE PROCARYOTIC WORLD

Mono-ADP-Ribosylation was originally discovered as the mechanism by which diphtheria toxin (DT) inactivates protein synthesis in human cells (1) (Fig. 2) (see Chapters 4 and

5). Since then, genes encoding mADPRTs have been cloned from many different bacteria and their phages. Most of these encode secreted enzymes that act on proteins of other organisms, including bacterial toxins acting on animal cell proteins (2) as well as a recently described mADPRT, halovibrin, which is secreted by the bacterial symbiont of luminescent fish and which is thought to trigger the development of the light organ in its eucaryotic host (3). Cytoplasmic, endogenous mADPRTs (DRAT) have been described in some photosynthetic bacteria, in which nitrogen fixation is regulated by reversible ADP-ribosylation of the key enzyme dinitrogenase reductase (4) (see Chapter 7). Moreover, some E.coli phages encode mADPRTs (Alt, Mod) which target the host cell RNA polymerase (5) (see Chapter 8).

Subfamilies of procaryotic mono(ADP-ribobsyl)transferases can be distinguished on the basis of similarities in structure and amino acid/target protein specificities (Fig. 2) (2). Members of each subfamily show moderate to strong amino acid sequence similarities. Although members of distinct subfamilies show almost no discernible amino acid se-

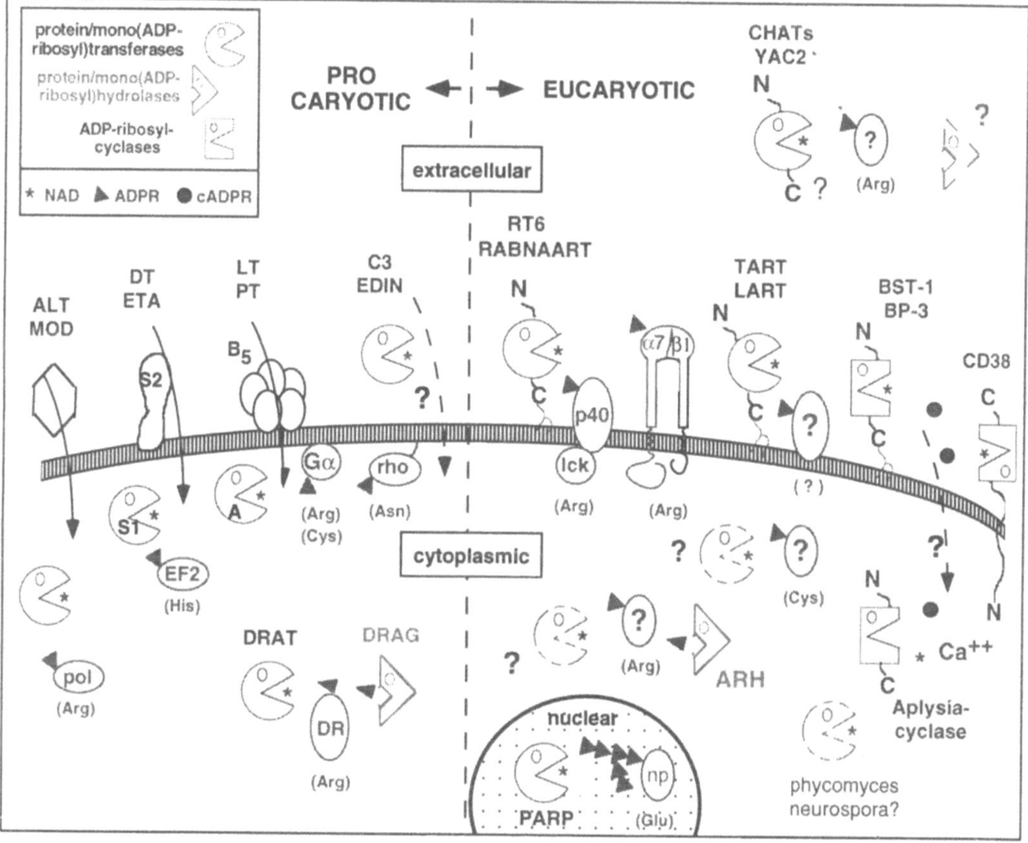

Figure 2. 'Group-picture' of cloned family members of pro- and eucaryotic mono(ADP-ribosyl)transferases and related enzymes. At least five subfamilies of mADPRTs acting in procaryotes and their phages can be distinguished on the basis of sequence similarities and target protein specificities (left hand side). An ADP-ribosylprotein hydrolase has been cloned from photosynthetic bacteria. Neither ADP-ribosyl cyclases nor poly(ADP-ribose)polymerases have yet been cloned from procaryotes. All eucaryotic mADPRTs cloned to date are predicted to exist as GPI-anchored membrane or seretory proteins (right hand side). In contrast, the single eucaryotic ADP-ribosylprotein hydrolase (ARH) identified to date is predicted to be a cytoplasmic protein. The nuclear poly(ADP-ribose)polymerase (PARP) has been cloned from many different eucaryotes, including yeast, fish, chicken and mammals. Another recently emerging eucaryotic family of NAD^+-metabolizing enzymes catalyzes the synthesis of cyclic ADP-ribose (far right).

quence identity, their core protein folds evidently are still remarkably similar: in the four known crystal structures of toxin mADPRTs, 44 amino acid residues in the same linear order of secondary structure units can be superimposed with a root mean square deviation of 1.6 Å (6–8) (see Chapters 4 and 12).

Diphtheria toxin and pseudomonas exotoxin A (DT and ETA) are made as part of a large secretory precursor protein which is proteolytically processed into two subunits (Fig. 2) (2) (see Chapters 4 and 5). The nonenzymatic S2 subunit mediates binding of the holo-toxin to a receptor on the target cell, the enzymatic S1 subunit subsequently penetrates into the host cell cytoplasm where it ADP-ribosylates elongation factor 2, thereby shutting off host cell protein synthesis. In case of DT the enzyme subunit is N-terminal in the holo-toxin, in case of ETA it is C-terminal.

E. coli heat labile enterotoxin, cholera toxin, pertussis toxin (LT, CT, PT) and related toxins have an $A(B_5)$ multimer structure in which the B pentamer mediates membrane binding to the target cell (2) (see Chapters 9–11). The enzymatic A subunit penetrates into the host cell cytoplasm where it catalyzes mono-ADP-ribosylation of the α subunit of regulatory heterotrimeric G-proteins. This affects the protein which is regulated by the respective G-protein (e.g. adenylate cyclase in case of LT and CT, several different signal-transducing receptors in case of PT).

Monomeric secretory enzymes from *Clostridium botulinum*, *Clostridium limosum*, *Bacillus cereus* and *Staphylococcus aureus* (C3, EDIN) catalyze mono-ADP-ribosylation of small GTP-binding proteins of the ras superfamily upon incubation of cell lysates with these enzymes (9). However, it is not clear how these secretory enzymes penetrate the membrane barrier to reach their cytoplasmic targets. A distinct family of clostridial toxins (C2, iota) mono-ADP-ribosylate actin in animal cells (10) (see Chapter 6).

Cytoplasmic mADPRTs so far have been molecularly characterized only in photosynthetic bacteria: the DRAT mono(ADP-ribosyl)transferases of *Rhodospirillium rubrum* and *Azospirillium braziliense* inactivate dinitrogenase reductase by mono-ADP-ribosylation (4). These photosynthetic bacteria are also the sole examples in the procaryotic world in which enzymes capable of catalyzing the reverse reaction, i.e. removal of ADP-ribose from target proteins have been cloned: the DRAG ADP-ribosylglycohydrolases reactivate nitrogenase by removal of the ADP-ribose moiety (see Chapter 7).

Most of the known bacterial mADPRTs transfer ADP-ribose onto arginine residues (LT, CT, halovibrin, DRAT, ACT, and C2) although some enzymes do target other amino acids such as cysteine (PT), diphthamide (DT and ETA), or asparagine (C3- and related exoenzymes).

To our knowledge, no poly(ADP-ribose)polymerases or ADP-ribosyl cyclases have yet been cloned in procaryotes.

MONO-ADP-RIBOSYLATION IN THE EUCARYOTIC WORLD

The idea that endogenous relatives of bacterial toxins may regulate protein functions also in vertebrates has long aroused the curiosity of biochemists and cellular biologists and ample evidence has been put forth to corroborate this hypothesis (11) (see Chapters 2 and 3). Adding exogenous NAD^+ or known inhibitors of mono-ADP-ribosylation profoundly modifies the activity of several distinct types of intact cells including muscle cells, neurons, activated cytotoxic T cells and macrophages (12–19) (see Chapters 24–26, 29, 31, 36–42). Morevoer, upon incubation of intact cells or cell lysates with $[^{32}P]NAD^+$, several proteins usually become covalently modified by mono-ADP-ribose (14, 15,

20–22) (see Chapters 24, 36–42). Identified target proteins for endogenous ADP-ribosylation reactions include membrane proteins such as integrins, Ca^{2+} ATPase, and p40, a regulator of tyrosine kinases, as well as intracellular proteins such as the house keeping enzyme G3PDH, the heat shock protein BiP, the cytoskeletal protein desmin, the neuronal phosphoprotein B-50/GAP-43, and proteins that are known to serve as targets also for bacterial enzymes: $G\alpha$, rho, and actin (14, 19, 23–27) (see Chapters 24, 36–46).

AN EMERGING GENE FAMILY OF VERTEBRATE mADPRTs

Mono-ADP-ribosylation in eucaryotes has been implicated to modulate cell-cell interactions, signal transduction, the architecture of the cellular cytoskeleton, and vesicular traffic (13–18, 28). The responsible enzymes, however, eluded molecular cloning until the groups of J. Moss at the NIH, USA, and M. Shimoyama at the Shimane Medical University, Japan, succeeded in purifying and sequencing proteins with enzyme activities akin to those of bacterial mADPRTs from rabbit skeletal muscle (designated RABNAART) and chicken bone marrow cells (designated CHAT-1 and CHAT 2), respectively (29, 30) (Figs. 2 and 3) (see Chapters 15, 16, 39).

Homology searches revealed significant sequence similarity of RABNAART and the CHATs to only one other previously known eucaryotic protein of unknown function: the T-cell differentiation antigen RT6 (31–33) (see Chapters 13, 26, 28). RT6, indeed, has been shown to possess the enzyme activities predicted by the structural homology (34–38) (see Chapters 13, 15, 20, 22, 23). With oligonucleotides derived from known mADPRTs as well as from mADPRT-related expressed sequence tags, additional mono(ADP-ribosyl)transferase gene family members have been cloned from chicken, mouse, human, and other species (39–41) (see Chapters 12–15, 17, 19). Presently, six distinct mADPRT gene family members have been fully cloned from mouse and/or human and three from chicken (Fig. 3) (see Appendix).

One of the murine mADPRT genes is expressed predominantly in cardiac and skeletal muscle, three are preferentially expressed in lymphatic tissues, and two in testis (see Chapters 13–15, 19, 39). Five of these enzymes are expressed as GPI-anchored membrane proteins, whereas one of the testis enzymes appears to be a secretory protein. Four of these enzymes specifically target arginine, the specificities of the two other enzymes still need to be established. Two of the chicken mADPRT genes are expressed predominantly in bone marrow, the third in erythroblasts (see Chapters 16 and 17). All of these have been shown to target arginine. The chicken enzymes are most closely related to the mammalian muscle enzyme (50% sequence identity). The different expression patterns and the absence of a GPI-anchor signal sequence in the chicken enzymes, however, make it seem unlikely that any of these represents the chicken homologue of the mammalian muscle transferase. It remains to be determined whether species-homologues of the cloned mammalian mADPRTs exist in chicken and vice versa. The isolation of a GPI-anchored mono(ADP-ribosyl)transferase from chicken spleen indicates that this, indeed, may be so (see Chapter 30).

WHERE AND WHO ARE THE TARGETS OF THE VERTEBRATE mADPRTs?

It is likely, though not yet firmly established that the cloned vertebrate mADPRTs or their respective species homologues account for at least some of the mono-ADP-ribosyla-

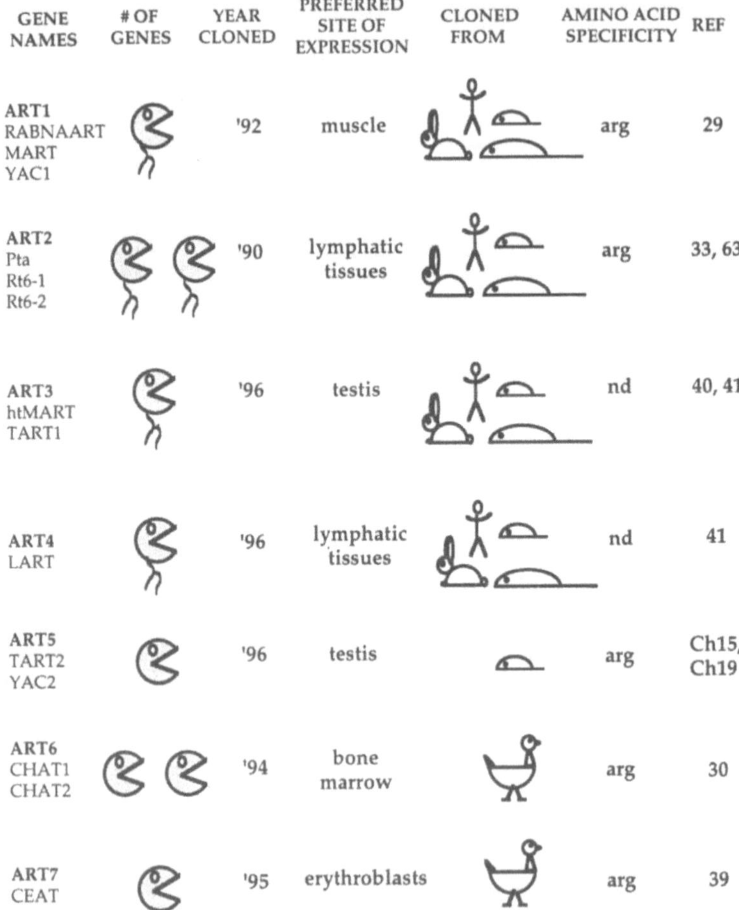

Figure 3. Cloned vertebrate members of the mono(ADP-ribosyl)transferase gene family. Nine distinct mADPRT-encoding genes have been fully cloned from vertebrates. The gene symbol "ART" for ADP-ribosyltransferase has been assigned provisionally for members of this gene family by the nomenclature committee of the human genome project and was approved by a panel of experts at the Hamburg workshop (see appendix). Arginine-specific mono(ADP-ribosyl)transferase activity has been demonstrated for ART1-homologues in mouse, rabbit and human, for ART2-homologues in rat and mouse, for ART5 in mouse, and for ART6 and ART7 in chicken (see Chapters 13, 15–17). Inclusion of ART3 and ART4 is on the basis of sequence similarity (see Chapters 13, 14, 19). The ART2 gene in the human is inactivated by premature stop codons (62). We have cloned species-homologues of ART1-ART4 from at least three mammalian lineages, indicating that the gene duplication generating the ancestral genes for ART1-ART4 predates the mammalian radiation (37, 41) (and F. K.-N., et al., in preparation). ART2 has been duplicated into two functional gene copies in the murine lineage, ART6 in the chicken lineage (30, 63).

tion reactions detected in animal tissues. A common feature of the vertebrate mADPRT-encoding genes is that their expression is restricted to one or a few tissues and that they show relatively low signal intensities in Northern Blot analyses with whole tissue RNA (33, 33, 40, 41). This may reflect a need to tightly control the site and level of expression of theses presumptive regulatory enzymes. There is, however, a slight degree of overlap in expression and low transcript levels are detectable also in tissues where other gene family members are most prominent. A clear correlation between enzyme activities detected in animal tissues and cell extracts with cloned mADPRT gene family members, thus, must be established by further experimental investigations.

It is intriguing that all of the eucaryotic mADPRTs cloned so far are predicted to be either secretory or GPI-anchored membrane proteins. The GPI-anchor inherently is subject to potential cleavage (see Chapters 47 and 48). Indeed, a GPI-anchored T cell mADPRT has been shown to be released from the cell surface into the surrounding medium upon triggering of the T cell receptor (see Chapter 24). This raises the question whether these enzymes have local and/or distant targets and whether their targets are extra- and/or intracellular (Fig. 4). If they have extracellular targets as some evidence seems to suggest, how is access to the required substrate NAD^+ assured? After all, NAD^+ is considered a classic intracellular metabolite and cell membranes are impermeable to NAD^+. A similar NAD-access problem is posed for the ecto-ADP-ribosyl cyclases (see below). It is conceivable that a secretory mechanism for NAD^+ exists akin to that for ATP, another classic intracellular metabolite for which cell surface receptors have been found on neurons, lymphocytes, and in testis (42, 43). If the vertebrate ecto-mADPRTs target intracellular proteins as do their bacterial toxin cousins, the question arises as to their mechanism of entry into the cytoplasm. Should they indeed be capable of translocating across the cell membrane, they might also account for some of the intracellular ADP-ribosylation reactions that have been observed in animal tissues.

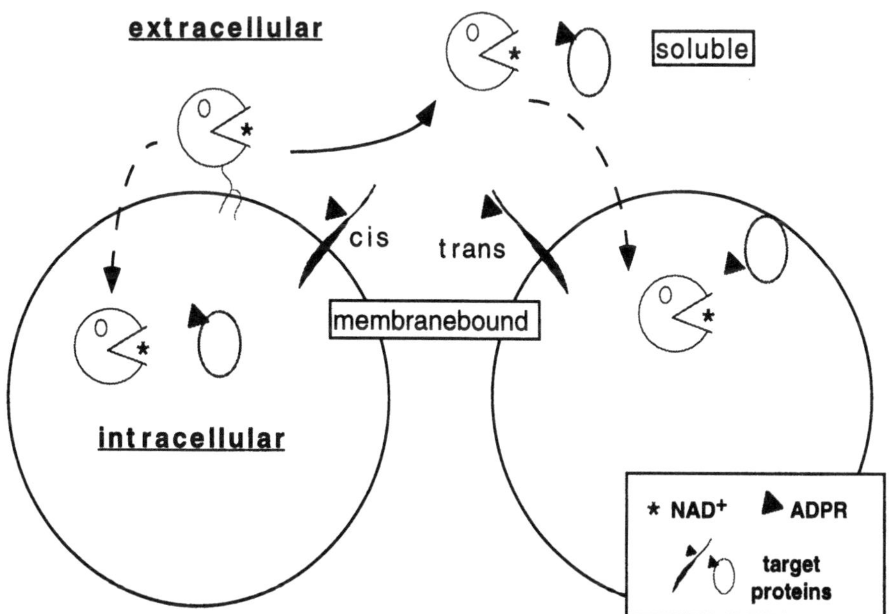

Figure 4. Potential target proteins for eucaryotic ecto-mADPRTs. All of the eucaryotic mADPRTs cloned so far are predicted to be either secretory or GPI-anchored membrane proteins. The GPI-anchor inherently is subject to potential cleavage. This raises the question whether these enzymes have local and/or distant targets and whether their targets are extra- and/or intracellular. Cytoplasmic mADPRT-activity has been observed in eucaryotic cells but it is not known whether eucaryotic ecto-mADPRTs can translocate across the cell membrane as some of the bacterial mADPRTs do.

RELATIONSHIP OF mADPRTs, PARP, ARH, AND ADP-RIBOSYL CYCLASES

The three-dimensional structure of a prototype vertebrate mADPRT has yet to be elucidated. Structure prediction analyses, however, strongly suggest that these enzymes exhibit a core fold similar to that of their bacterial toxin cousins (38) (see Chapters 12 and 23). These analyses as well as the results of site directed mutagenesis studies strongly support the hypothesis that the eu- and procaryotic mADPRTs share a common fold and ancestry (8, 38, 44) (see Chapters 15 and 23).

A single poly(ADP-ribose)polymerase gene (PARP) has been cloned from human and mouse. A species-homologue has been identified in many different eucaryotes, including mammals, birds, fish, plants and yeast (45). PARP is expressed in many different tissues and catalyzes the posttranslational polymerization and branching of ADP-ribose onto glutamic acid residues in nuclear proteins. PARP is much larger than the known mADPRTs and, in addition to the catalytic C-terminal region, also contains an N-terminal DNA-binding domain and a central regulatory segment. The crystal structure of the catalytic fragment of PARP has recently been elucidated and reveals a core fold that is remarkably similar to that of the bacterial toxins (46). Thus, pro- and eucaryotic mADPRTs together with PARP appear to constitute a super gene family that is related in evolution.

In contrast to the eucaryotic mADPRTs cloned so far, the single eucaryotic ADP-ribosylprotein hydrolase (ARH) cloned to date is predicted to be a cytoplasmic enzyme (47) like its counterpart in photosynthetic bacteria (48) (Fig. 2) (see Chapters 3 and 7). It still needs to be determined whether this enzyme participates in reversible ADP-ribosylation cycles with any of the known transferases. It is of note that ADP-ribosylprotein hydrolase activity has been demonstrated also on the surface of intact cells (36). However, it has yet to be established whether this activity originates from a cytoplasmic ARH that has translocated across the cell membrane or from a distinct extracellular isoform. Moreover, the three dimensional structures of pro- and eucaryotic ARH have yet to be elucidated. Although it is not unreasonable to predict that they will be similar to one another, it is open whether they will resemble mADPRTs.

One other gene family encoding NAD^+- metabolizing enzymes, the ADP-ribosylcyclases, has emerged with the cloning of related genes from snail, mouse and human (49–51) (see Chapters 51–58). These enzymes catalyze the cleavage of NAD^+ to nicotinamide and cyclic ADP-ribose (cADPR) (49–51) (Figs. 1 and 2). cADPR has attracted interest as a Ca^{2+} mobilizing second messenger. Cloned family members include a cytoplasmic enzyme from the snail Aplysia and two membrane proteins from human and mouse bone marrow and hematopoetic cells: CD38, a type II membrane protein, and GPI-anchored BST-1/BP3 (49–51). An ADPR cyclase-related gene fragment has been also been identified in the parasite *Schistosoma mansonii* (see Chapter 19). As in case of ecto-mADPRTs, it is still unclear whether and how CD38 and BST-1, their substrate NAD^+ or their product cADPR cross the cell-membrane barrier.

Despite their similar cellular localization and common substrate, the ADP-ribosyl cylases and ADP-ribosyltransferases evidently constitute two distinct gene families. Thus, it has not been possible to demonstrate cyclase activity for any of the known transferases or transferase activity for any of the known cyclases (see Chapters 21 and 52). (The 'transferase' activity recently reported for CD38 (52) is actually a consequence of its NADase activity and the reactivity of the generated free ADP-ribose with cysteine in serum albumin and other proteins and thus does not qualify as a true transferase activity). Moreover, preliminary data on the crystal structure of the Aplysia enzyme indicates that its NAD$^+$-binding fold is completely different from that of mono-ADP-ribosylating bacterial toxins and PARP (see Chapter 53). As some of the ADP-ribosyl cyclases and some of the mADPRTs exhibit NADase activity, they both could account for the NADase activities that have been detected biochemically in animal tissues (see Chapters 49–58). It remains to be seen whether any NAD-glycohydrolases exist that do not belong to either the ADPR-cyclase or the mADPRT gene families.

DISEASE ASSOCIATION OF DEFECTIVE mADPRTs

In different animal models, polygenetically determined autoimmune disorders such as juvenile diabetes and systemic lupus have been found to coincide with defects in the structure and function of the RT6 T cell ecto-mADPRT (53, 54) (see Chapters 13, 26, 27, 32–35). In the rat, susceptibility and resistance to diabetes correlate with absence and presence of RT6-expressing T cells, repsectively (53, 55–57). Moreover, treatment of animals with RT6-specific antibodies can induce autoimmune reactions up to full blown diabetes (58). In a mouse model for systemic lupus erythematosus, a gene locus influencing disease paramaters was recently mapped near the *Rt6* locus (59). Moreover, we have found *Rt6* gene-inactivating mutations in two distinct lupus prone mouse strains (54) (see Chapter 35). A causal relationship between RT6 defects and enhanced disease susceptibility still needs to be established. A possible mechanism for an RT6-mediated protective function is pointed out by the finding that the cytotoxic functions of activated T cells can be profoundly inhibited by cell surface ADP-ribosylation reactions (15, 19) (see Chapter 24).

PERSPECTIVES

Do the nine distinct mADPRT gene family members that have been cloned from mammals and birds (see Fig. 3) represent the tip of an iceberg? Considering that mono-ADP-ribosylation has been detected also in tissues where the identified genes do not appear to be transcribed (e.g. brain, peripheral nervous system, liver, erythrocytes, oocytes) (20, 60) (see Chapters 2, 3, 36–38), it is possible that additional mADPRTs exist in verebrates. The results of EST database searches (see Chapter 19), however, indicate that ADP-ribosyl transferases and ADP-ribosylprotein hydrolases constitute rather small gene families - in contrast to say those of protein kinases and phosphatases. It is quite possible that most if not all mammalian mADPRT genes have already been identified - unless subfamilies with only very little sequence similarities to the known mADPRTs, i.e. below the level of detection with state-of-the-art database search programs, exist. We eagerly await the molecular cloning of mADPRT genes from other eucaryotic species in which ADP-ribosylation reactions have been described, e.g. fungus and frogs (20) (see Chapter 18). Moreover, it will be interesting to see what role if any mADPRTs play in the world of plants.

An important endeavour is to determine which of the molecularly cloned mADPRTs is responsible for which of the mono-ADP-ribosylation reactions observed in animal tissues. Mono-ADP-ribosylation has been observed in the cell cytoplasm, but it is not known whether the eucaryotic ecto-mADPRTs can translocate across the cell membrane as some of the bacterial mADPRTs do. It is conceivable that cytoplasmic mADPRT isoforms exist in vertebrates in addition to the ecto-enzymes identified to date, as has been observed for other enzyme families, e.g. the carbonic anhydrases (61).

The availability of recombinant mADPRTs should facilitate the identification of the specific target protein(s) for each enzyme. The application of gene targeting technologies to the mADPRT-encoding genes should help clarify the physiological functions affected by the respective mADPRT-catalyzed target protein modifications.

Considering the association of RT6 defects with autoimmune diseases, it is not unreasonable to predict that inherited and acquired malfunctioning of other vertebrate mADPRTs will also be of clinical import. Moreover, recombinant versions of these enzymes as well as their inhibitors may prove to be of use for experimental and therapeutic interventions, in particular in case of the readily accessible ecto-mADPRTs. Perhaps some day mADPRTs may even be engineered to modify specified target proteins.

The molecular cloning of vertebrate mADPRTs and related enzymes has already caused a flurry of excitement, which reverberates throughout the following contributions to this volume. New fields of experimental investigation have been seeded at the interfaces of enzymology, microbiology, cell biology, immunology, signal transduction, and molecular biology. It is not unwarrented to expect some exiciting follow up discoveries in the near future.

REFERENCES

1. Honjo, T., Y. Nishizuka, O. Hayaishi and I. Kato. 1968. Diphtheria toxin/dependent adenosine diphosphate ribosylation of aminoacyl transferase II and inhibition of protein synthesis. *J. Biol. Chem. 243*: 3553–3555.

2. Passador, L. and W. Iglewski. 1994. ADP-ribosylating toxins. *Methods Enzymol. 235*: 617–631.

3. Reich, K. A. and G. K. Schoolnik. 1996. Halovibrin, screted from the light organ symbiont *Vibrio fischeri*, is a member of a new class of ADP-ribosyltransferases. *J. Bacteriol. 178*: 209–215.

4. Ludden, P. W. 1994. Reversible ADP-ribosylation as a mechanism of enzyme regulation in procaryotes. *Mol. Cell. Biochem. 138*: 123–129.

5. Koch, T. and W. Rüger. 1994. The ADP-ribosyltransferases (gpAlt) of bacteriophages T2, T4, and T6: sequencing of the genes and comparison of their products. *Virology 203*: 294–298.

6. Choe, S., M. J. Bennett, G. Fujii, P. M. Curmi, K. A. Kantardjieff, R. J. Collier and D. Eisenberg. 1992. The crystal structure of diphtheria toxin. *Nature 357*: 216–222.

7. Stein, P., A. Boodhoo, G. D. Armstrong, S. A. Cockle, M. H. Klein and R. J. Read. 1994. The crystal structure of pertussis toxin. *Structure. Curr. Biol. 2*: 45–57.

8. Domenighini, M., C. Magagnoli, M. Pizza and R. Rappuoli. 1994. Common features of the NAD-binding and catalytic site of ADP-ribosylating toxins. *Mol. Microbiol. 14*: 41–50.

9. Aktories, K. (ed.) 1991. *ADP-ribosylating toxins*. Springer Verlag, Berlin.

10. Aktories, K. and A. Wegner. 1992. Mechanisms of the cytopathic action of actin-ADP-ribosylating toxins. *Mol. Microbiol. 6*: 2905–8.

11. Althaus, F. R., H. Hilz and S. Shall. 1985. *ADP-ribosylation of proteins*. Springer Verlag, Berlin.

12. Kharadia, S. V., T. W. Huiatt, H. Y. Huang, J. E. Peterson and D. J. Graves. 1992. Effect of arginine-specific ADP-ribosyltransferase inhibitor on differentiation of embryonic chick skeletal muscle cells in culture. *Exp. Cell Res. 201*: 33–42.

13. McMahon, K. K., K. J. Piron, V. T. Ha and A. T. Fullerton. 1993. Developmental and biochemical characteristics of the cardiac membrane-bound arginine-specific mono-ADP-ribosyltransferase. *Biochem. J. 293*: 789–793.

14. Zolkiewska, A. and J. Moss. 1993. Integrin alpha 7 as substrate for a glycosylphosphatidylinositol-anchored ADP-ribosyltransferase on the surface of skeletal muscle cells. *J. Biol. Chem. 268*: 25273–25276.

15. Wang, J., E. Nemoto, A. Y. Kots, H. R. Kaslow and G. Dennert. 1994. Regulation of cytotoxic T cells by ecto-nicotinamide adenine dinucleotide (NAD) correlates with cell surface GPI-anchored/arginine ADP-ribosyltransferase. *J. Immunol. 153*: 4048–4058.

16. Hauschildt, S., P. Scheipers and W. G. Bessler. 1994. Lipopolysaccharide-induced change of ADP-ribosylation of a cytosolic protein in bone-marrow derived macrophages. *Biochem. J. 297*: 17–20.

17. Pellat, D. C., J. Wietzerbin and J. C. Drapier. 1994. Nicotinamide inhibits nitric oxide synthase mRNA induction in activated macrophages. *Biochem. J. 297*: 53–58.

18. Schuman, E. M., M. K. Meffert, H. Schulman and D. V. Madison. 1994. An ADP-ribosyltransferase as a potential target for nitric oxide action in hippocampal long-term potentiation. *Proc. Natl. Acad. Sci. USA 91*: 11958–11962.

19. Wang, J., E. Nemoto and G. Dennert. 1996. Regulation of CTL by ecto-NAD involves ADP-ribosylation of a p56[lck] associated protein. *J. Immunol. 156*: 2819–2827.

20. Godeau, F., D. Belin and S. S. Koide. 1984. Mono(adenosine diphoshpate ribosyl) transferase in *Xenopus* tissues. Direct demonstration by a zymographic localization in Sodium Dodecyl Sulfate-Polyacrylamide Gels. *Anal. Biochem. 137*: 287–296.

21. Brune, B., y. V. L. Molina and E. G. Lapetina. 1990. Agonist-induced ADP-ribosylation of a cytosolic protein in human platelets. *Proc. Natl. Acad. Sci. USA 87*: 3304–8.

22. Soman, G., A. Haregewoin, R. C. Hom and R. W. Finberg. 1991. Guanidine group specific ADP-ribosyltransferase in murine cells. *Biochem. Biophys. Res. Commun. 176*: 301–308.

23. Maehama, T. K., K. Takahashi, Y. Ohoka, T. Otsuka, M. Ui and T. Katada. 1991. Identification of a botulinum C3-like enzyme in bovine brain that catalyzes ADP-ribosylation of GTP-binding proteins. *J. Biol. Chem. 266*: 10062–10065.

24. Freiden, P. J., J. R. Gaut and L. M. Hendershot. 1992. Interconversion of three differentially modified and assembled forms of BiP. *EMBO J. 11*: 63–70.

25. Taniguchi, M., M. Tsuchiya and M. Shimoyama. 1993. Comparison of acceptor protein specificities on the formation of ADP-ribose.acceptor adducts by arginine-specific ADP-ribosyltransferase from rabbit skeletal muscle sarcoplasmic reticulum with those of the enzyme from chicken peripheral polymorphonuclear cells. *Biochim. Biophys. Acta 1161*: 265–71.

26. Huang, H. Y., D. J. Graves, R. M. Robson and T. W. Huiatt. 1993. ADP-ribosylation of the intermediate filament protein desmin and inhibition of desmin assembly in vitro by muscle ADP-ribosyltransferase. *Biochem. Biophys. Res. Commun. 197*: 570–7.

27. Quist, E. E., D. L. Coyle, R. Vasan, N. Satumtira, E. L. Jacobson and M. K. Jacobson. 1994. Modification of cardiac membrane adenylate cyclase activity and Gs alpha by NAD and endogenous ADP-ribosyltransferase. *J. Mol. Cell. Cardiol. 26*: 251–260.

28. De Matteis, M., et al. 1994. Stimulation of endogenous ADP-ribosylation by brefeldin A. *Proc. Natl. Acad. Sci .USA 91*: 1114–8.

29. Zolkiewska, A., M. S. Nightingale and J. Moss. 1992. Molecular characterization of NAD:arginine ADP-ribosyltransferase from rabbit skeletal muscle. *Proc. Natl. Acad. Sci. USA 89*: 11352–11356.

30. Tsuchiya, M., N. Hara, K. Yamada, H. Osago and M. Shimoyama. 1994. Cloning and expression of cDNA for arginine-specific ADP-ribosyltransferase from chicken bone marrow cells. *J. Biol. Chem. 269*: 27451–27457.

31. Thiele, H. G., F. Koch, A. Hamann and R. Arndt. 1986. Biochemical characterization of the T-cell alloantigen RT-6.2. *Immunol. 59*: 195–201.

32. Koch, F., H. G. Thiele and M. G. Low. 1986. Release of the rat T cell alloantigen RT-6.2 from cell membranes by phosphatidylinositol-specific phospholipase C. *J. Exp. Med. 164*: 1338–1343.

33. Koch, F., F. Haag, A. Kashan and H. G. Thiele. 1990. Primary structure of rat RT6.2, a nonglycosylated phosphatidylinositol-linked surface marker of postthymic T cells. *Proc. Natl. Acad. Sci. USA 87*: 964–967.

34. Takada, T., K. Iida and J. Moss. 1994. Expression of NAD glycohydrolase activity by rat mammary adenocarcinoma cells transformed with rat T cell alloantigen RT6.2. *J. Biol. Chem. 269*: 9420–9423.

35. Haag, F., V. Andresen, S. Karsten, F. Koch-Nolte and H.-G. Thiele. 1995. Both allelic forms of the rat T cell differentiation marker RT6 display nicotinamide adenine dinucleotide (NAD)-glycohydrolase activity, yet only RT6.2 is capable of automodification upon incubation with NAD. *Eur. J. Immunol. 25*: 2355–2361.

36. Maehama, T., H. Nishina, S. Hoshino, Y. Kanaho and T. Katada. 1995. NAD+-dependent ADP-ribosylation of T lymphocyte alloantigen RT6.1 reversibly proceeding in intact rat lymphocytes. *J. Biol. Chem. 270*: 22747–22751.

37. Koch-Nolte, F., F. Haag, R. Kastelein and F. Bazan. 1996. Uncovered: The family relationship of a T cell membrane protein and bacterial toxins. *Immunol. Today 17*: 402–405.

38. Koch-Nolte, F., D. Petersen, S. Balasubramanian, F. Haag, D. Kahlke, T. Willer, R. Kastelein, F. Bazan and H. G. Thiele. 1996. Mouse T cell membrane proteins Rt6–1 and Rt6–2 are arginine/protein mono ADP-ribosyltransferases and share secondary structure motifs with ADP-ribosylating bacterial toxins. *J. Biol. Chem. 271*: 7686–7693.

39. Davis, T. and S. Shall. 1995. Sequence of a chicken erythroblast mono(ADP-ribosyl)transferase-encoding gene and its upstream region. *Gene 164*: 371–372.

40. Lévy, I., Y. Q. Wu, N. Roeckel, F. Bulle, A. Pawlak, S. Siegrist, M. G. Mattéi and G. Guellaen. 1996. Human testis specifically expresses a homologue of the rodent T lymphocytes RT6 mRNA. *FEBS Lett. 382*: 276–280.

41. Koch-Nolte, F., F. Haag, R. Braren, M. Kühl, J. Hoovers, S. Balasubramanian, F. Bazan and H. G. Thiele. 1996. Two novel human members of an emerging mammalian gene family related to mono-ADP-ribosylating bacterial toxins. *Genomics, in press.*

42. Evans, R. J., V. Derkach and A. Surprenant. 1992. ATP mediates fast synaptic transmission in mammalian neurons. *Nature 357*: 503–505.

43. Valera, S., N. Hussy, R. J. Evans, N. Adami, R. A. North, A. Surprenant and G. Buell. 1994. A new class of ligand-gated ion channel defined by P2x receptor for extracellular ATP. *Nature 371*: 516–9.

44. Takada, T., K. Iida and J. Moss. 1995. Conservation of a common motif in enzymes catalyzing ADP-ribose transfer. Identification of domains in mammalian transferases. *J. Biol. Chem. 270*: 541–544.

45. DeMurcia, G. and J. MénissierDeMurcia. 1994. *Trends Biochem. Sci. 19*: 172–176.

46. Ruf, A., J. M. DeMurcia, G. M. DeMurcia and G. E. Schulz. 1996. Structure of the catalytic fragment of poly(ADP-ribose)polymerase from chicken. *Proc. Natl. Acad. Sci. USA 93*: 7481–7485.

47. Moss, J., S. J. Stanley, M. S. Nightingale, J. J. Murtagh, L. Monaco, K. Mishima, H. C. Chen, K. C. Williamson and S. C. Tsai. 1992. Molecular and immunological characterization of ADP-ribosylarginine hydrolases. *J. Biol. Chem. 267*: 10481–8.

48. Fitzmaurice, W. P., L. L. Saari, R. G. Lowery, P. W. Ludden and G. P. Roberts. 1989. Genes coding for the reversible ADP-ribosylation system of dinitrogenase reductase from Rhodospirillum rubrum. *Mol. Gen. Genet. 218*: 340–347.

49. Lee, H. C. and R. Aarhus. 1991. ADP-ribosyl cyclase: an enzyme that cyclizes NAD into a calcium-mobilizing metabolite. *Cell. Regulation 2*: 203.

50. Malavasi, F., A. Funaro, S. Roggero, A. Horenstein, L. Calosso and K. Mehta. 1994. Human CD38: a glycoprotein in search of a function. *Immunol. Today 15*: 95–97.

51. Howard, M., J. C. Grimaldi, J. F. Bazan, F. E. Lund, A. L. Santos, R. M. Parkhouse, T. F. Walseth and H. C. Lee. 1993. Formation and hydrolysis of cyclic ADP-ribose catalyzed by lymphocyte antigen CD38. *Science 262*: 1056–1059.

52. Grimaldi, J. C., S. Balasubramanian, N. H. Kabra, A. Shanafelt, J. F. Bazan, G. Zurawski and M. C. Howard. 1995. CD38-mediated ribosylation of proteins. *J. Immunol. 155*: 811–817.

53. Greiner, D. L., E. S. Handler, K. Nakano, J. P. Mordes and A. A. Rossini. 1986. Absence of the RT-6 T cell subset in diabetes-prone BB/W rats. *J. Immunol. 136*: 148–151.

54. Koch-Nolte, F., J. Klein, C. Hollmann, M. Kühl, F. Haag, H. R. Gaskins, E. H. Leiter and H. G. Thiele. 1995. Defects in the structure and expression of the genes for the T cell marker Rt6 in NZW and (NZB x NZW)F1 mice. *Int. Immunol. 7*: 883–890.

55. Burstein, D., J. P. Mordes, D. L. Greiner, D. Stein, N. Nakamura, E. S. Handler and A. A. Rossini. 1989. Prevention of diabetes in BB/Wor rat by single transfusion of spleen cells. Parameters that affect degree of protection. *Diabetes 38*: 24–30.

56. McKeever, U., J. P. Mordes, D. L. Greiner, M. C. Appel, J. Rozing, E. S. Handler and A. A. Rossini. 1990. Adoptive transfer of autoimmune diabetes and thyroiditis to athymic rats. *Proc .Natl. Acad. Sci. USA 87*: 7618–22.

57. Fowell, D. and D. Mason. 1993. Evidence that the T cell repertoire of normal rats contains cells with the potential to cause diabetes. Characterization of the CD4+ T cell subset that inhibits this autoimmune potential. *J. Exp. Med. 177*: 627–36.

58. Greiner, D. L., J. P. Mordes, E. S. Handler, M. Angelillo, N. Nakamura and A. A. Rossini. 1987. Depletion of RT6.1+ T lymphocytes induces diabetes in resistant biobreeding/Worcester (BB/W) rats. *J. Exp. Med. 166*: 461–475.

59. Drake, C. G., S. J. Rozzo, H. F. Hirschfeld, N. P. Smarnworawong, E. Palmer and B. L. Kotzin. 1995. Analysis of the New Zealand Black contribution to lupus-like renal disease. Multiple genes that operate in a threshold manner. *J. Immunol. 154*: 2441–2447.

60. Williamson, K. C. and J. Moss. 1990. Mono-ADP ribosyltransferases and ADP-ribosylarginine hydrolases: a mono-ADP-ribosylation cycle in animal cells. In: *ADP-ribosylating bacterial toxins and G-proteins. Insights into signal transduction.* (J. Moss and M. Vaughan, eds.). American Society for Microbiology, Washington D.C.

61. Okuyama, T., S. Sato, X. L. Zhu, A. Waheed and W. S. Sly. 1992. Human carbonic anhydrase IV: cDNA cloning, sequence comparison, and expression in COS cell membranes. *Proc. Natl. Acad. Sci. USA 89*: 1315–1319.

62. Haag, F., F. Koch-Nolte, M. Kühl, S. Lorenzen and H. G. Thiele. 1994. Premature stop codons inactivate the *RT6* genes of the human and chimpanzee species. *J. Mol. Biol. 243*: 537–546.
63. Hollmann, C., F. Haag, M. Schlott, A. Damaske, H. Bertuleit, M. Matthes, M. Kühl, H. G. Thiele and F. Koch-Nolte. 1996. Molecular characterization of mouse T-cell ecto-ADP-ribosyltransferase Rt6: cloning of a second functional gene and identification of the Rt6 gene products. *Mol. Immunol. 33*: 807–817.

2

ADP-RIBOSE

A Historical Overview

Helmuth Hilz

Institut für Physiologische Chemie
Universitätskrankenhaus Eppendorf, Martinistrasse 52
D-20246 Hamburg, Germany

ABSTRACT

ADP-ribosylation of proteins was first detected as a modification of nuclear proteins by polymeric ADP-ribose residues derived from NAD^+. Subsequently, the field developed into three destinct sections: (i) Possible function(s) of poly-ADP-ribosylation with increasing evidence for participation in DNA excision repair and perhaps in DNA recombination; (ii) Mono-ADP-ribosylation as a mechanism to inactivate (by some toxins) or to regulate enzymes/proteins (e.g. in bacterial nitrogen fixation, in protein traffic through membranes, in intercellular communication); (iii) Intramolecular ADP-ribosylation converting NAD^+ to cyclic ADP-ribose, a possible Ca^{2+}-mobilizing agonist. Thus, NAD^+ first known as a cofactor of oxidoreductases has experienced an impressive metamorphosis from a housekeeping coenzyme to a multifunctional group transfering metabolite involved in an increasing number of intracellular and intercellular regulatory loops.

INTRODUCTION

I fcel very much honored by the organizers' invitation to review the history of ADP-ribose research in front of such eminent scholars. In presenting this retrospection it is also a pleasure to draw a line from the beginnings, let's say from the detection of NAD^+, to Heinz-Günter Thiele to whom this meeting is devoted.

The mutual denominator of ADP-ribose transfer reactions, the basis of all the processes to be discussed during this meeting, the biochemical all of you have on the shelf or in the ice box is NAD^+, the coenzyme originally isolated from yeast as "Cozymase". The complete structure was identified in 1936 by Hans von Euler (1). However, I think, it was Otto Warburg, who laid the ground of our present day understanding of coenzyme functions. This man not only revolutionized biochemical methodology, invented devices to

ADP-Ribosylation in Animal Tissue, edited by Haag and Koch-Nolte
Plenum Press, New York, 1997

quantitate oxygen uptake by cells and by extracts, described the function of iron in biological oxidation, introduced procedures to crystallize enzymes. He was also the first to realize the existence and far-reaching significance of non-proteinaceous cofactors for the catalysis of biological processes. And it was his ingenious intuition which had him discover a nicotinamide-containing "Coferment", when he was analyzing erythrocytes brought to consume oxygen by the addition of methylene blue, and to oxidize Robison ester to phosphogluconic acid. He showed that the coferment acted as a reversible hydrogen acceptor, that the hydrogen taken from the substrate was used to reduce the nicotinamide portion of the coenzyme, and that this changed the UV spectrum in such a way as to allow the quantitation not only of the reduced form of the coenzyme, but of all reactions in which it could be involved (2). He thus became the originator of the optical test. How did this man get all the ideas? Perhaps, he has drawn the inspiration from the daily excursions with his mare "Nixe" (Fig. 1). In any case, the pyridine nucleotide story was supposed to be celebrated by a Nobel price in 1944, which would have been the second one for Otto Warburg. However, by a special edict of Hitler, a German was not allowed at that time to accept the Nobel prize.

During the following decades, biochemists were primarily concerned with the elucidation of NAD^+ as a coenzyme of oxidoreductases and the involvement in metabolic pathways. Developments in other fields were required to learn that NAD^+ is more than a general purveyor of reactive hydrogen, that additional potentials are buried in its structure. Especially the concept of group-transfering coenzymes and the existence of energy-rich bonds by Fritz Lipmann opened up new horizons (3). A step in that direction was the discovery by Olivera and Lehman that NAD^+ could serve as an energy donor in bacterial DNA ligation (4). Here, the energy of the pyrophosphate bond was transferred as AMP residue and finally used to form a phospho diester, thus ligating two pieces of DNA (Fig. 2).

Figure 1. Otto Warburg with his mare Nixe. [Leitbild of a biochemist ?]

POLY-ADP-RIBOSYLATION

At about the same time additional capacities of NAD^+ as a group transfering coenzyme were detected. They relate to the glycosidic bond linking ADP-ribose to the quaternary nitrogen of the pyridine ring, the second energy-rich bond in the molecule, which allows the use of "activated ADP-ribose" in a variety of transfer reactions (Fig.2).

The first reaction of this type of transfer was discovered in 1963 in Strassburg, where Paul Mandel and his coworkers looked for RNA and poly(A) synthesis in cell nuclei. When they observed a one thousand-fold stimulation of $[^{32}P]ATP$ incorporation into the acid-insoluble fraction by nicotinamide mononucleotide, they first misinterpreted the result as enhanced poly(A) formation (5). Nevertheless, the finding marks the beginning of the ADP-ribosylation era, and when Janine Doly joined the group, they identified the product as a polymer of ADP-ribose and the substrate as NAD^+ (Table I). At about the same time, the laboratories of Takashi Sugimura in Tokyo and of Osamu Hayaishi in Kyoto reported similar results (Table I).

The field of poly(ADP-ribose) rapidly expanded (Table I): The polymerizing enzyme was purified and characterized as a multifunctional catalyst of about 120 kD, able to transfer a first ADP-ribose residue from NAD^+ to Glu residues of the acceptor proteins, to elongate the ADP-ribose chain, and to branch it. The enzyme depends on nicked DNA for activity, and it possesses an auto ADP-ribosylated domain. Further landmarks were the determination, purification and characterization of degrading enzymes, the development of methods to quantify protein-bound polymeric and monomeric ADP-ribose residues which helped to clarify the possible function of this nuclear reaction.

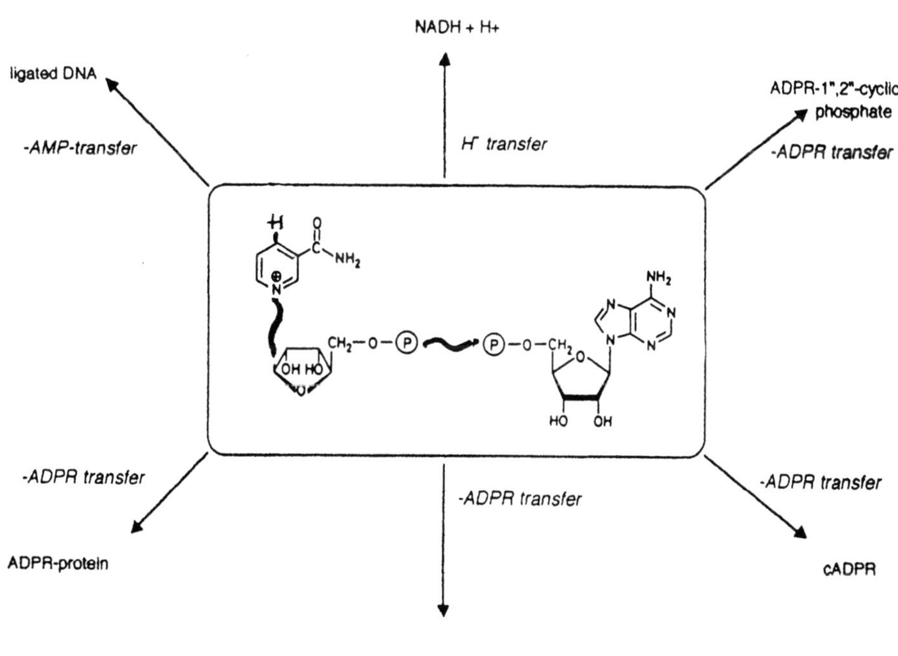

Figure 2. Reactions of NAD^+.

Table 1. Time table on poly-ADP-ribosylation

	Author	Publ.Y.	Ref.[a]
Detection and identification of poly(ADP-ribose)	Chambon et al.	1963/1966	(4,5)
	Fujimura et al.	1967	(4,6)
	Nishizuka et al.	1967	(4,6)
Branched structure of the polymer	Miwa et al.	1979	(8,9)
purification and characterization of	Yoshihara et al.	1978	(6,7)
poly(ADP-ribose)polymerase	Ueda et al.	1982	(6,7)
Three functional domains in	Shizuta et al.	1985	(7)
poly(ADP-ribose)polymerase			
poly(ADP-ribose)polymerase as principal acceptor of	Kawaichi et al.	1981	(7)
poly(ADP-ribose) *in vivo* (automodification)	Adamietz	1981	(7)
Cloning of poly(ADP-ribose)polymerase	Smulson et al.	1987	(8)
	Uchida et al.	1987	(8)
	Kurosaki et al.	1987	(8)
Degrading enzymes	Ueda et al.	1972	(8,9)
	Okayama et al.	1978	(8,9)
modification of chromatin structure	Poirier et al.	1982	(8)
	Althaus et al.	1985	(8)
Quantitation (polymer)	Kanai et al.	1974	(7)
(polymer)	Jacobson et al	1979	(7,11)
(polymer and monomer)	Bredehorst et al.	1978	(7)
	Wielckens et al.	1981	(7)
Participation in DNA excision repair	Durkacz et al.	1980	(10)
poly(ADP-ribose) turnover and NAD+ depletion	Hilz et al.	1982	(11)
after DNA damage	Jacobson et al.	1983	(11)
Similar poly(ADP-ribose)polymerase levels in			
proliferating and non-proliferating cells/ tissues	Ludwig et al.	1988	(12)
Zn finger in poly(ADP-ribose)polymerase	Menissier-de Murcia	1989	(13)
directs recognition of DNA strand breaks			
poly(ADP-ribose)polymerase-deficient cells are	Chatterjee&Berger	1994	(14)
hypersensitive to X-rays and alkylating agents	Küpper et al.	1995	(15)
Mice lacking poly(ADP-ribose)polymerase develop			
normally but are susceptible to skin disease	Wang et al.	1995	(16)

[a] see authors in corresponding sections of cited reviews

During these earlier years, virtually every process occuring in the nucleus was brought in connection with ADP-ribosylation, a special favorite being DNA and cell replication. However, little difference was noted between rapidly proliferating and non-proliferating cells and tissues with respect to polymer content (7). Furthermore, immuno-quantitation of the polymerase protein indicated again similar levels of the catalyst in a number of tissues in different states of growth and proliferation (12). So, the idea never gained momentum. Also, divergent reports on an involvement in differentiation processes may relate to artifactual activation of the polymerase as seen in the differentiation of pre-adipocytes while endogenous poly(ADP-ribose) did not change (8).

As to the possible function of poly(ADP-ribose): a breakthrough came from Shall and coworkers, their data demonstrating participation of poly(ADP-ribose) formation in DNA excision repair, esp. in the ligation step (10). Consonant with these findings was the observation of a dramatic rise of endogenous polymer in response to DNA fragmentation as reported by Juarez-Salinas in Jacobsons laboratory (7,11). A big issue at that time was the question, whether the loss of NAD^+ seen in cells treated with alkylating agents was due to poly(ADP-ribose) formation. Doubts were indeed justified by the fact that the amounts of ADP-ribose residues present in NAD^+ surpass those in mono(ADP-ribosyl) groups 40 fold, and those in poly(ADP-ribose) by orders of magnitude (7). And although in alkylated cells, polymer levels increase by a factor of 100 or so, there remains a vast discrepancy between substrate and product levels (11). Yet it is this fraction which can deplete cells of their entire reservoir of active ADP-ribose, of NAD^+, because DNA fragmentation not only induces a rapid accumulation of polymer but also leads to a dramatic increase of polymer turnover yielding half-life values near one min (11).

By recent analyses using mutant cell lines and mice lacking poly(ADP-ribose)polymerase, the main message comes out clear: poly-ADP-ribosylation is dispensable for normal development, but required for the repair of some damage (13–16).

MONO-ADP-RIBOSYLATING SYSTEMS

In those early days, the development of the ADP-ribosylation story appeared as going from the complex world of poly(ADP-ribose) to the beautiful simplicity of mono-ADP-ribosylation reactions, probably because the toxin transferases act unidirectionally, downhill, without the need for reversible regulation, their biochemical message meaning killing without resurrection. However, on close inspection, we can recognize in this system "complexity" as well. Cholera toxin for instance, exhibits catalytic properties which according to Moss and Vaughan (18), not only comprise ADP-ribosyl transferase activity. The toxin also acts as a NAD^+ glycohydrolase, and it is able to perform an automodification. Furthermore, choleragen responds to the help of GTP-binding proteins, the ARF's, with increased activity (8).

The first toxin found to produce its effect as an ADP-ribosyltransferase was actually diphtheria toxin. Apparently primed by the detection of the poly-ADP-ribosylating system in cell nuclei, the Hayaishi group made a good guess in assuming that the NAD^+ requirement for the toxin's action was due to an ADP-ribosyltransferase reaction modifying elongation factor 2 (19). Additional examples are listed in Tab II.

Complexity of mono-ADP-ribosylation systems increases further, when the covalent modification requires reversibility. It is the merit of Paul Ludden to have provided the first example of a truly regulatory mono-ADP-ribosylation cycle, involving the regulated enzyme, dinitrogenase reductase, an ADP-ribosyltransferase inducing inhibition, and a reactivating glycohydrolase (8).

When methods became availabe to quantitate endogenous protein-bound ADP-ribosyl residues, it became also apparent that in mammalian cells of various origin protein-bound mono(ADP-ribosyl) groups not only outnumbered the polymeric ADP-ribosyl residues by factors of 20–100, but also that different types of mono(ADP-ribosyl)polypeptides exist that can be distinguished by their sensitivity toward specific chemical agents, like mercury salts introduced by Thomas Meyer for the detachment of cysteine-linked ADP-ribose (8), or hydroxylamine indicating arginine-bound (and glutamic acid-linked polymeric) residues (20). Interesting changes of such conjugates with respect to amounts and composition were reported quite early, for instance during development of the liver (20).

More recently, defined acceptor proteins modified at specific amino acid residues have been described (21). The heterogeneity of such conjugates in vertebrate systems was further substantiated by the detection of acceptor-specific transferases as well as (linkage-specific) hydrolases. The brilliant work of Joel Moss and his associates, which includes themes like integrin a7 as a substrate of GPI-anchored transferase or the detection of sequence homologies with the T cell marker RT6 has set landmarks in this rapidly expanding field.

Arriving at RT6, I come to the line from NAD^+ to Heinz-Günter Thiele. I must remind you that he is originally a whole-hearted immunologist. And it meant for him to travel a rather curved road until he arrived at cloning surface proteins and determining ribosyl transferase activities. Therefore, when I draw that line originating at NAD$^+$, it has to wind itself around many obstacles before it reaches its goal, Heinz-Günter at his best (Fig. 3). Indeed, impressive results have been obtained by the group at the Abteilung für Immunologie (cf. 22 and Table II. Only two of their highlights are listed here).

Additional fascinating stories are emerging in the field of mono-ADP-ribosylation processes, like protein traffic through membranes, which includes the physiological function of ARF (cf. 23) and possibly the reversible ADP-ribosylation of a new class of G-proteins as discovered with the aid of Brefeldin A (23). Consider also the recent detection of an entire new class of ADP-ribosyl proteins by Jacobson et al. (24).

CYCLIC ADP-RIBOSE

In the third field, the ADP-ribose moiety of NAD$^+$ plays again a key role. The field has been opened by H.C. Lee. In a series of inspirational papers, he and his group have described the formation, structure and mechanism of action of the calcium releasing factor cyclic ADP-ribose (cADPR) (25,27). In the meantime, the function as a calcium mobilizing agent has been firmly established in various systems, at least *in vitro* (27).

In the synthetic reaction forming cADPR, the activated N-glycosidic linkage in NAD$^+$ is used to produce an intramolecular bond leading to the production of cADPR (Fig. 4). The link involves N1 of the adenine ring, as shown first by Jacobson (26). This is surprising since both linkages, in the substrate as well as in the product, are formally identical, both being glycosidic to a quarternary nitrogen, both being energy-rich. The cyclase reaction, therefore, should not go to completion, in contrast to the transferase reaction, where a low-energy bond is formed. Indeed, most ADP-ribosyl cyclases exhibit primarily NAD$^+$ glycohydrolase but very little (<5%) cyclase (and cADPR hydrolase) activities. Even the Aplysia cyclase, which is practically devoid of cADPR hydrolase activity, cata-

Figure 3. The line from NAD$^+$ to H. G. Thiele.

Table 2. Time table on mono-ADP-ribosylation

	Author	Publ.Y.	Ref.[a]
A. *Toxins*			
Diphtheria toxin an ADP-ribosyltransferase	Honjo et al.	1968	(17)
Cholera toxin ADP-ribosylates Gs α	Moss & Vaughan	1977/1988	(18,19)
	Cassel & Pfeuffer	1978	(19)
	Gill & Meren	1978	(19)
Activation of Cholera toxin by ARF	Kahn & Gilman	1984	(19)
	Tsai et al.	1987	(19)
Pertussis toxin ADP-ribosylates Gi α	Katada & Ui	1982	(19)
Clostr. botulinum toxins as ADP-ribosyltransferases	Aktories et al.	1986	(19)
	Rubin et al.	1988	(19)
B. *Nitrogen-fixing bacteria*			
Regulation of *Rhodospirillum* nitrogenase by ADP-ribosylation	Pope & Ludden	1985	(8)
C. *Vertebrate systems*			
Linkage heterogeneity of mono(ADP-ribosyl) proteins and quantitation	Bredehorst et al.	1978	(9)
	Adamietz et al.	1981	(9)
ADP-ribosyltransferase (Arg-spec.; erythrocytes)	Moss et al.	1980	(21)
ADP-ribosylprotein hydrolase	Moss,Jacobson	1985	(21)
ADP-ribosyltransferase (Cys-spec.; erythrocytes)	Tanuma et al.	1987	(21)
Integrin α7 as substrate of GPI-anchored ADP-ribosyltransferase from muscle	Zolkiewska et al.	1992/1993	(21)
Cloning of RT6	Koch, Haag & Thiele	1990	(21)
Cloning of Arg-spec. ADP-ribosyl transferase	Zolkiewska et al.	1992	(21)
RT6 is an ADP-ribosyltransferase	Koch-Nolte et al.	1996	(22)
Brefeldin A inhibits ARF binding to Golgi and stimulates endog. ADP-ribosylation	Girolamo et al.	1995	(23)
Acid-labile (acetal-linked) ADP-ribosyl proteins	Cervantes- et al.	1995	(24)

[a] see authors in corresponding sections of cited reviews

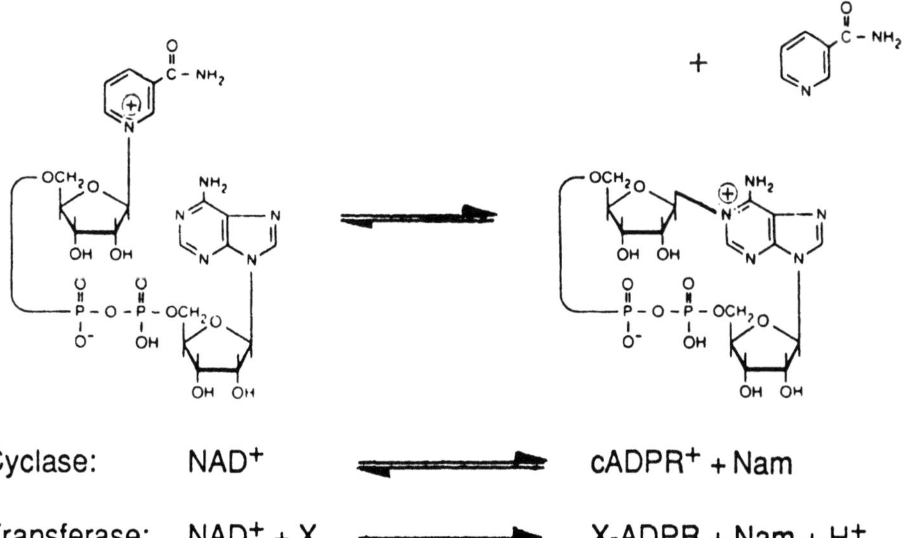

Cyclase: NAD^+ ⇌ $cADPR^+ + Nam$

Transferase: $NAD^+ + X$ ⟶ $X\text{-}ADPR + Nam + H^+$

Figure 4. Reversibility of the ADP-ribosyl cyclase reaction.

Table 3. Time table on ADP-ribosyl cyclases

	Author	Publ. Y	Ref.[a]
Discovery of cADPR as a Ca^{2+}-releasing factor	Lee et al.	1989	(25)
Structure of cADPR	Kim et al.	1993	(26)
	Lee et al.	1989/1994	(25,27)
Purification and cloning of Aplysia cyclase	Hellmich et al.	1991/1992	(27)
cADPR and insulin secretion	Takasawa et al.	1993	(26,27)
NADase from spleen is also a cADPR cyclase and cADPR hydrolase	Kim et al.	1993	(26)
CD38 acts as a cADPR cyclase and cADPR hydrolase	Zocchi et al.	1993	(26,27)
	Takasawa et al.	1993	(26,27)
2′-Phospho cADPR+ synthesized from $NADP^+$ by	Zhang et al.	1995	(28)
cyclases is another Ca^{2+} releasing agent	Jacobson et al.	1995	(29)
	Guse et al.	1996	
CD38 exhibits Cys-specific ADP-ribosyltransferase activity	Grimaldi et al.	1995	(30)

[a] see authors in corresponding sections of cited reviews

lyzes a reaction which comes to a halt long before the substrate is used up. It seems strange that a reaction producing a messenger-like metabolite should not go to completion. Things would change, if the equilibrium of the cyclase reaction would be shifted to the right by removal of nicotinamide, e.g. by methylation to N-methyl nicotinamide, or by deamidation to nicotinate.

More compounds of the pyridine nucleotide family may play a role in calcium signalling, like cyclic 2′-phospho $ADPR^+$ produced by ADP-ribosyl cyclases with $NADP^+$ as substrate (28,29), or $NAADP^+$, the nicotinic acid analog of $NADP^+$ (27)(Table III). Finally, one should not neglect the surprising observation of Grimaldi et al. (30) who reported that the cyclase CD38 exhibits also ADP-ribosyl transferase activity.

CONCLUSIONS AND PERSPECTIVES

When we compare the three groups of ADP-ribose-handling enzymes, it appears that they are characterized, among other features, by a rather unusual accumulation of divergent catalytic potentials. This applies especially to the poly-ADP-ribosylating enzyme, which initiates, elongates and branches polymer chains. It also exhibits significant NAD^+ glycohydrolase and auto modification activities. Transferase, hydrolase and automodifying activities are also apparent in toxins and in mammalian mono(ADP-ribosyl)transferases (19,21). Even the ADP-ribosyl cyclases have more than two catalytic capacities: They synthesize and hydrolize cADPR, they also hydrolize NAD^+ certainly not via free cADPR, and in the case of CD38 at least, they are able to act as an ADP-ribosyltransferase. The biological significance of the multicatalytic properties of these enzymes remains to be determined.

I come to the end of my historical review. Many impressive results have been achieved which indicate important biological functions of ADP-ribosylation processes. Yet, a number of problems remain unsolved. Not solved is the molecular mechanism whereby poly-ADP-ribosylation primes chromatin proteins for an active role in DNA excision repair; enigmatic is the supply of NAD^+ for ecto-ADP-ribosyltransferases, or the removal of ADP-ribose groups from external acceptor proteins. Other fields involving ADP-ribose are just too young to be prepared for complete answers. Thus, ADP-ribosylation emerges as one of the most fascinating fields open for many more surprises! Dr. Haag and Dr. Koch-Nolte have organized this timely meeting, which by virtue of an excellent selection of topics and experts promises to become one of the most stimulating sessions on ADP-ribose. It is the third one on this subject to be held in Hamburg.

REFERENCES

1. Von Euler, H., H. Albers, & F. Schlenk. 1936. Chemische Untersuchungen an hochgereinigter Co-Zymase. *Hoppe-Seyl. Z.Physiol.Chem. 240*: 113–126.

2. Warburg, O., W. Christian, & A. Griese. 1935. Wasserstoffübertragendes Co-Ferment, seine Zusammensetzung und Wirkungsweise. *Biochem. Z. 282*: 157–165; Warburg, O. 1948 . *Wasserstoff-übertragende Fermente*. W. Saenger-Verlag, Berlin.

3. Lipmann, F. 1941. Metabolic generation and utilization of phosphate bond energy. *Advances in Enzymology, 1*, 99–126.

4. Hilz, H. & P. Stone. 1976. Poly(ADP-Ribose) and ADP-Ribosylation of proteins. *Rev. Physiol. Biochem. Pharmacol. 76*:1–58.

5. Chambon, P., J.D. Weill, & P. Mandel. 1963. Nicotinamide mononucleotide activation of a new DNA-dependent polyadenylic acid synthesizing nuclear enzyme. *Biochem. Biophys. Res. Commun. 11*:39–43.

6. Hayaishi, O. & K. Ueda. 1977. Poly(ADP-Ribose) and ADP-Ribosylation of proteins. *Ann. Rev. Biochem. 46*: 95–116

7. Althaus, F.R., H. Hilz, & S. Shall. 1985. *ADP-Ribosylation of Proteins*. Springer Verlag, Berlin.

8. Jacobson, M.K. & E.L. Jacobson. 1989. *ADP-Ribose Transfer Reactions*. Springer Verlag, New York.

9. Hayaishi, O. & K. Ueda (Eds.) 1982. *ADP-Ribosylation Reactions*. Academic Press,Inc., New York.

10. Durkacz, B., O. Omidiji, D. Gray, & S. Shall. 1980. Poly (ADP-ribose) participates in DNA excision repair. *Nature 283*: 593–596.

11. Miwa, M., O. Hayaishi, S. Shall, M. Smulson, & T. Sugimura. 1983. *ADP-Ribosylation, DNA Repair and Cancer*. Japan Scientific Societies Press, Tokyo.

12. Ludwig, A., B. Behnke, J. Holtlund, & H. Hilz. 1988. Immunoquantitation and size determination of intrinsic poly(ADP-ribose)polymerase from acid precipitates. An analysis of the in vivo status in mammalian cells and in lower eukaryotes. *J.Biol.Chem. 263*: 6993–6999.

13. Menissier-de Murcia, J., M. Molinete, G. Gradwohl, F. Simonin, & G. de Murcia. 1989. Zinc-binding domain of poly(ADP-ribose)polymerase participates in the recognition of single strand breaks on DNA. *J.Mol.Biol. 210*: 229–233

14. Chatterjee S., M.F. Cheng, & S.J. Berger. 1991. Alkylating agent hypersensitivity in poly(adenosine diphosphate-ribose)polymerase deficient cell lines. *Cancer Commun. 3*: 71–75;

15. Küpper, J.H., M. Müller, M.K. Jacobson, J. Tatsumi, D.L. Coyle, E.L. Jacobson, & A. Bürkle. 1995. Transdominant inhibition of poly-ADP-ribosylation sensitizes cells against g-irradiation and NMNG but does not limit DNA replication of a polyomavirus replicon. *Mol.Cell.Biol. 15*: 3154–3163.

16. Wang, Z.Q., B. Auer, L. Stingl, H. Berghammer, D. Haidacher, M. Schweiger, & E.F. Wagner. 1995. Mice lacking ADPRT and poly-ADP-ribosylation develop normally but are susceptible to skin disease. *Genes and Development 9*: 509–520.

17. Honjo, T., Y. Nishizuka, O.Hayaishi, & I. Kato. 1968. Diphtheria toxin-dependent adenosine diphosphate ribosylation of aminoacyl transferase II and inhibition of protein sythesis. *J.Biol.Chem. 243*: 3553–3555.

18. Moss, J. & M. Vaughan.1977. Mechanism of action of choleragen. Evidence of ADP-ribosyltransferase activity with arginine as an acceptor. *J.Biol.Chem. 252*: 2455–2457.

19. Pekala, P.H. & B.M. Anderson. 1982. Non-oxidation-reduction reactions of pyridine nucleotides. in: Everse, J., B. Anderson, & K. You (eds.) *The pyridine nucleotide coenzymes*. Academic Press, New York.

20. Hilz, H., R. Bredehorst, P. Adamietz, & K. Wielckens. 1982. Subfractions and subcellular distribution of mono(ADP-ribosyl) proteins in eukaryotic cells. In: Hayashi O. & K. Ueda (eds.). *ADP-Ribosylation Reactions*. Academic Press, Inc.. New York.

21. Okazaki, I.J., A. Zolkiewska, T. Takada, & J. Moss. 1995. Characterization of mammalian ADP-ribosylation cycles. *Biochimie 77*: 319–325.

22. Koch-Nolte, F., D. Petersen, S. Balasubramanian, F. Haag, D. Kahlke, T. Willer, R. Kastelein, F. Bazan, & H.-G. Thiele. 1996. Mouse T cell membrane proteins Rt6–1 and Rt6–2 are arginine/protein mono(ADP-ribosyl)transferases and share secondary structure motifs with ADP-ribosylating bacterial toxins. *J.Biol.Chem. 271*: 7686–7693.

23. Girolamo, M.D., M.G. Siletta, M.A. De Matteis, A. Braca, A. Colanzi, D. Pawlak, M.M. Rasenick, A. Luini, & D. Corda. 1995. Evidence that the 50-kDa substrate of brefeldin A-dependent ADP-ribosylation binds GTP and is modulated by the G-protein βγ subunit complex. *Proc.Natl.Acad.Sci.USA 92*:7065–7069.

24. Cervantes-Laureant, D., P.T.Loflin, D.E.Minter, E.L.Jacobson,& M.K.Jacobson. 1995. Protein modification by ADP-ribose via acid-labile linkages. *J.Biol.Chem. 270*:7929–7936.

25. Lee,H.C., T.F.Walseth, G.T.Bratt, R.N.Hayes, & D.L.Clapper. 1989. Structural determination of a cyclic metabolite of NAD$^+$ with intracellular Ca^{2+}-mobilizing activity. *J.Biol.Chem. 264*:1608–1615.

26. Jacobson, M.K., J. Amé, W. Lin, D.L. Coyle, & E.L. Jacobson. 1995. Cyclic ADP-Ribose. *Receptor 5*: 43–49.

27. Lee, H.C., R. Graeff, & T.F.Walseth. 1995. Cyclic ADP-ribose and its metabolic enzymes. *Biochimie 77*:345–355.

28. Zhang F, Q. Gu, P. Jing,& C.J.Sih. 1995. Enzymatic cyclization of nicotinamide adenine dinucleotide phosphate (NADP). *Bioorg.Med.Chem. Letters 5*: 2267–2272;

29. Jacobson M.K., Vu C.Q., P. Lu, C. Chen,& M.K.Jacobson. 1995. 2′-Phospho cyclic ADP-ribose, a calcium-mobilizing agent derived from NADP. *J.Biol.Chem. 271*:4747–4754.

30. Grimaldi, J.C., S. Balasubramanian, N.H.Kabra, A. Shanafelt, J.F.Bazan, G. Zerawski, & M.C.Howard. 1995. CD38-Mediated Ribosylation of Proteins. *J. Immunol. 155*: 812–817.

3

ADP-RIBOSYLARGININE HYDROLASES AND ADP-RIBOSYLTRANSFERASES

Partners in ADP-Ribosylation Cycles

Joel Moss, Anna Zolkiewska, and Ian Okazaki

Pulmonary-Critical Care Medicine Branch
National Heart, Lung, and Blood Institute
National Institutes of Health
Bethesda, Maryland 20892

ABSTRACT

Mono-ADP-ribosylation is a reversible modification of arginine residues in proteins, with NAD:arginine ADP-ribosyltransferases and ADP-ribosylarginine hydrolases constituting opposing arms of a putative ADP-ribosylation cycle. The enzymatic components of an ADP-ribosylation cycle have been identified in both prokaryotic and eukaryotic systems. The regulatory significance of the cycle has been best documented in prokaryotes. As shown by Ludden and coworkers, ADP-ribosylation controls the activity of dinitrogenase reductase in the phototropic bacterium *Rhodospirillum rubrum*. ADP-ribosylation of other amino acids, such as cysteine, has also been demonstrated, lending credence to the hypothesis that this modification is heterogeneous.

In eukaryotes, the functional relationship between ADP-ribosyltransferases and ADP-ribosylarginine hydrolases is less well documented. The transferase-catalyzed reaction results in stereospecific formation of α-ADP-ribosylarginine from β-NAD; ADP-ribosylarginine hydrolases specifically cleave the α-anomer, leading to release of ADP-ribose and regeneration of the free guanidino group of arginine. The two reactions can thus be coupled *in vitro*. Coupling *in vivo* is dependent on cellular localization. The deduced amino acid sequences of ADP-ribosyltransferases from avian and mammalian tissues have common consensus sequences involved in catalytic activity but, in some instances, enzyme-specific cellular localization signals. The presence of amino- and carboxy-terminal signal sequences is consistent with the glycosylphosphatidylinositol(GPI)-anchoring to the cell surface. The muscle and lymphocyte transferases ADP-ribosylate integrins. Some transferases lack the carboxy- terminal signal sequence needed for GPI-anchoring. Most ADP-ribosylarginine hydrolase activity is cytosolic, although perhaps some is located at the cell surface. Deduced amino acid sequences of hydrolases from a number of mammal-

ADP-Ribosylation in Animal Tissue, edited by Haag and Koch-Nolte
Plenum Press, New York, 1997

ian species are consistent with their cytoplasmic localization. Katada and coworkers have determined, however, that auto-ADP-ribosylated RT6, a GPI-linked protein, is metabolized by a hydrolase-like activity, consistent with the existence of an ADP-ribosylation cycle. ADP-ribosyl RT6 may be internalized, thereby coming in contact with the cytosolic hydrolase; alternatively, a novel form of the hydrolase may be located at the surface. The mechanism of coupling of ADP-ribosyltransferases and hydrolases in eukaryotic ADP-ribosylation cycles has yet to be clarified.

IDENTIFICATION AND CHARACTERIZATION OF ADP-RIBOSYLTRANSFERASES

ADP-ribosylation and ADP-ribosyltransferases have been reviewed recently[1,2]. The NAD:arginine ADP-ribosyltransferases were first isolated from avian tissues. In turkey erythrocytes, there is a family of transferases that exhibits different kinetic, regulatory and physical properties and intracellular localizations[3–6]. Erythrocyte ADP-ribosyltransferase A, a soluble protein, is activated by histone, NaCl, and certain phospholipids, and detergents with the activity of the protein determined by its state of self-association; the monomeric transferase is active, the aggregated protein, inactive[7–9]. ADP-ribosyltransferase B is membrane-associated and inhibited by salt[5]. Both enzymes are minor components of the cell and required extensive purification to achieve homogeneity. In mammalian cells, muscle and lymphocyte transferases appear to be anchored to the external surface through a phosphatidylinositol-specific phospholipase C (PI-PLC)-sensitive linkage identified by immunoreactivity and radiolabeling studies as a glycosylphosphatidylinositol anchor[10,11].

The transferases catalyze a stereospecific reaction (Fig. 1), utilizing β-NAD as a substrate and forming the α-anomeric ADP-ribosylarginine[12]. The S_N2-like reaction is similar to that catalyzed by the cholera toxin NAD:arginine ADP-ribosyltransferase[13]. Although the α-anomeric product can anomerize *in vitro*, leading to the formation of the β-anomer, it is not clear whether anomerization occurs *in vivo*, where the ADP-ribose attached to a protein may be physically constrained. As noted below, the ADP-ribosylarginine hydrolase cleaves only the α-anomer[14].

To understand the structure and expression of the ADP-ribosyltransferases in more detail, the cDNAs were cloned from a number of sources including humans, mouse, rabbit, and chicken[10,11,15–18]. Deduced amino acid sequences of the avian and mammalian eukaryotic transferases have hydrophobic amino termini, compatible with signal sequences

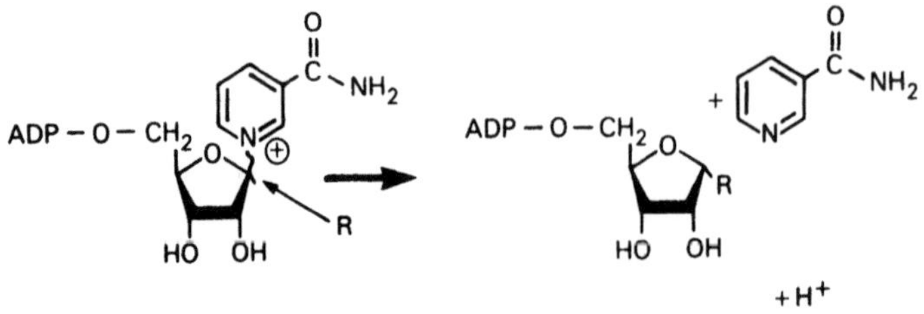

Figure 1. Stereochemistry of the NAD:arginine ADP-ribosyltransferase-catalyzed reaction. In this reaction, β-NAD is the substrate; the S_N2-like reaction proceeds with inversion of configuration and formation of α-anomeric ADP-ribosylarginine.

used for the transport of proteins into the endoplasmic reticulum. In some instances, notably the muscle and lymphocyte transferases, there are hydrophobic carboxy-termini, compatible with signal sequences used for the addition of a glycosylphosphatidylinositol anchor[10,15]. When the cDNAs have been expressed in mammalian cells, the mature enzyme is present on the cell surface and can be released by incubation with phosphatidylinostiol-specific phospholipase C[10,11]. With a testis and chicken transferases, however, the carboxy-terminus does not exhibit the characteristic hydrophobicity, leading to the conclusion that the protein may be secreted[16,18]. Models for the transferases discussed thus far are shown schematically in Fig. 2. Clearly further information obtained through cloning or other means is needed for a fuller understanding of the intracellular transferases in avian tissues.

The mammalian and avian transferases contain consensus sequences, similar to regions found in the bacterial toxins[2]. Based on a large number of mutagenesis and cross-linking experiments, it appears that some of these regions are involved in the formation of the catalytic site (Fig. 2). One region appears to have a histidine or arginine and may participate in hydrogen bonding. The active site region has a key glutamate, which can be cross linked to the nicotinamide moiety of NAD[2].

The ADP-ribosyltransferases that have been sequenced thus far, appear to be localized, based on the presence of specific signal sequences, either to secretory compartments or to the cell surface through the GPI anchor. Localization would appear to play an important role in the choice of substrates for modification. In intact muscle cells, integrin α_7, part of the heterodimeric cell surface protein $\alpha_7\beta_1$, is ADP-ribosylated by the muscle ADP-ribosyltransferase[19]. In cytotoxic T cells (CTL), a GPI-linked transferase modifies a cell surface 40-kDa protein involved in the regulation of p56[lck] (Ref. 20). The NAD:arginine ADP-ribosyltransferases are capable of using as substrates, simple guanidino compounds (e.g., arginine, agmatine) as well as proteins apparently unrelated to their regulatory role in cells, e.g., histones[1,2]. Thus, the fact that a protein is modified *in vitro* does not necessarily mean that it is part of *in vivo* signalling pathways involving a trans-

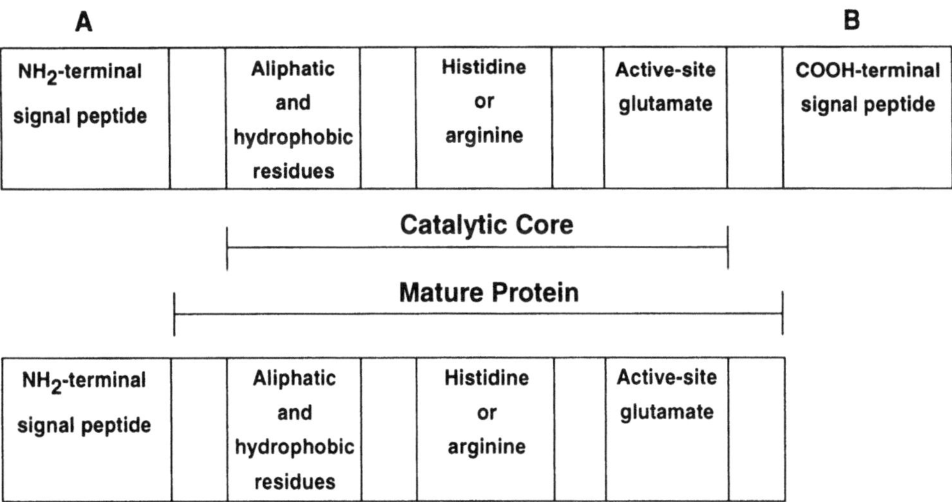

Figure 2. Schematic of proposed consensus domains for mammalian and avian ADP-ribosyltransferases. Regions A and B denote hydrophobic amino and carboxy terminal signal sequences. Some common consensus regions are noted.

ferase. Disruption of muscle cell integrity results in the modification of proteins other than integrin α_7, presumably due to the presence of accessible arginine(s). Thus, protein ADP-ribosylation needs to be interpreted with care.

Most of the mono-ADP-ribosylation studies to be discussed at this conference involve proteins that catalyze the ADP-ribosylation of arginine or similar simple guanidino compounds. To a certain extent, this may reflect the ease of the assay rather than the predominance of the ADP-ribosylated substrates or of the specific ADP-ribosyltransferases. The eukaryotic NAD:arginine ADP-ribosyltransferases catalyze reactions similar to the bacterial toxins, cholera toxin and *E. coli* heat-labile enterotoxin[21]. The initial observations that the toxins modified arginine and related compounds, in addition to the ADP-ribosylation of the heterotrimeric guanine nucleotide-binding protein, G_s, led to the obvious question as to whether the toxins could be mimicking, in part, a normal eukaryotic regulatory mechanism[3].

ADP-RIBOSYLATION OF CYSTEINE AND OTHER AMINO ACIDS

There is strong evidence that mono-ADP-ribose is attached to many amino acids *in vivo*. Dr. Hilz pioneered these studies, by developing mechanisms for differentiating the different ADP-ribose-amino acid linkages based on chemical stability[22]. Further advances in the identification of ADP-ribosylated amino acids were made by the Jacobsons[23-26]. It would appear that cysteine and serine/threonine are ADP-ribosylated, in addition to arginine.

NAD:cysteine ADP-ribosyltransferases, characterized, based, in part, on assays using the free amino acid, and ADP-ribosylcysteine hydrolases, have also been described[27,28]. Cysteine is much more reactive than arginine; free ADP-ribose readily forms thiazolidones with cysteine[29]. In these molecules, ADP-ribose is linked to cysteine in a cyclic compound involving the amino and sulfhydryl groups. The nonenzymatic ADP-ribosylation of free cysteine results from its reaction with free ADP-ribose generated from NAD by the action of NAD glycohydrolases. Both the thiazolidone product and the enzymatically formed ADP-ribosyl-cysteine are sensitive to mercuric chloride[29]. The thiazolidone can, however, be differentiated from an enzymatically formed ADP-ribosyl-cysteine based on its sensitivity to hydroxylamine[29]. The enzymatic ADP-ribosylation of cysteine in the guanine nucleotide-binding protein transducin or G_t by pertussis toxin results in a bond that is labile to $HgCl_2$, but not to hydroxylamine[30]. A further complication was introduced into the analysis when it was observed that free ADP-ribose can react with cysteines in proteins leading to the formation of $HgCl_2$-, but not hydroxylamine-sensitive, linkages[31]. Thus, in these instances, enzymatic and nonenzymatic ADP-ribosylation cannot, at least on this superficial level, be differentiated. Other experimental procedures can be used to differentiate between enzymatic and nonenzymatic ADP-ribosylation (e.g., addition of excess free ADP-ribose to dilute the radiolabeled product of NAD breakdown). Unfortunately, there are alternative explanations for some of the data generated by these procedures (e.g., inhibition of the transferase by free ADP-ribose).

ADP-RIBOSYLARGININE HYDROLASES

ADP-ribosylarginine hydrolases release ADP-ribose from modified proteins (arginine); hydrolase activity has been detected in mammalian, avian and bacterial tissues[1,2].

A regulatory ADP-ribosylation cycle was proposed based on the presence of NAD:arginine ADP-ribosyltransferases and ADP-ribosylarginine hydrolases (Fig. 3). In the photosynthetic bacterium *Rhodospirillum rubrum*, dinitrogenase reductase, which is part of a nitrogen-reducing enzyme complex, is inactivated by ADP-ribosylation of a critical arginine catalyzed by dinitrogenase reductase ADP-ribosyltransferase (DRAT) in response to environmental stimuli such as darkness, or a source of fixed nitrogen[32]. Dinitrogenase reductase ADP-ribose glycohydrolase (DRAG) removes the ADP-ribose moiety after exposure of the bacteria to light or depletion of the nitrogen source, regenerating free arginine and activating dinitrogenase reductase, completing the ADP-ribosylation cycle[32]. ADP-ribosylarginine hydrolases have been purified from several animal tissues including turkey erythrocytes[33], and hydrolase cDNAs have been cloned from rat[34], mouse[35], and human[35] brain.

TURKEY ERYTHROCYTE ADP-RIBOSYLARGININE HYDROLASES

A 39-kDa ADP-ribosylarginine hydrolase was extensively purified from turkey erythrocytes[33]. The enzyme required Mg^{2+} (5–10 mM) and dithiothreitol (5–10 mM) for maximal activity, and was inhibited by NaF (5 mM) or NaCl (200 mM)[36]. The ADP-ribose moiety, but not the guanidino-containing compound, was crucial for substrate recognition[14]. The hydrolase was inhibited competitively by ADP-ribose > ADP > AMP; there was no effect of arginine, agmatine, or guanidine. Arginine, generated from ADP-ribosyl-

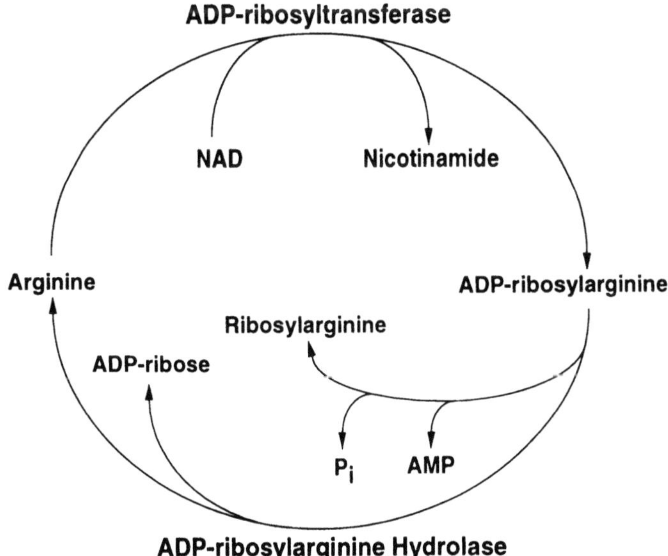

Figure 3. An ADP-ribosylation cycle. Within the framework of this model, NAD:arginine ADP-ribosyltransferases catalyze the stereospecific modification of arginine residues in proteins, leading to formation of the α-anomer of ADP-ribosylarginine. ADP-ribosylarginine hydrolases catalyze cleavage of α-ADP-ribose-arginine, resulting in the formation of ADP-ribose and regeneration of the guanidino group of arginine. In some instances the ADP-ribosyl-protein is cleaved by a pyrophosphatase leading to release of AMP and formation phosphoribosyl-protein.

arginine by treatment with the ADP-ribosylarginine hydrolase, served as a substrate for the turkey erythrocyte ADP-ribosyltransferase, indicating that the guanidino group of arginine was intact following removal of ADP-ribose[36]. Moreover, the hydrolase, like the transferase, catalyzed a stereospecific reaction; the α-anomer of ADP-ribosylarginine, generated stereospecifically in the transferase-catalyzed reaction, was utilized as a substrate by the hydrolase; the β-anomer was not hydrolyzed[14]. This finding is consistent with the notion that the ADP-ribosyltransferases and the ADP-ribosylarginine hydrolases serve as opposing arms of an ADP-ribosylation cycle.

MAMMALIAN ADP-RIBOSYLARGININE HYDROLASES

An ADP-ribosylarginine hydrolase was purified ~20,000-fold from rat brain[34]. Hydrolases were also partially purified from other rat tissues and brains from several other species for structural and functional comparisons[34]. Among rat tissues, hydrolase activity was greatest in brain, spleen, and testis. Hydrolase activities in rat and mouse brain were higher than those in guinea pig, rabbit, sheep, pig, and calf brain. The rat and mouse brain hydrolases were stimulated by Mg^{2+} and dithiothreitol, properties also of the turkey hydrolase, whereas pig and calf hydrolases were stimulated by Mg^{2+} but not dithiothreitol; the human hydrolase appeared to be dithiothreitol independent as well. Additionally, polyclonal rabbit anti-rat brain hydrolase antibodies reacted on immunoblot with 39-kDa proteins from turkey erythrocytes, and mouse, rat and calf brain, despite differences in enzymatic characteristics[34].

The rat brain hydrolase coding region cDNA, cloned from a cDNA library, has an open reading frame of 1089 bp that encodes a 363-amino acid protein[34]. The coding region cDNAs of the mouse and human brain hydrolases were cloned by PCR-based techniques[35]. The nucleotide and deduced amino acid sequences of the mouse hydrolase cDNA were, respectively, 92 and 94% identical to those of the rat; human and rat hydrolases were 82 and 83% identical in nucleotide and deduced amino acid sequences, respectively.

A rat brain hydrolase cDNA probe hybridized on Northern analysis with a 1.7-kb band in total RNA from rat tissues[34]. Levels were highest in brain, spleen, and testis, which correlated with levels of enzyme activity. The rat brain hydrolase cDNA did not hybridize with poly(A)$^+$RNA from chicken, rabbit, or bovine brain. A human hydrolase-specific oligonucleotide probe hybridized with a 4-kb band in poly(A)$^+$RNA from human brain, lung, and placenta, but not IMR-32, or undifferentiated HL-60 cells[35].

In rat and mouse hydrolases, the positions of five cysteines were identical, whereas only four of the five cysteines were present in the human hydrolase[35]. In the human hydrolase, serine 103 corresponds to cysteine 108 in the rat and mouse sequences. Since the hydrolases from rat and mouse, but not human tissues, were stimulated by dithiothreitol, site-directed mutagenesis was utilized to determine whether this difference in thiol-dependence was related to the fifth cysteine[35]. The rat hydrolase cDNA, synthesized in *E. coli*, demonstrated Mg^{2+}- and dithiothreitol-dependence, identical to that observed with the native enzyme, whereas the wild-type recombinant human hydrolase was, as expected, remained thiol-independent. Expression in *E. coli* of a mutant human hydrolase, in which serine 103 was replaced by a cysteine, generated a thiol-dependent enzyme similar to the native rat hydrolase. On the other hand, a mutant rat hydrolase containing serine at position 108 instead of cysteine, was thiol-independent, whereas the similarly expressed wild-type rat hydrolase was thiol-dependent. The mutant rat and human hydrolases were identical to their respective native enzymes in Mg^{2+} requirements and specific activities.

Anti-hydrolase antibodies reacted on immunoblot with the native and recombinant wild-type rat hydrolases, but only weakly with the mutant rat hydrolase and not at all with the wild-type or mutant human hydrolases[35]. The differences in dithiothreitol requirements, and perhaps immunoreactivity, observed between species among the wild-type and mutant recombinant hydrolases may be due to the replacement of a cysteine, which is identical to that occurring between native rat and human hydrolases.

The rat brain hydrolase, synthesized as a GST-fusion protein, hydrolyzed the ADP-ribose group from $G_{s\alpha}$ that had been ADP-ribosylated by cholera toxin[37]. The recombinant hydrolase similarly removed the ADP-ribose moiety from an auto-ADP-ribosylated cholera toxin A subunit. Nonmuscle actin modified by botulinum C2 toxin was also a substrate for the recombinant hydrolase enzyme[37]. On the other hand, elongation factor 2 ADP-ribosylated by diphtheria toxin, $G_{o\alpha}$ modified by pertussis toxin, and the Rho protein modified by C3 exoenzyme were not substrates for the recombinant hydrolase[37], consistent with the fact that the ADP-ribosylarginine hydrolase specifically hydrolyzes the ADP-ribose arginine bond.

The ADP-ribosylarginine hydrolases appear to be soluble, intracellular proteins. If they participate in an ADP-ribosylation cycle, the question arises as to how they interact with cell surface proteins modified in their extracellular domains by GPI-linked transferases. In muscle cells, it appears that the ADP-ribosylated integrin α_7 is processed by a pyrophosphatase, rather than a hydrolase, resulting in the formation of phosphoribosylarginine, and perhaps, ribosylarginine, leaving, obviously, a blocked guanidino group[38]. Katada and coworkers[39] observed that auto-ADP-ribosylated rat RT6.2 is de-ADP-ribosylated with release of a ADP-ribose and regeneration of arginine on the protein. The mechanism of that reaction is not certain; although it was postulated that a cell surface hydrolase may participate. Based on these studies, it appears that an ADP-ribosylation cycle may in fact be composed of alternative pathways based on whether the intact ADP-ribose moiety is removed or only AMP is removed, leaving the phosphoribose attached to the protein[38]. Presumably this residue would then be subjected to cleavage by phosphatases. These data would support the concept of a modified ADP-ribosylation cycle, with either a free guanidino group of arginine or phosphoribosylarginine being generated (Fig. 3).

REFERENCES

1. Williamson K. C., & J. Moss. 1990. Mono-ADP-ribosyltransferases and ADP-ribosylarginine hydrolases: A mono-ADP-ribosylation cycle in animal cells. In: *ADP-ribosylating Toxins and G Proteins: Insights into Signal Transduction*, J. Moss, and M. Vaughan (eds). American Society for Microbiology, Washington, D.C., p. 493–510.
2. Okazaki, I. J., & J. Moss. 1996. Mono-ADP-ribosylation: A reversible posttranslational modification of proteins. In: *Advances in Pharmacology*. T. J. August, M. W. Anders, F. Murad, & J. T. Coyle (eds). Academic Press, San Diego, CA, *35*: 247–280.
3. Moss, J., & M. Vaughan. 1978. Isolation of an avian erythrocyte protein possessing ADP-ribosyltransferase activity and capable of activating adenylate cyclase. *Proc. Natl. Acad. Sci. USA. 75*: 3621–3624.
4. Moss, J., S. J. Stanley, & P. A. Watkins. 1980. Isolation and properties of an NAD- and guanidine-dependent ADP-ribosyltransferase from turkey erythrocytes. *J. Biol. Chem. 255*: 5838–5840.
5. Yost D. A., & J. Moss. 1983. Amino acid-specific ADP-ribosylation. Evidence for two distinct NAD:arginine ADP-ribosyltransferases in turkey erythrocytes. *J. Biol. Chem. 258*: 4926–4929.
6. West, R. E. Jr, & J. Moss. 1986. Amino acid specific ADP-ribosylation: Specific NAD:arginine mono-ADP-ribosyltransferases associated with turkey erythrocyte nuclei and plasma membranes. *Biochem. 25*: 8057–8062.
7. Moss, J., & S. J. Stanley. 1981. Histone-dependent and histone-independent forms of an ADP-ribosyltransferase from human and turkey erythrocytes. *Proc. Natl. Acad. Sci. USA. 78*: 4809–4812.

8. Moss, J., S. J. Stanley, & J. C. Osborne Jr. 1981. Effect of self-association on activity of an ADP-ribosyltransferase from turkey erythrocytes. *J. Biol. Chem. 256*: 11452–11456.

9. Moss, J., J. C. Osborne Jr, & S. J. Stanley. 1984. Activation of an erythrocyte NAD:arginine ADP-ribosyltransferase by lysolecithin and nonionic and zwitterionic detergents. *Biochem. 23*: 1353–1357.

10. Okazaki, I. J., A. Zolkiewska, M. S. Nightingale, & J. Moss. 1994. Immunological and structural conservation of mammalian skeletal muscle glycosylphosphatidylinositol-linked ADP-ribosyltransferases. *Biochem. 33*: 12828–12836.

11. Okazaki, I. J., H-. J. Kim, N. G. McElvaney, & J. Moss. 1996. Molecular characterization of a glycosylphosphatidylinositol-linked ADP-ribosyltransferase from lymphocytes. *Blood*, in press.

12. Moss, J., S. J. Stanley, & N. J. Oppenheimer. 1979. Substrate specificity and partial purification of a stereospecific NAD-and guanidine-dependent ADP-ribosyltransferase from avian erythrocytes. *J. Biol. Chem. 254*: 8891–8894.

13. Oppenheimer. N. J. 1978. Structural determination and stereospecificity of the choleragen-catalyzed reaction of NAD$^+$ with guanidines. *J. Biol. Chem. 253*: 4907–4910.

14. Moss, J., N. J. Oppenheimer, R. E. West Jr, & S. J. Stanley. 1986. Amino acid specific ADP-ribosylation: Substrate specificity of an ADP-ribosylarginine hydrolase from turkey erythrocytes. *Biochem. 25*: 5408–5414.

15. Zolkiewska, A., M. S. Nightingale, & J. Moss. 1992. Molecular characterization of NAD:arginine ADP-ribosyltransferase from rabbit skeletal muscle. *Proc. Natl. Acad. Sci. USA. 89*: 11352–11356.

16. Tsuchiya, M., N. Hara, K. Yamada, H. Osago, & M. Shimoyama 1994. Cloning and expression of cDNA for arginine-specific ADP-ribosyltransferase from chicken bone marrow cells. *J. Biol. Chem. 269*: 27451–27457.

17. Davis, T., & S. Shall. 1995. Sequence of a chicken erythroblast mono(ADP-ribosyl)transferase-encoding gene and its upstream region. *Gene 164*: 371–372.

18. Okazaki, I. J., H-. J. Kim, & J. Moss. 1996. A novel membrane-bound lymphocyte ADP-ribosyltransferase cloned from Yac-1 cells. Abstr., *Am. Thoracic Soc.* 608078.

19. Zolkiewska, A., & J. Moss. 1993. Integrin α7 as substrate for a glycosylphosphatidylinositol-anchored ADP-ribosyltransferase on the surface of skeletal muscle cells. *J. Biol. Chem. 268*: 25273–25276.

20. Wang, J., E. Nemoto, & G. Dennert. 1996. Regulation of CTL by ecto-nicotinamide adenine dinucleotide (NAD) involves ADP-ribosylation of a p56^{lck}-associated protein. *J. Immunol. 156*: 2819–2827.

21. Moss, J., & M. Vaughan. 1988. ADP-ribosylation of guanyl nucleotide-binding proteins by bacterial toxins. *Adv. Enzymol. 61*: 303–379.

22. Hilz, H., R. Bredehorst, P. Adamietz, & K. Wielckens. 1982. Subfractionations and subcellular distribution of mono(ADP-ribosyl)proteins in eukaryotic cells. In: *ADP-ribosylation Reactions. Biology and Medicine.* O. Hayaishi, & K. Ueda (eds). Academic Press, San Diego, CA, p. 207–219.

23. Payne, M. D., E. L. Jacobson, J. Moss, & M. K. Jacobson. 1985. Modification of proteins by mono(ADP-ribosylation) in vivo. *Biochem. 24*: 7540–7549.

24. Jacobson, M. K., P. T. Loflin, N. Aboul-Ela, M. Mingmuang, J, Moss, & E. L. Jacobson. 1990. Modification of plasma membrane protein cysteine residues by ADP-ribose *in vivo. J. Biol. Chem. 265*: 10825–10828.

26. Cervantes-Laurean, D., D. E. Minter, E. L. Jacobson, & M. K. Jacobson. 1993. Protein glycation by ADP-ribose: Studies of model conjugates. *Biochem. 32*: 1528–1534.

26. Cervantes-Laurean, D., P. T. Loflin, D. E. Minter, E. L. Jacobson, & M. K. Jacobson. 1995. Protein modification by ADP-ribose via acid-labile linkages. *J. Biol. Chem. 270*: 7929–7936.

27. Tanuma, S., K. Kawashima, & H. Endo. 1988. Eukaryotic mono(ADP-ribosyl)transferase that ADP-ribosylates GTP-binding regulatory G$_i$ protein. *J. Biol. Chem. 263*: 5485–5489.

28. Tanuma, S., & H. Endo. 1990. Identification in human erythrocytes of mono(ADP-ribosyl) protein hydrolase that cleaves a mono(ADP-ribosyl) G$_i$ linkage. *FEBS Lett. 261*: 381–384.

29. McDonald, L. J., L. A. Wainschel, N. J. Oppenheimer, & J. Moss. 1992. Amino acid-specific ADP-ribosylation: Structural characterization and chemical differentiation of ADP-ribose-cysteine adducts formed nonenzymatically and in a pertussis toxin-catalyzed reaction. *Biochem. 31*: 11881–11887.

30. McDonald, L. J., & J. Moss. 1994. Enzymatic and nonenzymatic ADP-ribosylation of cysteine. *Mol. Cell. Biochem. 138*: 221–226.

31. McDonald, L. J., & J. Moss. 1993. Nitric oxide-independent, thiol-associated ADP-ribosylation inactivates aldehyde dehydrogenase. *J. Biol. Chem. 268*: 17878–17882.

32. Ludden, P. W. 1994. Reversible ADP-ribosylation as a mechanism of enzyme regulation in procaryotes. *Mol. Cell. Biochem. 138*: 123–129.

33. Moss, J., S-. C. Tsai, R. Adamik, H-. C. Chen, & S. J. Stanley. 1988. Purification and characterization of ADP-ribosylarginine hydrolase from turkey erythrocytes. *Biochem. 27*: 5819–5823.

34. Moss, J., S. J. Stanley, M. S. Nightingale, J. J. Murtagh Jr, L. Monaco, K. Mishima, H-. C. Chen, K. C. Williamson, & S-. C. Tsai. 1992. Molecular and Immunological characterization of ADP-ribosylarginine hydrolases. *J. Biol. Chem. 267*: 10481–10488.

35. Takada, T., K. Iida, & J. Moss. 1993. Cloning and site-directed mutagenesis of human ADP-ribosylarginine hydrolase. *J. Biol. Chem. 268*: 17837–17843.

36. Moss, J., M. K. Jacobson, & S. J. Stanley. 1985. Reversibility of arginine-specific mono(ADP-ribosyl)ation: Identification in erythrocytes of an ADP-ribose-L-arginine cleavage enzyme. *Proc. Natl. Acad. Sci. USA. 82*: 5603–5607.

37. Maehama, T., H. Nishina, & T. Katada. 1994. ADP-ribosylarginine glycohydrolase catalyzing the release of ADP-ribose from the cholera toxin-modified α-subunits of GTP-binding proteins. *J. Biochem. 116*: 1134–1138.

38. Zolkiewska, A., & J. Moss. 1995. Processing of ADP-ribosylated integrin α7 in skeletal muscle myotubes. *J. Biol. Chem. 270*: 9227–9233.

39. Maehama, T., H. Nishina, S. Hoshino, Y. Kanaho, & T. Katada. 1995. NAD$^+$-dependent ADP-ribosylation of T lymphocyte alloantigen RT6.1 reversibly proceeding in intact rat lymphocytes. *J. Biol. Chem. 270*: 22747–22751.

CRYSTAL STRUCTURE OF DIPHTHERIA TOXIN BOUND TO NICOTINAMIDE ADENINE DINUCLEOTIDE[*]

Charles E. Bell and David Eisenberg

UCLA-DOE Lab of Structural Biology and Molecular Medicine
Molecular Biology Institute and Department of Chemistry and Biochemistry
Box 951569, UCLA, Los Angeles, California 90095-1569

ABSTRACT

The crystal structure of diphtheria toxin (DT) in complex with nicotinamide adenine dinucleotide (NAD) has been determined by x-ray crystallography to 2.3Å resolution. NAD binds to a cleft on the surface of the catalytic (C) domain of DT, interacting closely with the side chains of Tyr54, Tyr65, His21, Thr23, and Glu 48. The carboxylate group of Glu148 of DT lies approximately 4Å from the scissile, N-glycosidic bond of NAD, suggesting a possible catalytic role for Glu148 in stabilizing a positively charged oxocarbonium intermediate. Residues 39–46 of the active-site loop of the C-domain become disordered upon NAD-binding, suggesting a potential role for these residues in binding to elongation facor-2 (EF-2). Structural alignments of the DT-NAD complex with the structures of other ADP-ribosylating toxins suggest how NAD may bind to these other enzymes.

BACKGROUND

Several bacterially secreted toxins, including DT, cholera toxin (CT), pertussis toxin (PT), *Pseudomonas aeruginosa* exotoxin A (ETA), and *Escherichia coli* heat-labile enterotoxin (LT)[1,2], catalyze the transfer of an ADP-ribose group from NAD to a specific residue within a host protein, in order to mediate toxic activity. Although the crystal structures of four of these ADP-ribosylating toxins, DT[3], ETA[4], LT[5], and PT[6] have revealed a common fold within the catalytic domains of these toxins, structural data of the NAD complex of these enzymes has been lacking, until recently[7].

[*] This work was supported by an NIH Grant GM31299 and a USPHS National Research Service Award GM07185 (C.E.B.), and is based on a previous publication (Bell, C.E. & Eisenberg, D. 1996. *Biochemistry 35*, 1137–1149).

ADP-Ribosylation in Animal Tissue, edited by Haag and Koch-Nolte
Plenum Press, New York, 1997

DT was the first of the ADP-ribosylating toxins to be studied in detail and over the course of several decades the basic mechanism by which DT elicits its toxic effect has been unraveled. DT is a 58 kD protein secreted by strains of *Corynebacterium diphtheriae* which have been infected by a phage that carries the DT gene[8]. DT causes the disease diphtheria in humans by gaining entry into the cytoplasm of host cells and inhibiting protein synthesis[9]. DT consists of three domains[3]: a C-terminal receptor-binding (R) domain, a central translocation (T) domain, and an N-terminal catalytic (C) domain. While the R, and T domains aid in the delivery of the C domain to the cytoplasm of a host cell, the C-domain catalyzes the transfer of an ADP-ribose group from NAD to a diphthamide (post-translationally modified histidine) residue of EF-2[10]. When the diphthamide residue of EF-2 is ADP-ribosylated, protein synthesis is inhibited and cell death occurs. It has been shown that a single C-domain of DT can kill a cell[11]. In the absence of EF-2, the C-domain, as well as whole DT, catalyzes the slow hydrolysis of NAD at the N-glycosidic bond to ADP-ribose and nicotinamide[12], a reaction for which no biological significance has been demonstrated.

Crystal structures of DT in complex with an inhibitor, adenylyl 3'-5' uridine monophosphate (ApUp) have previously been determined[3,13–15]. ApUp is similar in structure to NAD, and its binding site on DT has been assumed to mark the NAD-binding site. This site within the C-domain of DT is a prominent cleft between two sub-domains. The side chains of residues within this cleft, including His21 and Tyr65, have been shown by site-directed mutagenesis[16,17] and chemical modification studies[18,19] (Papini et al., 1989; Papini et al., 1991) to be involved in NAD binding. UV-crosslinking and site-directed mutagenesis studies have implicated Glu148 as being involved in catalysis of the ADP-ribosylation reaction[20,21]. In order to determine more precisely how the C-domain of DT binds to NAD, and gain further insight into the mechanism by which the C-domain catalyzes the ADP-ribosylation of EF-2, we have determined the crystal structure of DT in complex with NAD to 2.3Å resolution.

EXPERIMENTAL PROCEDURES

The details of DT purification, crystallization, x-ray data collection, structure determination and crystallographic refinement have been reported previously[7]. Briefly, the structure of the dimeric DT-NAD complex was solved by (1) crystallizing nucleotide-free dimeric DT, (2) soaking the crystals in an excess of NAD, and (3) immediately flash freezing under liquid nitrogen-cooled nitrogen vapor (-180 °C) in order to prevent NAD hydrolysis. The x-ray reflection intensities were measured while the crystal was maintained at -180 °C, and the crystal structure was solved by molecular replacement (X-PLOR)[22] using the refined dimeric DT-ApUp structure (PDB identification code 1DDT)[13] without ApUp as a starting model. After adding NAD to the model, crystallographic refinement of atomic positions and temperature factors using X-PLOR yielded a final model with an R-factor and free R-factor of 22.7% and 30.7%, respectively. The atomic coordinates and structure factors for the structure of the DT-NAD complex have been deposited in the Brookhaven Protein Data Bank (identification code 1TOX).

RESULTS

After crystallographic refinement of the structure of DT in the DT-NAD crystal, a 2Fo-Fc electron density map contoured at 1σ (Fig. 1) showed clear, readily interpretable electron density for NAD in the active sites of both DT molecules in the asymmetric unit of the crystal.

NAD binds to a prominent cleft on the surface of the C-domain, as shown in the ribbon representation of the C-domain-NAD complex in Figure 2 (the R and T-domains have been omitted in the figure, although they are present in the structure). Residues of the C-domain which line the inside of this cleft and make direct contacts with NAD extend primarily from α-helices CH2 & CH3 and β-strands CB2 and CB3.

Atomic interactions between the C-domain and NAD are summarized in Figure 3. NAD forms roughly a U-shape such that the adenine and nicotinamide rings at each end of the U project into the active-site cleft of the C-domain making extensive hydrophobic and hydrogen bonding interactions, while the phosphates at the base of the U stick out of the cleft and are more exposed to solvent. The adenine ring of NAD binds to a hydrophobic pocket formed by Trp153, Pro38, Tyr27, Ile31, Ile35, and Phe53, and also forms a pair of hydrogen bonds to the backbone carbonyl of Gly34 and amide of Gln36. The side chains of His21 and Thr23 of the C-domain hydrogen bond to the O2′ hydroxyl of the adenosine ribose. At the other end of the active-site pocket, the nicotinamide mononucleotide

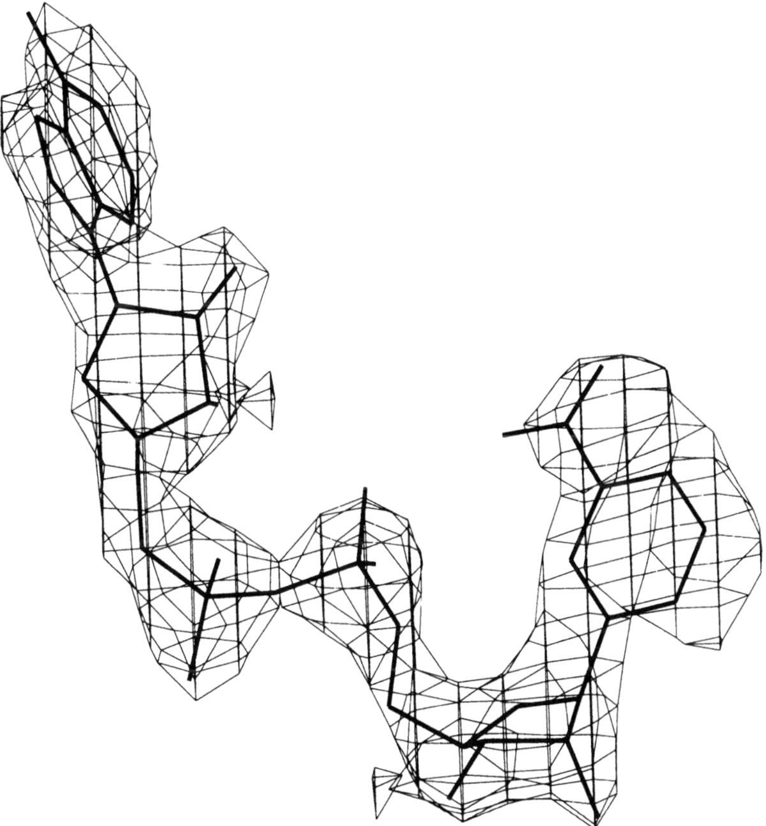

Figure 1. Electron density for NAD in the structure of the DT-NAD complex. A 2Fo-Fc electron density map (thin lines) is shown superimposed on the final, refined model for NAD (thick lines) for one of the two DT-NAD complexes in the asymmetric unit of the crystal. The density for the other NAD molecule is of equal quality. The electron density map was calculated before NAD was included in crystallographic refinement, and is therefore unbiased. The figure was generated using the program FRODO[28], and is reprinted with permission[7].

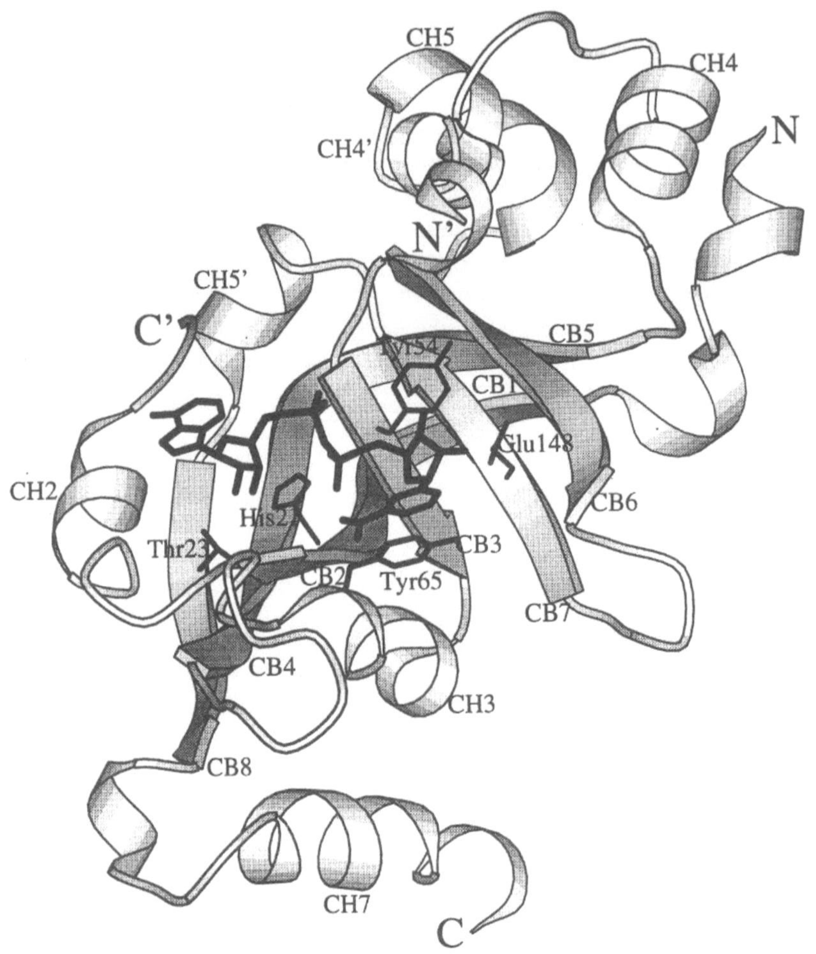

Figure 2. Ribbon representation of the structure of the C-domain of DT bound to NAD. The R and T domains of DT have been omitted from the figure, although they are present in the crystal structure. NAD and side chains of the C-domain which contact NAD are shown in black in stick form. Residues 39–47 of the active-site loop, indicated by C' and N', are disordered in the crystal structure and are not included in the model. Secondary structures are labeled as previously[13]. The figure was generated using MOLSCRIPT[29], and is reprinted with permission[7].

(NMN) portion of NAD binds to a groove formed by Tyr54, which packs against the NMN ribose, and Tyr65, which forms stacking interactions with the nicotinamide ring. The carboxamide group of the nicotinamide ring forms a pair of hydrogen bonds with the backbone carbonyl and amide groups of Gly22. The carboxylate group of Glu148, a residue which has been shown to be critical for catalysis of the ADP-ribosylation of EF-2, is positioned 4Å from the N1N and C1´N atoms of NAD which form the scissile, N-glycosidic bond.

Surprisingly, the negatively charged phosphates of NAD make no direct contacts to atoms of the C-domain, and are not neutralized by nearby positively charged side-chains. Instead, the phosphates, together with the NMN ribose hydroxyls, form a solvent-exposed surface that faces away from the active-site cleft of the C-domain.

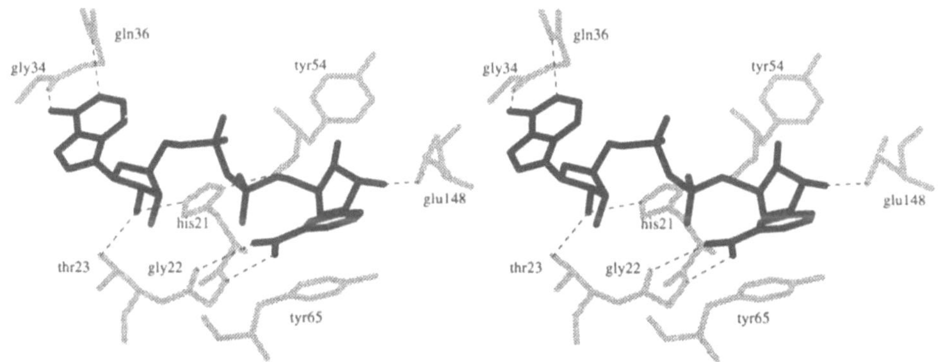

Figure 3. Stereo view of atomic interactions between NAD and the C-domain of DT. NAD is shown in grey, and residues of the C-domain are shown in black. All hydrogen bonds are shown as dashed lines. Two water molecules, which hydrogen bond to NAD and to residues of the C-domain, have been omitted for clarity. The figure was generated using MOLSCRIPT[29].

Residues 39–48 of the active-site loop of the C-domain (the loop between CH2 and CB3 as shown in Fig. 2) are disordered in the structure of the DT-NAD complex, for both molecules in the asymmetric unit of the crystal. In the DT-ApUp and nucleotide free-DT crystal structures, these residues are well-ordered and form the upper lip of the active-site cleft. Thus, upon NAD-binding, the active-site loop of the C-domain is displaced from its position over the active-site cleft to a disordered state.

DISCUSSION

Previous studies, in which the side chain of Glu148 of DT was UV-crosslinked to the nicotinamide ring of NAD, suggested that the Glu148 carboxylate group may be important in catalysis of the ADP-ribosylation reaction[20]. This was later confirmed by site-directed mutagenesis studies at this position[21]. Based on these studies, and on the known inversion of configuration at the C1′N atom during the ADP-ribosylation of EF-2 (substrate is the β-anomer of NAD, product is the α-anomer of ADP-ribosyl diphthamide)[23] a catalytic role for Glu148 was proposed in which the carboxylate group activates the diphthamide imidazole of EF-2 for a nucleophilic attack on the C1′N atom of NAD, in an S_N-2 type displacement reaction[21,24].

In agreement with the crosslinking study, the crystal structure of the DT-NAD complex reported here shows that Glu148 and the nicotinamide ring of NAD are in close proximity. More specifically, the carboxylate group of Glu148 is positioned ~4Å from the positively charged N1N atom of the nicotinamide ring and ~4Å from the C1′N atom of the NMN ribose (Fig. 3). These two atoms form the N-glycosidic bond that is cleaved in the ADP-ribosylation reaction. This geometry suggests an S_N-1 type mechanism for the ADP-ribosylation reaction in which the negatively charged carboxylate group of Glu148 could potentially stabilize a positively charged oxocarbonium intermediate generated by dissociation of the nicotinamide ring. The observed inversion of configuration at C1′N could be explained by the structure of the DT-NAD complex in which only the opposite side of the C1′N atom, relative to nicotinamide, is exposed and able to react with the incoming diphthamide imidazole (Figure 3).

Kinetic studies have shown that the ADP-ribosylation of EF-2 proceeds through a sequential mechanism, in which the C-domain of DT binds to NAD first and then binds to EF-2[25]. Moreover, it has been shown that the C-domain binds to EF-2 only in the presence of NAD: when NAD is absent, there is no detectable interaction between the C-domain and EF-2[25,26]. This can be explained if (1) NAD-binding induces some significant structural change in the C-domain which is important for binding to EF-2 and/or (2) NAD is bound to the C-domain such that some significant portion of NAD is exposed and important for binding to EF-2.

The crystal structure of the DT-NAD complex offers a possible explanation for the sequential mechanism of ADP-ribosylation observed biochemically. Upon NAD-binding, residues 39–48 of the active site loop become disordered. This loop may act as an arm that is dislodged from the active-site cleft upon NAD-binding and is then free to bind EF-2. In addition, the negatively charged phosphates and NMN-ribose hydroxyls of NAD form an exposed surface in the DT-NAD complex which could also be important for binding to EF-2.

The crystal structure of the DT-NAD complex may be useful as a structural framework for mapping the NAD-binding site onto the structures of other ADP-ribosylating toxins, such as LT and PT, for which the structure of the NAD-complex has not been determined. The ADP-ribosylating toxins whose structures are known (DT, LT, PT, ETA) share a common core fold of approximately 100 amino acids which contains the NAD-binding site. This is illustrated by least-squares alignments of the structures the catalytic domains of LT, PT, and ETA with the structure of the C-domain of DT in complex with NAD (Fig. 4). A sequence alignment, derived from the structural alignments (Fig. 5) shows the residues of LT, PT, and ETA which map to the predicted NAD-binding site, and thus are implicated as possibly being important for NAD-binding. As seen in Figure 5, only two residues, Gly34 and Glu148 of DT, are strictly conserved in all four toxins. The DT-NAD structure would suggest that the conserved Glycine residue is important for binding to the adenine ring of NAD, while the conserved glutamate binds near the scissile, N-glycosidic bond of NAD and is involved in catalysis.

Interestingly, alignment of the structure of the DT-NAD complex with the structures of PT and LT results in the overlap of NAD with an active-site segment, in both cases. The predicted NAD binding site on LT overlaps with residues 48–56 of the active-site loop of LT. This loop is topologically similar to the active-site loop of DT that becomes disordered upon NAD-binding. This indicates that removal of residues 48–56 of LT from the active-site cleft would be required for NAD-binding, as has been proposed previously[5]. Similarly, the predicted NAD binding site on PT overlaps with residues 199–207 of PT which form a short α-helix. Cys201 of this helix and Cys41 of PT form a disulphide bond, the reduction of which is known to be required for activity[27]. Thus, the predicted NAD binding site on PT, based on alignment with the DT-NAD complex, suggests that reduction of the Cys41-Cys201 disulphide bond and displacement of residues 199–207 of PT from the active-site cleft are required for NAD-binding, as has been previously proposed[6]. This feature of an active-site segment that is displaced from the active-site cleft upon NAD-binding could be conserved for the ADP-ribosylating family of toxins, and may be important for binding to the ADP-ribose acceptor substrate.

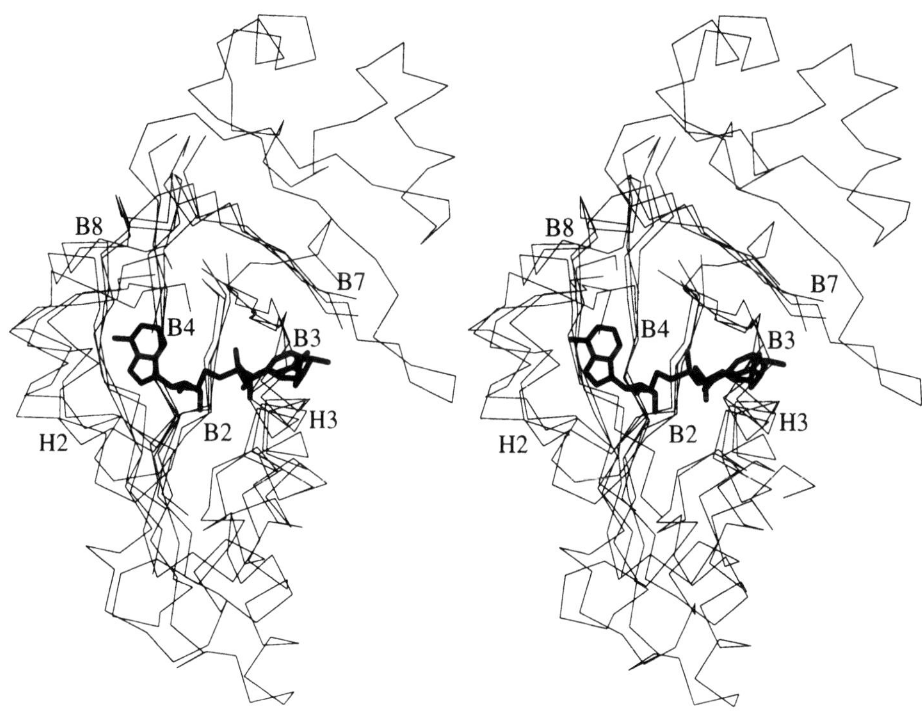

Figure 4. Stereo view showing the NAD-binding core structure common to four ADP-ribosylationg toxins. The structures of the catalytic domains of LT (PDB code 1LTS), PT (PDB code 1PRT), and ETA (Davie Davies, personal communication), were superimposed on the structure of the DT-NAD complex using only Cα atoms and the algorhithm of Kabsch[30]. Only the conserved segments are shown; these are residues 437–457, 469–505, and 552–573 of ETA, residues 4–25, 60–92, and 111–168 of LT, and residues 6–26, 51–96, and 127–148 of PT. Residues 18–187 of DT are shown. The figure was generated using MOLSCRIPT[29] and reprinted with permission[7].

```
                CB2               CH2
                ___________       ___________
DT      18      SSYHGTKPGYVD.SIQK...GIQKP        38
ETA     437     VGYHGTFLEAAQ.SIVF..GGVRAR        458
PT      6       TVYRYDS..RPPEDVFQN..GFTAW         26
LT      4       RLYRADS..RPPDEI.KRSGGLMPR         25


                CB3       CH3                                       CB4
                ______    ___________                              __________
DT      53      FYSTDNKYDAA.GYS..VDNEN..PLS..........GK.AGGVVKVT.YPG      87
ETA     469     FYIAGDPALAY.GYA..QDQAPDARGR..........IR.NGALLRVY.VPR      505
PT      51      VSTSSSRRYTE.VYLEHRM..Q..EAVEAERAGRGTGHF.IGYIYEVR.AD.      94
LT      60      VSTSLSLRSAHLAGQSIL......S...........GYSTYYIYVIATA..      91


                CB7               CB8
                __________        ___________
DT      146     SVEYINNWEQAKALSV.ELE.INF         167
ETA     551     RLETILGWPLAERTVV.IPSAIPT         573
PT      127     QSEYLAHRR.IPPENI.RRV.TRV         147
LT      110     EQEVSALGG.IPYSQIYGW..YRV         130
```

Figure 5. Sequence alignment of the catalytic domains of four ADP-ribosylating toxins, based on an alignment of their structures. Residues of DT that contact NAD are shown in boldface. Residues of ETA, PT, and LT, which align close to the predicted NAD binding site, and thus are implicated as possibly being important for NAD binding, are also shown in boldface. Secondary structures of DT are indicated by lines.

REFERENCES

1. Passador, L.,& W. Iglewski. 1994. ADP-ribosylating toxins. *Methods Enzymol. 235*: 617–630.

2. Burnette, W.N. 1994. AB$_5$ ADP-ribosylating toxins: compariative anatomy and physiology. *Structure 2*: 151–158.

3. Choe, S., M.J. Bennett, G. Fujii, P.M.G. Curmi, K.A. Kantardjieff, R,J. Collier,& D. Eisenberg. 1992. The crystal structure of diphtheria toxin. *Nature 357*: 216–222.

4. Allured, V.S., R.J. Collier, S.F. Carroll, & D.B. McKay. 1986. Structure of exotoxin A of *Pseudomaonas aeruginosa* at 3.0 Ångstrom resolution. *Proc. Natl. Acad. Sci. U.S.A. 83*:1320–1324.

5. Sixma, T.K., K.H. Kalk, B.A.B. van Zanten, Z. Dauter, J. Kingma, B. Witholt, & W.G.J. Hol. 1993. Refined structure of *Escherichia coli* heat-labile enterotoxin, a close relative of cholera toxin. *J. Mol. Biol. 230*: 890–918.

6. Stein, P.E., A. Boodhoo, G.D. Armstrong, S.A. Cockle, M. H. Klein,& R.J. Read. 1994. The crystal structure of pertussis toxin. *Structure 2*: 45–57.

7. Bell, C.E., & D. Eisenberg. 1996. Crystal structure of diphtheria toxin bound to nicotinamide adenine dinucleotide. *Biochemistry 35*: 1137–1149.

8. Freeman, V.J. 1951. Studies on the virulence of the bacteriophage-infected strains of *Corynebacterium diphtheriae. J. Bacteriol. 61*: 675–678.

9. Collier, R.J. 1975. Diphtheria toxin: mode of action and structure *Bacteriol. Rev. 39*: 54–85.

10. Van Ness, B.G., J.B. Howard, & J.W. Bodley. 1980. ADP-ribosylation of elongation factor 2 by diphtheria toxin. NMR spectra and proposed structures of ribosyl-diphthamide and its hydrolysis products. *J. Biol. Chem. 255*: 10710–10716.

11. Yamaizumi, M., E. Mekada, T. Uchida, & Y. Okada. 1978. One molecule of diphtheria toxin fragment A introduced into a cell can kill the cell. *Cell 15*:245–250.

12. Lory, S., S.F. Carroll, P.D. Bernard, & R.J. Collier. 1980. Ligand interaction of diphtheria toxin I. Binding and hydrolysis of NAD. *J. Biol. Chem. 255*: 12011–12015.

13. Bennett, M.J., S. Choe, & D. Eisenberg. 1994. Refined structure of dimeric diphtheria toxin at 2.0Å resolution. *Protein Sci. 3*: 1444–1463.

14. Bennett, M.J., & D. Eisenberg. 1994. Refined structure of monomeric diphtheria toxin at 2.3Å resolution. *Protein Sci. 3*: 1464–1475.

15. Weiss, M.S., S.R. Blanke, R.J. Collier, & D. Eisenberg. 1995. Structure of the isolated catalytic domain of diphtheria toxin. *Biochemistry 34*: 773–781.

16. Blanke, S.R., K. Huang, B.A. Wilson, E. Papini, A. Covacci, & R.J. Collier. 1994. Active-site mutations of the diphtheria toxin catalytic domain: role of histidine-21 in nicotinamide adenine dinucleotide binding and ADP-ribosylation of elongation factor 2. *Biochemistry 33*: 5155–5161.

17. Blanke, S.R., K. Huang, & R.J. Collier. 1994. Active-site mutations of diphtheria toxin: role of tyrosine-65 in NAD binding and ADP-ribosylation. *Biochemistry 33*: 15494–15500.

18. Papini, E., G. Schiavo, D. Sandona, R. Rappuoli, & C. Montecucco. 1989. Histidine 21 is at the NAD binding site of diphtheria toxin. *J. Biol. Chem. 264:* 12385–12388.

19. Papini, E., A. Santucci, G. Schiavo, M. Domenighini, P. Neri, R. Rappuoli, & C. Montecucco. 1991. Tyrosine 65 is photolabeled by 8-azidoadenine and 8-azidooadenosine at the NAD binding site of diphthera toxin. *J. Biol. Chem. 266*: 2494–2498.

20. Carroll, S.F., & R.J. Collier. 1984. NAD binding site of diphtheria toxin: identification of a residue within the nicotinamide subsite by photochemical modification with NAD. *Proc. Natl. Acad. Sci. U.S.A. 81*: 3307–3311.

21. Wilson, B.A., K.A. Reich, B.R. Weinstein, & R.J. Collier. 1990. Active-site mutations of diphtheria toxin: effects of replacing glutamic acid-148 with aspartic acid, glutamine, or serine. *Biochemistry 29*: 8643–8651.

22. Brünger, A.T. 1990. *X-PLOR Manual Version 3.1*. Yale University, New Haven, CT.

23. Oppenheimer, N.J., & J.W. Bodley. 1981. Diphtheria toxin: site and configuration of ADP-ribosylation of diphthamide in elongation factor 2. *J. Biol. Chem. 256*: 8579–8581.

24. Wilson, B.A., & R.J. Collier. 1992. Diphtheria toxin and *Pseudomaona aeruginosa* exotoxin A: active-site structure and enzymic mechanism. *Curr. Top. Microbiol. Immunol. 175*: 27–39.

25. Chung, D.W., & R.J. Collier. 1977. The mechanism of ADP-ribosylation of elongation factor 2 catalyzed by fragment A from diphtheria toxin. *Biochem. Biophys. Acta 483*: 248–257.

26. Kessler, S.P., & D.R. Galloway. 1992. *Pseudomonas aeruginosa* exotoxin A interaction with eucaryotic elongation factor 2. *J. Biol. Chem. 267*: 19107–19111.

27. Moss, J., et al., & E.L. Hewlett. 1983. Activation by thiol of the latent NAD glycohydrolase and ADP-ribo-syltranferase activities of *Bordetella pertussis* toxin (islet-activating protein). *J. Biol. Chem. 258*: 11879–11882.

28. Jones, T.A. 1978. Graphics model building and refinement system for macromolecules. *J. Appl. Crystallogr. 11*: 268–272.

29. Kraulis, P. 1991. MOLSCRIPT: a program to produce both detailed and schematic plots of proteins. *J. Appl. Crystallogr. 24*, 946–950.

30. Kabsch, W. 1978. A discussion of the solution for the best rotation to relate two sets of vectors. *Acta Crystallogr 34*: 827–828.

5

SELECTION OF DIPHTHERIA TOXIN ACTIVE-SITE MUTANTS IN YEAST

Rediscovery of Glutamic Acid-148 as a Key Residue

Haian Fu, Steven R. Blanke, Larry C. Mattheakis, and R. John Collier

Department of Microbiology and Molecular Genetics
200 Longwood Avenue
Boston, Massachusetts 02115

ABSTRACT

Saccharomyces cerevisiae was transformed with expression plasmids carrying the DTA gene under control of the *GAL1* promoter; colonies that formed under inducing conditions were selected; and plasmids from these colonies were screened for mutations in DTA that failed to block expression of the protein. Substitutions at three sites were identified, all of which are in the active-site cleft; and each of the substitutions reduced ADP-ribosyltransferase activity by $>10^5$. The substitutions include a charge reversal mutation of a catalytically important residue (Glu148Lys) and replacements of either of two glycines (Gly22 and Gly52) with bulky residues. The fact that multiple mutations were identified in these same residues implies that there are relatively few sites at which substitutions ablate ADP-ribosyltransferase activity without blocking expression of the full-length protein. Incorporation of a primary attenuating mutation into the DTA gene allowed *S. cerevisiae* also to be used to select complementary secondary mutations which altered activity less drastically. Besides elucidating structure-activity relationships, mutations identified by these approaches may be useful in designing new vaccines.

INTRODUCTION

Diphtheria toxin (DT) has been studied extensively in recent years as a model for understanding active-site structure and function in bacterial toxins that act by ADP-ribosylation mechanisms (1–4). Despite only weak sequence similarities between DT and most other ADP-ribosylating toxins, it is clear from x-ray crystallographic studies that most, if not all, toxins in this family have similar active-site folds (5–8). Moreover, there is in-

ADP-Ribosylation in Animal Tissue, edited by Haag and Koch-Nolte
Plenum Press, New York, 1997

creasing evidence that the catalytic domains of eukaryotic ADP-ribosyltransferases resemble those of the bacterial toxins (9–12).

Various approaches have been used to identify mutations that affect the ADP-ribosyltransferase activity of DT. The first such mutations came from studies in which *Corynebacterium diphtheriae* infected with randomly mutagenized Corynephage carrying the DT gene *(tox)* were screened for production of inactive forms of DT (13,14). Later, Carroll *et al.* used photoaffinity-labeling with NAD^+ to identify Glu148 (1,15). Among the various functionally important residues that have been identified in the ADP-ribosyltransferases, the active-site glutamate corresponding to Glu148 of DT is the only one that may prove to be universally conserved in this class of enzymes (1,16–19) (see also article by F. Bazan, this volume).

In the current studies we searched for attenuating mutations in DTA that had escaped detection by earlier approaches. The approach chosen involved random mutagenesis of the DTA gene, followed by expression in yeast, a eukaryotic organism that has diphthamide-containing elongation factor-2 (EF-2) and is therefore highly sensitive to the introduction of DTA into the cytosol (20). Mutations in DTA that allowed survival and colony formation by yeast were then selected and characterized. DT-resistant mutants of *Saccharomyces cerevisiae*, including ones deficient in the synthesis of diphthamide, had been isolated earlier using ADP-ribosylation as the basis of selection (21). In the current work, selection in yeast permitted us to identify a highly restricted set of amino acid substitutions in DTA which reduced ADP-ribosyltransferase activity by $>10^5$, a significantly higher factor than that of substitutions characterized earlier. The substitutions identified may be useful in developing new vaccines against diphtheria and possibly other diseases involving ADP-ribosylating toxins.

RESULTS

We constructed yeast expression vectors carrying DTA's coding sequence under the control of the yeast *GAL1* promoter. Healthy colonies were seen within two to three days after yeast transformants carrying the vectors were streaked onto medium containing glucose (to suppress expression of *GAL1*), but no colonies formed when the same transformants were streaked onto medium containing galactose(to induce DTA synthesis). Yeast carrying the vectors lacking the DTA gene formed healthy colonies in medium containing either glucose or galactose.

The plasmids were randomly mutagenized by passage through an *E. coli* mutator strain (Fig. 1), to elevate the mutation frequency in DTA above that of spontaneous host mutations conferring resistance to DTA. From 2×10^5 yeast transformants, 250 colonies formed on galactose-selection medium; and from these colonies, we identified by Western-blotting with anti-DT serum 8 plasmid-borne DTA mutants that produced full-length DTA-cross-reacting-material. Sequencing of the DTA gene from the isolated plasmids revealed mutations in the codons for Gly22 (mutated to Trp, Arg, and Glu), Gly52 (to Arg) and Glu148 (to Lys, alone or in combination with Lys51Glu) (Table 1). Mutations in Gly22 and Glu148 were isolated more than once from independently mutagenized pools, suggesting that this selection procedure approached saturation.

The repeated isolation of mutants at a limited number of sites suggested that DTA's activity must be drastically reduced for colony formation to occur, and that substitutions at only a limited number of sites produce the required level of reduction in activity without destabilizing the protein. To probe this possibility, we cloned three well characterized ac-

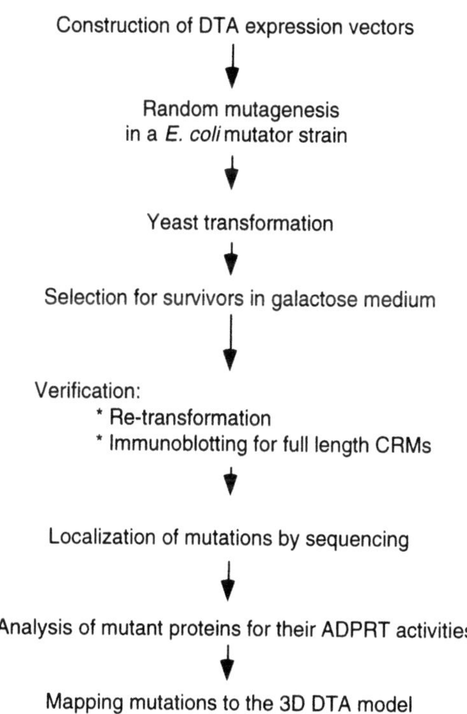

Figure 1. Genetic selection protocols for isolating DTA mutants.

tive-site mutations into the vector-borne DTA gene: Glu148Asp, Glu148Ser and Glu148Δ (deletion)—which decrease ADP-ribosyltransferase activity relative to wild-type by ca. 100-, 300-, and ~10^5-fold, respectively. Surprisingly, all three mutant DTAs prevented colony formation. Thus even the Glu148 deletion mutation, which reduced ADP-ribosyl-transferase activity by five orders of magnitude blocked growth and colony formation.

To decrease the stringency of the selection so that mutations causing more modest reductions in activity could be identified, yeast was transformed with plasmids encoding attenuated forms of DTA, instead of the wild type. DTA-Glu148Lys, which permitted only small colonies to form, and DTA-Glu148Δ, which was just potent enough to block colony formation, were chosen as primary attenuating mutations, and separate plasmids

Table 1. Mutants and their phenotypes

Mutant	Base change	Amino acid substitution	Frequency	Colony morphology
mplw133	gGa > gAg	Gly22Glu	1	small
mpllmm5-3	Ggg > Tgg	Gly22Trp	1	small
mplllw5-2	Gtt > Att, Ggg > Agg	Val5Ile, Gly22Arg	1	large & healthy
mplVw3-14	Ggg > Agg	Gly52Arg	2	small
mplw64	Gaa > Aaa	Glu148Lys	2	small
mplw33	Aaa > Gaa	Lys51Glu, Glu148Lys	1	large & healthy
mpllm2-2*	(Gaa > Aaa), Tgg > Ggg	(Glu148Lys), Trp153Gly	1	large &healthy
mplVm9-1**	(GAA deleted), Tgg>Agg	(Glu148Δ), Trp50Arg	1	large & healthy

*The starting DTA plasmid contained the Glu148Lys mutation. Yeast expressing DTA-Glu148Lys gave rise to small colonies.

**The starting DTA plasmid contained Glu148Δ. Expression of DTA-Glu148Δ arrested yeast growth.

carrying DTA-Glu148Lys and DTA-Glu148Δ, respectively, were randomly mutagenized and transformed as described above. From 10^5 transformants, two point mutations were identified that allowed formation of healthy colonies: Trp153Gly from DTA-Glu148Lys, and Trp50Arg from DTA-Glu148Δ.

To quantify the biochemical effects of the identified mutations, we measured the NAD^+:EF-2 ADP-ribosyltransferase activity of the proteins expressed in *E. coli*. Activities in extracts from *E. coli* were measured and normalized on the basis of DTA concentrations estimated from Western blots. As predicted, all of the mutant DTAs from trials originating with the wild-type gene exhibited drastically reduced levels of ADP-riboslyltransferase activity. The activity of the Glu148Lys mutant, which gave small colonies, was reduced by $\sim 2 \times 10^5$. The Lys51Glu mutation decreased ADP-ribosyltransferase activity by 10-fold, and combined with Glu148Lys, reduced activity to undetectable levels, consistent with the healthy colonies seen with the double mutant. Individual mutations in Gly22 and Gly52 also reduced activity to undetectable levels. The secondary mutation, Trp50Arg, decreased enzymic activity 2400-fold, consistent with the fact that the DTA-Glu148Δ /Trp50Arg double mutant had undetectable activity and gave healthy colonies. The activity of DTA-Trp153Gly, derived from the double mutant Glu148Lys/Trp153Gly, was undetectable. Trp153Gly may have destabilized the protein, given that little crossreacting material was detected for this mutant. DTA containing the Glu148Asp, Glu148Ser, and Glu148Δ mutations showed activity levels $\sim 1.6\%$, 0.5% and 0.0017% that of wild-type, respectively, values that are similar to ones reported earlier for these mutants.

DISCUSSION

It is known that introduction of a single molecule of DTA into the cytosol of CHO cells is lethal for those cells (22), and yeast cells show apparently similar sensitivity. Thus, our finding that only mutations that diminished the protein's ADP-ribosyltransferase activity by greater than $\sim 10^5$ permitted *S. cerevisiae* to form readily visible colonies is not surprising. It is also not surprising that only three sites were found at which individual amino acid substitutions meet this criterion without truncating the protein or rendering it susceptible to rapid degradation. While our search for such mutations was not exhaustive, the fact that mutations at two of the three sites were isolated repeatedly from pools of independently mutagenized plasmids suggests that our selection/screening protocol approached saturation.

The three loci at which single amino acid replacements allowed colony formation—Gly22, Gly52, and Glu148—all lie within the perimeter of the active-site cleft of DTA (Fig. 2). Glu148 lies on β-strand CB7 with its side chain projecting into the cleft. In the crystallographic model of NAD^+ bound to DT, recently reported by Bell and Eisenberg (23), the side-chain carboxylate group of Glu148 lies 4 Å from the N1N and C1'N atoms, which form the scissile N-glycosidic bond of NAD^+. Even conservative changes in Glu148 reduce ADP-ribosyltransferase activity by ≥ 100-fold, and given the evidence for functionality of a negatively charged side chain at this position, it is not surprising that substitution of a positively charged lysine for Glu148 virtually abolished DTA's enzymatic activity (Table 2). The fact that the Glu148Lys mutation was isolated twice—once as a single mutation exhibiting a sick colony phenotype and a second time as a double mutation (with Lys51Glu) exhibiting a healthy colony phenotype—provides independent confirmation of the critical role of Glu148 in catalyzing ADP-ribosylation. The small size of the colonies

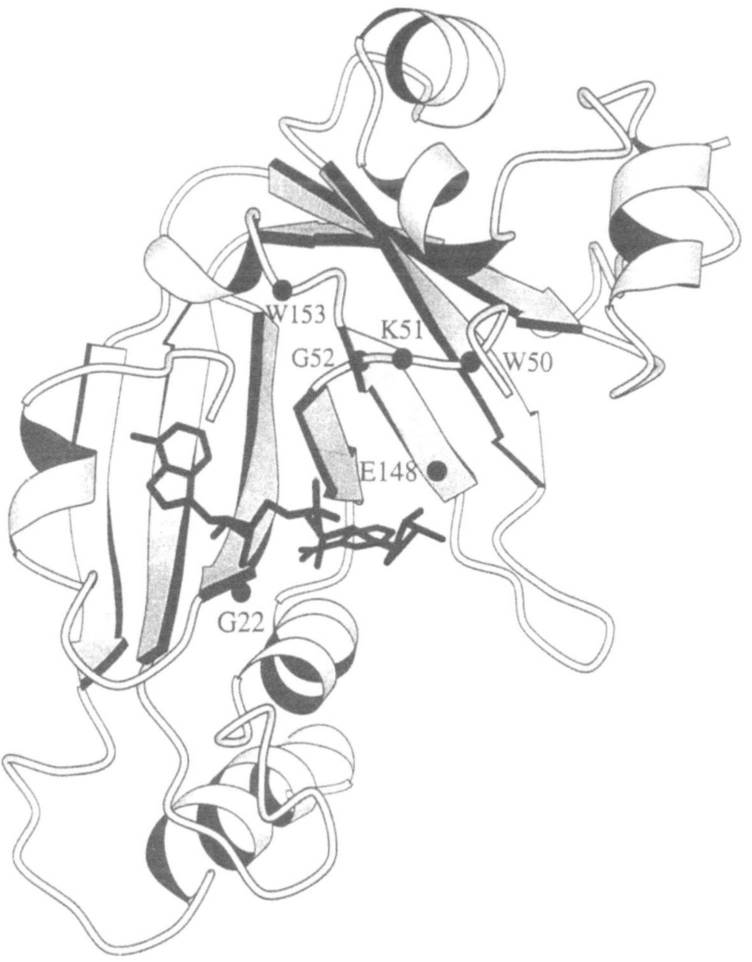

Figure 2. Mapping of isolated mutants on the three-dimensional crystallographic structural model of DTA with bound NAD$^+$. The approximate locations of the α-carbon atoms of residues identified in the current study are highlighted. The figure was generated using MOLSCRIPT (25) from atomic coordingates generously provided by C. Bell and D. Eisenberg (23).

in which the single mutant, DTA-Glu148Lys, was expressed apparently reflects slight growth inhibition due to residual ADP-ribosyltransferase activity of this mutant.

Three independent amino acid replacements allowing colony formation were found at Gly22 (Gly22Glu, Gly22Trp, and Gly22Arg). Gly22 is located between H21 and T23 near the carboxyl terminus of a 7-residue β strand (CB2). In the crystallographic structure of DT-bound NAD$^+$, the backbone carbonyl oxygen and amide nitrogen atoms of Gly22 are hydrogen bonded to the carboxamide group of nicotinamide. In addition, the backbone carbonyl oxygen of Gly22 makes an indirect hydrogen bond to the NMN phosphate, via a well-ordered water molecule (WAT743)—which in turn makes additional hydrogen bonds to NAD$^+$. Each of the three mutations identified at Gly22 (to Arg, Trp, and Glu) represents the introduction of a bulky side chain, which would be predicted to project away from the cleft and sterically overlap with adjacent polypeptide segments. This would likely produce

Table 2. Comparison of specific activities of DTA mutants relative to wild-type DTA*

Protein	Specific Activity**	Relative specific activity
DTA-WT	39±7.7	100%(1/1)
Lys51Glu	3.9 ±1.4	10%(1/10)
Glu148Asp	0.61 ±0.073	1.6% (1/64)
Glu148Ser	0.19 ±0.059	0.49%(1/210)
Trp50Arg	0.017 ±0.0047	0.041%(1/2400)
Glu148Δ	0.00065 ±0.000099	0.0017%(1/60,000)
Glu148Lys	0.00019 ±0.000078	0.00049%(1/210,000)
Glu148Lys/Lys51Glu	Not measurable***	
Gly22Trp	Not measurable	
Gly22Arg/Val5Ile	Not measurable	
Gly52Arg	Not measurable	
Trp153Gly	Not measurable	
Glu148Lys/Trp153Gly	Not measurable	
Glu148D/Trp50Arg	Not measurable	

*All assays were performed under the following conditions: 50 mM Tris-HCl, pH8.0, 1 mM EDTA, 10 mM DTT, 50 mM NAD+, and 0.5 mM EF-2, at 25oC. The values shown were generated from at least two assays, each performed in duplicate. Toxin concentrations were estimated from the average of at least three separate Western blots.

**Defined as mol EF-2 ADP-ribosylated per minute per mol enzyme.

***Limit of sensitivity of assay employed, ca. 10-6 that of wild-type DTA.

distortions of the main chain in and around Gly22 and disrupt the hydrogen bonding interactions of this residue and probably also of the flanking residues, H21 and T23. Derivatization of H21 with diethylpyrocarbonate causes a major disruption in ADP-ribosylation activity, as does replacement of this residue with most other residues tested. In DT-bound NAD^+, the adenosine ribose is hydrogen bonded to both H21 and T23. Perturbing the interactions of H21 and T23 with NAD^+ would be expected to compound the disruptions of Gly22's interactions.

The single substitution at Gly52 (to Arg) that allowed colony formation reduced ADP-ribosylation activity below the detection limit of our assay. Gly52 is also the site of the inactivating mutation (to Glu) in a well studied mutant form of DT, CRM197, which was isolated by random mutagenesis and screening. This residue lies near the carboxyl terminal end of a loop region between helix CH2 and β strand CB3. Part of this loop becomes disordered upon NAD^+ binding and has been proposed to be involved in binding EF-2. Gly52 is centrally located within a cluster of aromatic residues (Trp50, Phe53, Tyr54, Trp153) on one face of the active-site cleft. The side chain of Y54 packs against the NMN ribose, and the backbone carbonyl of Tyr54 hydrogen bonds both to His21, which is important in NAD^+ binding, and to the well ordered water (WAT743), which is also hydrogen bonded to the carbonyl of Gly22. Phe53 and Trp153 form part of a hydrophobic pocket that binds the adenine ring of NAD^+. Thus perturbation of local structure resulting from substitution of Gly52 with Arg would likely disrupt contacts of several adjacent residues with NAD^+, and might well also affect interactions with EF-2. Substitution of Trp50 with Ala has been shown to decrease activity by nearly 10^5-fold, and, as demonstrated here, replacement of Lys51 with Glu was found to cause a 10-fold reduction in activity.

One of the goals in understanding the structure-function relationships of toxins is to create inactive forms for vaccine development. An enzymically attenuated form of pertussis toxin, in which Glu129 and Arg9 (the homolog of Glu148 and His 21, respectively, of

DT) were mutated, has recently been licensed in Italy for use in a new parenterally administered vaccine against pertussis (24). Studies are also underway to identify mutations in DT that are optimal for such purposes. Some of the mutations identified in the current study may be particularly useful in this regard, particularly for reducing ADP-ribosyltransferase activity to the level that may be required for development of DNA vaccines, which require that the protein synthesis machinery of the host cells into which the modified *tox* gene is introduced remain active, in order that the antigen be synthesized.

ACKNOWLEDGMENTS

We thank Robert Deresiewicz and Sims Kochi for helpful discussions.

REFERENCES

1. Carroll, S.F. & R.J. Collier. 1984. NAD binding site of diphtheria toxin: identification of a residue within the nicotinamide subsite by photochemical modification with NAD. *Proc.Natl.Acad.Sci.USA 81*: 3307–3311.
2. Wilson, B.A., K.A. Reich, B.R. Weinstein & R.J. Collier. 1990. Active-site mutations of diphtheria toxin: effects of replacing glutamic acid-148 with aspartic acid, glutamine, or serine. *Biochemistry 29*: 8643–8651.
3. Blanke, S.R., K. Huang, B.A. Wilson, E. Papini, A. Covacci & R.J. Collier. 1994. Active-site mutations of the diphtheria toxin catalytic domain: Role of histidine-21 in nicotinamide adenine dinucleotide binding and ADP-ribosylation of elongation factor 2. *Biochemistry 33*: 5155–5161.
4. Blanke, S.R., Collier, R.J., Covacci, A., et al. Mutations affecting ADP-ribosyltransferase activity of diphtheria toxin. In: *Bacterial Protein Toxins, Fifth European Workshop*, edited by Witholt, B., Alouf, J.E., Boulnois, G.J., et al. New York: Gustav Fischer Verlag, 1992, p. 349–354.
5. Allured, V.S., R.J. Collier, S.F. Carroll & D.B. McKay. 1986. Structure of exotoxin A of Pseudomonas aeruginosa at 3.0-Angstrom resolution. *Proc.Natl.Acad.Sci.USA 83*: 1320–1324.
6. Sixma, T.K., S.E. Pronk, K.H. Kalk, et al. 1991. Crystal structure of a cholera toxin-related heat-labile enterotoxin from *E. coli. Nature 351*: 371–377.
7. Choe, S., M.J. Bennett, G. Fujii, et al. 1992. The crystal structure of diphtheria toxin. *Nature 357*: 216–222.
8. Stein, P.E., A. Boodhoo, G.D. Armstrong, S.A. Cockle, M.H. Klein & R.J. Read. 1994. The crystal structure of pertussis toxin. *Structure 2*: 45–57.
9. Takada, T., K. Iida & J. Moss. 1995. Conservation of a common motif in enzymes catalyzing ADP-ribose transfer. Identification of domains in mammalian transferases. *J.Biol.Chem. 270*: 541–544.
10. Marsischky, G.T., B.A. Wilson & R.J. Collier. 1995. Role of glutamic acid 988 of human poly-ADP-ribose polymerase in polymer formation. Evidence for active site similarities to the ADP-ribosylating toxins. *J.Biol.Chem. 270*: 3247–3254.
11. Domenighini, M., C. Magagnoli, M. Pizza & R. Rappuoli. 1994. Common features of the NAD-binding and catalytic site of ADP-ribosylating toxins. *Mol.Microbiol. 14*: 41–50.
12. Koch-Nolte, F., D. Petersen, S. Balasubramanian, et al. 1996. Mouse T-cell membrane proteins RT6–1 and RT6–2 are arginine protein mono(ADPribosyl)transferases and share secondary structure motifs with ADP-ribosylating bacterial toxins. *J.Biol.Chem. 271*: 7686–7693.
13. Uchida, T., A.M.Jr. Pappenheimer & R. Greany. 1973. Diphtheria toxin and related proteins. I. Isolation and properties of mutant proteins serologically related to diphtheria toxin. *J.Biol.Chem. 248*: 3838–3844.
14. Uchida, T., D.M. Gill & A.M. Pappenheimer,Jr.. 1971. Mutation in the structural gene for diphtheria toxin carried by temperate phage. *Nature New.Biol. 233*: 8–11.
15. Carroll, S.F., J.A. McCloskey, P.F. Crain, N.J. Oppenheimer, T.M. Marschner & R.J. Collier. 1985. Photoaffinity labeling of diphtheria toxin fragment A with NAD: structure of the photoproduct at position 148. *Proc.Natl.Acad.Sci.USA 82*: 7237–7241.
16. Carroll, S.F., J.T. Barbieri & R.J. Collier. 1988. Diphtheria toxin: purification and properties. *Methods In Enzymology 165*: 68–76.

17. Aktories, K., M. Jung, J. Bohmer, G. Fritz, J. Vandekerckhove & I. Just. 1995. Studies on the active-site structure of C3-like exoenzymes: involvement of glutamic acid in catalysis of ADP-ribosylation. [Review]. *Biochimie 77*: 326–332.

18. Carroll, S.F. & R.J. Collier. 1987. Active site of Pseudomonas aeruginosa exotoxin A. Glutamic acid 553 is photolabeled by NAD and shows functional homology with glutamic acid 148 of diphtheria toxin. *J.Biol.Chem. 262*: 8707–8711.

19. Liu, S.Y., S.M. Kulich & J.T. Barbieri. 1996. Identification of glutamic acid 381 as a candidate active site residue of *Pseudomonas aeruginosa* exoenzyme S. *Biochemistry 35*: 2754–2758.

20. Murakami, S., J.W. Bodley & D.M. Livingston. 1982. Saccharomyces cerevisiae spheroplasts are sensitive to the action of diphtheria toxin. *Mol.Cell Biol. 2*: 588–592.

21. Chen, J.Y., J.W. Bodley & D.M. Livingston. 1985. Diphtheria toxin-resistant mutants of Saccharomyces cerevisiae. *Mol.Cell Biol. 5*: 3357–3360.

22. Yamaizumi, M., E. Mekada, T. Uchida & Y. Okada. 1978. One molecule of diphtheria toxin fragment A introduced into a cell can kill the cell. *Cell 15*: 245–250.

23. Bell, C.E. & D. Eisenberg. 1996. Crystal structure of diphtheria toxin bound to nicotinamide adenine dinucleotide. *Biochemistry 35*: 1137–1149.

24. Rappuoli, R., G. Douce, G. Dougan & M. Pizza. 1995. Genetic detoxification of bacterial toxins - a new approach to vaccine development. *International Archives of Allergy & Immunology 108*: 327–333.

25. Kraulis, P. 1991. MOLSCRIPT: A program to produce both detailed and schematic plots of protein structures. *J.Appl.Crystallogr. 24*: 946–950.

6

IDENTIFICATION OF THE CATALYTIC SITE OF CLOSTRIDIAL ADP-RIBOSYLTRANSFERASES

Klaus Aktories

Institut für Pharmakologie und Toxikologie der
Albert-Ludwigs-Universität Freiburg
Hermann-Herderstr. 5, D-79104 Freiburg
Germany

ABSTRACT

The catalytic sites of clostridial ADP-ribosyltransferases were studied by photoaffinity-labelling with [carbonyl-^{14}C]NAD$^+$. In C3-like transferases, which are known to modify low molecular mass GTP-binding Rho proteins, Glu-174 was identified to be essential for catalysis. In *C. perfringens* iota toxin, Glu-380 and Glu-378 may have pivotal roles in the active site of this actin-ADP-ribosylating toxin.

INTRODUCTION

Clostridial ADP-ribosyltransferases can be divided into two families. One family consists of the Rho-GTPase ADP-ribosylating toxins, members of which are various isoforms of C3-like transferases produced by *C. botulinum* and by *C. limosum* (1–5). Moreover, ADP-ribosyltransferases from *B. cereus* and *S. aureus* were shown to belong to this family of transferases (6,7). These transferases are very basic proteins (pI >9), have molecular masses of about 25000 Da and are 30 to 75% identical at their amino acid sequence. The enzymes share the same protein substrates and modify low molecular mass GTP-binding proteins of the Rho-family (RhoA, B, C) at asparagine-41 (8). Rho proteins are involved in the regulation of the actin cytoskeleton and ADP-ribosylation inactivates the regulatory proteins and causes depolymerization of the actin cytoskeleton (9,10). Furthermore, these proteins are involved in a variety of cellular signal transduction processes including control of phosphoinositide-3-kinase (11), phosphatidylinositol-4-phosphate-5-kinase (12), phospholipase D (13), smooth muscle contraction (14), cell-cell contact (15), endocytosis (16) and transcriptional activation (17,18). All these pathways are blocked by ADP-ribosylation of Rho proteins.

ADP-Ribosylation in Animal Tissue, edited by Haag and Koch-Nolte
Plenum Press, New York, 1997

The second family of clostridial ADP-ribosylating toxins comprises toxins that attack eukaryotic cells by specific ADP-ribosylation of actin (19–21). Among these toxins are *Clostridium botulinum* C2 toxin (22), *Clostridium perfringens* iota toxin (23–25), *Clostridium spiroforme* toxin (26) and a transferase produced by *Clostridium difficile* (27). These toxins are binary in structure consisting of a biologically active enzyme component and a separate binding component, which is not linked to the enzyme component. Attachment of the binding component to the cell surface induces a binding site for the enzyme component and, subsequently, results in endocytosis and translocation of the transferase into the cytosol. In the cytosol, the transferases ADP-ribosylates monomeric actin at arginine-177 (28). This modification inhibits actin polymerisation most likely by sterical hindrance. Additionally, ADP-ribosylated actin acts like a capping protein to inhibit polymerisation of non-modified actin at the fast polymerising end of F-actin filaments (29). Furthermore, ADP-ribosylated actin inhibits the nucleation activity of gelsolin-actin complexes (30). All these effects may participate in the dramatic redistribution of the actin cytoskeleton typically observed after treatment of intact cells with the toxins.

Photoaffinity labelling has been proved to be a powerful method to identify the catalytic site of ADP-ribosylating toxins. It was first shown by Carroll and Collier (31) that UV-irradiation of diphtheria toxin with [carbonyl-^{14}C]NAD$^+$ causes covalent modification of a glutamic acid residue later unequivocally shown to be crucial for catalysis of ADP-ribosyltransferase activity of this toxin. Recently, the same method was applied to identify the catalytic site of various clostridial transferases (32,33). These studies largely increased our knowledge about the structure-function relationships of clostridial ADP-ribosylating toxins.

RESULTS AND DISCUSSION

The active-site structure of C3-like exoenzymes were identified by UV-induced covalent linkage of NAD$^+$ to *C. limosum* transferase (32). To this end, *C. limosum* exoenzyme was UV-irradiated in the presence of [carbonyl-^{14}C-]NAD$^+$ or [adenyl-^{32}P]NAD$^+$ for up to 3 h at 4° C. After precipitation of the photolabeled protein by trichloroacetic acid and filtration on nitro-cellulose filters, the incorporated radioactivity was analysed by liquid scintillation counting. In the presence of [^{14}C]NAD$^+$, about 1 mol radiolabel per mol enzyme was incorporated. In contrast, in the presence of [^{32}P]NAD$^+$ only up to 0.2 mol radiolabel pro mol enzyme was bound. Concomitantly with incorporation of radiolabel, NAD$^+$ glycohydrolase and ADP-ribosyltransferase activities were inhibited. To identify amino acids involved in photoaffinity binding of NAD$^+$, the radiolabelled *C. limosum* enzyme was digested by V8 protease and trypsin. Subsequently, peptides formed were isolated by reversed-phase HPLC and the radiolabelled peptide was sequenced. Thereby Glu-174 of *C. limosum* transferase was identified as acceptor for photoaffinity labelling.

To corroborate the hypothesis that Glu-174 is involved in catalysis of the ADP-ribosyltransferase reaction, mutants of *C. limosum* exoenzyme were constructed in which Glu-174 was replaced by glutamine or aspartic acid (34). Comparison of the enzyme activity of wild-type and mutant transferases revealed reduction in enzyme activity by about 1000fold with the E174D mutant and even less enzyme activity with the E174Q mutant (Table I). In contrast, the K_m values for NAD$^+$ of mutant transferases were only slightly increased, indicating that decrease in enzyme activity by exchange of Glu-174 was not caused by loss in NAD$^+$-binding capacity of the enzyme. Similar results were obtained by studying the quenching of intrinsic protein fluorescence by NAD$^+$ for mutant proteins

Table 1. Kinetic parameters of wild type and mutant *C. limosum* transferases

	WT	E174D	E174Q
K_m (μM)	52.2	102.6	N.D.
k_{cat} (1/min)	0.276	3.2×10^{-4}	N.D.
relative k_{cat}	1	1.2×10^{-3}	N.D.
k_{cat}/Km (M^{-1}/min^{-1})	5.3×10^3	3.2	N.D.
relative k_{cat}/K_m	1	5.7×10^{-4}	N.D.

Data from 34. ND, not detectable. Standard deviations were not greater than +/- 15%.

(34). Moreover, photoaffinity-labelling with $[^{14}C]NAD^+$ showed that the E174D and E174Q mutants incorporated much less or no radioactivity, respectively. All these data are consistent with the hypothesis that glutamic acid at position 174 is involved in the catalytic activity of *C. limosum* transferase. Similar results were obtained with *C. botulinum* exoenzyme C3 (34,35), which is about 70 % identical with the *C. limosum* transferase. (Note that two C3 isoforms exist: in the isoform produced by *C. botulinum* type C strain C468, the pivotal glutamic acid is at position 174 whereas C3 from strain C003–9 has the equivalent glutamic acid residue at position 173. Recently, another C3-like transferase, which is produced by *B. cereus*, was identified (36). This enzyme shows no immunological relationship with any other Rho-modifying transferase and appears to be even more distantly related to C3 than other Rho-ADP-ribosylating transferases (6). Again, modification of *B. cereus* transferase by UV-irradiation in the presence of NAD^+ caused inhibition of ADP-ribosyltransferase and NAD^+ glycohydrolase activity. When the exoenzyme was photolabeled with $[carbonyl-^{14}C]NAD^+$, subsequent sequencing of a radioactively labelled peptide revealed modification of a glutamic acid residue

Comparison of the amino acid sequence of the enzyme component *of C. perfringens* iota toxin with the region adjacent to Glu-174 of C3 toxins revealed a high degree of similarity (Fig. 1). As observed in C3 transferases, in iota toxin a glutamic acid residue (Glu-380) is followed by a hydrophobic tripeptide suggesting a similar function as Glu-174 of C3 toxins. In order to confirm the role of Glu-380 in the catalysis, iota toxin was radiolabelled in the presence of $[^{14}C]NAD^+$ by UV-irradiation. To determine the amino acid residue modified, the labelled transferase was first cleaved by CNBr and subsequently digested by trypsin. Peptides formed were isolated by reverse-phase HPLC. Analysis of the radiolabelled peptide showed the following sequence: Gly-Ser-Pro-Gly-Ala-Tyr-Leu-Ser-Ala-Ile-Pro-Gly-Tyr-Ala-Gly-X-Tyr-Glu-Val-Leu-Leu-Asn-His-Gly-Ser-Lys corresponding with the region Gly-363 through Lys-388 in the *C. perfringens* iota toxin (Fig. 2). Residue X that carried the radiolabel eluted as an unconventional PTH-derivative with a retention time between dimethylphenylthiourea and PTH-Ala. A similar unconventional PTH -derivative was found in the course of the analysis of *C. limosum* transferase where the DNA sequence (as in the case of iota toxin) predicted a glutamic acid residue at this position. However, sequence analysis of the radiolabelled iota toxin peptide showed that Glu-378 and not Glu-380 was radiolabelled.

It was reported that photoaffinity labelling of ADP-ribosylating toxins in the presence of NAD^+ results in the incorporation of the complete nicotinamide moiety of NAD^+ into the transferase. As analysed with diphtheria toxin first, nicotinamide covalently links through C-6 to the decarboxylated γ-methylene carbon of the catalytic glutamic acid residue thereby increasing the mass of the peptide modified by about 76 Da (37). Similarly, in the case of iota toxin, mass spectrometric data revealed that photoaffinity labelling with

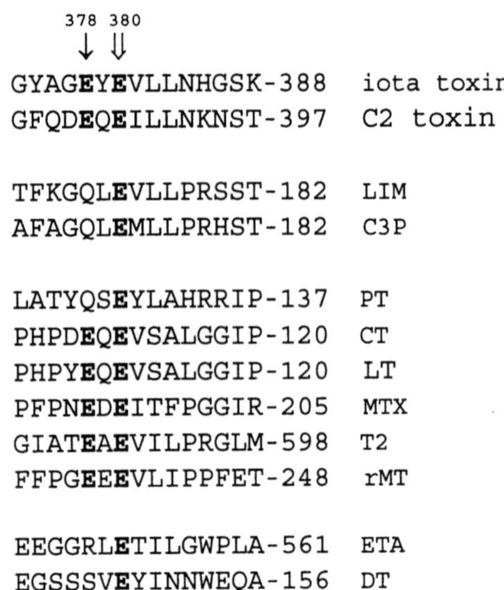

Figure 1. Comparison of the amino aicd sequences of the region adjacent to a glutamic acid residue involved in catalysis of the ADP-ribosylation reaction by various transferases (iota toxin, *C. perfringens* iota toxin; C2 toxin, *C. botulinum* C2 toxin; Lim, *C. limosum* ADP-ribosyltransferase; C3P, *C. botulinum* (strain C468) C3 transferase; CT, cholera toxin; LT, *E. coli* heat-labile toxin; MTX, mosquitocidal toxin from *B. sphaericus*; PT, pertussis toxin; T2, bacteriophage ADP-ribosyltransferase; rMT, rabbit muscle ADP-ribosyltransferase, ETA, *Pseudomonas* exotoxin A; DT, diphtheria toxin.

NAD^+ caused increase in the mass of the above mentioned iota toxin peptide by 76 Da corresponding to the known modification of a glutamic acid side chain. These findings indicated that an identical type of reaction was responsible for covalent labelling of *C. perfringens* iota toxin and favoured the hypothesis that Glu-378 of iota toxin interacts with NAD^+ in the same mode as analysed for the labelling of Glu-148 of diphtheria toxin.

The above described studies on the photoaffinity-labelling of clostridial ADP-ribosyltransferases are in line with the view that a glutamic acid residue is crucial for the enzyme activity of these transferases. Thus, Glu-174 of *C. limosum* and of *C. botulinum* C3 transferase appear to be equivalent to Glu-148 of diphtheria toxin (31), Glu-553 of *Pseudomonas* exotoxin A (38) and Glu-129 of pertussis toxin (39). In all these cases, specific glutamic acid residues were modified after UV-irradiation of the toxins in the presence of NAD^+, modifications that resulted in inhibition of enzyme activity. It was confirmed by crystal structure analysis of diphtheria toxin (40), *Pseudomonas* exotoxin A (41) and pertussis toxin (42) that the radiolabelled glutamic acid residues are located within the active site cleft of the toxins. Furthermore, crystal structure analysis of *E coli* heat labile enterotoxin revealed equivalence of Glu-112 of the cholera toxin-related transferase with the glutamic acid residues Glu-148 and Glu-553 of diphtheria toxin and *Pseudomonas* toxin, respectively (43). As described above with C3-transferases, the role of glutamic acids in catalysis was confirmed by site-directed mutagenesis in various ADP-ribosylating toxins (40,44–46). Moreover, studies indicate that a glutamic acid residue is basically involved not only in ADP-ribosylation reactions catalysed by toxins but also by several prokaryotic and eukaryotic mono(ADP-ribosyl)transferases identified recently, and even by the eukaryotic poly(ADP-ribosyl)polymerase. In most of these transferases the catalytic glutamic acid is located within a region of significant similarity with a

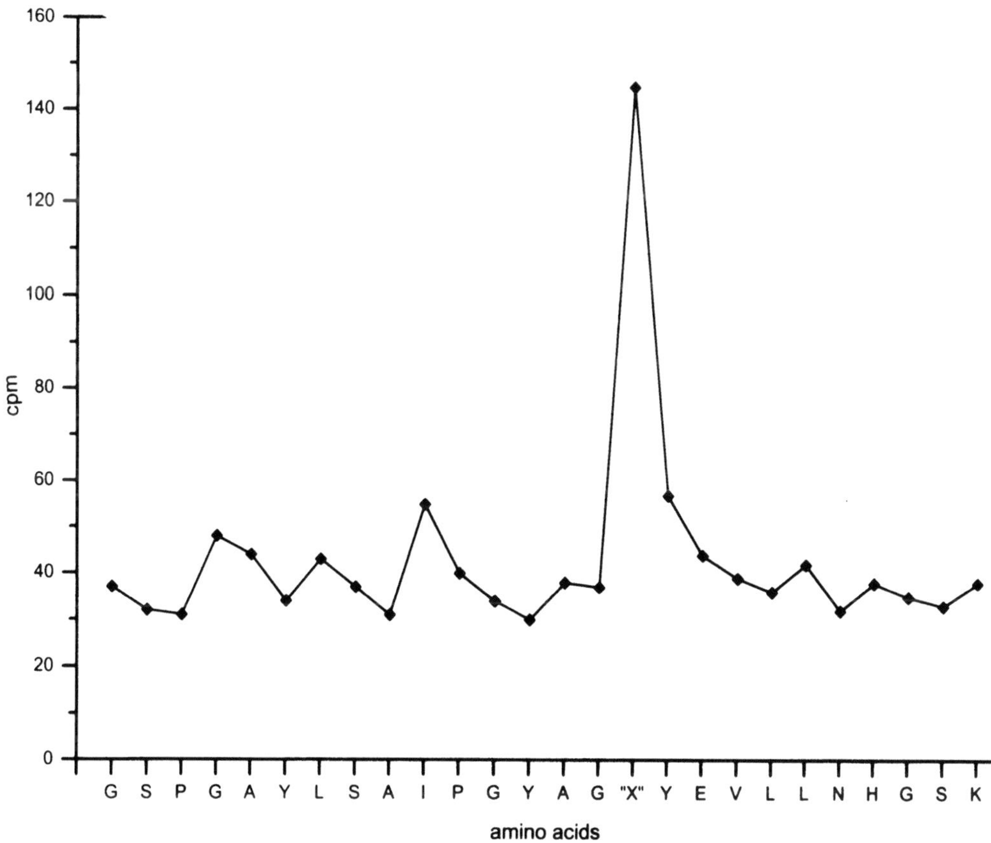

Figure 2. Amino acid sequence analysis and measurements of radioactivity in the residues of a ^{14}C-labelled proteolytic peptide from *C. perfringens* iota toxin previously photoaffinity labelled with [carbonyl-^{14}C]NAD$^+$.

number of other ADP-ribosyltransferases including the family of *Rhodospirillum*-like transferases (47), the T2, T4 and T6 bacteriophage transferases (48) and various eukaryotic glycosylphosphatidyl-anchored mono(ADP-ribosyl)transferases (49,50).

Surprisingly, in iota toxin the photoaffinity-labelled glutamic acid residue was not Glu-380, which residue perfectly aligns with the conserved catalytic glutamic acid residues of other transferases, but Glu-378 was labelled. Transferases with a similarly spaced biglutamic acid sequence are the related *C. botulinum* C2 toxin, cholera toxin, the *E. coli* heat-labile toxins, the mosquitocidal toxin from *B. sphaericus*, *Rhodospirillum* transferases and various eukaryotic mono(ADP-ribosyl)transferases. Interestingly, it has been reported that mutant *E. coli* heat-labile toxin in which glutamic acid residues at either position 110 and 112 are changed with aspartic acid have severely reduced transferase activity (51). Moreover, it was shown with rabbit muscle ADP-ribosyltransferase that the glutamic acid residues Glu-238 and Glu-240, which are equivalent with Glu-378 and Glu-380 of iota toxin, respectively, are both essential for enzyme activity (49,50). Thus it appears that in cases of "biglutamic" transferases, which are all ADP-ribosyltransferases modifying arginine residues, both glutamic acid residues are essential for catalysis.

REFERENCES

1. Aktories, K., U. Braun, B. Habermann, & S. Rösener. 1990. *ADP-ribosylating Toxins and G Proteins: Botulinum ADP-ribosyltransferase C3.* American Society for Microbiology. Washington. 97–115.

2. Aktories, K., C. Mohr, & G. Koch. 1992. Clostridium botulinum C3 ADP-ribosyltransferase. *Curr. Top. Microbiol. Immunol. 175*: 115–131.

3. Just, I., C. Mohr, G. Schallehn, L. Menard, J.R. Didsbury, J. Vandekerckhove, J. van Damme, & K. Aktories. 1992. Purification and characterization of an ADP-ribosyltransferase produced by *Clostridium limosum. J. Biol. Chem. 267*: 10274–10280.

4. Aktories, K., U. Weller, & G.S. Chhatwal. 1987. Clostridium botulinum type C produces a novel ADP-ribosyltransferase distinct from botulinum C2 toxin. *FEBS Lett. 212*: 109–113.

5. Rubin, E.J., D.M. Gill, P. Boquet, & M.R. Popoff. 1988. Functional modification of a 21-Kilodalton G protein when ADP-ribosylated by exoenzyme C3 of Clostridium botulinum. *Mol. Cell. Biol. 8*: 418–426.

6. Just, I., J. Selzer, M. Jung, J. van Damme, J. Vandekerckhove, & K. Aktories. 1995. Rho-ADP-ribosylating exoenzyme from *Bacillus cereus* -purification, characterization and identification of the NAD-binding site. *Biochemistry, 34*: 334–340.

7. Sugai, M., K. Hashimoto, A. Kikuchi, S. Inoue, H. Okumura, K. Matsumota, Y. Goto, H. Ohgai, K. Morishi, B. Syuto, K. Yoshikawa, H. Suginaka, & Y. Takai. 1992. Epidermal cell differentiation inhibitor ADP-ribosylates small GTP-binding proteins and induces hyperplasia of epidermis. *J. Biol. Chem. 267*: 2600–2604.

8. Sekine, A., M. Fujiwara, & S. Narumiya. 1989. Asparagine residue in the rho gene product is the modification site for botulinum ADP-ribosyltransferase. *J. Biol. Chem. 264*: 8602–8605.

9. Paterson, H.F., A.J. Self, M.D. Garrett, I. Just, K. Aktories, & A. Hall. 1990. Microinjection of recombinant p21[rho] induces rapid changes in cell morphology. *J. Cell Biol. 111*: 1001–1007.

10. Chardin, P., P. Boquet, P. Madaule, M.R. Popoff, E.J. Rubin, & D.M. Gill. 1989. The mammalian G protein rho C is ADP- ribosylated by Clostridium botulinum exoenzyme C3 and affects actin microfilament in Vero cells. *EMBO J. 8*: 1087–1092.

11. Zhang, J., W.G. King, S. Dillon, A. Hall, L. Feig, & S.E. Rittenhouse. 1993. Activation of platelet phosphatidylinositide 3-kinase requires the small GTP-binding protein Rho. *J. Biol. Chem. 268*: 22251–22254.

12. Chong, L.D., A. Traynor-Kaplan, G.M. Bokoch, & M.A. Schwartz. 1994. The small GTP-binding protein Rho regulates a phosphatidylinositol 4-phosphate 5-kinase in mammalian cells. *Cell, 79*: 507–513.

13. Malcolm, K.C., A.H. Ross, R.-G. Qiu, M. Symons, & J.H. Exton. 1994. Activation of rat liver phospholipase D by the small GTP-binding protein RhoA. *J. Biol. Chem. 269*: 25951–25954.

14. Hirata, K.-i., A. Kikuchi, T. Sasaki, S. Kuroda, K. Kaibuchi, Y. Matsuura, H. Seki, K. Saida, & Y. Takai. 1992. Involvement of *rho* p21 in the GTP-enhanced calcium ion sensitivity of smooth muscle contraction. *J. Biol. Chem. 267*: 8719–8722.

15. Tominaga, T., K. Sugie, M. Hirata, N. Morii, J. Fukata, A. Uchida, H. Imura, & S. Narumiya. 1993. Inhibition of PMA-induced, LFA-1-dependent Lymphocyte aggregation by ADP-ribosylation of the small molecular weight GTP binding protein, *rho. J. Cell Biol. 120,No.6*: 1529–1537.

16. Schmalzing, G., H.P. Richter, A. Hansen, W. Schwarz, I. Just, & K. Aktories. 1995. Involvement of the GTP binding protein Rho in constitutive endocytosis in *Xenopus laevis* oocytes. *J. Cell Biol. 130*: 1319–1332.

17. Olson, M.F., A. Ashworth, & A. Hall. 1995. An essential role for Rho, Rac, and Cdc42 GTPases in cell cycle progression through G_1. *Science, 269*: 1270–1272.

18. Hill, C.S., J. Wynne, & R. Treisman. 1995. The Rho family GTPases RhoA, Rac1, and CDC42Hs regulate transcriptional activation by SRF. *Cell, 81*: 1159–1170.

19. Aktories, K. & A. Wegner. 1989. ADP-ribosylation of actin by clostridial toxins. *J. Cell Biol. 109*: 1385–1387.

20. Aktories, K. & I. Just. 1993. *GTPases in biology I: GTPases and actin as targets for bacterial toxins.* Springer-Verlag. Berlin-Heidelberg. 87–112.

21. Aktories, K. & A. Wegner. 1992. Mechanisms of the cytopathic action of actin-ADP-ribosylating toxins. *Mol Microbiol. 6*: 2905–2908.

22. Aktories, K., M. Bärmann, I. Ohishi, S. Tsuyama, K.H. Jakobs, & E. Habermann. 1986. Botulinum C2 toxin ADP-ribosylates actin. *Nature, 322*: 390–392.

23. Simpson, L.L., B.G. Stiles, H.H. Zapeda, & T.D. Wilkins. 1987. Molecular basis for the pathological actions of Clostridium perfringens Iota toxin. *Infect. Immun. 55*: 118–122.

24. Stiles, B.G. & T.D. Wilkens. 1986. Purification and characterization of Clostridium perfringens iota toxin: dependence on two nonlinked proteins for biological activity. *Infect. Immun. 54*: 683–688.

25. Stiles, B.G. & T.D. Wilkins. 1986. Clostridium perfringens iota toxin: Synergism between two proteins. *Toxicon, 24*: 767–773.

26. Simpson, L.L., B.G. Stiles, H. Zepeda, & T.D. Wilkins. 1989. Production by Clostridium spiroforme of an iotalike toxin that possesses mono(ADP-ribosyl)transferase activity: Identification of a novel class of ADP-ribosyltransferases. *Infect. Immun. 57*: 255–261.

27. Popoff, M.R., E.J. Rubin, D.M. Gill, & P. Boquet. 1988. Actin-specific ADP-ribosyltransferase produced by a clostridium difficile strain. *Infect. Immun. 56*: 2299–2306.

28. Vandekerckhove, J., B. Schering, M. Bärmann, & K. Aktories. 1988. Botulinum C2 toxin ADP-ribosylates cytoplasmic β/g-actin in arginine 177. *J. Biol. Chem. 263*: 696–700.

29. Wegner, A. & K. Aktories. 1988. ADP-ribosylated actin caps the barbed ends of actin filaments. *J. Biol. Chem. 263*: 13739–13742.

30. Wille, M., I. Just, A. Wegner, & K. Aktories. 1992. ADP-ribosylation of the gelsolin-actin complex by clostridial toxins. *J. Biol. Chem. 267*: 50–55.

31. Carroll, S.F. & R.J. Collier. 1984. NAD binding site of diphtheria toxin: identification of a residue within the nicotinamide subsite by photochemical modification with NAD. *Proc. Natl. Acad. Sci. USA, 81*: 3307–3311.

32. Jung, M., I. Just, J. van Damme, J. Vandekerckhove, & K. Aktories. 1993. NAD-binding site of the C3-like ADP-ribosyltransferase from *Clostridium limosum. J. Biol. Chem. 268*: 23215–23218.

33. van Damme, J., M. Jung, F. Hofmann, I. Just, J. Vandekerckhove, & K. Aktories. 1996. Analysis of the catalytic site of the actin ADP-ribosylating *Clostridium perfringens* iota toxin. *FEBS Lett.* 380:291–295.

34. Böhmer, J., M. Jung, P. Sehr, G. Fritz, M. Popoff, I. Just, & K. Aktories. 1996. Active site mutation of the C3-like ADP-ribosyltransferase from *Clostridium limosum* - Analysis of glutamic acid 174. *Biochemistry, 35*: 282–289.

35. Saito, Y., Y. Nemoto, T. Ishizaki, N. Watanabe, N. Morii, & S. Narumiya. 1995. Identification of Glu[173] as the critical amino acid residue for the ADP-ribosyltransferase activity of *Clostridium botulinum* C3 exoenzyme. *FEBS Lett. 371*: 105–109.

36. Just, I., G. Schallehn, & K. Aktories. 1992. ADP-ribosylation of small GTP-binding proteins by *Bacillus cereus. Biochem. Biophys. Res. Commun. 183*: 931–936.

37. Carroll, S.F., J.A. McCloskey, P.F. Crain, N.J. Oppenheimer, T.M. Marschner, & R.J. Collier. 1985. Photoaffinity labeling of diphtheria toxin fragment A with NAD:structure of the photoproduct at position 148. *Proc. Natl. Acad. Sci. USA, 82*: 7237–7241.

38. Carroll, S.F. & R.J. Collier. 1987. Active site of Pseudomonas aeruginosa exotoxin A. Glutamic acid 553 is photolabeled by NAD and shows functional homology with glutamic acid 148 of diphtheria toxin. *J. Biol. Chem. 262*: 8707–8711.

39. Barbieri, J.T., M. Mende-Mueller, R. Rappuoli, & R.J. Collier. 1989. Photolabeling of glu-129 of the S-1 subunit of pertussis toxin with NAD. *Infect. Immun. 57*: 3549–3554.

40. Choe, S., M.J. Bennett, G. Fujii, P.M.G. Curmi, K.A. Kantardjieff, R.J. Collier, & D. Eisenberg. 1992. The crystal structure of diphtheria toxin. *Nature, 357*: 216–222.

41. Allured, V.S., R.J. Collier, S.F. Carroll, & D.B. McKay. 1986. Structure of exotoxin A of Pseudomonas aeruginosa at 3,0-Angström resolution. *Proc. Natl. Acad. Sci. USA, 83*: 1320–1324.

42. Stein, P.E., A. Boodhoo, G.D. Armstrong, S.A. Cockle, M.H. Klein, & R.J. Read. 1994. The crystal structure of pertussis toxin. *Structure, 2*: 45–57.

43. Sixma, T.K., S.E. Pronk, K.H. Kalk, E.S. Wartna, B.A.M. van Zanten, B. Witholt, & W.G.J. Hol. 1991. Crystal structure of a cholera toxin-related heat-labile enterotoxin from *E. coli. Nature, 351*: 371–377.

44. Barbieri, J.T. & R.J. Collier. 1987. Expression of a mutant, full-length form of diphtheria toxin in *Escherichia coli. Infect. Immun. 55*: 1647 1651.

45. Douglas, C.M. & R.J. Collier. 1987. Exotoxin A of *Pseudomonas aeruginosa* substitution of glutamic acid 553 with aspartic acid drastically reduces toxicity and enzymatic activity. *J. Bacteriol. 169*: 4967–4971.

46. Locht, C., C. Capian, & L. Feron. 1989. Identification of amino acid residues essential for the enzymatic activities of pertussis toxin. *Proc. Natl. Acad. Sci. USA, 86*: 3075–3079.

47. Fitzmaurice, W.P., L.L. Saari, R.G. Lowery, P.W. Ludden, & G.P. Roberts. 1989. Genes coding for the reversible ADP-ribosylation system of dinitrogenase reductase from *Rhodospirillum rubrum. Mol. Gen. Genet. 218*: 340–347.

48. Koch, T. & W. Rüger. 1994. The ADP-ribosyltransferases (gpAlt) of bacteriophages T2, T4, and T6: Sequencing of the genes and comparison of their products. *Virology, 203*: 294–298.

49. Zolkiewska, A., M.S. Nightingale, & J. Moss. 1992. Molecular characterization of NAD:arginine ADP-ribosyltransferase from rabbit skeletal muscle. *Proc. Natl. Acad. Sci. USA, 89*: 11352–11356.

50. Takada, T., K. Iida, & J. Moss. 1995. Conservation of a common motif in enzymes catalyzing ADP-ribose transfer. *J. Biol. Chem. 270*: 541–544.

51. Lobet, Y., C. Cluff, & J.W. Cieplak. 1991. Effect of site-directed mutagenic alterations on ADP-ribosyl-transferase activity of the A subunit of *Escherichia coli* heat-labile enterotoxin. *Infect. Immun. 59,No.9*: 2870–2879.

7

A PROPOSED ROLE FOR PROTEIN

Protein Complexes in the Regulation of the Reversible ADP-Ribosylation of Dinitrogenase Reductase

Sandra K. Grunwald,[1] Yaoping Zhang,[3] Cale Halbleib,[1] Gary P. Roberts,[2] and Paul W. Ludden[1]

[1]Departments of Biochemistry
[2]Departments of Bacteriology
[3]The Center for the Study of Nitrogen Fixation
University of Wisconsin-Madison
420 Henry Mall
Madison, Wisconsin 53706

INTRODUCTION

Regulation of Dinitrogenase Reductase by Reversible ADP-Ribosylation

The regulation of dinitrogenase reductase by ADP-ribosylation is depicted in Fig. 1 and is described in detail in a review.[1] The ADP-ribose moiety of β-NAD$^+$ is transferred to arginine-101 of dinitrogenase reductase by dinitrogenase reductase ADP-ribosyltransferase (DRAT, the *draT* gene product).[2,3] Modification of one of the two identical subunits of dinitrogenase reductase renders the protein inactive[4] and modification of both subunits of the same dimer has not been observed. The structure of dinitrogenase reductase shows arg-101 to be at the surface of the protein near the FeS center.[5] ADP-ribose is attached to arginine-101 by an α-specific bond to the guanidinium N.[6,7] The ADP-ribosylated dinitrogenase reductase in its reduced, native form can be activated by dinitrogenase reductase activating glycohydrolase (DRAG, the *draG* gene product) in an MgATP- and Mn^{2+}-dependent reaction. DRAG removes ADP-ribose and regenerates the site of modification in the process.[8] ADP-ribose can be removed from both oxidized and oxygen-denatured dinitrogenase reductase in the absence of MgATP, but a divalent metal (Mn^{2+}) is still required.[9,10] Likewise, DRAG can hydrolyze low molecular weight substrates such as dansyl arginine methylester ADP-ribose in the absence of nucleotide.[9] DRAT is unable to ADP-ribosylate small molecule substrates.[3]

ADP-Ribosylation in Animal Tissue, edited by Haag and Koch-Nolte
Plenum Press, New York, 1997

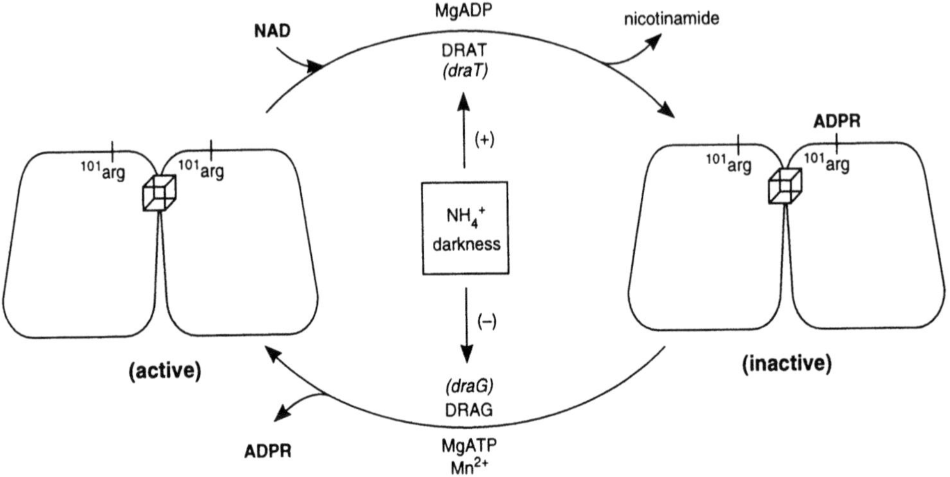

Figure 1. Reversible ADP-ribosylation of Dinitrogenase Reductase.

Factors that lead to modification of dinitrogenase reductase *in vivo* include darkness and fixed nitrogen sources such as ammonium, glutamine and asparagine.[11–14] The mechanisms by which these effectors stimulate modification are not known. It is known that the ADP-ribose attached to dinitrogenase reductase does not turn-over during switch-off conditions, and thus the activity of DRAG is regulated *in vivo*.[12] DRAT activity is also regulated *in vivo*.[15] This conclusion is based on the observation that mutants that lack DRAG accumulate dinitrogenase reductase in the active (not ADP-ribosylated) form until the cells are signalled to ADP-ribosylate the protein.

The effect of fixed N sources on ADP-ribosylation and the fact that the glutamine synthetase inhibitor, methionine sulfoximine, blocks ADP-ribosylation of dinitrogenase reductase *in vivo* in response to fixed N, suggest a role for glutamine or glutamine synthetase in the regulation.[16–19] However, analysis of changes in glutamine pools during switch-off initiated by fixed N, darkness, or phenazine methosulfate does not allow the conclusion that changes in glutamine concentration *in vivo* are either necessary or sufficient to initiate switch-off.[20,21] No effect of glutamine or ammonium has been observed on activity of purified DRAG or DRAT *in vitro*. Similarly, the requirements for nucleotides in the *in vitro* assays of DRAG and DRAT as well as the effect of darkness suggest a role for energy charge in the regulation. Again, results of analyses of adenine nucleotide pools do not allow the conclusion that changes in these pools are necessary or sufficient to initiate switch-off.[22,23] Nordlund and coworkers have shown that NAD⁺ added to cells can stimulate modification.[24]

DRAT and DRAG are encoded by the *draT* and *draG* genes, respectively.[25] While the expression of DRAT and DRAG are not co-regulated with the *nif* genes,[26] the two genes are arranged sequentially on the *R. rubrum* genome next to the *nifKDH* region. There is an open reading frame immediately downstream of *draG* that encodes a protein of 15 kD (ORF) and is an obvious candidate for involvement in the regulation of DRAT and DRAG activity.

RESULTS AND DISCUSSION

Development of a Model for DRAT and DRAG Regulation

The following observations guide our thinking about the regulation and are used to develop the working hypothesis for DRAT and DRAG regulation shown in Fig. 2:

- Both DRAT and DRAG activities are regulated *in vivo*.
- Neither DRAT nor DRAG is stably covalently modified.

All attempts to observe a shift in migration of DRAT or DRAG proteins on 2-D gels in response to treatment of cells and all attempts to observe ^{32}P incorporation into DRAT or DRAG have proven negative.

- The regulation of both DRAT and DRAG is lost upon extraction from the cell, as both enzymes are active *in vitro*, regardless of the ADP-ribosylation status of the dinitrogenase reductase in the cells from which DRAT and DRAG are isolated.
- The regulation of both DRAT and DRAG *in vivo* must be by effectors that inhibit the activity of DRAT or DRAG, because positive effectors are not required for activity of pure DRAT or pure DRAG *in vitro*. (And the activities observed *in vitro*

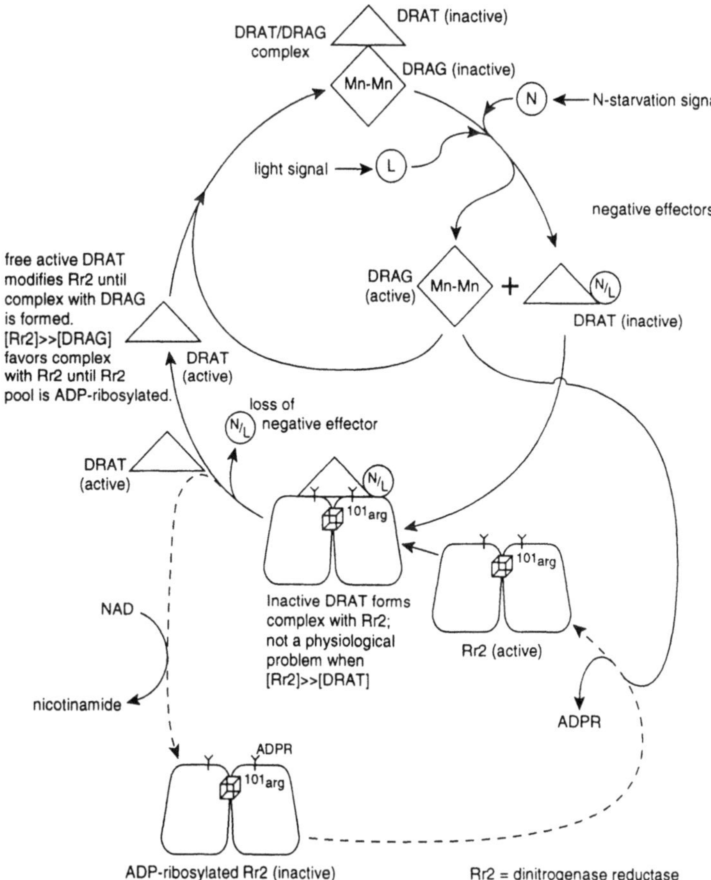

Figure 2. Working hypothesis for the regulation of DRAT and DRAG.

are sufficient to perform their roles at the rate observed *in vivo*). A mutant form of DRAT that fails to be inhibited *in vivo* is described below.

- When DRAT is overexpressed, the whole-cell nitrogenase activity is greatly decreased (5% of normal) even though unmodified dinitrogenase reductase accumulates to a nearly normal level. An explanation for this observation is that DRAT might bind to dinitrogenase reductase even when it is not signaled to ADP-ribosylate dinitrogenase reductase. Because the site of ADP-ribosylation is at the presumed site of interaction of dinitrogenase and dinitrogenase reductase, the binding of DRAT at the ADP-ribosylation site prevents the interaction of the two components of nitrogenase, thus inhibiting its activity.

- The target protein, dinitrogenase reductase, must play some role in the signal transduction pathway. Different species of dinitrogenase reductase (the *nifH*-encoded dinitrogenase reductase or the *anfH*-encoded dinitrogenase reductase) respond differently to different signals *in vivo*. The *anfH*-encoded dinitrogenase reductase becomes modified *in vivo* in response to darkness, but not in response to ammonium.

- DRAT is necessary for the negative regulation of DRAG. When DRAG is overexpressed, its regulation *in vivo* is lost and it is not possible to accumulate dinitrogenase reductase in the ADP-ribosylated form in response to darkness or ammonium. When DRAG and DRAT are overexpressed together, regulation of DRAG is regained. The simplest explanation for this observation is that DRAT binds to DRAG, and this simple idea has been incorporated into our working model for the regulation of DRAG (Fig. 2).

- The regulation of DRAT and DRAG is not a simple, reciprocal on/off in which one of the two is off while the other is on. While DRAT appears to be off when DRAG is on, there also appears to be a state in which both activities are off. This "both off" state is proposed because dinitrogenase activity is never completely eliminated *in vivo*, even in a *draG⁻* strain that cannot produce DRAG. The extent to which the pool of dinitrogenase reductase is ADP-ribosylated *in vivo* can be manipulated, depending on conditions. DRAT activity is transiently observed in response to the signal, and then returns to an inactive form (while DRAG remains inactive).

These observations have been used to develop the working hypothesis for DRAT and DRAG regulation shown in Figure 2. In this model, DRAT and DRAG are seen as catalytically active when present as free molecules with no macromolecular or small molecule effectors bound. DRAT and DRAG are also proposed to form a complex in which both are inactive by virtue of being in the complex. When the cells are exposed to light or are starved for N, negative effectors for DRAT accumulate in the cell and bind to DRAT, removing DRAT from the DRAT/DRAG complex. DRAT with its negative effector(s) bound is inactive, while the DRAG released from the complex is now active and proceeds to remove ADP-ribose from dinitrogenase reductase, thereby activating it. DRAT, with its negative effector bound, binds to active dinitrogenase reductase. Under normal physiological conditions, this complex is not a problem for nitrogen fixation because the dinitrogenase reductase concentration exceeds the DRAT concentration by 100-fold. The negative effector(s) for DRAT generated in response to N-starvation or light (designated (N/L) in Fig. 2) are shown binding at the interface of DRAT and dinitrogenase reductase because dinitrogenase reductase is thought to play a role in perceiving the signal. When cells are exposed to darkness or ammonium, the negative effectors for DRAT are consumed or otherwise removed from the cell and DRAT becomes active and ADP-ribosylates the dinitrogenase reductase molecule to which it is bound. DRAT then releases from the ADP-ribosylated molecule of dinitrogenase reductase and either binds to another molecule of dinitrogenase reductase (and ADP-ribosylates it) or to a molecule of DRAG, thereby inactivating DRAG and DRAT. Because the dinitrogenase reductase concentration exceeds both the DRAT and DRAG concentrations by 100-fold, DRAT will modify many molecules of dinitrogenase before it is bound in the stable, inactive complex with DRAG. For simplicity, DRAG is shown to be inactivated only by binding to DRAT; but it is likely that the real situation is more complicated. For example, it is quite possible that negative effectors for DRAG activity accumulate in the cell when fixed N is available or when cells are placed in the dark.

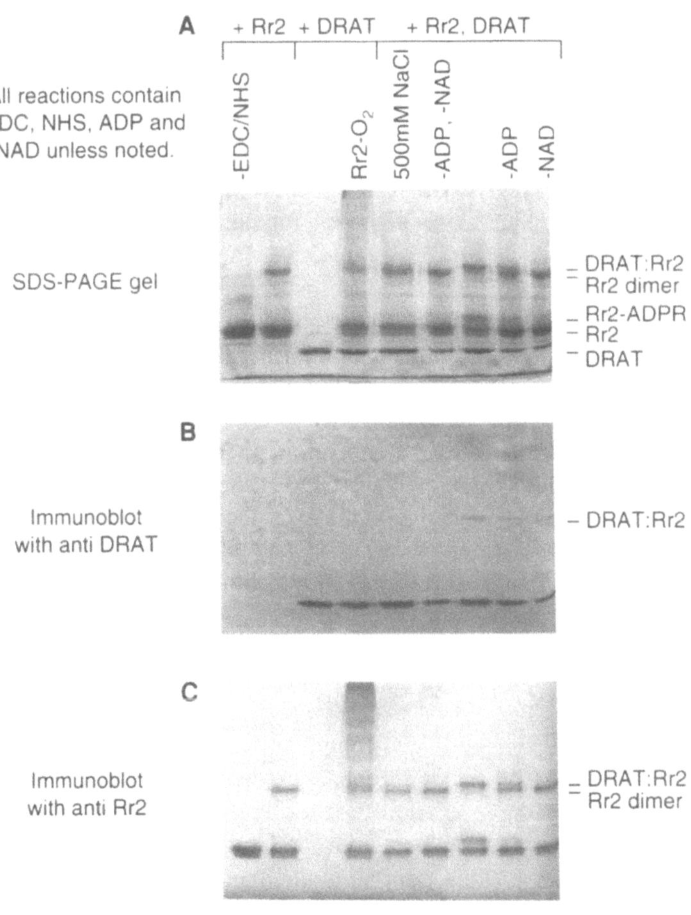

Figure 3. SDS-PAGE gel and immunoblots of DRAT:Dinitrogenase Reductase complex formation.

Crosslinking of DRAT and Dinitrogenase Reductase

A model that predicts complexes of DRAT with dinitrogenase reductase and with DRAG has been developed to explain the results obtained with *draTG* mutants, *draTG* overexpressers, and with the alternative dinitrogenase reductase as described above. In order to test the model, protein crosslinking experiments have been performed. Figure 3 shows a gel and immunoblots from an experiment using the crosslinking agent EDC (1-ethyl-3-(3-dimethylaminopropyl)carbodiimide). This reagent is effective at crosslinking proximal carboxyl groups to amino groups and has been used by Howard and coworkers to study the formation of complexes between the nitrogenase components.[27] Two types of crosslinked complexes have been observed in these experiments: 1) the subunits of *R. rubrum* dinitrogenase are crosslinked to form a complex that electrophoreses on SDS PAGE at the position expected for a dimer of subunits (i.e. lane 2, Fig. 3; this band is labelled "Rr2 dimer" in the figure). This complex can be detected with anti-dinitrogenase reductase antibodies but not with anti-DRAT antibodies. 2) A second complex that crossreacts with antibodies against both dinitrogenase reductase and DRAT migrates on SDS PAGE at a position just above the position of the dimer of dinitrogenase reductase subunits (i.e. lane 7, labelled "DRAT:Rr2"). This is the position that would be expected for a complex of DRAT and a single subunit of dinitrogenase reductase. This crosslinked complex is barely observed if the crosslinking reaction is attempted in the presence of 500 mM salt (lane 5), suggesting that formation of the complex is inhibited by salt, just as the DRAT-catalyzed ADP-ribosylation of dinitrogenase reductase is inhibited by high salt. Also, no DRAT:Rr2 complex is seen if Rr2 is O_2-denatured (lane 4), suggesting that the native form of dinitrogenase reductase is required for complex formation. Formation of the DRAT:dinitrogenase reductase complex is dependent on the presence of NAD^+ and is stimulated by the presence of MgADP (lanes 8 & 9). (MgADP is required for the DRAT-catalyzed ADP-ribosylation of dinitrogenase reductase.)

The NAD^+-dependence of the complex formation is intriguing. Sandra Grunwald in the lab has attempted to demonstrate the binding of NAD^+ to purified DRAT using a number of methods, and all results to date are negative (the positive control, NAD^+ binding to alcohol dehydrogenase, works well in all methods tested). This result, taken together with the crosslinking data, suggests EITHER: that NAD^+ binds to dinitrogenase reductase, followed by DRAT binding to the DRAT:dinitrogenase reductase complex; OR that NAD^+ binds to a weak complex of DRAT and dinitrogenase reductase (or a different complex that lacks the proximal carboxyl group:amino group pair that EDC crosslinks). The results show that a complex of DRAT:dinitrogenase reductase:NAD^+ is formed in the absence of ADP. In the absence of ADP, essentially no ADP-ribosylation of *R. rubrum* dinitrogenase reductase takes place and it would be expected that DRAT would be inactive in this complex.

Reduced pyridine nucleotide was completely ineffective in promoting the formation of a complex between DRAT and dinitrogenase reductase. This is not surprising as neither NADH nor NADPH will serve as a substrate for the ADP-ribosylation of dinitrogenase reductase. Surprisingly, $NADP^+$ promotes the formation of the DRAT:dinitrogenase reductase complex, even though it does not allow the phospho-ADP-ribosylation of the target protein. This is a useful observation, as it allows the complex to be studied in the absence of DRAT enzymatic activity. ADPR and nicotinamide, alone or in combination, did not promote complex formation.

NADP$^+$ Serves as a Substrate for DRAT when *Azotobacter vinelandii* Dinitrogenase Reductase Is Used as the Target Substrate

A surprising result was the NADP$^+$-dependent phospho-ADP-ribosylation of dinitrogenase reductase from *A. vinelandii*. This result is surprising for several reasons: 1) DRAT is unable to use NADP$^+$ as a substrate for the modification of its physiological target, the dinitrogenase reductase from *R. rubrum*;[3] 2) the toxin ADP-ribosyltransferases are highly specific for NAD$^+$;[28, 29] and 3) the turkey erythrocyte ADP-ribosyl transferase, which will use NADP$^+$ in place of NAD$^+$ is not target selective with respect to NAD$^+$ vs. NADP$^+$.[29] This result is consistent with the hypothesis that the target protein plays a role in forming the functional complex that results in modification of dinitrogenase reductase.

Isolation of a Mutation in *draT* which Results in Unregulated DRAT

A strain of *R. rubrum* in which a substantial portion of dinitrogenase reductase is always modified has been isolated. This strain contains unregulated DRAT and was obtained by imprecise PCR of the *draT* gene. Following imprecise PCR of the *draT* gene, the mutated gene was packaged in a plasmid that will stably transform *R. rubrum*. A *R. rubrum* strain deficient in *draT* (strain UR212; *draT*::kan) is used as a recipient for plasmids containing altered *draT* genes; the plasmid also encodes gentamycin and tetracycline resistance, and transformed strains are selected by growth on plates containing kanamycin, gentamycin and tetracycline. Each colony is then grown in a test tube culture under nitrogenase derepressing conditions and tested for growth under a regimen of light/dark/light. Under these conditions, cells that contain normal levels of appropriately regulated DRAT grow with a doubling time of approximately 20 hours. Cells that lack DRAT activity grow with a doubling time of about 15 hours (many of this sort of mutant are obtained). Cells with unregulated DRAT grow with a doubling time of approximately 40 hours because most of their nitrogenase is switched-off most of the time.

Using this approach, Yaoping Zhang has obtained 2 mutants in which dinitrogenase accumulates in the ADP-ribosylated form. These mutants accumulate normal levels of DRAT protein and dinitrogenase reductase protein, but the DRAT appears to be unregulated. One of these, a mutant with lysine 103 converted to glutamate (K103E), shows less than 10% of normal whole cell nitrogenase activity and upon treatment with ammonium or darkness, that activity decreases to less than 1% of full activity that is observed with wild type cells.

DRAG Contains a Binuclear Mn Cluster

The enzyme arginase has been shown to contain a binuclear Mn cluster at its active site.[30] Based on the similarity of the DRAG and arginase sequences and on the similarity of the reactions they catalyze, DRAG was examined for the presence of a Mn-Mn cluster. DRAG has long been known to require Mn^{2+} or a similar divalent metal, but the stoichiometry of Mn per DRAG monomer has not been examined. The signature of a binuclear Mn cluster is an 11 line signal in the EPR spectrum; generally, these signals can only be observed at low temperatures (below 20 K). Such a signal is observed for *R. rubrum* DRAG and it is clear that DRAG contains a Mn binuclear cluster.

ACKNOWLEDGMENTS

We wish to thank Russ Poyner and George Reed at the Enzyme Institute at UW-Madison for their assistance with Mn EPR. This work was supported by NIH grant GM54910 to PWL and NRI grant 93–37305–9237 to GPR.

REFERENCES

1. Ludden, P.W., & G.P. Roberts. 1989. Regulation of nitrogenase activity by reversible ADP-ribosylation. *Current Topics in Cellular Regulation. 30*: 23–56.
2. Lowery, R.G., L.L. Saari, & P.W. Ludden. 1986. Reversible regulation of the nitrogenase iron protein from *Rhodospirillum rubrum* by ADP-ribosylation *in vitro, J. Bacteriol. 166*: 513–518.
3. Lowery, R.G., & P.W. Ludden. 1988. Purification and properties of dinitrogenase reductase inactivating ADP-ribosyltransferase from the photosynthetic bacterium *Rhodospirillum rubrum, J. Biol. Chem. 263*: 16714–16719.
4. Preston, G.G., & P.W. Ludden. 1982. Change in subunit composition of the iron protein of nitrogenase from *Rhodospirillum rubrum* during activation and inactivation of iron protein. *Biochem J. 205*: 489–494.
5. Georgiadis, M.M., H. Komiya, P. Chakrabarti, D. Woo, J.J. Kornuc, & D.C. Rees. 1992. Crystallographic structure of the nitrogenase iron protein from *Azotobacter vinelandii, Science 257*: 1653–1659.
6. Pope, M.R., S.A. Murrell, & P.W. Ludden. 1985. Covalent modification of the iron protein of nitrogenase from *Rhodospirillum rubrum* by adenosine diphosphoribosylation of a specific arginyl residue. *Proc. Natl. Acad. Sci. USA 82*: 3173–3177.
7. Pope, M.R., S.A. Murrell, & P.W. Ludden. 1985. Purification and properties of the heat released nucleotide modifying group from the inactive Fe protein of nitrogenase from *Rhodospirillum rubrum. Biochemistry. 24*: 2374–2380.
8. Saari, L.L., M.R. Pope, S.A. Murrell, & P.W. Ludden. 1986. Studies on the activating enzyme for iron protein of nitrogenase from *Rhodospirillum rubrum. J. Biol. Chem. 261*: 4973–4977.
9. Pope, M.R., L.L. Saari, & P.W. Ludden. 1986. *N*-glycohydrolysis of adenosine diphosphoribosyl arginine linkages by dinitrogenase reductase activating glycohydrolase (activating enzyme) from *Rhodospirillum rubrum, J. Biol. Chem. 261*: 10104–10111.
10. Pope, M.R., L.L. Saari, & P.W. Ludden. 1987. Fluorometric assay for ADP-ribosylarginine cleavage enzymes. *Anal. Biochem. 160*: 68–77.
11. Kamen, M.D., & H. Gest. 1949. Evidence for a nitrogenase system in the photosynthetic bacterium *Rhodospirillum rubrum. Science . 109*: 560.
12. Kanemoto, R.H., & P.W. Ludden. 1984. Effect of ammonia, darkness, and phenazine methosulfate on whole-cell nitrogenase activity and Fe protein modification in *Rhodospirillum rubrum. J. Bacteriol. 158*: 713–720.
13. Neilson, A.H., & S. Nordlund. 1975. Regulation of nitrogenase synthesis in intact cells of *Rhodospirillum rubrum*: inactivation of nitrogen fixation by ammonia, L-glutamine and L-asparagine. *J. Gen. Microbiol. 91*: 53–62.
14. Schick, H.-J. 1971. Substrate and light dependent fixation of molecular nitrogen in *Rhodospirillum rubrum. Arch. Mikrobiol. 75*: 89–101.
15. Liang, J., G.M. Nielsen, D.P. Lies, R.H. Burris, G.P. Roberts, & P.W. Ludden. 1991. Mutations in the *draT* and *draG* genes of *Rhodospirillum rubrum* result in loss of regulation of nitrogenase by reversible ADP-ribosylation. *J. Bacteriol. 173*: 6903–6909.
16. Meyer, J., & P.M. Vignais. 1979. Effects of L-methionine-DL-sulfoximine and B-N-oxayl;-L-α, β-diaminopropionic acid on nitrogenase and activity in *Rhodospseudomonas capsulata. Biochem. Biophys. Res. Commun. 89*: 353–359.
17. Sweet, W.J., & R.H. Burris. 1981. Inhibition of nitrogenase activity by NH_4^+ in *Rhodospirillum rubrum. J. Bacteriol. 145*: 824–831.
18. Sweet, W.J., & R.H. Burris.1982. Effects of *in vivo* treatments on the activity of nitrogenase isolated from *Rhodospirillum rubrum. Biochim. Biophys. Acta 680*: 17–21.
19. Weare, N.M., & K.T. Shanmugam. 1976. Photoproduction of ammonium ion from N_2 in *Rhodospirillum rubrum. Arch. Microbiol. 110*: 207–213.
20. Kanemoto, R.H., & P.W. Ludden. 1987. Amino acid concentrations in *Rhodospirillum rubrum* during expression and switch-off of nitrogenase activity. *J. Bacteriol. 169*: 3035–3043.

21. Li, J., C.-Z. Hu, & D.C. Yoch. 1987. Changes in amino acid and nucleotide pools of *Rhodospirillum rubrum* during switch-off of nitrogenase activity initated by NH_4^+ or darkness. *J. Bacteriol. 169*: 231–237.

22. Paul, T.D., & P.W. Ludden. 1984. Adenine nucleotide levels in *Rhodospirillum rubrum* during switch-off of whole-cell nitrogenase activity. *Biochem J. 224*: 961–969.

23. Nordlund, S., & L. Hogland. 1986. Studies of the adenylate and pyridine nucleotide pools during nitrogenase "switch-off" in *Rhodospirillum rubrum.. Plant and Soil. 90*: 203–209.

24. Soliman, A., & S. Nordlund. 1992. Studies on the effect of NAD(H) on nitrogenase activity in *Rhodospirillum rubrum. Arch. Microbiol. 157*: 431–435.

25. Fitzmaurice, W.P., L.L. Saari, R.G. Lowery, P.W. Ludden, & G.P. Roberts. 1989. Genes coding for the reversible ADP-ribosylation system of dinitrogenase reductase from *Rhodospirillum rubrum. Mol. Gen. Genet. 218*: 340–347.

26. Triplett, E.W., J.D. Wall, & P.W. Ludden. 1982. Expression of the activating enzyme and Fe protein of nitrogenase from *Rhodospirillum rubrum. J. Bacteriol. 152*: 786–791.

27. Willing, A.H., M.M. Georgiadis, D.C. Rees, & J.B. Howard. 1989. Cross-linking of nitrogenase components: Structure and activity of the covalent complex. *J. Biol. Chem. 264*: 8400–8503.

28. Collier, R.J., "Diphtheria toxin: Structure and function of a cytocidal protein," *ADP-Ribosylating Toxins and G Proteins*, ed. J. Moss, and M. Vaughan. (Washington, DC: American Society for Microbiology, 1990) 3–19.

29. Vaughan, M., & J. Moss. 1981. Mono(ADP-ribosyl)transferases and their effects on cellular metabolism. *Current Topics in Cellular Regulation. 20*: 205–246.

30. Reczkowski, R.S., & D.E. Ash. 1992. EPR evidence for binuclear Mn(II) centers in rat liver arginase. *J. Am. Chem. Soc. 114*: 10992–10994.

8

ADP-RIBOSYLATION AND EARLY TRANSCRIPTION REGULATION BY BACTERIOPHAGE T4

Kai Wilkens,[1] Bernd Tiemann,[1] Fernando Bazan,[2] and Wolfgang Rüger[1]

[1]Arbeitsgruppe Molekulare Genetik
Lehrstuhl für Biologie der Mikroorganismen
Ruhr-Universität Bochum
44780 Bochum, Germany
[2]DNAX Research Institute of Molecular and Cellular Biology, Inc.
901 California Avenue
Palo Alto, California 94304–1104

SUMMARY

Bacteriophage T4 codes at least for two ADP-ribosylating activities, the 76 kDa Alt and the 24 kDa Mod gene products. The main target for both enzymes is the host RNA polymerase. We cloned and sequenced the *alt* gene and overexpressed the corresponding enzyme. The recombinant protein shows ADP-ribosylating activities *in vitro*, as had been described earlier for the native enzyme isolated from phage heads. The native as well as the recombinant protein ADP-ribosylate the α-subunit of RNA polymerase, but also subunits ß, ß′ and σ^{70} and perform an autoribosylation reaction.

Taking advantage of the pKWIII test system, constructed to measure promoter strengths *in vivo*, it was found that ADP-ribosylation of RNA polymerase leads to an increase of transcription from T4 early promoters up to a factor of two. In an infected host cell this should cause an enhanced expression of T4 genes. Depending on whether RNA polymerase was ADP-ribosylated or not, it initiated transcription at T4 promoters with different sequence characteristics: unribosylated RNA polymerase recognizes the early T4 promoters by an extended -10 region, whereas the ribosylated enzyme selects for T4 early promoters with an extended T4-specific and highly conserved -35 region. These results may reflect how the virus, step by step imposes its genetic program on the host cell, and in part they give a rationale for the extension of the consensus sequence observed with these promoters.

We also sequenced the genomic region of the T4 *mod* gene and found two open reading frames coding both for proteins of approximately 24 kDa. Up to now none of the reading frames could be cloned into *E. coli* in an active form, making it highly probable

ADP-Ribosylation in Animal Tissue, edited by Haag and Koch-Nolte
Plenum Press, New York, 1997

that the ADP-ribosylation pattern inflicted by gene product Mod on host RNA polymerase is deleterious to these bacteria. Comparisons of the amino acid sequences showed significant homologies among the two reading frames. Computer analysis reveals that both Mod sequences and also the sequence of the Alt protein exhibit a structural concordance with the catalytic domains of other prokaryotic ADP-mono-ribosyltransferases such as the *Pseudomonas aeruginosa* exotoxin A, the cholera labile enterotoxin, the diphteria toxin, the heat labile enterotoxin A of *E. coli*, and the pertussis toxin. We present a detailed model for T4 transcription regulation.

INTRODUCTION

The 169 kb genome of bacteriophage T4 is completely sequenced and encodes about 150 genes well described and identified with respect to their functions. Some 150 open reading frames remain to be characterized[1]. Only 69 genes have been shown to be essential under standard laboratory conditions[1]. Surprisingly, a large number of functions, seemingly important to the infection cycle, turned out to be non-essential in genetic analysis. The rationale for this finding is that either important pathways are redundantly secured, or host proteins may in part supplement for the genetic defect, possibly at the expense of the number of progeny phage. Examples for redundantly secured pathways are the glucosylation of the viral DNA which is catalyzed by two independent glucosyltransferases. Even with both transferases missing, DNA still remains protected against cleavage by most restriction enzymes, since it contains 5-hydroxymethylcytosine instead of cytosine. Likewise the phage codes for at least two ADP-mono-ribosyltransferases, gene products Alt and Mod, presumably arginine-directed, and both non-essential, even in the combination alt⁻ mod⁻[2]. It is known that these enzymes trigger regulatory functions and both ADP-ribosylate at least Arg^{265} of the α-subunit of RNA polymerase. While Alt (RNAP *alt*eration) is reported to ribosylate only one of the two α-subunits, Mod (RNAP *mod*ification) is supposed to ribosylate both. It is argued that the 76 kDa Alt protein may not have access to both α-subunits in contrast to the much smaller Mod protein of approximately 24 kDa. Our present investigations are designed to find an answer to the following questions: why do both enzymes target at the same residue, what are the differences in the target specificities and what are the consequences of these ADP-ribosylations for host RNA polymerase and for viral transcription regulation.

RESULTS AND DISCUSSION

T4 DNA and Transcription

T4 DNA has several structural peculiarities. The A-T content is 64%, and cytosine is replaced by 5-hydroxy-methylcytosine (HMC). All 5-HMC residues carry a mono glucosyl group attached in α- or β-linkage. The latter modifications render phage DNA refractory to most restriction endonucleases and to the *rgl* restriction system of its host *E. coli*. Phage specific transcription is rifampicin sensitive up to the lysis of host cells. From this it is deduced that T4 relies on the host RNA polymerase throughout the infection cycle. However, at least 7 phage encoded proteins act as polymerase subunits or auxiliary factors to co-ordinate the timely ordered gene expression and direct the enzyme to early[3], middle[4] or late promoters[5]. The first step of numerous modifications is the ADP-ribosylation of

RNA polymerase by the gene product Alt, prior to any phage specific gene expression. Therefore, ADP-ribosylation exerted by an internal protein of the phage head that enters the host cell together with the viral DNA, might push host RNA polymerase to favour transcription from T4 early promoters. A general reduction of transcription on non-T4 DNA as a result of ADP-ribosylation of the α-subunits of the host RNA polymerase was reported previously by several authors[6,7,8,9,10].

T4 Early Promoters

Analysis of early transcripts allowed identification and mapping of 31 promoters[11]. Their strengths posed considerable difficulties in cloning and sequencing. The construction of a special vector carrying a transcription termination signal immediately downstream from the multiple cloning site, helped to overcome this obstacle. With this phage vector, M13HDL17, 29 early promoter sequences were compiled and a consensus sequence was deduced[12]. Additional eight promoters were detected when screening T4 sequences for particular features of these sites, bringing the total number of T4 early promoters on the T4 genome to 37. Their consensus sequence is particular in length and sequence from that of *E. coli* promoters. The most prominent deviations are a characteristic and absolutely conserved -35 region, an extended -10 region, and a significant conservation around positions -40 to -52. This leads to a high information content of these promoters[13], much higher than would be required to recognize T4 promoters when screening host and phage DNA. T4 early promoters thus provide a number of characteristics for a putative modulation of promoter recognition[14]. The high information content might indicate the existence of promoter-overlapping binding sites for additional, phage coded proteins and/or an ADP-ribosylated host RNA polymerase. Figure 1 shows the consensus sequences of *E. coli* and T4 early promoters for comparison.

Isolation of the T4 ADP-Ribosyltransferase Alt

It had been shown earlier that the capsid of bacteriophage T4 contains up to 50 copies of the ADP-ribosyl transferase Alt which enter the host cell together with the infecting DNA[15]. Alt is a 76 kDa protein and could be distinguished easily from a second T4-specific ADP-ribosylating activity of about 24 kDa, designated as Mod. While the gene coding for the Mod-activity could be traced to position 12.500 on the genomic map[17], the position of *alt* was identified at about 123.000[16]. All map positions indicated refer to the genomic map worked out by Kutter et al.[1].

Attempts to express and isolate a recombinant Alt protein started with restriction fragment *Xho*I-10 (map positions 121.677–127.792). This fragment of 6115 bp was recovered from preparative agarose gels and then treated with restriction endonucleases *Sph*I and *Cla*I. The *Sph*I-*Cla*I restriction fragment, representing the DNA between 123.114 and 126.151, was isolated and cloned into vector pBluescript KS. The entire nucleotide sequence was determined from both strands[18] (EMBL Data Library Accession No. X15811). The 2 kb reading frame of *alt* appeared to be under control of a late promoter immediately upstream of the gene. We over-expressed the cloned gene with the phage T7 expression system[19] and purified the protein by affinity chromatography on a Blue Sepharose column (Pharmacia Biotech). The high affinity of the ribosyltransferase to Blue Sepharose could be competed with NAD⁺. For comparison we also purified the native Alt protein from the T4 capsid. The protein was detected on polyacrylamide gels, comparing T4 wildtype and T4 alt⁻ preparations. The corresponding protein band could be isolated from preparative

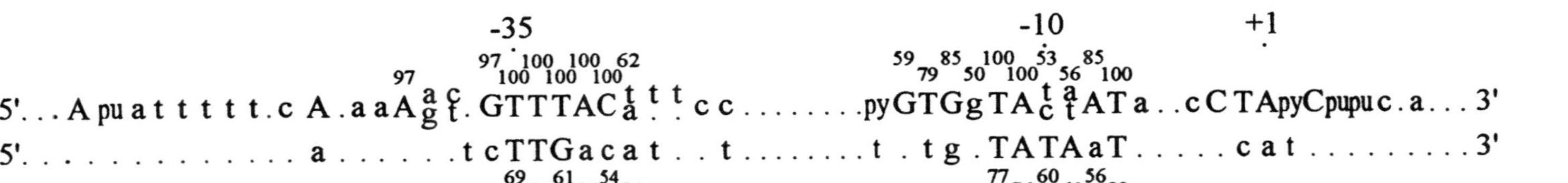

Figure 1. The consensus sequence of T4 early promoters[3] in the upper line is compared to that of *E. coli* in the lower line[32,33]. Upper case letters represent highly conserved bases while lower case letters correspond to positions less conserved. Two bases at the same position represent either one of two alternatives (pu = purines; py = pyrimidines). The percentage of the most frequent base at each position is given above and below the corresponding consensus sequence.

gels and the first 15 NH_2-terminal amino acids were determined. Their sequence confirmed the start of the reading frame of the *alt* gene and revealed that the amino terminus of the protein is processed by cleaving off 6 residues. This processing may be catalyzed by T4 gene product 21 in the course of head assembly[20].

We also constructed and over-expressed a deletion clone carrying only the N-terminal 250 amino acids of the polypeptide. This truncated protein still binds to the Blue Sepharose column. Its binding may be prevented by an incubation with a surplus of NAD^+ prior to loading the enzyme onto the column. Therefore, the nucleotide binding domain of Alt seems to reside in the amino terminal part of the protein. Secondary structure prediction suggests a nucleotide binding fold within the first 150 amino acids[21,22].

The recombinant protein readily ADP-ribosylates RNA polymerase *in vitro*. Not only the α-subunit served as an acceptor in these experiments but the radioactive label was also transferred from NAD^+ to the RNA polymerase subunits ß, ß′ and σ^{70}. Moreover, the Alt protein showed an autoribosylation reaction[23]. The radioactive label transferred to the RNA polymerase subunits could not be chased by adding large amounts of cold NAD^+. Thus, the Alt-catalyzed ADP-ribosylation does not seem to be reversible. The same was reported for Alt isolated from phage capsids[24]. These observations are supported by the fact that up to date no ADP-ribosyl hydrolase activity has been detected in T4 infected cells.

Influence of Alt on Promoter Recognition *in Vivo*

To investigate the effect of the ADP-ribosylation on RNA polymerase and promoter utilization *in vivo*, we constructed a vector to constitutively express Alt. The *alt* gene was taken from the pBKS-clone described above. It was combined with the P15A-origin and kanamycin-resistance of plasmid pGP1–2[19] and promoter P_{lac} derived from plasmid pT7/T3 (Pharmacia-LKB). The resulting vector pTKRI[23] is shown in Figure 2a. Cells harboring this vector grew slower than cells without.

The promoter-test system, vector pKWIII, principally consists of two independent transcription units (Fig. 2b). The first one serves as an internal standard. The weak and constitutive promoter P_{bla} directs transcription of *lacG*. The corresponding gene product, 6-phospho-β-galactosidase[25,26], an enzyme activity not present in *E. coli*, is easy to quanti-

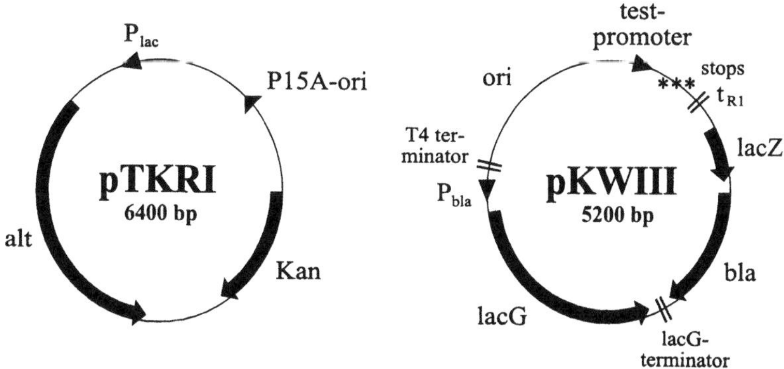

Figure 2. The arrangement of the different DNA elements on plasmids pTKRI and pKWIII. For further details see text.

tate in a photometric assay. The second transcriptional unit on vector pKWIII is controlled by the promoter to be tested. In order to allow cloning and analysis of very strong promoters the initiated transcription is attenuated by insertion of a termination signal, immediately downstream of the promoter cloning site. Reading through this termination site from a strong test promoter with a frequency of approximately 20–30%, results in sufficient expression of the reporting gene *bla*, encoding β-lactamase. This enzyme activity also may be quantitated photometrically. The ratio of the two enzyme activities in crude cell extracts reflects promoter strength. In this system promoter activity also may be quantitated in parallel at the level of transcription. Both experimental approaches result in very similar promoter activities. The development of the test system and further details are described elsewhere[14].

$E.$ *coli* cells transformed either with pKWIII alone, or alternatively, with pKWIII plus pTKRI may serve to test ADP-ribosylated RNA polymerase, with respect to promoter recognition. The data of these experiments are compiled in Table 1. In this system we tested 3 control promoters, P_{A1}, P_{A2} and P_{207}, isolated from phages, T7 (P_A) and T5. These promoters showed no difference in strengths in absence or presence of pTKRI. The 21 T4 early promoters tested behaved less uniformly. Transcription from some promoters did not seem to be influenced by the presence of Alt, transcription initiated at a few others seemed

Table 1. *In vivo* activities of T4 early promoters in the presence and absence of Alt. The values are averaged from three independent tests for each promoter. The strategy that can be deduced from these values will be published elsewhere (Tiemann et al.)

Promoter Tested	Without Alt			With Alt		
	β-lactamase (ΔE/min×ml)	6-P-Gal. (ΔE/min×ml)	Ratio	β-lactamase (ΔE/min×ml)	6-P-Gal. (ΔE/min×ml)	Ratio
Control promoters						
PA1 (T7)	3.15	0.14	22.50	2.26	0.10	22.60
PA2 (T7)	1.82	0.18	10.11	1.75	0.19	9.21
P207 (T5)	3.41	0.14	24.36	1.90	0.08	23.75
T4 early promoters						
P3.6	1.67	0.16	10.44	2.43	0.11	22.09
P128.2	0.60	0.07	8.57	0.55	0.07	7.86
P12.8	1.11	0.10	11.10	3.33	0.13	25.62
P65.6	0.89	0.15	5.93	1.34	0.19	7.05
P131.7	1.60	0.15	10.66	4.30	0.19	22.63
P148.6	2.09	0.23	9.09	3.91	0.24	16.29
P50.0	2.42	0.11	22.00	3.64	0.15	24.27
P1.4-3.9	0.71	0.15	4.73	0.59	0.11	5.36
P57.9	2.56	0.14	18.29	1.86	0.11	16.91
P8.1	2.27	0.11	20.64	7.08	0.14	50.57
P54.4	2.11	0.22	9.59	5.13	0.21	24.43
P161.1	0.96	0.06	16.00	3.42	0.14	24.43
P164.5	1.67	0.06	27.83	0.70	0.10	7.00
P73.0	1.91	0.14	13.64	1.95	0.13	15.00
P41.0	4.88	0.23	21.22	3.69	0.15	24.60
P11.5	1.63	0.16	10.19	1.93	0.19	10.15
P128.6	1.75	0.15	11.66	3.67	0.12	30.58
P46.7	0.72	0.12	6.00	1.12	0.15	7.46
P40.4	0.78	0.13	6.00	0.97	0.16	6.06
P5.9-7.3	0.67	0.14	4.79	0.95	0.16	5.93
P35.3	0.81	0.19	4.26	1.26	0.13	4.39

to be weaker, however, at most of the promoters transcription was initiated more frequently under the influence of ADP-ribosylation of host polymerase.

Sorting of the T4 promoter sequences which had been tested in the presence or absence of the Alt gene product and according to their corresponding strength, stresses the importance of poly-A tracks centered around -42 and -52 for the interaction with ADP-ribosylated RNA polymerase. This finding is in agreement with the fact that Arg[265] resides in the carboxyterminus of the α-subunit that was reported to make DNA contacts in this upstream promoter region[27,28,29,30]. Figure 3 shows the sequences of the strongests T4 early promoters in the two hierarchic orders.

T4 ADP-Ribosyltransferase Mod

ADP-ribosylation of host RNA polymerase by the Alt gene product leads to a controlled increase in T4-specific transcription and this seems to be a reasonable step on the way to impose the viral genetic program on host cell metabolism. It is more difficult to understand why the phage codes for a second ADP-ribosylating activity, seemingly targeting at the same protein. To find a solution to this question we intended to clone and overexpress *mod* and to characterize the gene product biochemically. By analyzing several deletion mutants of T4, some lacking the Mod activity, Horvitz mapped at least one gene encoding an essential component of Mod, between genes 39 and 56 at about 12.500 kb of the T4 genomic map[17]. We amplified the corresponding part of the genome between T4 early promoters P12.8 and P11.5 by PCR and subdivided the DNA product into three restriction fragments. Cloning and sequencing of these fragments revealed two ORFs (Accession No X98695), linked by an ATGA sequence. In this constellation the start codon ATG of the downstream gene and stop codon TGA of the upstream gene overlap by one base pair. It is believed that this arrangement of two reading frames leads to a rapid and coordinate translation of both messages. The reading frames code for polypeptides of 200 and 207 aminoacids, respectively. Figure 4 illustrates this part of the genomic map. Our results are in agreement with the data given by Kutter et al.[1] and the peptide sequences for these reading frames submitted recently by G. Mosig [SwissProt Database, Accession No. P39421 (1995) and PIR Database Accession No. JZ0007 (1995)]. The Mod-activity earlier had been traced to a fraction of proteins of 18 - 28 kDa[31]. Therefore, with respect to the molecular masses of the encoded proteins and their position on the genome, both reading

Cells grown without pTKRI/Alt

```
                                  -35                                          -10        +1
P8.1     TATCTCTTCCAACAAAACGATAAAAAGTT GTTTACT TCCTCGGTTAGTTGTGG TACTAT AACACCATAGCTACTGAGGATAAT
P164.5   TCTTTCCATTTAAAAAATTTCACAAAACA GTTTACA TACCACAAGGACCGTGG TACTAT ACAACTATCAACTACTGATACGGA
P73.0    CAGATGATTTAAAATAATTTCACAAAGTT GTTTACA TACTGATGAGGTAGTGA TACTAT TACCTCATCAAAATTAATTAGGAA
P57.9    CCTTCGGGAGACTTTTTTTCATTTTACCG GTTTACT TTCCGTTTGAGCTGTGG TACTAT ACAACCATCGGATAAAGAGGAGAA
P128.6   TTTTTATTTTCCAAAGATTGCACAAAGTT GTTTACA GTATAGTTCCTTTGTGA TAGTAT TATCTTACACAAACAAAGGAGAAT
P50.0    CTGAACAATGCACTAAATGCACTGAACTA GTTTACT TTGCCACAAGGATGTGG TATAAT GTTCTTACTTTCTACTGAGGAGAT
P131.7   CAAACTGCGGAATTCAAAGCGATTTAATG GTTTACT TTACGGTAGAGTTGTGA TATTAT AGCTCTACCAAAACAAATGAGGAA
```

Cells grown with pTKRI/Alt

```
                                  -35                                          -10        +1
P8.1     TATCTCTTCCAACAAAACGATAAAAAGTT GTTTACT TCCTCGGTTAGTTGTGG TACTAT AACACCATAGCTACTGAGGATAAT
P128.6   TTTTTATTTTCCAAAGATTGCACAAAGTT GTTTACA GTATAGTTCCTTTGTGA TAGTAT TATCTTACACAAACAAAGGAGAAT
P12.8    GCTCCCTGGGCAAAATAATTCAAAAAGTT GTTTACT TTCCTTTCTAACGATGA TATGAT AGCTTCTGAAGTATACGGAGGCTA
P41.0    TCATTCCCTAAAATATTTTTCACAAAGTT GTTTACA ACAAGTTCAAACCGTGG TATTAT TAACATATGAATTACCTTTGAGGA
P161.1   TTTTATTTTAAAAATTTTTTCACAAAACG GTTTACA ACCAAAGCATACTGTGG TACTAT ACAACTATCAACTACTGATACAGA
P54.4    TTTTTATTTCAAAATTTTTTCACAAAACG GTTTACA AGCATAAAGCTTTATGG TACTAT ACAACTATCAACTGATACGGATTT
P50.0    CTGAACAATGCACTAAATGCACTGAACTA GTTTACT TTGCCACAAGGATGTGG TATAAT GTTCTTACTTTCTACTGAGGAGAT
```

Figure 3. Compilation of promoter sequences most frequently recognized by *E. coli* RNA polymerase and its ADP-ribosylated form. The promoters are sorted according to their respective strengths.

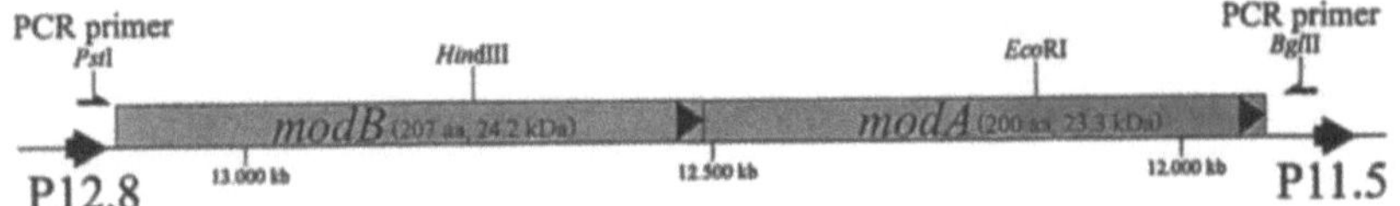

Figure 4. The arrangement of reading frames *modB* and *modA* between promoters P12.8 and P11.5. The sizes of the corresponding gene products are indicated. The PCR primers for DNA amplification and the restriction cleavage sites for cloning of the three subfragments are given. The cleavage sites within oligonucleotides were generated by PCR. Values in kilobases correspond to the positions on the T4 genomic map[1].

frames are candidates to represent gene *mod*. The two proteins show significant amino acid homologies among each other. Homology searches in the databases as well as the comparison with other arginine directed ADP-ribosyltransferases make it highly probable that both proteins belong to this enzyme family. Further evidence for a structural link is the observation that the Alt protein and the two Mod proteins can be threaded into the catalytic domains of ADP-ribosylating bacterial toxins and display the expected catalytic residues (Fig. 5). It is interesting to note that the *modB/modA* operon ranges among those, most stimulated by ADP-ribosylation.

The appearance of a putative third ADP-ribosylating activity on the T4 genome is somewhat unexpected. The two Mod polypeptides coordinately expressed may either act separately and possibly in sequence, ADP-ribosylating each selected Arg residues which would drive transcription regulation, or they might form a heterodimer, ensuring a highly stereo-specific access to regulation-relevant Arg residues, e.g. on both α-subunits of RNA polymerase. In contrast to the conditions found with the *alt* gene, the *mod* reading frames could not be cloned in their entirety, especially *modA* showing more homologies with other ADP ribosyltransferases seems to be toxic: even with the ribosomal binding site deleted, this reading frame withstands cloning in *E. coli*. Work is in progress to overcome these difficulties and to isolate recombinant Mod proteins supporting the characterization of these ADP-ribosyltransferases and their importance for transcription regulation.

The Mechanisms Possibly Underlying T4 Transcription Regulation

Pulling together the experimental data presented in this publication as well as the facts already known about T4 transcription regulation in general, then the following model may explain the different modes of transcription:

T4 DNA entering the host cell is transcribed by host cell RNA polymerase. The T4 early promoters are recognized efficiently by their extended and rather conserved -10 regions. Their -35 region might contribute only little or not at all to the recognition process, since the consensus sequence is away from what host RNA polymerase is used to find in *E. coli* cells. Increasing ADP-ribosylation of one the α-subunits (and possibly others) leads to a rearrangement of the α-subunit and especially of its carboxy terminal domain. This rearrangement hampers the recognition of *E. coli* specific -35 regions, but at the same time the recognition of the highly conserved T4-specific -35 and upstream regions becomes perfect, T4 early transcription now is at its maximum. Addressing certain promoters very effectively (Table 1), T4 enhances trancription of most promoters serving genes with transcription factors like *mot*A, gp55, and *asi*A. The operon coding for genes *modB* and *modA* is transcribed strongest.

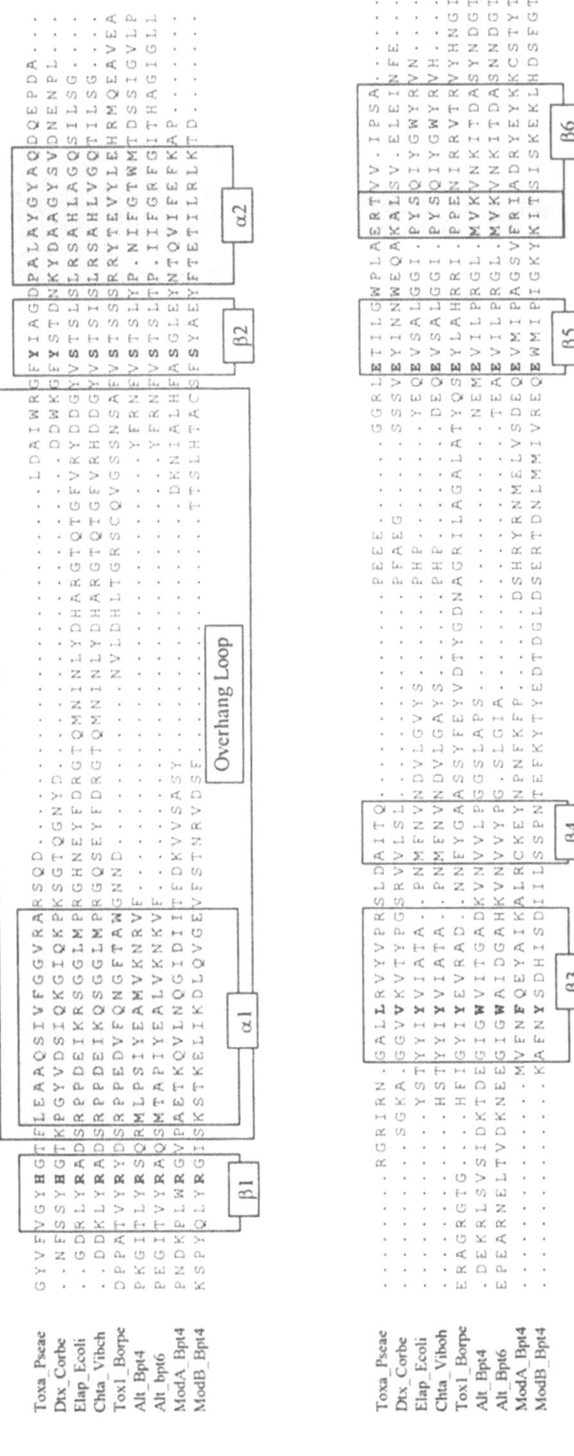

Figure 5. The alignment of the Alt and Mod ADP ribosyltransferases of bacteriophage T4 (and Alt T6) to the catalytic domains of bacterial toxins (and other eukaryotic arginine directed mADPRTs, not shown) according to threading programs. For further details see F. Bazan as well as C.E. Bell and D. Eisenberg (this volume). The enzyme codes used are: Toxa_pseae = exotoxin A of *Pseudomonas aeruginosa*, Dtx_Corbe = diphtheria toxin of *Corynebacterium diphtheriae*, Elap_Ecoli = heat-labile enterotoxinA of *E. coli*, Chta_Viboh = cholera enterotoxin of *Vibrio cholerae*, Tox1_Borpe = pertussis toxin of *Bordetella pertussis*, Alt_Bpt4 = Alt protein of phage T4, Alt_Bpt6 = Alt protein of phage T6, ModA_Bpt4 = ModA protein of phage T4, ModB_Bpt4 = ModB protein of phage T4.

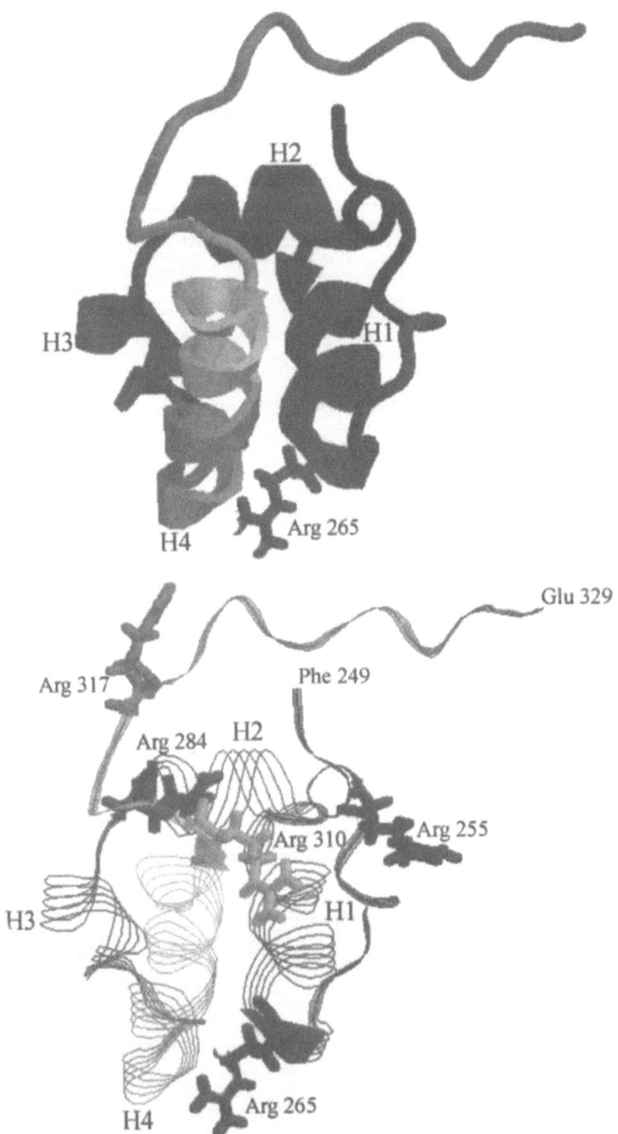

Figure 6. The carboxy terminus of the α-subunit of RNA polymerase according to Igarashi et al.[27], as taken from the Brookhaven data library (access code 1COO) and processed with the RasMol program. The top part of the figure shows the 4 α-helices (H1 - H4), constituting this domain. The Arg[265] residue is highlighted. The bottom presentation shows the α-helices for clarity as "strands" with the remaining Arg residues as putative ribosylation targets together with the NH_2 and COO^- terminal amino acids. The 3D structure of the carboxy terminal domain of the alpha subunit is composed of four α-helices[27] which were reported to be highly resistant to endoproteinases and therefore, must form a compact structure[34]. The Arg[265] residue sits in an exposed position at the N-terminal portion of helix 1 and it is known that helix 1, helix 4 and the proximal loop region contact the *E. coli rrn*B P1 promoter up element. ADP-ribosylation of Arg[265] abolishes the binding to the up element. ADPR not only replaces the positive charge of the Arg residue by two negative oxygen charges but it also renders the arginine side chain a rather bulky construct with new bonding capacities, possibly causing rearrangements within the carboxy terminal domain and allowing for more flexibility.

The ModA protein or possibly a ModA/B heterodimer now starts with the second step ribosylation of the α-subunit not reached previously by the Alt protein. The structural consequences of this reaction are rather similar to those described before, but now with identical structures, both carboxy termini of the α-subunits act synchronously, possibly in a dimeric configuration. These doublet helix turn helix motifs may function as a "built in repressor" which shuts off transcription from all early promoters. Since polymerase would be blocked at the step of promoter clearance, even the high affinity to the -10 regions does not help the enzyme to continue early transcription.

The RNA polymerase may be reactivated and the repressor action abolished if the carboxy terminus will engage in protein-protein interactions as e.g. by contacts with other transcription factors and/or auxiliary proteins like motA and possibly asiA for middle mode transcription, or gene products 45, 33 and possibly 55 in late transcription. In both cases transcription will proceed with new recognition specificities as conferred e.g. by motA (mot box, ttTGCTTcA) or in cooperation with the factors for late transcription (late promoter region, TATAAATA).

This model does not exclude that other proteins not mentioned here might come into the game, or that ADP-ribosylation proceeds, either to further modify RNA polymerase and/or modify auxiliary factors in a way that they are released from or enabled to enter the transcription complex, as necessary at the switch from middle mode to late transcription. The ModB activity may be a candidate to support such reactions on the way to built e.g. "late modified" RNA polymerase. The isolation of the Mod activities would help to solve several of the open questions on T4 transcription regulation.

ACKNOWLEDGMENT

Our thanks are due to Betty Kutter, Evergreen State College, Olympia, WA. for critically reading this manuscript and for valuable discussion. This work was supported financially by the Deutsche Forschungsgemeinschaft with grants Ru 123/22–1 and 22–2.

REFERENCES

1. Kutter E., T. Stidham, B. Guttman, D. Batts, S. Peterson, T. Djavakhishvili, F. Arisaka, V. Mesyanzhinov, W. Rüger, & G. Mosig. 1994. In: *Molecular biology of bacteriophage T4*; J.D. Karam, J.W. Drake, K.N. Kreuzer, G. Mosig, D.H. Hall, ASM Press, Washington D.C. pp. 491–519.
2. Goff, C.G., & J. Setzer. 1980. ADP-ribosylation of *Escherichia coli* RNA polymerase is nonessential for bacteriophage T4 development. *J. Virol. 33*: 547–549.
3. Wilkens K., & W. Rüger. 1994. In: *Molecular biology of bacteriophage T4*; J.D. Karam, J.W. Drake, K.N. Kreuzer, G. Mosig, D.H. Hall, ASM Press, Washington D.C. pp. 132–141.
4. Stitt B., & D. Hinton. 1994. In: *Molecular biology of bacteriophage T4*; J.D. Karam, J.W. Drake, K.N. Kreuzer, G. Mosig, D.H. Hall, ASM Press, Washington D.C. pp. 142–160.
5. Williams K.P., G.A. Kassavetis, D.R. Herendeen, & E.P. Geiduschek. 1994. In: *Molecular biology of bacteriophage T4*; J.D. Karam, J.W. Drake, K.N. Kreuzer, G. Mosig, D.H. Hall, ASM Press, Washington D.C. pp. 161–175.
6. Mailhammer, R., H.L. Yang, G. Reiness, & G. Zubay. 1975. Effects of bacteriophage T4-induced modification of *Escherichia coli* RNA polymerase on gene expression *in vitro*. *Proc. Natl. Acad. Sci. U. S. A. 72*: 4928–4932.
7. Goldfarb, A. 1981. Changes in the promoter range of RNA polymerase resulting from bacteriophage T4-induced modification of core enzyme. *Proc. Natl. Acad. Sci. U. S. A. 78*: 3454–3458.
8. Goldfarb, A., & P. Palm. 1981. Control of promoter utilization by bacteriophage T4-induced modification of RNA polymerase alpha subunit. *Nucleic Acids Res. 9*: 4863–4878.

9. Goldfarb, A., & S. Malik. 1984. Changed promoter specificity and antitermination properties displayed in vitro by bacteriophage T4-modified RNA polymerase. *J. Mol. Biol. 177*: 87–105.

10. Drivdahl, R.H., & E.M. Kutter. 1990. Inhibition of transcription of cytosine-containing DNA *in vitro* by the alc gene product of bacteriophage T4. *J. Bacteriol. 172*: 2716–2727.

11. Gram, H., H.D. Liebig, A. Hack, E. Niggemann, & W. Rüger. 1984. A physical map of bacteriophage T4 including the positions of strong promoters and terminators recognized in vitro. *Mol. Gen. Genet. 194*: 232–240.

12. Liebig, H.D., & W. Rüger. 1989. Bacteriophage T4 early promoter regions. Consensus sequences of promoters and ribosome-binding sites. *J. Mol. Biol. 208*: 517–536.

13. Schneider, T.D., G.D. Stormo, L. Gold, & A. Ehrenfeucht. 1986. Information content of binding sites on nucleotide sequences. *J. Mol. Biol. 188*: 415–431.

14. Wilkens, K., & W. Rüger. 1996. Characterization of bacteriophage T4 early promoters *in vivo* with a new promoter probe vector. *Plasmid 35*: 108–120.

15. Onorato, L., B. Stirmer, & M.K. Showe. 1978. Isolation and characterization of bacteriophage T4 mutant preheads. *J. Virol. 27*: 409–426.

16. Goff, C.G. 1979. Bacteriophage T4 alt gene maps between genes *30* and *54*. *J. Virol. 29*: 1232–1234.

17. Horvitz, H.R. 1974. Bacteriophage T4 mutants deficient in alteration and modification of the *Escherichia coli* RNA polymerase. *J. Mol. Biol. 90*: 739–750.

18. Koch, T., & W. Rüger. 1994. The ADP-ribosyltransferases (gpAlt) of bacteriophages T2, T4, and T6: sequencing of the genes and comparison of their products. *Virology 203*: 294–298.

19. Tabor, S., & C.C. Richardson. 1985. A bacteriophage T7 RNA polymerase/promoter system for controlled exclusive expression of specific genes. *Proc. Natl. Acad. Sci. USA 82*: 1074–1078.

20. Coppo, A., A. Manzi, J.F. Pulitzer, & H. Takahashi. 1973. Abortive bacteriophage T4 head assembly in mutants of *Escherichia coli*. *J. Mol. Biol. 76*: 61–87.

21. Rossmann, M.G., D. Moras, & K.W. Olsen. 1974. Chemical and biological evolution of nucleotide-binding protein. *Nature 250*: 194–199.

22. Chou, P.Y., & G.D. Fasman. 1978. Empirical predictions of protein conformation. *Annu Rev Biochem 47*: 251–276.

23. Koch, T., A. Raudonikiene, K. Wilkens, & W. Rüger. 1995. Overexpression, purification, and characterization of the ADP-ribosyltransferase (gpAlt) of Bacteriophage T4: ADP-ribosylation of *E. coli* RNA polymerase modulates T4 "early" transcription. *Gene Expression 4*: 253–264.

24. Rohrer, H., W. Zillig, & R. Mailhammer. 1975. ADP-ribosylation of DNA-dependent RNA polymerase of *Escherichia coli* by an NAD+: protein ADP-ribosyltransferase from bacteriophage T4. *Eur. J. Biochem. 60*: 227–238.

25. Morse, M.L., K.L. Hill, J.B. Egan, & W. Hengstenberg. 1968. Metabolism of lactose by *Staphylococcus aureus* and its genetic basis. *J. Bacteriol. 95*: 2270–2274.

26. Breidt, F.J., W. Hengstenberg, U. Finkeldei, & G.C. Stewart. 1987. Identification of the genes for the lactose-specific components of the phosphotransferase system in the *lac* operon of *Staphylococcus aureus*. *J. Biol. Chem. 262*: 16444–16449.

27. Igarashi, K., N. Fujita, & A. Ishihama. 1991. Identification of a subunit assembly domain in the alpha subunit of *Escherichia coli* RNA polymerase. *J. Mol. Biol. 218*: 1–6.

28. Ross, W., K.K. Gosink, J. Salomon, K. Igarashi, C. Zou, A. Ishihama, K. Severinov, & R.L. Gourse. 1993. A third recognition element in bacterial promoters: DNA binding by the alpha subunit of RNA polymerase. *Science 262*: 1407–1413.

29. Blatter, E.E., W. Ross, H. Tang, R.L. Gourse, & R.H. Ebright. 1994. Domain organization of RNA polymerase alpha subunit: C-terminal 85 amino acids constitute a domain capable of dimerization and DNA binding. *Cell 78*: 889–896.

30. Busby, S., & R.H. Ebright. 1994. Promoter structure, promoter recognition, and transcription activation in prokaryotes. *Cell 79*: 743–746.

31. Skorko, R., W. Zillig, H. Rohrer, H. Fujiki, & R. Mailhammer. 1977. Purification and properties of the NAD+: protein ADP-ribosyltransferase responsible for the T4-phage-induced modification of the alpha subunit of DNA-dependent RNA polymerase of *Escherichia coli*. *Eur. J. Biochem. 79*: 55–66.

32. Hawley, D.K., & W.R. McClure. 1983. Compilation and analysis of *Escherichia coli* promoter DNA sequences. *Nucleic Acids Res. 11*: 2237–2255.

33. Lisser, S., & H. Margalit. 1993. Compilation of *E. coli* mRNA promoter sequences. *Nucleic Acids Res. 21*: 1507–1516.

34. Negishi, T., Fujita, N. & Ishihama, A. 1995. Structural map of the alpha subunit of *Escherichia coli* RNA polymerase: structural domains identified by proteolytic cleavage. J. Mol. Biol. *248*: 723–728.

9

PERTUSSIS TOXIN

Entry into Cells and Enzymatic Activity

Ali el Bayâ, Ruth Linnemann, Lars von Olleschik-Elbheim, and
M. Alexander Schmidt

Institut für Infektiologie
Zentrum für Molekularbiologie der Entzündung (ZMBE)
Universität Münster
von Esmarch-Str. 56, 48149 Münster
Germany

ABSTRACT

In this study we present data supporting the concept of a retrograde transport mechanism for pertussis toxin to the Golgi complex. We also describe a novel GTP-binding 32 kDa cellular target protein, which is ADP-ribosylated upon exposure to PT and different from the classical PT substrate, the α-subunit of Gi proteins.

INTRODUCTION

Pertussis toxin (PT) is the main virulence factor of *Bordetella pertussis*, the causative agent of whooping cough. Analogous to other bacterial toxins, like cholera toxin and shiga toxin, the exotoxin consists of an enzymatically active A-protomer (S1 subunit) and a B-oligomer (composed of the subunits S2, S3, (2x) S4, and S5) which is responsible for target cell-binding. S1 exhibits an ADP-ribosyltransferase activity and is known to inactivate the α-subunit of Gi proteins, thus deranging the adenylate cyclase regulatory system (1). So far the Giα-subunits are the only PT substrates known.

In recent studies we could show binding of PT to membrane glycoproteins of a ß-cell derived insulin-secreting cell line of the hamster pancreas (HIT-T15) as well as CHO cells (2). EM studies with gold-labelled pertussis toxin and subcellular fractionation experiments demonstrated that the toxin enters the PT-sensitive HIT-T15 and CHO cells (PT enhances the insulin secretion of HIT-T15 cells and induces a "clustered growth pattern" in CHO cells), by endocytosis. Further, we obtained conclusive evidence for a retrograde transport mechanism for the toxin to the Golgi complex (3). These results are in accord-

ADP-Ribosylation in Animal Tissue, edited by Haag and Koch-Nolte
Plenum Press, New York, 1997

ance with the findings of Xu and Barbieri who used the Golgi complex-disturbing agent Brefeldin A to suggest a participation of the Golgi complex in PT intoxication (4).

RESULTS AND DISCUSSION

Retrograde Transport of PT into Eukaryotic Target Cells

In previous studies we analyzed the uptake of PT into CHO cells by electron microscopy (EM) (with gold-labelled toxin) and subcellular fractionation experiments, and obtained convincing evidences for retrograde transport of PT to the Golgi complex (3). To demonstrate the necessity of the Golgi route for PT action, we conducted a series of experiments using a CHO mutant cell line (V24.1) in which the Golgi complex can be disrupted by culturing the cells at elevated temperature (5). When these cells were grown at 41°C a significant protection from PT-mediated modification of Gi proteins was observed as compared to V24.1 cells grown at 34°C or CHO wild type cells. The PT uptake kinetic of the V24.1 cells grown at 41°C resembles that of the PT receptor-deficient CHO-lec1 cell line (Fig. 1 **A**).

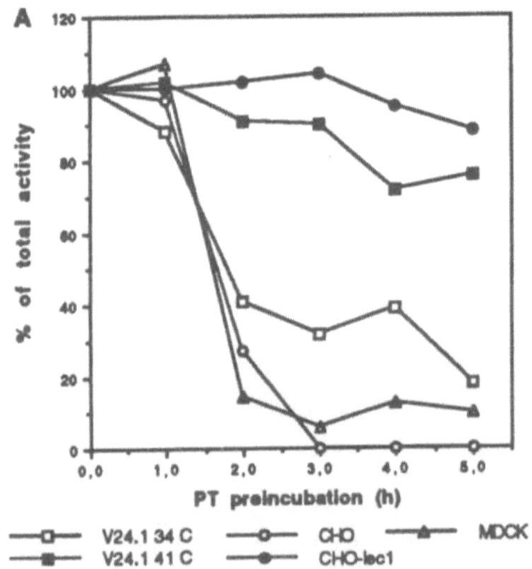

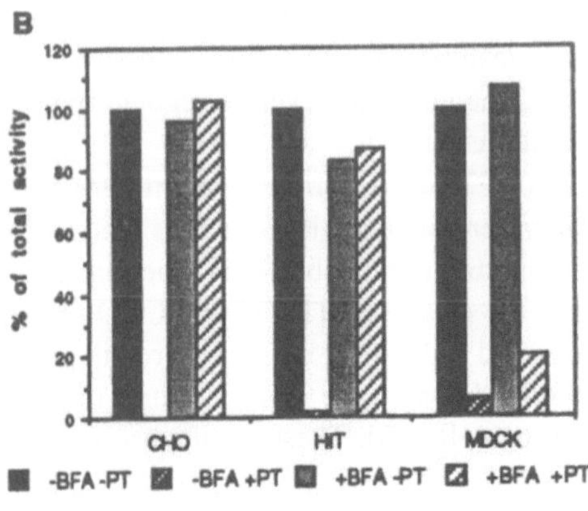

Figure 1. In vitro ADP-ribosylation of G-proteins after PT preincubation in culture. A. V24.1 CHO mutant cells grown at 34°C or 41°C and CHO, CHO-lec1, and MDCK cells were preincubated with 200 ng PT/ml for the periods indicated. Subsequently membrane proteins were solubilized with 0.2 M n-octylglucoside/2 mM PMSF/50 mM K_3PO_4 and used as the substrate in an *in vitro* ADP-ribosylation assay with 140 ng activated PT/ml (1 hour at RT in 100 mM Tris-HCl, pH 8.0, 25 mM DTT, 2 mM ATP, and 1 µCi ^{32}P-NAD; PT activation with 50 mM DTT in 100 mM Tris-HCl, pH 8,2). B. Cells were preincubated with or without 1 µg BFA/ml and afterwards with or without 200 ng PT/ml. The solubilized membrane proteins were then used as the substrate in an *in vitro* ADP-ribosylation assay with 140 ng activated PT/ml as described in A. After separation on a 15% SDS-PAGE binding of ^{32}P-ADP ribose was visualized and quantified using a Bioimager (Fuji BAS 1000).

We also could demonstrate that MDCK cells which are resistent to the Golgi complex-disrupting agent Brefeldin A (BFA) could not be protected from PT intoxication by preincubation with BFA while BFA-sensitive cells (e.g. CHO and HIT cells) could be protected by this treatment (Fig. 1 **B**). Control experiments showed that PT is indeed taken up by MDCK cells (Fig. 1 **A**). As MDCK cells are only resistant against the Golgi-disrupting potency and not against other effects of BFA (e.g. on endosomes) these findings underline the importance of an intact Golgi complex for PT intoxication.

PT Not Only ADP-Ribosylates Gi Proteins but also a Protein of Approximately 32 kDa

CHO cells are known to exhibit a clustered growth pattern upon incubation with PT (6), but the cellular events upon PT intoxication that cause these alterations are not known. We therefore looked for potential cytoskeleton rearrangements and found a disrupting effect of PT on actin stress fibers by staining with Phalloidine-FITC (data not shown). Similar effects had been observed by Scapigliati et al. (7). As other bacterial toxins directly ADP-ribosylate actin or small GTP-binding proteins (e.g. Rho) that are involved in the organization of the actin cytoskeleton (8), we looked for PT substrates smaller than the 42 kDa α-subunit of Gi proteins. In a cell extract enriched in cytoskeleton components we detected a protein of approximately 32 kDa which is ADP-ribosylated by PT *in vitro* (Fig. 2, lanes 1 and 2) and, in contrast to the 42 kDa α-Gi, could not be protected from ADP-ribosylation by preincubating the cells with PT prior to extraction. In a ^{32}P-GTP overlay experiment with the blotted proteins of the same cell extract we could further demonstrate GTP-binding to a single protein of the same size (Fig. 2, lanes 3 and

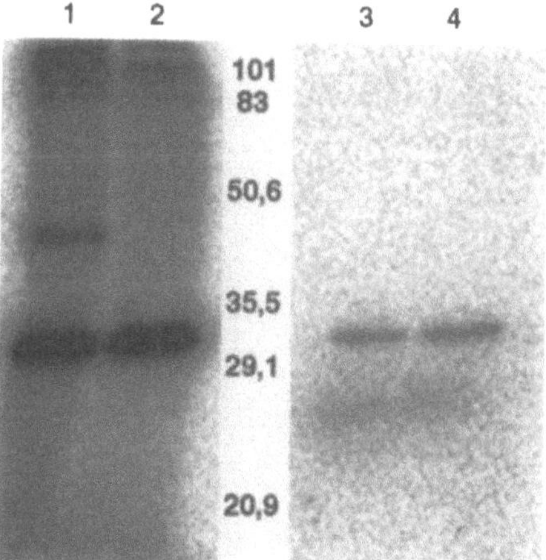

Figure 2. ADP-ribosylation of and GTP-binding to a 32 kDa protein of CHO cells. CHO cells were incubated without (lanes 1 & 3) or with 200 ng PT/ml (lanes 2 & 4) for 3 h. Then the cells were mechanically disrupted by Ultraturrax treatment and centrifuged. The pelleted proteins were solubilized with 0.2 M n-octylglucoside/2 mM PMSF/50 mM K_3PO_4. After centrifugation the remaining insoluble pellet was resuspended in PBS/0.5% Tween 20 and used in a *in vitro* ADP-ribosylation assay as described in Fig.1 (lanes 1 & 2) and in a GTP overlay assay (lanes 3 & 4). After SDS-PAGE the proteins were blotted on nitrocellulose filters and incubated with 100 μM ATP in 50 mM Tris-HCl, pH 7.6, 10 μM MgCl2, and 0.03% Tween 20 for 10 minutes. Afterwards 1μCi ^{32}P-GTP/ml was added. After 1 hour the blots were rinsed and binding of GTP was visualized with a bioimager (Fuji BAS 1000).

4). Thus, we propose that this 32 kDa protein represents a novel G-protein and could be involved in the action of PT on the actin cytoskeleton. Experiments are underway to further characterize this 32 kDa protein.

An effect of PT on the cytoskeleton of target cells could potentially be another characteristic of the toxic activity of PT besides the well-known influence on the intracellular cAMP level of target cells. Moreover, pertussis toxin could develop into an interesting tool for cell biologists investigating the regulation of the (actin) cytoskeleton.

REFERENCES

1. Katada, T., and M. Ui. 1982. Direct modification of the membrane adenylate cyclase system by islet-activating protein due to ADP-ribosylation of a membrane protein. Proc. Natl. Acad. Sci. USA *79*: 3129–3133.
2. El Bayâ, A., R. Linnemann, L. von Olleschik-Elbheim, and M. A. Schmidt. 1995. Identification of binding proteins for pertussis toxin on pancreatic ß cell-derived insulin-secreting cells. Microb. Pathogenesis. *18*: 173–185.
3. El Bayâ, A., R. Linnemann, L. von Olleschik-Elbheim, H. Robenek, and M. A.. Schmidt. Endocytosis and retrograde transport of Pertussis toxin to the Golgi complex as a prerequisite for cellular intoxication. Eur. J. Cell Biol., in revision.
4. Xu, Y. and J. T. Barbieri. 1995. Pertussis toxin-mediated ADP-ribosylation of target proteins in Chinese Hamster Ovary cells involves a vesicle trafficking mechanism. Infect. Immun. *63*, 825–832.
5. Kao, C.-Y., and R. K. Draper. 1992. Retention of secretory proteins in an intermediate compartment and disappearance of the Golgi complex in an END4 mutant of Chinese hamster ovary cells. J. Cell Biol. *117*: 701–715.
6. Hewlett, E. L., K. T. Sauer, G. A. Myers, J. L. Cowell, and R. L. Guerrant. 1983. Induction of a new morphological response in Chinese hamster ovary cells by pertussis toxin. Infect. Immun. *40*: 1198–1203.
7. Scapigliati, G., R. Rappuoli, and S. Silvestri. 1988. Cytoskeletal alterations as a parameter for assessment of toxicity. Xenobiotica *18*: 715–724.
8. Aktories, K. 1994. Clostridial ADP-ribosylating toxins: effects on ATP and GTP-binding proteins. Mol. Cell. Biochem. *138*: 167–176.

10

MEMBRANE ANCHORED SYNTHETIC PEPTIDES AS A TOOL FOR STRUCTURE–FUNCTION ANALYSIS OF PERTUSSIS TOXIN AND ITS TARGET PROTEINS

Lars von Olleschik-Elbheim, Ali el Bayâ, and M. Alexander Schmidt

Institut für Infektiologie
Zentrum für Molekularbiologie der Entzündung (ZMBE) Universität Münster
von-Esmarch-Str. 56, 48149 Münster
Germany

ABSTRACT

Solid phase fixed peptides are useful tools for the investigation of enzyme substrate interactions. We could show that interactions of target sequences, like e.g. the acceptor domain of G_i-proteins for the of mono ADP-ribosylation by pertussis toxin and the toxin itself can be studied by this method. Reaction specifity is identical to modification of free peptides. The method is extremly versatile and has enourmous potential if the focus is on the effects of substrate modification or substrate length.

INTRODUCTION

The major factor for virulence and antigenicity of the whooping cough-causing agent *Bordetella pertussis* is the exotoxin pertussis toxin (PT). The A-B toxin consists of the S1 subunit (A-protomer) which harbours an ADP-ribosyltransferase activity and the B-oligomer (subunits S2, S3, (2x) S4, and S5) which is responsible for target cell-binding. The α-subunits of a couple of heterotrimeric (inhibitory) G-proteins serve as natural substrates for PT. PT inhibits the function of these proteins by transfering an ADP-ribosyl moiety from NAD to a cysteine residue at position -4 of the C-terminus. 20mer soluble peptides corresponding to the C-terminal regions of the α-subunits function as substrates for PT (1). Employing the spot synthesis technique (2) we could show that PT specifically also modifies these residues when the peptides are fixed on cellulose filters via their C-ter-

ADP-Ribosylation in Animal Tissue, edited by Haag and Koch-Nolte
Plenum Press, New York, 1997

minus. Thus, this opens the possibility for fast and efficient "mutational analysis" of the amino acid sequences needed for ADP-ribosylation.

MATERIAL AND METHODS

C-terminally fixed synthethic 5–25mer peptides revealing C-terminal sequences of G-Protein α–subunits of G_{i3}, G_i, G_s, G_{o1}, G_{o2}, G_{oX1}, T_{rod}, G_z, $G_{q/11}$ and G_h (Table 1) were synthesized by conventional Fmoc synthesis methods on specifically aktivated cellulose sheets assisted by a pipeting-robot (AMS 422; Abimed Analysentechnik, Langenfeld, Germany).

During synthesis peptides grew on β-alanine anchors on acid stable cellulose sheets (Whatman 540) of 9x13 cm. The membranes primarily had been functionalized by coupling of β-alanine via esterisation to hydroxyl groups of cellulose filaments. A second β-alanine was coupled to the deprotected NH_2-function of the first alanine. Thus, these anchors were used for the synthesis of the specific peptides as described by Frank (2).

For ADP-ribosylation experiments sheets of 9x13cm with fixed peptides were soaked in 15ml ehanol; if not specially mentioned all procedures were done at room temperature. This was done to prevent the formation of stable hydrophobic aggregates. The buffer was exchanged against 1xTBS pH 7.0 by sequential washings.

Pertussis toxin was preactivated for one hour (450 µg/ml PT, 50 mM DTT, 0.1 M Tris pH 8.2) and afterwards ATP and ^{32}P-NAD were added.

Filters were incubated with 5 ml of ADP-ribosylation solution (0.9 µg/ml aktivated PT, 25 mM DTT, 100 mM Tris pH 7.0, 2 mM ATP, 10 mCi/ml ^{32}P-NAD) for 2 hours. Membranes were washed with H_2O and 5% Tween-TBS (3x 5% Tween-TBS 2 min. / 1x 60°C 5% Tween-TBS 30 min. /3x H_2O 5 min.)

Remaining ^{32}P-activity was detected by exposing filters on a Fuji imaging plate (Type: BAS-IIIs) for 1.5 hours and subsequent scanning with a Bio-Imager (Fuji BAS 1000). Quantification was done with Fuji MacBAS® imaging and quantification software and results were graphically visualized using the arbitrary units produced by the quantification software.

Table 1. C-terminal sequences of G-protein α-subunit
(last 20 aa). The sequences are taken from the
corresponding Swiss-Prot database entries

G protein α-subunit	C-terminal sequence (last 20 aa)
Gi1/Gi2	VFDAV TDVII KNNLK DCGLF
Gi3	VFDAV TDVII KNNLK ECGLY
Go1	VFDAV TDVII AKNLR GCGLY
Go2/GoX1	VFDAV TDVII AKNLR GCCLY
Trod	VFDAV TDIII KENLK DCGLF
Gs	VFNDC RDIIQ RMHLR QYELL
Gq/11	VFAAV KDTIL QLNLK EYNLV
Gz	VFDAV TDVII QNNLK YIGLC

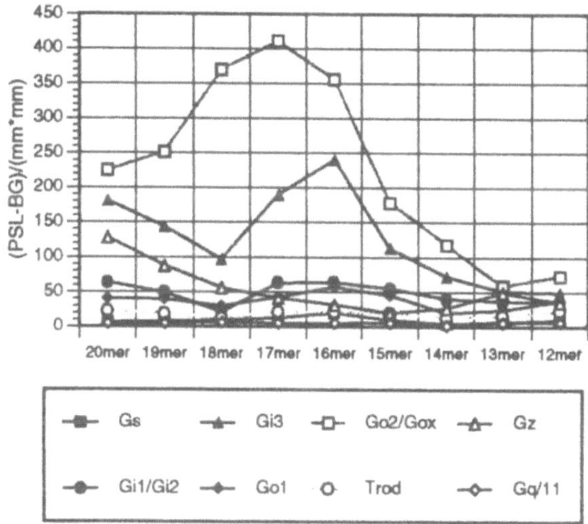

Figure 1. C-terminal peptides of G-protein α-subunits (20 - 12mer) C-terminally fixed on cellulose sheets were ADP-ribosylated by pertussis toxin as described. The graph shows the influence of peptide length on ADP-ribosylation afficiency.

RESULTS AND DISCUSSION

ADP ribosylation of fixed peptides was compared with the results of Graf *et.al.* (1) who investigated ADP-ribosylation of soluble 20mer and 15mer peptides using the same sequences and peptide lengths. Both approaches detect the same sequences as ADP ribosylation substrates. In our study we investigated a number of different C-terminal sequences of G_i-proteins and it turned out that $G_{o2/oX1}$ and G_{i3} exhibit the most efficient acceptor sequences for ADP-ribosylation of all PT substrates tested (Fig. 1). This is in ac-

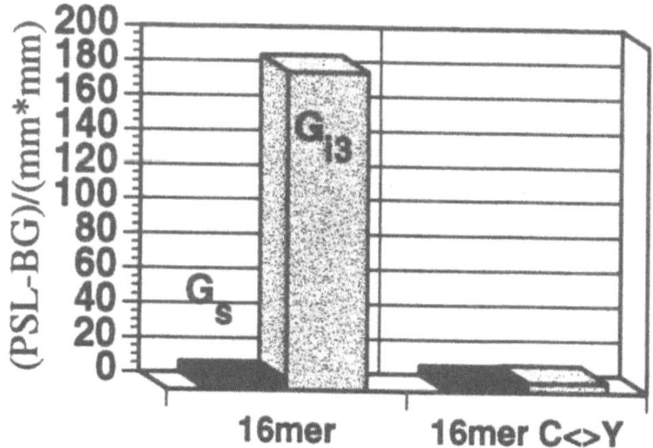

Figure 2. C-terminal fixed 16mer C-terminal peptides of G_{i3} and G_s were ADP-ribosylated by pertussis toxin as described. Exchange of cysteine at position -4 against tyrosine and *vice versa* leads to a complete loss of ADP-ribosylation efficiency.

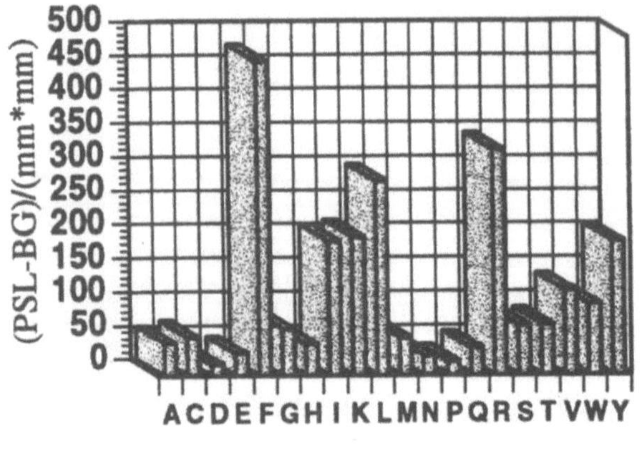

Figure 3. ADP ribosylation of C-terminal fixed 16mer C-terminal peptides of mutated G_{i3} α-subunit. Amino acids at position -1 have been exchanged against all 20 natural amino acid residues. The native aminoacid at this positon is tyrosine (Y).

cordance with the results of Graf *et.al* (1) who suggested G_{i3} as *bona fide* sequence for their approach. Not surprisingly the C-terminal peptides of the cholera toxin-substrates G_s, G_z, and $G_{q/11}$ did not contain acceptor sites for PT-mediated ADP-ribosylation (Fig. 1) as the -4 cysteine is missing.

Positional amino acid exchange of the peptide sequences was used as an easy way for mutational analysis. We determined the effect of amino acid exchanges at at position-4 of G_{i3}, the position where ADP-ribosylation is supposed to take place. As expected any other amino acid than the authentic cysteine lead to a complete loss of the substrate characteristic (Fig. 2).

To investigate whether amino acids in the surrounding positions play an important role for the efficiency of ADP-ribosylation by PT we exchanged them at several positions. Here we present the results of amaino acid exchange at position -1 (Fig. 3). Substitution of tyrosine at position -1 to asparagine or prolin for instance, lead to a loss of the PT-mediated modification. Phenylalanine or arginine at position -1, on the other hand, even lead to enhancement of ADP-ribosylation at position -4.

The method of membrane anchored synthetic peptides turned out to be a very versatile and useful tool to investigate interactions between PT and its substrates.

Compared to similar appoaches using free peptides (1) this technique allows many parallel experiments in combinations with highly conserved reaction conditions. Thus, it should be applicable for further investigation of this and other enzyme/substrate interactions and should be helpful in the process of identifying inhibitory peptides.

REFERENCES

1. Graf, R., Codina, J., and Birnbaumer, L. 1992. Peptide inhibitors of ADP-ribosylation by Pertussis toxin are substrates with affinities comparable to those of the trimeric GTP-binding proteins. *Molecular pharmacology* 42: 760 - 764.

2. Frank, R. 1992. Spot synthesis: An easy technique for the positionally addressable, parallel chemical synthesis on membrane support. *Tetrahedron* 48: 9217 - 9232.

3. von Olleschik-Elbheim, L., el Bayâ, A., and Schmidt, MA. 1996. Positional analysis of the C-terminal acceptor site of G_i proteins for pertussis toxin mediated ADP-ribosylation. *Journal of Biological Chemisty.* submitted.

11

ENHANCED DEGRADATION OF STIMULATORY G-PROTEIN (Gsα) BY CHOLERA TOXIN IS MEDIATED BY ADP-RIBOSYLATION OF Gsα PROTEIN BUT NOT BY INCREASED CYCLIC AMP LEVELS

Bukhtiar H. Shah

Department of Physiology and Pharmacology
The Aga Khan University
Karachi, Pakistan

ABSTRACT

Cholera toxin (CT) catalyses ADP-ribosylation of the α-subunit of stimulatory protein (Gs) leading to stimualtion of adenylyl cyclase and elevated intracellular cAMP. Persistent treatment (24–48 h) of C6 glioma cells with cholera toxin (100 ng/ml) caused marked downregulation of Gsα (75–80%) which could not be mimicked by dibutyryl cAMP (1 mM) and forskolin (10 µM) over the same time periods suggesting that CT-mediated Gsα downregulation is independent of cAMP production. However, CT increased the expression of Gq/11α proteins at 24 and 48 h of treatment. The increase in mRNA levels of Gq/11α proteins preceded the increase in Gq/11 proteins. Such stimulatory effects of CT were mimicked by forskolin and dibutyryl-cAMP. These results suggest that CT-mediated downregulation of Gsα is independent of cAMP but CT upregulates the expression of Gq/11α proteins in a cAMP-dependent mannner.

INTRODUCTION

Cholera toxin (CT) is secreted by virulent strains of *Vibrio cholera* and is responsible for massive diarrhoea associated with the disease (1). The majority of the effects of cholera toxin on the regulation of cellular signalling cascades are produced via modification of guanine nucleotide binding proteins (G-proteins) by the ADP-ribosyltransferase activity of the toxin (2). CT catalyses ADP-ribosylation of the α-subunit of stimulatory G-protein (Gs) leading to inhibition of intrinsic GTPase activity thereby causing constitutive stimulation of adenylyl cyclase activity and elevated intracellular cAMP levels (2).

ADP-Ribosylation in Animal Tissue, edited by Haag and Koch-Nolte
Plenum Press, New York, 1997

Recent studies have shown that persistent activation of the stimulatory adenylyl cyclase pathway desensitizes and downregulates the Gsα protein and this effect is cAMP-independent (3). However very little is known about the interaction of CT-induced increase in cAMP with pertussis toxin-insensitive G-proteins (Gq/11α). These G-proteins mediate their effects by activating phospholipase C and thus generating the second messengers, inositol 1,4,5-triphosphate (IP_3) and diacylglycerol (DAG) (4). Here we studied the effects of CT, dibutyryl cAMP (dB-cAMP) and forskolin (FSK) on the CT-sensitive stimulatory (Gs) and pertussis toxin-sensitive (Gi) and -insensitive G-proteins (Gqα and G11α) and their mRNA levels in C6 glioma cells.

EXPERIMENTAL PROCEDURES

Cells were grown in DMEM supplemented with 10% fetal-calf serum, in 5% CO2 at 37 °C in 75 cm^2 tissue-culture flasks and were harvested just before confluency. The treatment of cells with CT (100 ng/ml), dB-cAMP (1mM) or FSK (10 µM) for different times was carried out as described in figure legends. Membranes were prepared from the treated cells by homogenization and differential centrifugation. The immunoblots for various G-proteins (Gsα, Gq/11α and Gi2α) were done by using antisera specific against the α-subunit of these proteins as described (4). The mRNA levels of G-proteins in control and drug treated cells were measured by reverse transcription/polymerase chain reaction (RT/PCR) as described previously (4). Reaction products were separated by 1.5–1.75% agarose gel electrophoresis and transcripts quantitated (5).

RESULTS AND DISCUSSION

The membranes prepared from cells exposed to CT (100 ng/ml) for different time periods were subjected to immunoblotting with antipeptide antiserum which was generated against extreme carboxy terminus of α-subunit of Gs protein. CT treatment rapidly decreased the long form (45 kDa) of Gsα-protein without any significant effect on short form (42 kDa) which is quite less abundant in C6 cells. CT-mediated decrease in Gsα immunoreactivity was rapid initially for 8 h and then slowed down thereafter till 48 h. The levels of immunoreactive Gsα-membrane (long form) at 24 and 48 hours of CT treatment were 40 and 28 per cent, respectively (Fig. 1). Measurment of mRNA levels in C6 cells by RT/PCR showed that CT-induced reduction in the levels of Gsα was not related to alterations in its mRNA, which confirms previous findings (6) in C6 glioma cells but is in contrast to results of Lin et. al. (7) who observed a decrease in the levels of both Gsα protein and its mRNA in GH3 lactotoph cells treated with CT.

Since CT-mediated ribosylation of Gsα results in marked production of cAMP within the cells (3,6), we tested if the increased intracellular cAMP levels are responsible for decreased expression of membrane Gsα protein. For that cells were treated with agenst that increase intracellular cAMP (FSK) and a cAMP analogue (db-cAMP). In contrast to the effects of CT, FSK (10 µM) and db-cAMP (1 mM) treatment of C6 glioma cells for 24 and 48 h had no effect on the levels of either form of Gsα indicating that loss of these polypeptides was independent of the generation of cAMP (Fig. 2). In fact recent studies have shown that activation of Gs protein by CT (3,8), by constitutive activating mutation (9) and by prolonged exposure to neurotransmitter agonists (10) induces a loss of Gsα

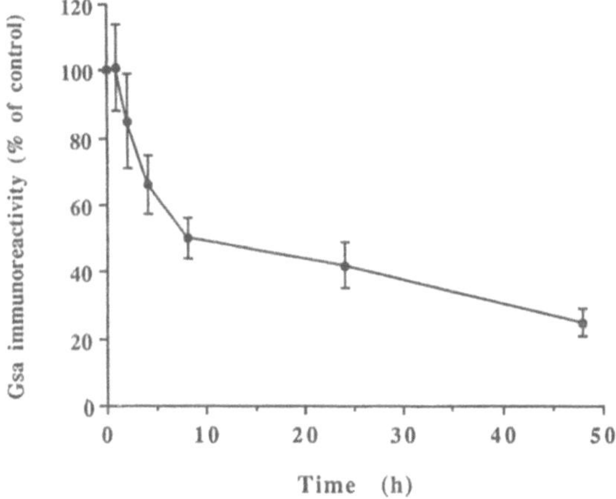

Figure 1. Time-course effect of cholera toxin (100 ng/ml) on the membrane Gsα levels in C6 glioma cells. Membranes were prepared and immunoblots done using antisera specific to the α-subunit of Gs. The data is average and SEM of 5 experiments.

from the plasma membrane. This process occurs mainly through depalmitoylation as it causes loosening of Gsα attachement from the cell membrane (11).

C6 glioma cells are considered to be a useful model for studying G-protein linked signal transduction mechanisms (6). Since cAMP is known to affect other signalling pathways (12), we checked whether CT-mediated elevation of intracellular cAMP affects the pertussis toxin- sensitive (Gi) and -insensitive (Gq/11) G-proteins. Both CT and FSK treatment of C6 glioma cells for 24–48 h had no effect on levels of the pertussis toxin-sensitive G-protein, Gi2α, an observation similar to that seen by Boehm et al (8) recently in chick sympathetic neurons. However, persistent stimulation with CT, in contrast to its

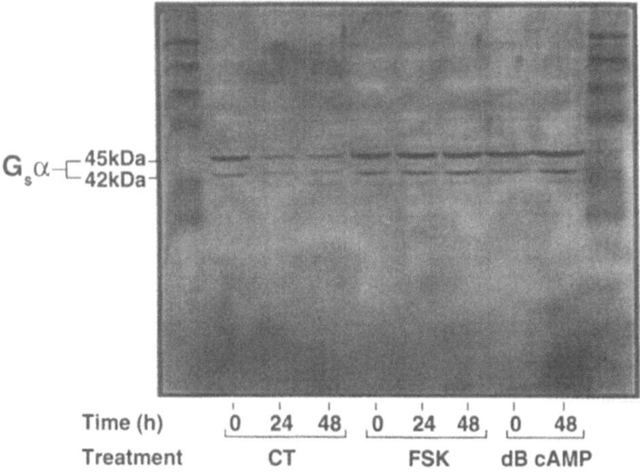

Figure 2. Immunoblots of Gsα in C6 cells treated with CT (100 ng/ml), FSK (10 μM) and db-cAMP (1 mM) for 24 and 48 h. CT downregulates Gsα protein up to 80% without any effect of FSK and db-cAMP indicating CT-induced Gsα reduction is independent of cAMP levels.

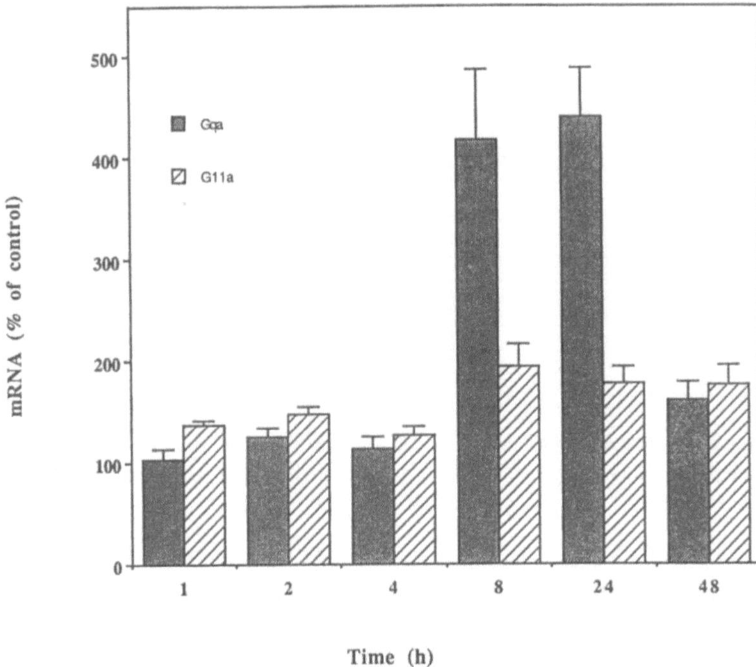

Time (h)

Figure 3. Effect of cholera toxin (100 ng/ml) on Gqα and G11α mRNA levels in C6 glioma cells at various times of treatment. The mRNA levels in control and treated cells were measured by RT/PCR as described in Experimental Procedures.

downregulatory effect on Gsα, increased the level of pertussis toxin-insensitive G-proteins, Gq/11α. The CT-mediated increase in Gq/11α-proteins was accompanied by enhanced expression of Gqα (4-fold) and G11α (2-fold) transcripts as measured through RT/PCR. Time course studies showed that the maximum level of Gqα and G11α mRNA levels were found at 8 and 24 h of CT treatment (Fig. 3). These effects were mimicked by treatment with both forskolin and db-cAMP indicating the likely involvement of cAMP-dependent regulation of the Gqα/ and G11α genes. The cAMP mediated transcriptional regulation of G-proteins (13) and linked receptors (14) is well documented. Few recent studies have shown that cAMP-mediated regulation of adrenergic receptors occur at the level of gene transcription not mRNA stability (8, 15). These studies demonstrate dual effects of CT on G-proteins and provides an evidence of long-term cross regulation between stimulatory cascades involving Gs and Gq/11 proteins.

REFERENCES

1. Middlebrook, J.L., & R.B. Dorland. 1984. Bacterial toxins: cellular mechamisms to action. *Microbiol. Rev.,* *48*: 199–221.
2. Lobban, M.D., & G. Milligan. 1993. Structure and function of bacterial exotoxins which regulate second messenger generation. In: Natural and Synthetic Neurotoxins. pp: 47–63.
3. Chang, Fu-H., & H. R. Bourne. 1989. Cholera toxin induces cAMP-independent degradation of Gs. *J. Biol. Chem., 264*: 5352–5357.

4. Shah, B.H., & G. Milligan. 1994. The gonadotrophin releasing hormone receptor of αT3–1 cells interacts with and regulates phosphoinositidase C-linked G-proteins, Gqα and G11α equally. *Mol. Pharmacol.*, *46*:1–7.

5. Shah, B.H., D.J. McEwan, & G. Milligan. 1995. Gonadotrophic releasing hormone receptor agonist-mediated down-regulation of Gqα/G11α G-proteins in αT3–1 gonadotroph cells reflects increased G-protein turnover but not alterations in mRNA levels. *Proc. Natl. Acad. Sci. USA*, *92*: 1886–1890.

6. Carr, C., C. Loney, C. Unson, J. Knowler, & G. Milligan. 1990. Chronic exposure of rat glioma C6 cells to cholera toxin induces loss of the α-subunit of the stimulatory guanine-nucleotide binding protein (Gs). *Eur. J. Pharmacology, Mol. Pharm. section*, *188*: 203–209.

7. Lin, J. H., H.Y. Wang, J.C. Fong, J.T. Pan, & F.F. Wang. 1993. Correlation between prolactin secretion and Gs protein expression during sustained cholera-toxin stimulation. *Biochem. J.*, *296*:335–340.

8. Boehm S., S. Huck, A. Motejlek, H. Brobny, E.A. Singer, & M. Freissmuth. 1996. Cholera toxin induces cyclic AMP-independent down-regulation of Gsα and sensitization of α2-autoreceptors in chick sympathetic neurons. *J. Neurochem.*, *66*: 1019–1026.

9. Levis, M.J., & H.R. Bourne. 1992. Activation of the α-subunit of Gs in intact cells alters its abundance, rate of degradation, and membrane avidity. *J. Cell Biol.*, *119*: 1297–1307.

10. Wedegaertner, P.B., & H.R. Bourne. 1994. Activation and depalmitoylation of Gsα. *Cell*, *77*: 1063–1070.

11. Wedegaertner, P.B., P.T. Wilson, & H.R. Bourne. 1995. Lipid modifications of G proteins. *J. Biol. Chem.*, *270*, 503–506.

12. Hadcock, J.R., M. Ros, D.C. Watkins, & C.C. Malbon. 1990. Cross regulation between G-protein mediated pathways. Stimulation of adenylyl cyclase increases expression of the inhibitory G-protein Giα2. *J. Biol. Chem.*, *265*; 14784–14790.

13. Loganzo, Jr., F. & P.W. Fletcher. 1993. Follicle-stimulating hormone increases the turnover of G-protein αi1-and αi2-subunit messenger RNA in sertoli cells by a mechanism that is independent of protein synthesis. *Mol. Endocrinology*, *7*:434–440.

14. Collins, S., M. Bouvier, M.A. Bolanowski, M.G. Caron, & R.J. Lefkowitz. 1989. cAMP stimulates transcription of the α2-adrenergic receptor gene in response to short term agonist exposure. *Proc. Natl. Acad. Sci. USA*, *86*: 4853–4857.

15. Hosoda, K., G.K. Feussner, L. Rydelek-Fitzgerald, P.H. Fishman, & R. S. Duman. 1994. Agonist and cyclic AMP-mediated regulation of β1-adrenergic receptor mRNA and gene transcription in rat C6 glioma cells. *J. Neurochem.*, *63*: 1635–1645.

12

SEQUENCE AND STRUCTURAL LINKS BETWEEN DISTANT ADP-RIBOSYLTRANSFERASE FAMILIES

J. Fernando Bazan[1] and Friedrich Koch-Nolte[2]

[1]Department of Molecular Biology
DNAX Research Institute
901 California Ave.
Palo Alto, California 94304–1104
[2]Department of Immunology
University Hospital
D-20246 Hamburg
Germany

ABSTRACT

The low resolution structure of the *Pseudomonas aeroginosa* exotoxin A (ETA) presented in 1986 provided the first tantalizing three-dimensional view of an ADP-ribosyltransferase (ADPRT) catalytic domain. The major features of this protein fold have recurred in the more recently solved crystal structures of the cholera toxin-related heat-labile enterotoxin (LT), diphtheria toxin (DT) and pertussis toxin (PT). A core set of $\alpha+\beta$ elements define a minimal, conserved scaffold with remarkably plastic sequence requirements - only a single glutamic acid residue critical to catalytic activity is invariant. Other interchangeable residues in locations important for catalysis and binding are suggested by the cocrystal structures of DT with the inhibitor ApUp, ETA with bound AMP and nicotinamide, and DT with substrate NAD - in close accord with labeling and mutagenic data. Faint sequence resemblances that were earlier noticed among prokaryotic ADPRTs have now been securely extended by the structural concordance between toxin folds; more recently, eukaryotic ADPRTs have surfaced and their sequences can be reliably threaded into the conserved core fold. We will briefly summarize efforts in Palo Alto and Hamburg to explore these latter relationships, and to mount a rigorous search for new ADPRT families in the growing sequence databases.

ADP-Ribosylation in Animal Tissue, edited by Haag and Koch-Nolte
Plenum Press, New York, 1997

BACKGROUND

A new discovery effort in molecular biology has been spurred by the explosive growth in DNA sequence information and the development of sensitive computational tools that can tease notions of structure and function from sequence threads (1). These approaches lend speed and purpose to experiments, and yield a greater understanding of evolutionarily interconnected biological systems and protein families (2). We have specifically targeted ADP-ribosyltransferases (ADPRTs) because they form a functionally compelling class of enzymes with intriguing, established roles in prokaryotic systems -as potent bacterial toxins or metabolic controllers- and have been found with increasing frequency as regulatory components of eukaryotic cells (3). In the present work, we show that ADPRTs have a well conserved protein fold and catalytic geometry, and that this unique architecture facilitates their identification from sequence data.

The three-dimensional shapes of protein molecules tend to recur in Nature (4); at a practical level, this means that (i) there is a limited repertoire of folds for protein sequences -which simplifies the 'fold recognition' problem- (5), and (ii) protein sequence is much less well conserved than structure over large evolutionary time scales -a fact that plagues the 'sequence comparison' game (6). Yet, with a sufficiently diverse ensemble of related sequences (or folds) that conserve some aspect of biochemical function, it is possible to extract unapparent sequence and structural patterns that have great diagnostic utility (7). These computational methods, alongside more recent 'threading' approaches, have been quite successful in significantly extending protein families and revealing interesting catalytic divergences (8). The comparative analysis of related protein folds further allows the definition of an invariant 'core' as well as surface features -crevices or planes- that have been maintained by evolution for functional purpose (9,10). Excellent example of this analysis are the families of N-terminal-nucleophile hydrolases (11) or protein-tyrosine phosphatases (12).

We have applied these reductionist ideas and analytical tools to the ADPRT superfamily. The ADPRT fold in the catalytic domain of the *Pseudomonas* exotoxin A (ETA) structure (13) was soon recognized in the X-ray structures of the equivalent enzymes from an E. coli heat-labile enterotoxin (LT) (14), diphtheria toxin (15,16) and pertussis toxin (17); more recently, the ETA structure has been redetermined at higher resolution (18). In all cases, what appears to be a similar, core ADPRT module is partly obscured by significant chain insertions (17,19,20); at the sequence level, there is a strong realization that few amino acids survive unchanged in the growing menagerie of known or suspected ADPRTs, and that their correct, meager alignment is at the mercy of the structural variability surrounding the common cores (21,22,23).

RESULTS AND DISCUSSION

A structurally accurate sequence alignment of the toxin ADPRTs was obtained by progressively superimposing their protein scaffolds (obtained from the SCOP database) (24,25) and 'pruning' away their excess domains and loops -unique to each structure-, until only a conserved core (at a ~3 Å cutoff) remained. This core ADPRT module (~ 70 aa) comprised six β-strands, two α-helices and a single helical turn (Figures 1–2). Reasoning that these structural features were likely to be kept in other ADPRT folds, we employed a variety of computational tools (pattern matching, profiles, fold recognition and secondary

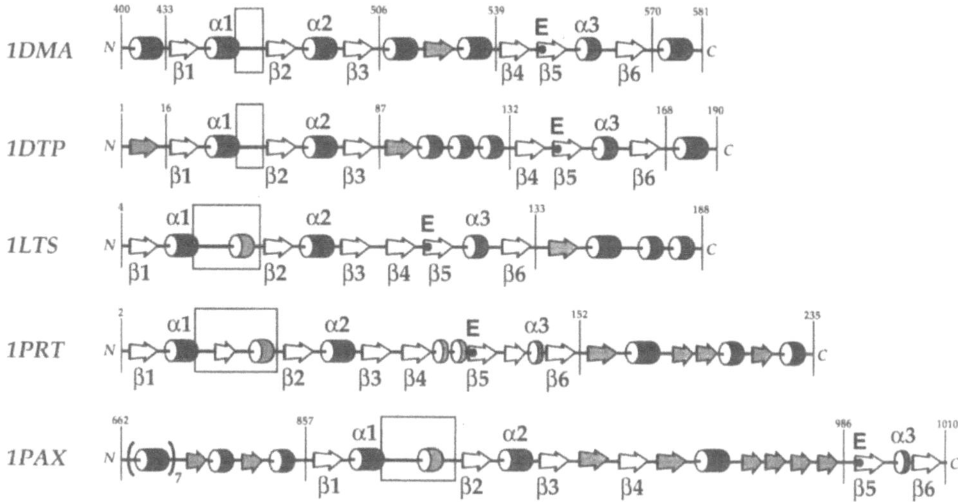

Figure 1. Structural decomposition of known ADPRT folds. The component β-strands and α-helices of the ETA (PDB code 1DMA) (18), DT (1DTP) (16), LT (1LTS) (14) and PT (1PRT) (17) toxin ADPRT folds are atop the elements of the recent PARP structure (1PAX) (20); figure is drawn to approximate scale. α-helices (core members labeled α1-α2) are modeled by cylinders and β-strands (β1-β6) are arrows. Shaded box marks the 'flap' region that covers the active site crevice. ADPRT domain boundary points along the chain are marked by appropriate residue numbers.

structure prediction) (5,7,26) to iteratively harvest and align a large number of known or suspected ADPRT sequences (Figure 3).

The six β-strands of the core ADPRT fold form two β-sheets (in complex strand order 6–3–1 and 2–5–4) that abut each other at an angle; the two α-helices in turn principally form the 'walls' of a bilobed crevice formed by the curled sheets (Figure 2). Three principal sequence and length-variable loops are evident: the first, linking α1 to β2, serves

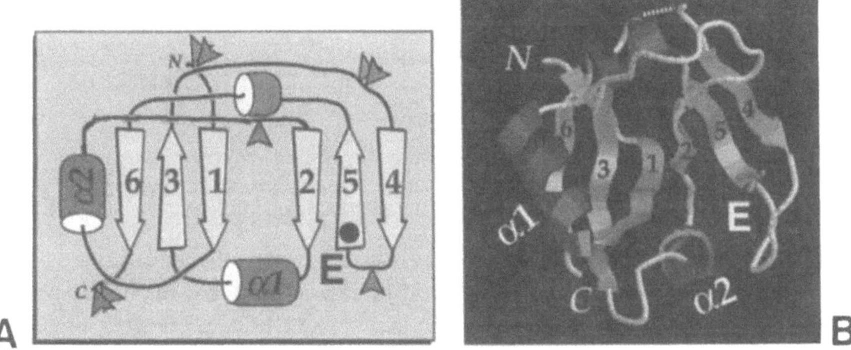

Figure 2. Core features of the ADPRT fold. A) Schematic topology diagram of the ADPRT core fold showing the elements of B). Known points of chain insertion in toxin and eukaryotic sequences are marked by arrowheads. The location of the catalytic Glu is marked by a black circle. A) Representative core of the (ETA) ADPRT fold once stripped of extraneous loops and structure, showing the six β-strands (labeled 1–6), two α-helices (α1 and α2) and a helical turn (at top) conserved in known toxin structures. The β4-β5 length-variable connection is marked by a dotted line; the catalytic Glu residue on β5 is shown. Graphics are by RASMOL (27).

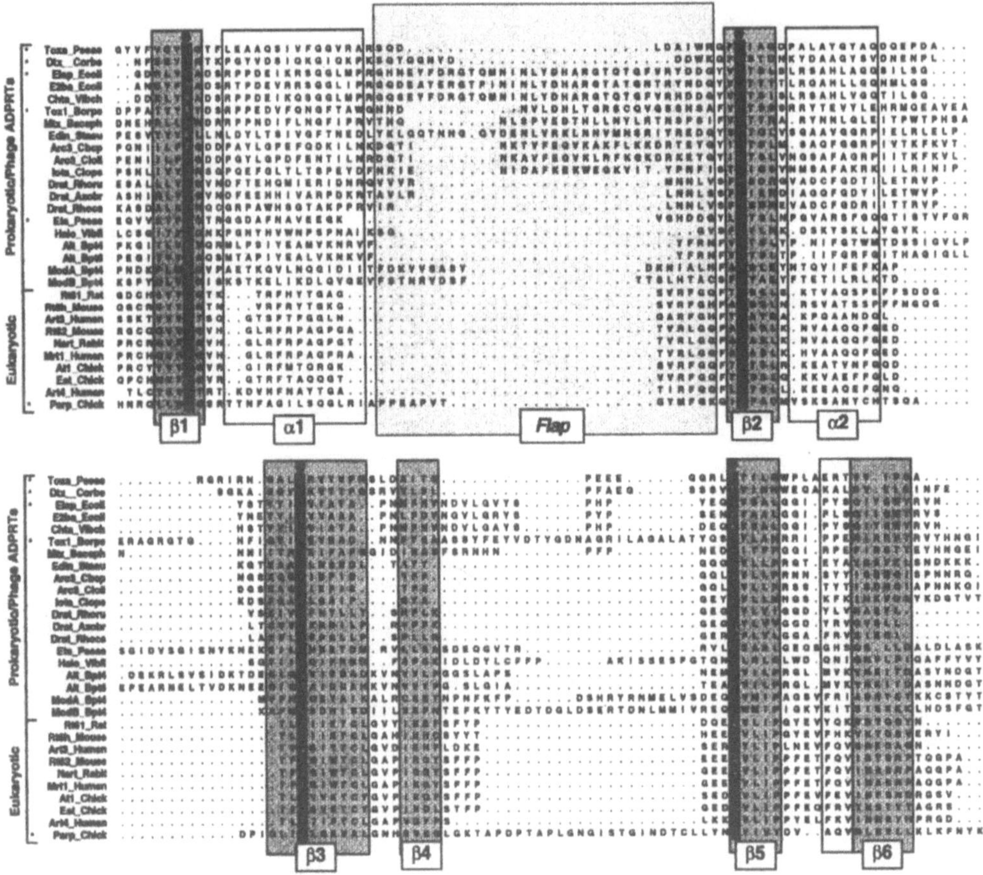

Figure 3. Structure-based alignment of representative prokaryotic and eukaryotic ADPRT cores. Aligned segments corresponding to known (parent sequences marked by an asterix) and predicted core β-strands (β1-β6) and α-helices (α1-α2) are boxed and differently shaded. The 'flap' segment is boxed and in grey. Fingerprint residues in β1 (an Arg or a His), β2 (Ser or Tyr), β3 (aromatic or Leu) and β5 (catalytic Glu) are noted by a black circle. The aligned sequences are noted by their Swissprot identifiers (format: protein_organism) and grouped into prokaryotic and phage (top), and eukaryotic (bottom) families. The seed of the alignment was in an earlier paper by our groups (22) which also serves as the principal reference for the sequences. Other sequences are presented in Ruger *et al.* (this volume) (ModA and ModB), Braren *et al.* (this volume) (assorted ARTs); PARP and halovibrin are taken from Refs. 20 and 28.

as a loose 'flap' that overhangs the active site crevice and may change structure upon substrate binding (19) ; two other loops connect α2 to β3 and β4 to β5 -the latter one perhaps affecting specificity due to its proximity to the active site Glu on β5 (Figure 1–3). It is notable that the longest chain insertions in the ADPRT fold are suffered by bacterial toxin sequences; by contrast, the vertebrate ADPRTs (homologs of RT6) (22) are relatively economical and make rather direct links between core elements. This evolutionary 'baggage' (along with the presence of additional domains flanking the ADPRT module) may protect the toxin's enzymatic machinery during its perilous journey to the besieged cell's interior. The derived evolutionary tree for the greater ADPRT superfamily securely places the prokaryotic/phage members in distinct branches from the eukaryotic enzymes (Figure 4).

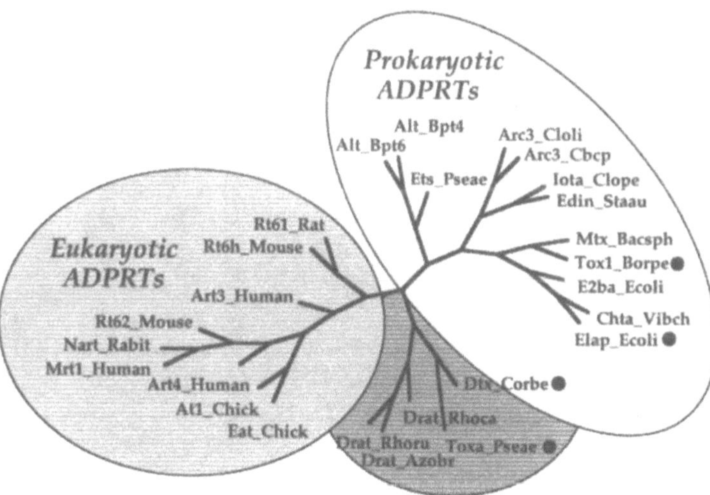

Figure 4. ADPRT evolutionary tree. ClustalW was used to create an evolutionary branching diagram from the alignment of Figure 3 using the Neighbor-joining method (7); the unrooted tree was radially drawn with Treeview. Three major branches are distinguished: the principal prokaryotic/phage group (with the sequences of PT and LT - black circles) and a smaller cluster of DT/ETA (black circles)-like sequences. All the eukaryotic sequences (save PARP -not shown) form a distinct branch.

Several conserved sequence motifs are apparent within the 'blocks' of structure that compose the ADPRT core fold (Figure 3). Three notable cases: the single invariant residue lies at the N-terminus of β5: a Glu residue implicated in catalysis whose carboxylate group is spatially close to the scissile, N-glycosidic bond of the bound NAD (20,29); a conserved His or Arg residue in β1 appears far from the catalytic Glu, and instead hydrogen-bonds to a conserved Tyr or Ser in β2, together cradling the bound NAD (20,29). Indeed, most chemically conserved residues in the Figure 3 alignment map to the inner surface of the active site crevice, particularly -or perhaps curiously- to the indentation that recognizes the adenine moiety (to the left in Figure 2's core pictures) : for example, an oft-conserved Gly at the end of α1 that closely approaches the bound ring (20). A loose correlation between sidechain specificity for both the toxins and vertebrate ADPRTs (22) was noticed with the amino acid type in a loop preceding the catalytic Glu in β5. A more detailed description of the residue topography and function in the active site pocket is relegated to another study (J.F.B and F.K-N., in preparation).

As a test of the sequence-to-structure mapping of Figure 3, we employed homology modeling techniques (30) to derive the three-dimensional structure of a representative vertebrate ADPRT. The sequence of mouse RT6 (22) was parsimoniously threaded atop the most economical toxin scaffold (ETA) (18) that served as the modeling template. Figure 5 shows the results: the RT6 fold is relatively unencumbered by long loops so typical in toxin structures, particularly in the flap segment (Figure 3). The catalytic machinery of the toxins is fully conserved in this model structure. Accurate model-building of the sidechain positions may certainly be hindered by the fact that distantly related sequences that adopt similar folds often evolve distinct packing arrangements in their hydrophobic cores (32). A different model-building approach, employing the full range of available ADPRT structures to form a sort of composite template (30), is perhaps better suited to future models of ADPRT homologs.

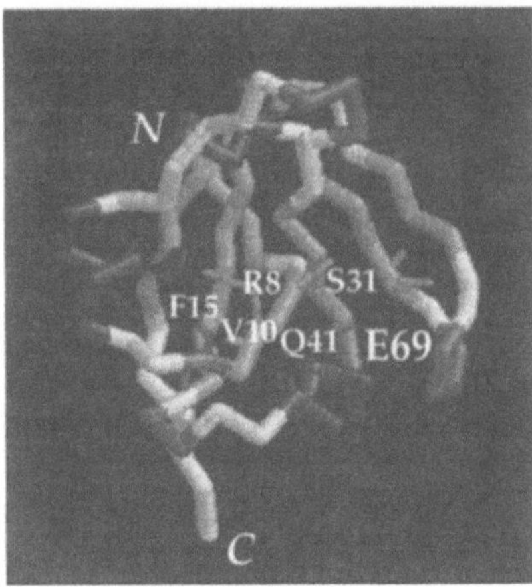

Figure 5. Molecular model of a mouse RT6 ADPRT fold. The alignment of Figure 3 was used to generate a full atomic model of mouse RT6 (22) by taking the most economical toxin fold -1DMA from ETA (18)- as the template structure for homology modeling. The program SEGMOD (within the LOOK 2.0 modeling package (Molecular Applications Group, Palo Alto, CA) uses a fragment library to progressively build up matching segments of the target structure upon the template backbone (31); 500 rounds of energy minimization by ENCAD were performed to improve the stereochemistry of the final RT6 model. Key amino acids positions in the active site crevice are noted on the fold. The model is displayed with RASMOL (27). Coordinates of the structure are available by email request to J.F.B.

CODA

Rumors of the imminent release of the first eukaryotic ADPRT three-dimensional structure in the guise of a poly-ADP-ribose polymerase (PARP) catalytic domain quietly percolated among the scientists at the 1996 Hamburg Workshop. This structural homology had been anticipated by the work of two groups that had noticed a faint resemblance of sequence patterns between the enzymatic region of PARPs and toxin ADPRT sequences (33,34). The Collier group elegantly tested their catalytic model by mutating the predicted, critical Glu988 residue in the human PARP chain to Asp, Gln, or Ala, and showed that these changes produced significant reductions in the rate of polymer formation by PARP (34). Indeed, the publication of the X-ray structure of the 40 kDa. catalytic fragment of chicken PARP bound to a nicotinamide-analog inhibitor by the de Murcia and Schulz groups (PDB code 1PAX) settled the matter: PARP has an ~150 aa ADPRT module at the tail end of its chain that most closely resembles the DT and ETA enzyme folds (20). The divergent PARP fold and catalytic function -ADPR units are progressively added to a chain rather than singly to a target protein sidechain- significantly extend the evolutionary diversity of the ADPRT superfamily.

CD38 heads another family of NAD-metabolizing ectoenzymes that may opportunistically be able to ribosylate proteins (35) but otherwise appears unrelated to ADPRTs. Nature has fashioned an alternative solution for the CD38 fold as was revealed during a Workshop talk by Lee (this volume) who presented an outline structure for the homologous *Aplysia* ADPR cyclase. The recently published X-ray work by the Stout group shows in detail that the cyclase fold is based on a flavodoxin-like α/β core sprouting a large helical domain stabilized by disulfide bridges (36). However, the cyclase fold, like the ADPRT structure, appears to also bind the substrate NAD in an extended manner, with separate pockets to cradle the adenine and nicotinamide moieties -before clipping off the latter (36).

More recent work in Palo Alto and Hamburg has branched along two lines: (i) investigating the structural basis for amino-acid-ribosylation specificity, and (ii) carrying out a broad-based search for new ADPRT sequence families. The first project makes use of the wealth of available, high-resolution X-ray structures for ADPRT folds, increasingly so in the form of complexes with substrates or inhibitors that thoughtfully detail the conserved constellation of residues that differentially act in binding and catalysis (16,18,19, 20). ADPRTs appear to recognize their exposed target sidechains without any preference for flanking residues, suggesting that the recognition 'code' for ribosylation may have a more topographic or even electrostatic basis (37). An end goal of this effort is to manipulate the specificity of any given ADPRT scaffold, or even design -or evolve- enzymes with novel recognition capabilities (38).

The second project has already yielded fruit in the form of novel cell surface ADPRTs harvested from the public EST database by vigilant sequence searches (Braren *et al.*, this volume). Because of the low degree of sequence similarity between ADPRT families -DT/ETA vs. LT/PT toxin groups vs the mammalian ART vs. PARP classes-, it may not be a simple matter to recognize new ADPRT sequence clusters. To circumvent this problem, we have focused on the likely invariant feature of ADPRTs, their unique fold, as the most reliable search tool. Assorted fold recognition methods (5), as well as more 'classical' pattern and profile-based algorithms (7), have fared well in this respect. For example, three new additions to the ADPRT sequence grouping in Figure 3 are the ModA and ModB enzymes from the T4 genome (Ruger *et al.*, this volume) and the halovibrin ADPRT from *Vibrio fischeri* (28). Other 'hits' from the human genome databases are currently undergoing experimental development in Hamburg and Palo Alto. To foster public access to these results and methods, and also serve as a continuously updated repository for ADPRT sequences, structures and biology, we are committed to setting up a dedicated Internet Web site at DNAX by the middle of 1997 - stay tuned!

ACKNOWLEDGMENT

Work at DNAX was supported by Schering-Plough; F.K-N.'s lab was supported by grant No310 from the Deutsche Forschungsgemeinschaft.

REFERENCES

1. Bork, P., C. Ouzounis, C. Sander. 1994. From genome sequences to protein function. *Curr. Opin. Struct. Biol. 4*: 393–403.
2. Pawson, T. 1995. Protein modules and signalling networks. *Nature 373*: 573–580.
3. Koch-Nolte, F., F. Haag, R. Kastelein, J.F. Bazan. 1996. Uncovered - the family relationship of a T-cell membrane protein and bacterial toxins. *Immunol. Today 17*: 402–405.
4. Holm, L., C. Sander. 1996. Mapping the protein universe. *Science 273*: 595–603.
5. Fischer, D., D. Rice, J.U. Bowie, D. Eisenberg. 1996. Assigning amino acid sequences to 3-dimensional protein folds. *FASEB J. 10*: 126–136.
6. Rost, B., A. Valencia. 1996. Pitfalls of protein sequence analysis. *Curr. Opin. Biotech. 7*: 457–461.
7. Bork, P., T. Gibson. 1996. Applying motif and profile searches. *Meth. Enzym. 266*: 162–184.
8. Firestine, S.M., A.E. Nixon, S.J. Benkovic. 1996. Threading your way to protein function. *Chem. and Biol. 3*: 779–783.

9. Casari, G., C. Sander, A. Valencia. 1995. A method to predict functional residues in proteins. *Nature Struct. Biol. 2*: 171–178.

10. Lichtarge, O., H.R. Bourne, F.E. Cohen. 1996. An evolutionary trace method defines binding surfaces common to protein families. *J. Molec. Biol. 257*: 342–358.

11. Brannigan, J.A., G. Dodson, H.J. Duggleby, P.C. Moody, J.L. Smith, D.R. Tomchick, A.G. Murzin. 1995. A protein catalytic framework with an N-terminal nucleophile is capable of self-activation. *Nature 378*: 416–419.

12. Fauman, E.B., M.A. Saper. 1996. Structure and function of the protein tyrosine phosphatases. *Trends Biochem. Sci. 21*: 413–417.

13. Allured, V.S., R.J. Collier, S.F. Carroll, D.B. McKay. 1986. Structure of exotoxin A of *Pseudomonas aeruginosa* at 3.0-Å resolution. *Proc. Natl. Acad. Sci. USA 83*: 1320–1324.

14. Sixma, T.K., S.E. Pronk, K.H. Kalk, E.S. Wartna, B.A.M. Van Zanten, B. Witholt, W.G.J. Hol. 1991. Crystal structure of a cholera toxin-related heat-labile enterotoxin from *E. coli*. *Nature 351*: 371–377.

15. Choe, S., M.J. Bennett, G. Fujii, P.M. Curmi, K.A. Kantardjieff, R.J. Collier, D. Eisenberg. 1992. The crystal structure of diphtheria toxin. *Nature 357*: 216–222.

16. Weiss, M.S., S.R. Blanke, R.J. Collier, D. Eisenberg. 1995. Structure of the isolated catalytic domain of diphtheria toxin. *Biochem. 34*: 773–781.

17. Stein, P.E., A. Boodhoo, G.D. Armstrong, S.A. Cockle, M.H. Klein, R.J. Read. 1994. The crystal structure of pertussis toxin. *Structure 2*: 45–57.

18. Li, M., F. Dyda, I. Benhar, I. Pastan, D.R. Davies. 1996. The crystal structure of *Pseudomonas aeruginosa* exotoxin domain III with nicotinamide and AMP: conformational differences with the intact exotoxin. *Proc. Natl. Acad. Sci. USA 92*: 9308–9312.

19. Bell, C.E., D. Eisenberg. 1996. Crystal structure of diphtheria toxin bound to nicotinamide adenine dinucleotide. *Biochem. 35*: 1137–1149.

20. Ruf, A., J. Mennissier de Murcia, G. de Murcia, G.E. Schulz. 1996. Structure of the catalytic fragment of poly(ADP-ribose) polymerase from chicken. *Proc. Natl. Acad. Sci. USA 93*: 7481–7485.

21. Takada, T., K. Iida, J. Moss. 1995. Conservation of a common motif in enzymes catalyzing ADP-ribose transfer. Identification of domains in mammalian transferases. *J. Biol. Chem. 270*: 541–544.

22. Koch-Nolte, F., D. Petersen, S. Balasubramanian, F. Haag, D. Kahlke, T. Willer, R. Kastelein, J.F. Bazan, H.G. Thiele. 1996. Mouse T cell membrane proteins Rt6–1 and Rt6–2 are arginine/protein mono(ADPribosyl)transferases and share secondary structure motifs with ADP-ribosylating bacterial toxins. *J. Biol. Chem. 271*: 7686–7693.

23. Domenighini, M., R. Rappouli. 1996. Three conserved consensus sequences identify the NAD-binding site of ADP-ribosylating enzymes, expressed by eukaryotes, bacteria and T-even bacteriophages. *Molec. Microbiol. 21*: 667–674.

24. Murzin, A.G., S.E. Brenner, T. Hubbard, C. Chothia. 1995. SCOP - A Structural Classification Of Proteins database for the investigation of sequences and structures. *J. Molec. Biol. 247*: 536–540.

25. Holm, L., C. Sander. 1995. Dali - A network tool for protein structure comparison. *Trends Biochem. Sci. 20*: 478–480.

26. Rost, B., C. Sander. 1996. Bridging the protein sequence-structure gap by structure predictions. *Ann. Rev. Biophys. Biomolec. Struct. 25*: 113–136.

27. Sayle, R.A., E.J. Milner-White. 1995. Rasmol - Biomolecular graphics for all. *Trends Biochem. Sci. 20*: 374–376.

28. Reich, K.A., G.K. Schoolnik. 1996. Halovibrin, secreted from the light organ symbiont Vibrio fischeri, is a member of a new class of ADP-ribosyltransferases. *J. Bacteriol. 178*: 209–215.

29. Li, M., F. Dyda, I. Benhar, I. Pastan, D.R. Davies. 1996. Crystal structure of the catalytic domain of *Pseudomonas* exotoxin A complexed with a nicotinamide adenine dinucleotide analog: implications for the activation process and for ADP ribosylation. *Proc. Natl. Acad. Sci. USA 93*: 6902–6906.

30. Sali, A. 1995. Modeling mutations and homologous proteins. *Curr. Opin. Biotech. 6*: 437–451.

31. Levitt, M. 1992. Accurate modeling of protein conformation by automatic segment matching. *J. Molec. Biol. 226*: 507–533.

32. Chung, S.Y., S. Subbiah. 1996. A structural explanation for the twilight zone of protein sequence homology. *Structure 4*: 1123–1127.

33. Domenighini, M., C. Montecucco, W.C. Ripka, R. Rappuoli. 1991. Computer modeling of the NAD binding site of ADP-ribosylating toxins - active site structure and mechanism of NAD binding. *Molec. Microbiol. 5*: 23.

34. Marsischky, G.T., B.A. Wilson, R.J. Collier. 1995. Role of glutamic acid 988 of human poly-ADP-ribose polymerase in polymer formation. Evidence for active site similarities to the ADP-ribosylating toxins. *J. Biol. Chem. 270*: 3247–3254.

35. Grimaldi, J.C., S. Balasubramanian, N.H. Kabra, A. Shanafelt, J.F. Bazan, G. Zurawski, M.C. Howard. 1995. CD38-mediated ribosylation of proteins. *J. Immunol. 155*: 811–817.
36. Prasad, G.S., D.E. McRee, E.A. Stura, D.G. Levitt, H.C. Lee, C.D. Stout. 1996. Crystal structure of *Aplysia* ADP ribosyl cyclase, a homologue of the bifunctional ectoenzyme CD38. *Nature Struct. Biol. 3*: 957–964.
37. Honig, B., A. Nicholls. 1995. Classical electrostatics in biology and chemistry. *Science 268*: 1144–1149.
38. Shao Z., F.H. Arnold. 1996. Engineering new functions and altering existing functions. *Curr. Opin. Struct. Biol. 6*: 513–518.

13

MOLECULAR CLONING AND CHARACTERIZATION OF THE T-CELL MONO(ADP-RIBOSYL)TRANSFERASE RT6

Relationships to Other mADPRTs and Possible Functions

Heinz-Günter Thiele, Friedrich Haag, and Friedrich Koch-Nolte

Department of Immunology
University Hospital Hamburg-Eppendorf
Martinistr. 52, D - 20246
Hamburg, Germany

1. INTRODUCTION

The RT6 system (1) was first detected and serologically defined on peripheral T cells of different inbred strains of the Norway rat (2–6). It gained particular interest when it was reported that the diabetes-prone BioBreeding rat (DP-BB), which develops insulin-dependent diabetes mellitus (IDDM) between postnatal day 60 and 120 in a high frequency, does not appropriately express RT6 on the surfaces of its T cells (7). Subsequently *RT6* -homologous genes were also detected in mice, primates and some other species (8, 9).

A new aspect arose in this field when it became apparent that a mono(ADP-ribosyl)transferase (mADPRT) isolated from rabbit skeletal muscle shares a conspicuous homology of its primary structure with that of rat RT6.2 (10).

Our review concerns the present knowledge of the RT6 system and related structures in the rat and some other species in structural, biochemical and biological terms.

2. THE RT6 SYSTEM IN THE NORWAY RAT

Rat RT6 was independently detected by different laboratories and initially appeared under various names in the literature (Pta (2), ART-2 (3), Ag-F (4), RT-Ly.2 (5), L-21 (11)). It exists in at least two alloantigenic forms (RT6.1 and RT6.2), which can be distinguished serologically by a number of poly- and monoclonal alloanti-sera/bodies. The rat *RT6* gene allotypes (*RT6^a* and *RT6^b*) were mapped to the short arm of chromosome 1

ADP-Ribosylation in Animal Tissue, edited by Haag and Koch-Nolte
Plenum Press, New York, 1997

where they are found in close linkage to the genes for the β-chain of hemoglobin (1q22) and tyrosinase (12). Both allelic forms of RT6 are coexpressed on the T cell surface membrane of adult heterozygous rats. However, the expression of the two alleles is asymmetric in that a surplus of 16 - 20% of single RT6.2 positive T cells is regularly found in such F_1 animals (13). This, indeed, is not a particularity of *RT6* heterozygous rats, since homozygous *RT6^b* rats consistently have a higher percentage of RT6$^+$ cells within their T cell compartment than age-matched homozygous *RT6^a* rats (13). This phenomenon, which points to a different regulation of the two *RT6* alleles, may explain why different authors report various percentages of RT6$^+$ T cells .

Rat RT6 is a differentiation antigen which appears on the surface membrane of maturing CD4$^+$ and CD8$^+$ T cells during early postthymic ontogeny (14, 15) and precedes the appearence of the tyrosine phosphatase CD45R (16, 17). Detailed FACS analyses show that the expression profile of both RT6 forms among peripheral T cells is not uniform but that there exist two distinct RT6$^+$ cell populations: one, which expresses low levels of RT6, is the predominant population in very young animals and another one of higher RT6 density prevails in older rats (Thiele et al., Transplant. Proc. in press). Whether these cells represent members of different T cell generations, which are exported from the thymus in a cyclic chronological order, as was reported in avians (18, 19) and mice (20), or whether the RT6^{low} cells are precursors of the RT6^{high} population, is presently under study in our laboratory. With respect to its possible biological function, which we shall discuss later in this review, it is noteworthy that intestinal intraepthelial lymphocytes (IEL) display a significantly higher expression of RT6 than peripheral lymph node lymphocytes (21).

Physicochemical analyses revealed that RT6 is relatively resistant to solubilization with nonionic detergents such as NP40 and Triton X-100 at 4°C, a phenomenon which may be explained by the fact that RT6 is anchored within the membrane by a glycosyl-phosphatidyl-inositol (GPI) moiety (22, 23). At 37° C both RT6 variants are solubilized by Triton X-100 within 20 minutes. In SDS PAGE analyses both RT6 forms appear as a doublet with M_rs of 26/28 kD (24, 25). RT6.1, however, appears additionally in at least five further bands (M_r 30–35 kD), which, in contrast to the 26/28 kD forms of both alloantigens, proved to be sensitive to endoglycosidase-F (25). This observation led already early to the suggestion that RT6.2 may not be glycosylated (24), whereas RT6.1 exists in non- and a number of differently glycosylated forms (25). Molecular cloning showed that both rat RT6 forms are encoded by a single copy gene. Deciphering of the primary structures of *RT6^a* and *RT6^b* cDNA clones revealed that the former, indeed, does not have any potential N-glycosylation site whereas the latter contains a single one (26, 27), which, however, is obviously differently glycosylated (25). The overall identity of RT6.1 and RT6.2 is 95% on the amino acid level, i.e. they are remarkably divergent, differing in 10 coding mutations, most of which result in nonconservative amino acid substitutions (28). In comparison, the two alloantigenic forms of the "classical" T cell marker Thy-1 (29) differ only in a single amino acid in position 89 (Thy-1.1: R, Thy-1.2:Q).(30). The functional meaning of this RT6 polymorphism is presently unknown. Uncovering the genomic organization of the *RT6* gene revealed the existence of two transcription start sites and two independent promoters. The gene contains 8 exons, which may be spliced alternatively (31, 32). Moreover, the 5′ prime untranslated region is strikingly heterogeneous. The major transcript begins in exon 2 and contains an open reading frame of 825 bp with classical N- and C-terminal signal sequences (encoded by exons 5 and 8, respectively), typical for GPI-anchored membrane proteins. The natively expressed RT6 protein is encoded by exon 7, which comprises 694 nucleotides and, thus, is unusually large (33).

A new era of RT6 research started when Zolkiewska et al. (10), submitting the nucleotide sequence of a mADPRT purified from rabbit skeletal muscle, noticed that this enzyme shares sequence identity of 30–40% with rat *RT6^b* (10), which had been cloned and sequenced some years before (8). In particular, amino acid residues 126 (R), 147 (S) and 189 (E) are conserved not only between RT6 and the rabbit muscle mADPRT, but also among many other pro- and eukaryotic mADPRTs (34) (see also Bredehorst et al. this volume). These amino acid residues apparently are essential parts of the active centres of these enzymes since site directed mutagenesis of any one of these three amino acids is sufficient to abolish the enzymatic activity of such molecules (34), (see also Bredehorst et al. this volume).

Indeed, *RT6^b* genes which were cloned and expressed in a rat mamma carcinoma cell line (10), as well as RT6.1 and RT6.2 of T cells expressing these structures natively (35–37) proved to cleave NAD$^+$ in nicotinamide and ADPR. While RT6.2 clearly automodifies itself in a hydroxylamine sensitive manner, RT6.1 lacks this potency (35, 37).

3. THE RT6 SYSTEM IN *MUS MUSCULUS*

The first evidence for the existence of the RT6 system in mice came from Southern "zoo-blot" analyses with a rat *RT6^b* probe (38). Using consensus oligonucleotides deduced from rat *RT6^a* and *RT6^b*, the cDNA of the translated region was prepared from Balb/c mouse spleen and lymph node lymphocytes and sequenced (8).

These experiments revealed that mouse Rt6, in contrast to rat RT6, is encoded by two closely linked genes (designated *Rt6–1* and *Rt6–2*) that are located on the long arm of chromosome 7 (7F1-ter) (39) which is syntenic to chromosome 1 of the rat. Comparative Southern blot analyses of *Rt6* genes of different inbred mouse strains showed that, as in the rat, the mouse *Rt6* genes are polymorphic and tightly linked to the gene of the β-chain of hemoglobin and the tyrosinase (albino) gene (39). Moreover, these analyses indicated further that the loci for *Rt6* are closely linked or even identical to the *H1* minor histocompatibility locus (40). The two *Rt6* genes have conserved open reading frames and an identity of 78.5% concerning their predicted amino acid sequences. The amino acid similarity to rat RT6 amounts to about 67%. Interestingly enough, *mus spretus*, which is a distinct wild type mouse species, and, more curiously, the house shrew, which is phylogenetically much more distant to *mus musculus* than the latter to the rat, also have two *Rt6* gene copies. On the other hand, some inbred strains of *mus musculus* have only a single copy of the *Rt6* gene, i.e. *Rt6–1* (s. Matthes this volume). As in the rat, the mouse Rt6 sequences contain three clusters of conserved amino acid residues typical for NAD$^+$-dependent arginine-specific GPI-anchored mADPRTs (34). Proceeding from the deduced amino acid sequences a model of the secondary structure of mouse Rt6 was predicted, which tightly resembles the pattern of the secondary structure motifs of bacterial mADPTs (34, Bazan et al. and Bredehorst et al. this volume). Functional analyses using recombinant soluble mouse Rt6–1 and Rt6–2 confirmed that both are NAD-dependent arginine specific mADPR-transferases (34).

4. THE RT6 SYSTEM IN MAN AND OTHER PRIMATES

In order to search for RT6 homologues in man and primates degenerate oligonucleotides from conserved regions of rat and mouse *RT6* were used to amplify a fragment of the RT6 homologous gene from a (caucasian) genomic DNA (9). This fragment was used to isolate the gene from a human genomic DNA library (9). By subcloning and sequencing appropriate restriction fragments a nucleotide sequence was obtained, which is 50% identical to those of rat and mouse *RT6*, with which it shares a conserved exon/intron structure. By PCR screening of human/rodent cell hybrids as well as by fluorescence in situ hybridization the human *RT6* gene was mapped to the long arm of chromosome 11 (11q13) centromeric to the tyrosinase gene (11q21), whereas the HBB locus is located on the short arm of chromosome 11 (11p15.4) (41). This, indeed, indicates an intrachromosomal reshuffling since the divergence of primates and rodents. Surprisingly, the human *RT6* nucleotide sequence contains three in-frame stop codons corresponding to amino acid residues 47, 141 and 193 of native rat RT6 (9). Stop codons of identical localization were also found in *RT6* sequences obtained from genomic DNAs of humans with different ethnic backgrounds (African, Native American) as well as of hominoids (chimpanzee, orang utan, gorilla). Further studies revealed that also the homologous gene of Old World (baboon, maccacus), New-World monkeys (cotton top, squirrel monkey) and of lemures (*lemur fulvus*) contain stop codons, though some of these are at different positions within the *RT6* nucleotide sequences (Fig. 1). There is, indeed, good evidence that the human *RT6* gene exists only as a single copy pseudogene (9) and thus is not translated. It is conceivable that this is also valid for the RT6 gene of the other primates mentioned.

5. RT6-RELATED mADPR-TRANSFERASES

Among an increasing number of RT6 related genes (42–44; and Davis et al. this volume) we wish here to point to a mouse gene which we recently cloned in our laboratory, and which may be the homologue of the rabbit muscle mADPRT gene (10) mentioned above. The expression of this single copy gene is restricted to mouse skeletal and cardiac muscle, is structurally distinct from both mouse Rt6 genes, which so far were found to be expressed only in T cells, but shares with the two Rt6 forms a similar exon/intron organisation, the characteristic structural feature of GPI-anchored membrane proteins and is located on chromosome 7 (45).

Moreover, by screening the EST (expressed sequence tags) data base we found 13 *RT6*-related gene fragments, which derive from two novel human genes encoding GPI-anchored membrane proteins and are expressed in testis and spleen (for details s. Braren et al. this volume).

6. CONSIDERATIONS ON THE FUNCTION OF THE RT6 SYSTEM

Although it recently became apparent that rat and mouse RT6 structurally and functionally are members of the mono(ADP-ribosyl)transferase gene family, their biological meaning in T-cell development and function still is imperfectly understood. Revisiting a number of data which were published by different laboratories, we here attempt to approach the biological function of the T-cell-associated RT6 system. First, there is clear evidence that rodent RT6 is a differentiation antigen (14, 15) which appears during the postthymic maturation of T cells (14). In this context it is of interest that the rabbit skeletal muscle mADPRT also is a differentiation antigen (46). It, indeed, would be important to learn whether further members of this gene family having found to be expressed by a variety of other tissues (43, 44) likewise are differentiation structures.

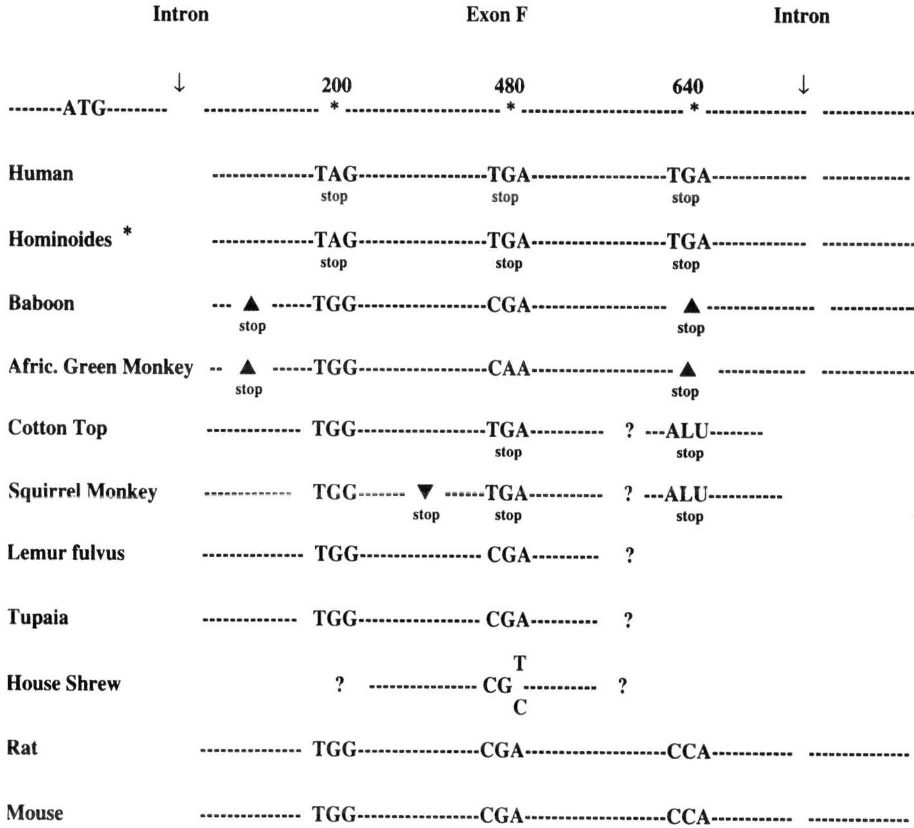

Figure 1. Gene-inactivity mutation in the RT6 genes of primates. The positions of three in-frame stop codons (TAG or TGA) occurring in humans and hominoids within exon 7 ("Exon F") of the RT6 gene are indicated. Homologous codons, as well as frameshift mutations found in other species are shown. ALU denotes the insertion of an *Alu* element into the corresponding region.

Another aspect, which we here wish to reconsider is the suggestion that the RT6 system may play a role in the prevention of autoimmune diseases. This supposition first arose when it was reported that DP-BB rats, which spontaneously generate IDDM, are not only lymphopenic (47), but lack RT6$^+$ cells (7).

In this context it is of interest that (i) transfer of RT6$^+$, CD4$^+$ T cells from non-diabetes prone inbred DR-BB rats, which are variants of the DP-BB rat that generate normal numbers of RT6$^+$ T cells, to congenic pre-diabetic animals of the DP-BB(RT1^u) strain (48, 49) or of PVG (RT1^u) rats to syngenic animals which had been made lymphopenic and susceptible for IDDM experimentally by thymectomy and irradiation (50), protect the recipients from IDDM and that (ii) experimental depletion of RT6$^+$ cells in the DR-BB rat led to a high incidence of IDDM (51). Remarkably, this effect was observed only in young DR-BB rats (30 days), but failed to appear when the depletion of RT6$^+$ cells was set in advanced age (60 days). From these experiments it may be concluded that an RT6$^+$, CD4$^+$

cell population, which was found to be CD45RC^{low} (50) has an autoimmune disease-preventing regulative function that (at least in the DR-BB rat model) seems to be operative mainly during the early postnatal life. This suggestion is consistent with the observation that CD4 single positive thymic lymphocytes, which, however, are RT6$^-$, prevent IDDM more efficiently in lymphopenic rats than do peripheral T cells of adult rats (50, 52). The fact that the protective effect of thymic lympocytes in these experiments was nonetheless only partial suggests that the respective cells used were functionally heterogeneous (50). The age of the donor of the thymocytes in these experiments was between 6 and 10 weeks (D. Mason, personal communication). Because thymocytes perinatally may emigrate from the thymus in a cyclic pattern (20) and since, as discussed below, cells of the first perinatal emigration wave may be more uniform and may have a distinct regulatory function, the question of the age of the donor of the thymic lymphocytes might have some bearing. Concerning the protective effect of these cells it should be noted that RT6$^-$, Thy-1$^+$ recent thymic migrants (RTM) of normal rats quickly display RT6 *in vitro* (53, 54) and *in vivo*, where the first RT6^{low} cells are found in the periphery before postnatal day 3 (55). This suggests that the protective effect in the thymocyte transfer experiment mentioned above (50, 52) eventually may be mediated by a CD4$^+$cell population which, although sharing some of the phenotypical characteristics with the "classical" TH2 cells represents a distinct cell population that predominantly emigrates from the thymus perinatally in a first wave and is the prevailing T cell category during the first 7 to 14 postnatal days (20, 56). These cells, which in a reduced number may also be exported from the thymus in adult life, differ, indeed, concerning their surface marker (CD4$^+$, RT6$^+$, CD45R^{low}, L-Selectin$^+$; Refs. 57, 58 and our own unpublished data) and cytokine (IL-4^{high}, IL-2$^{neg/low}$, IL-10^{low,} IFN-γ^{high}; Ref. 56) profile from TH2 cells (CD4$^+$, RT6$^+$, CD45R^{low}, L-Selectin$^-$; IL-4^{high}, IL-2^{low}, IL-10^{high}, IFNγ^{low}). It should, however, be noticed that these marker/cytokine profiles are highly versatile, so that it cannot be ruled out that the effective regulator cell may display other profiles.

The assumption that in rodents the regulative CD4 cell population is exported from the thymus particularly (but not exclusively !) in the immediate postnatal period is not necessarily contradictory to the observation of Fowell and Mason (50) that IDDM-protective cells also exist in long term thymectomized adult rats and therefore hardly can be RTM´s, if one supposes that the pertaining cells may be long living but outnumbered in the adult organism by functionally different cells of later emigration waves (18).

In line with the suggestion that in the early postnatal period the thymus exports a CD4$^+$ cell population which *a priori* has (or quickly acquires) a suppressive or regulatory potency on potentially autoreactive CD4$^+$ cells (the latter being present also in normal animals (52, 59, 60) is the observation that thymectomy of 2 - 5 day old mice may lead to organ-specific autoimmune disease (60–65), whereas thymectomy performed immediately after birth or on postnatal day 7 does not have such an effect (66–68). The outcome of thymectomy on postnatal day 2 -5 depends, however, greatly on the genetic background of the mouse strain used (63), thus indicating that further genetic factors should be involved in generation of autoimmune disease in such animals.

The susceptibility of DP-BB rats to develop an IDDM is essentially determined by three genetic traits (69), i.e. the lymphopenia resistance gene (*Lyp;* = *Iddm 1*), which was mapped to chromosome 4 in tight linkage to the neuropeptide Y gene (*Npy*) (69–71) and obviously is structurally or functionally defective in DP-BB rats, the MHC haplotype RT1^u (= *Iddm 2*) and a third gene (*Iddm 3*) (69), which reportedly is located on chromosome 18, about 22 +/- 7 cM from the *Olf* locus (72). Although the *Lyp* gene-dependent T cell lymphopenia, the molecular basis of which still is obscure but seems to be the result of apoptosis, is an absolute prerequisite (69), coincidence of all three traits seems to be necessary for full IDDM susceptibility in this model (69). This, indeed can be concluded from genetic crosses between DP-BB- and DR-BB-, Lewis- or Fischer- rats, which revealed, according to the Mendelian laws, a respective percentage of lymphopenic but not diabetes prone animals (69). Although the T cell lymphopenia (in DP-BB rats) coincides with defective RT6 expression (7) the question in which way both these defects are correlated to each other presently is discussed controversely. Southern blot analyses revealed that the RT6 expression defect does not concern the *RT6* gene itself because even DP-BB rats display normal *RT6^a* RFLP (restriction fragment length polymorphism) profiles indicating that the gene structurally is not grossly altered (73). In accord with this finding, and supplementary to the first pertaining report (7), it subsequently has been found that the deficiency of RT6$^+$ cells in DP-BB rats is not complete, but that there exists in the spleen of these animals a residual expression of RT6-specific mRNA and RT6.1 protein (74, 75). Densitometric analyses revealed that the quantity of RT6.1 protein displayed by DP-BB rat lymph node cells was below 10% of that expressed by an equal number of lymph node cells of normal (DR-BB) rats (74). By FACS analyses it became apparent that this diminuation of transcription does not reflect a global defect that affects all T cells equally. Thus, a reduced number of constitutively RT6.1-displaying T cells is found in DP-BB rats (14, 16, 76), e.g. in 32 day old DP animals about 7% of the OX19$^+$ T cells vs. 70% in age matched Lew rats (54). This implies that the regulation of *RT6* gene expression may be more sophisticated. The RT6 expression profile of these few cells (98% RT6^{low}, 2% RT6^{high}) is distinct from that of age matched Lew rats (80% RT6^{low}, 20% RT6^{high}). T cells of DP-BB rats have a shortened life span *in vivo* (77) and die quickly when cultured *in vitro* (54). It should, however, be noticed that in such cultures the percentage of surviving OX19$^+$ cells that express RT6 increases in the course of time. Although it is tempting to suggest that this relative increase reflects a higher survival potency of RT6$^+$ cells in comparison to those that (for what reason so ever) in DP-BB rats are unable to display RT6, it cannot definitively be excluded that this effect could reflect an ephemeral RT6 expression by RT6$^-$,Thy$^+$ cells or by cells the RT6 expression of which initially is below the sensitivity of immunofluorescence cytometry.

Since the *RT6^a* gene is existent and structurally not altered (73) in these rats and, moreover, was found to be expressed normally in (DP-BBxWF^b)F$_1$ (78) and (DP-BBxLew^b)F$_1$ (13) rats, the drastical diminuation of RT6 expressing cells in the DP-BB rat obviously is not concerned to the *RT6* gene itself but is the consequence of some other mechanism.

Two working hypotheses concerning lymphopenia and defective RT6 expression are up to discussion:

1. Because of the *Lyp* gene defect the T cells die before they attain the stage at which they constitutively turn on the *RT6* gene (74);
2. The majority of T cells dies prematurely since they are unable to turn on the *RT6* gene.

We here wish to present a model that pleas for the second hypothesis. This model supposes that the normal *Lyp* resistance gene codes for a factor that is essential for transcription of the main, but not of the minor RT6 transcript (s. above and Kuhlenbäumer et al. this volume; as well as Refs. 31 and 32). If this supposition would prove to be correct, *Lyp* gene defects such as those existing in the DP-BB rat should result in non transcription or in not detectable transcription- and translation-rates of RT6 in most T cells, but spare RT6 expression by a minor population that constitutively uses the other transcription start site. This hypothesis would be consistent with the suggestion that the restricted number of RT6 expressing T cells in DP-BB rats may have a higher survival rate *in vitro* than those cells that do not turn on the *RT6* gene. It is tempting to suggest that this latter cell population contains the diabetogenic effector cells in the DP-BB rat. Experiments to study this question are presently under way in our laboratory.

In this context it should be noted that the mere observation that an obviously distinct RT6-expressing cell population proved to have a protective effect against IDDM in different rat models (48, 50) does not necessarily mean that this function is mediated directly by the RT6 molecule itself. By contrast, this seems rather unlikely because RT6 is not restricted to a distinct T cell subset but is expressed by the majority of mature $CD4^+$ and $CD8^+$ T cells.

With regard to the questions dicussed above it may be helpful to consider the Rt6 system of the NOD mouse. This inbred mouse strain also develops spontaneous IDDM which, in contrast to the DP-BB rat, is not accompanied by lymphopenia but shows an accumulation of T- lymphocytes in peripheral lymphatic organs (see Leiter, this volume). Analyses of the expression rate of *Rt6* -specific mRNA in spleens of diabetes-prone NOD- and non diabetes-prone NON- (39) or NOR- (Leiter, this volume) mice revealed a reduced *Rt6* mRNA expression in young NOD mice (< 6 weeks of age). Moreover, it should be noted that the NZW mouse completely lacks *Rt6–2* gene transcripts whereas the *Rt6–1* genes of C57BL/6- and BxSB- mice are interrupted by premature stop codons (Matthes, this volume). Whether these defects, however, contribute to the susceptibility to autoimmune diseases in these mouse strains remains to be elucidated.

Two recently published observations, both referring to the NAD-converting activity of RT6 may contribute to a better understanding of its biological function. Wang et al. (79) found that the proliferative and the cytotoxic capacities of *in vitro* preactivated $CD8^+$ mouse T cells significantly were reduced when the cells were re-activated in presence of NAD. The effective enzymatic activity was found to be linked to a GPI-anchored membrane protein with a M_r of 35 kD. Whether this structure corresponds to one of the two mouse Rt6 forms is presently not known because deciphering of the primary structure of this 35 kD protein still awaits clarification and mouse Rt6-specific antibodies are not yet available. Comparable experiments recently were published by Rigby et al. (37), who showed that the proliferation of rat T cells which were stimulated *in vitro* with a number of different stimulants was significantly inhibited by NAD^+.

It is well known that mature ($RT6^+$) rodent T cells are distinctly less sensitive to activation-induced apoptosis than their $RT6^-$ immature precursors. This and the data of Wang et al. (80), and Rigby et al. (37) in mind, it is tempting to speculate that rodent RT6 may be involved in the complex mechanisms lowering the susceptibility of peripheral T cells towards activation-induced apoptosis by limiting their reactive potency. If this idea is correct, the high RT6 expression rate of IELs (81) would make sense because these cells in particular are confronted with an overload of conventional and super antigens, which would be expected to drive them to overactivation and programmed cell death unless their reactivity would be dampened.

An observation, which at first glance entangles the understanding of the biological function of RT6, is the fact that the RT6 homologues of humans of different races as well as of all primates tested so far will not be expressed because of the existence of premature stop codons, indicating that the function of these genes in these species is no longer used. Considering what is different between those species expressing RT6 and those not doing so, one point is conspicuous, i.e. the different gestation periods and, tightly correlated to it, the respective different state of maturity of the T cell system at birth. Thus, in rats and mice, which have a gestation period of 21 days, the peripheral circulation and the peripheral lymphatic tissues are settled first postnatally by functionally and phenotypically still immature T cells, whereas in species with a longer gestation period the settling process happens already during intrauterine life. For instance, in the sheep, which has a gestation period of 150 days and which at birth is more mature concerning its T cell system than rodents, the first T cells leave the thymus before gestation day 54 (82). This in mind, one is tempted to speculate that this difference may have some bearing concerning the requirement for RT6 expression. Alternatively it is conceivable that the primates, by running a different phylogenetic differentiation way, may have replaced the biological function of T cell RT6 by another molecule or mechanism. We have recently cloned a novel human member of the ADPRT gene family (LART/ART 4), whose expression pattern in the human is remarkably similar to that of the RT6 system in rodents (Braren et al. this volume). Like RT6, LART is a GPI-anchored membrane protein, and thus an attractive candidate for a molecule of the human immune system that may fulfil a function analogous to that of RT6 in rodent T cells.

7. ACKNOWLEDGMENTS

This work was supported by grant Ha 2369 from the Deutsche Forschungsgemeinschaft DFG to F.H. and H.G.T and by a stipend from the Sandoz Foundation for therapeutic research to F.H.

8. REFERENCES

1. Lubaroff, D. M., G. W. Butcher, C. DeWitt, T. Gill, E. Günther, J. Howard & K. Wonigeit. 1983. Standardized nomenclature for the rat T-cell alloantigens - report of the commitee. *Transplant. Proc. 15*: 1683.

2. Butcher, G. W. & J. C. Howard. 1977. An alloantigenic system on rat peripheral T cells. *Rat Newsletter 1*: 12.

3. Lubaroff, D. M., D. L. Greiner & C. W. Reynolds. 1979. Investigations of T-Lymphocyte subpopulations in the rat using alloantigenic markers. *Transplant. Proc. 11*: 1092.

4. DeWitt, C. W. & M. McCullough. 1975. Ag-F: serological and genetic identification of a new locus in the rat governing lymphocyte membrane antigens. *Transplantation 19*: 310.

5. Wonigeit, K. & E. Günther. 1977. Differentiation antigens of T lymphocytes. *Rat Newsletter 2*: 12.

6. Thiele, H. G., F. Koch & M. Low. 1987. Partial biochemical characterization of the rat T cell alloantigen RT6.2 as exposed by the T cell Line EpD3. In *Membrane Proteins*, S.C. Cohen, eds. Publ. Bio-Rad. Lab., p. 131.

7. Greiner, D. L., H. E.S., K. Nakanko, J. P. Mordes & A. A. Rossini. 1986. Absence of the RT6 T cell subset in diabetes-prone BB/W rats. *J. Immunol. 136*: 148.

8. Koch, F., F. Haag & H. G. Thiele. 1990. Nucleotide and deduced amino acid sequence for the mouse homologue of the rat T-cell differentiation marker RT6. *Nucleic Acids Res 18*: 3636.

9. Haag, F., F. Koch-Nolte, M. Kühl, S. Lorenzen & H.-G. Thiele. 1994. Premature stop codons inactivate the RT6 genes of the human and chimpanzee species. *J. Mol. Biol. 243*: 537.

10. Zolkiewska, A., M. S. Nightingale & J. Moss. 1992. Molecular characterization of NAD:arginine ADP-ribosyltransferase from rabbit skeletal muscle. *Proc. Natl. Acad. Sci. USA 89*: 11352.

11. Thiele, H. G., R. Arndt, R. Stark & K. Wonigeit. 1979. Detection and partial molecular characterization of the rat-T-lymphocyte surface protein L_{21} by allo-(anti-RT Ly 2.2) and xeno-(anti-RT-LN-Ly^{lg})sera. *Transplant. Proc. 11*: 1636.

12. Butcher, G. W., S. Clarke & E. M. Tucker. 1979. Close linkage of peripheral T-lymphocyte antigen A (PtaA) to the hemoglobin variant Hbb on linkage group I of the rat. *Transplant. Proc 11*: 1629.

13. Thiele, H.-G., F. Haag & F. Nolte. 1993. Asymmetric Expression of RT6.1 and RT6.2 Alloantigens in (RT6^{a}xRT6^b)F^1 Rats Is Due to a Pretranslational Mechanism. *Transpl. Proc. 25*: 2786.

14. Thiele, H. G., F. Koch & A. Kashan. 1987. Postnatal distribution profiles of Thy-1+ and RT6+ cells in peripheral lymph nodes of DA rats. *Transplant. Proc. 19*: 3157.

15. Mojcik, C. F., D. L. Greiner, E. S. Medlock, K. L. Komschlies & I. Goldschneider. 1988. Characterization of RT6 bearing rat lymphocytes - ontogeny of the RT6[+] subset. *Cell. Immunol. 114*: 336.

16. Haag, F., F. Nolte, A. Lernmark, C. Simrell & H.-G. Thiele. 1993. Analysis of T cell surface marker profiles during the postnatal ontogeny of normal and diabetes prone-rats. *Transpl. Proc. 25*: 2831.

17. Hosseinzadeh, H. & I. Goldschneider. 1993. Recent Thymic Emigrants in the Rat Express a Unique Antigenic Phenotype and Undergo Post-Thymic Maturation in Peripheral Tissues. *J. Immunol. 150*: 1670.

18. Jotereau, F. V., Le Douarin, N.M. 1982. Demonstration of a cyclic renewal of the lymphocyte precursor cells in quail thymus during embryonic and perinatal life. *J. Immunol. 129*: 1869.

19. Dunon, D., Imhof, B.A. 1996. T cell migration during ontogeny and T cell repertoire generation. *Current Topics Microbiol Immunol. 212*: 79.

20. Jotereau, F., Heuze, Francoise, Salomon-Vie, Véronique, Gascan, Hughes. 1987. Cell Kinetics in the fetal mouse thymus: precursor cell input, proliferation, and emigration. *J. Immunol. 138*: 1026.

21. Fangmann, J., R. Schwinzer & K. Wonigeit. 1991. Unusual phenotype of intestinal intraepithelial lymphocytes in the rat: predominance of T cell receptor alpha/beta+/CD2- cells and high expression of the RT6 alloantigen. *Eur J Immunol 21*: 753.

22. Thiele, H.-G. 1988. Rat Thy-1 antigen and its relation to RT6. In *Cell surface antigen Thy-1. Immunology, Neurology, and Therapeutic Applications*, A.E. Reif and M. Schlesinger, eds. Marcel Dekker, New York, p. 149.

23. Koch, F., H. G. Thiele & M. Low. 1987. Phosphatidylinositol is the membrane-anchoring domain of the rat T cell antigens RT6.2 and Thy.1. *Transplant. Proc. 19*: 3140.

24. Thiele, H.-G., R. Arndt & K. Wonigeit. 1983. RT6–2 is a nonglycosylated T-Lymphocyte surface membrane antigen anchored to the cytoskeleton. *Transplant. Proc. 15*: 1635.

25. Koch, F., A. Kashan & H. G. Thiele. 1988. The rat T-cell differentiation marker RT6.1 is more polymorphic than its alloantigenic counterpart RT6.2. *Immunology 65*: 259.

26. Koch, F., F. Haag, A. Kashan & H. G. Thiele. 1989. Construction of a rat T-cell hybridoma cDNA expression library and isolation of a cDNA clone for the rat T-cell differentiation marker RT6.2. *Immunology 67*: 344.

27. Haag, F., F. Koch & H. G. Thiele. 1990. Nucleotide and deduced amino acid sequence of the rat T-cell alloantigen RT6.1. *Nucleic Acids Res 18*: 1047.

28. Haag, F., F. Koch & H. G. Thiele. 1990. Polymorphism between rat T-cell alloantigens RT6.1 and RT6.2 is based on multiple amino acid substitutions. *Transplant Proc 22*: 2541.

29. Reif, A. E., Allen, J.M.V. 1963. The AKR thymic antigen and its distribution in leukemias and nervous tissues. *J. Exp. Med. 120*: 413.

30. Williams, A. F. 1989. *The Structure of Thy-1 Antigen*. Vol. 45. Immunology series "Cell surface Antigen Thy-1", ed. M. Schlesinger A. E. Reif. New York and Basel: Marcel Decker, Inc.

31. Haag, F., F. Nolte, C. Hollmann & H.-G. Thiele. 1993. Analysis of the Gene for the Rat T Cell Alloantigen RT6: Evidence for Alternative Splicing in the 5′ Region. *Transpl. Proc. 25*: 2884.

32. Haag, F., G. Kuhlenbäumer, F. Koch-Nolte, E. Wingender & H.-G. Thiele. 1996. Structure of the gene encoding the rat T cell ecto-ADP-ribosyltransferase RT6.2. *J. Immunol. 157*: 2022.

33. Hawkins, J. D. 1988. A survey on exon and intron lengths. *Nucl. Acids Res. 16*:

34. Koch-Nolte, F., D. Petersen, S. Balasubramanian, F. Haag, D. Kahlke, T. Willer, R. Kastelein, F. Bazan & H.-G. Thiele. 1996. The mouse T cell membrane proteins Rt6–1 and Rt6–2 show ADP-ribosyltransferase activity and share predicted secondary structure motifs with bacterial toxins. *J.Biol.Chem. 271*: 7686.

35. Haag, F., V. Andresen, S. Karsten, N. F. Koch & H. Thiele. 1995. Both allelic forms of the rat T cell differentiation marker RT6 display nicotinamide adenine dinucleotide (NAD)-glycohydrolase activity, yet only RT6.2 is capable of automodification upon incubation with NAD. *Eur J Immunol 25*: 2355.

36. Maehama, T., H. Nishina, S. Hoshino, Y. Kanaho & T. Katada. 1995. NAD+-dependent ADP-ribosylation of T lymphocyte alloantigen RT6.1 reversibly proceeding in intact rat lymphocytes. *J. Biol. Chem. 270*: 22747.

37. Rigby, M., Bortell, R., Stevens, L.A., Moss, J., Kanaitsuka, T., Shigeta, H., Mordes, J.P., Greiner, D., L., Rossini, A.A. 1996. Rat RT6.2 and mouse Rt6 locus 1 are NAD+: Arginine ADP ribosyltransferases with auto-ADP ribosylation activity. *J. Immunol. 156*: 4259.

38. Koch, F., F. Haag, A. Kashan & H. G. Thiele. 1990. Primary structure of rat RT6.2, a nonglycosylated phosphatidylinositol-linked surface marker of postthymic T cells. *Proc Natl Acad Sci USA 87*: 964.

39. Prochazka, M., H. R. Gaskins, E. H. Leiter, F. Koch-Nolte, F. Haag & H. G. Thiele. 1991. Chromosomal localization, DNA polymorphism, and expression of Rt-6, the mouse homologue of rat T-lymphocyte differentiation marker RT6. *Immunogenetics 33*: 152.

40. Koch-Nolte, N. F., C. Hollmann, M. Kuhl, F. Haag, M. Prochazka, E. Leiter & H. G. Thiele. 1995. Molecular polymorphism in the Rt6 genes of laboratory mice correlates with the allotypes of the H1 minor histocompatibility system. *Immunogenetics 41*: 152.

41. Koch-Nolte, F., F. Haag, M. Kühl, V. van Heyningen, J. Hoovers, K. H. Grzeschik, S. Singh & H. G. Thiele. 1993. Assignment of the human RT6 gene to 11q13 by PCR screening of somatic cell hybrids and *in situ* hybridization. *Genomics 18*: 404.

42. Davies, J. L., Y. Kawaguchi, S. T. Bennett, J. B. Copeman, H. J. Cordell, L. E. Pritchard, P. W. Reed, S. C. L. Gough, S. C. Jenkins, S. M. Palmer, K. M. Balfour, B. R. Rowe, M. Farrall, A. H. Barnett, S. C. Bain & J. A. Todd. 1994. A genome-wide search for human type 1 diabetes susceptibility genes. *Nature 371*: 130.

43. Lévy, I., Wu, Y.Q., Roeckel, N., Bulle, F., Pawlak, A., Siegrist, S., Mattéi, M.G., Guellaen, G. 1996. Human testis specifically expresses a homologue of the rodent T lymphocyte RT6 mRNA. *FEBS Lett. 382*: 276.

44. Koch-Nolte, F., Haag, F., Braren, R., Kühl, M., Hoovers, J., Balesubramanian, S., Bazan, F., Thiele, H.-G. 1996. Two Novel Human Members of an emerging Mammalian Gene Family in mono-ADP-ribosylating Bacterial Toxins. *Genomics in press*:

45. Koch-Nolte, F., Kühl, M., Haag, f., Cetkovic-Cvrijle, M., Leiter, E.H., Thiele, H.-G. 1996. Assignment of the human and mouse genes for muscle ecto mono (ADPribosyl) transferase to a conserved linkage group on human chromosome 11p15 and mouse chromosome 7. *Genomics 36*: 215.

46. Zolkiewska, A. & J. Moss. 1993. Integrin alpha 7 as substrate for a glycosylphosphatidylinositol-anchored ADP-ribosyltransferase on the surface of skeletal muscle cells. *J Biol Chem 268*: 25273.

47. Elder, M. E., Maclaren, N.K. 1983. Identification of profound peripheral T cell lymphocyte immunodeficiencies in the spontaneously diabetic BB rat. *J. Immunol. 130*: 1723.

48. Rossini, A. A., D. Faustman, B. A. Woda, A. A. Like, I. Szymanski & J. P. Mordes. 1984. Lymphocyte transfusions prevent diabetes in the Bio-Breeding/Worcester (BB/W) rat. *J. Clin. Invest. 74*: 39.

49. Burstein, D., J. P. Mordes, D. L. Greiner, D. Stein, N. Nakamura, E. S. Handler & A. A. Rossini. 1989. Prevention of diabetes in BB/Wor rat by single transfusion of spleen cells. Parameters that affect degree of protection. *Diabetes 38*: 24.

50. Fowell, D. & D. Mason. 1993. Evidence that the T Cell Repertoire of Normal Rats Contains Cells with the Potential to Cause Diabetes. Characterization of the CD4$^+$ T Cell Subset That Inhibits This Autoimmune Potential. *J. Exp. Med. 177*: 627.

51. Greiner, D. L., J. P. Mordes, E. S. Handler, M. Angelillo, N. Nakamura & A. A. Rossini. 1987. Depletion of RT6.1+ T lymphocytes induces diabetes in resistant biobreeding/Worcester (BB/W) rats. *J Exp Med 166*: 461.

52. Saoudi, A., Seddon, B., Heath, V., Fowell, D., Mason, D. 1996. The physiological role of regulatory T cells in the prevention of autoimmunity: the function of the thymus in the generation of the regulatory T cell subset. *Immunol. Rev.* 195.

53. Hunt, H. D. & D. M. Lubaroff. 1992. Identification of functional T cell subsets and surface antigen changes during activation as they relate to RT6. *Cell Immunol 143*: 194.

54. Thiele, H.-G., Haag, F., Nolte, F. in press. *in vitro* survival and RT6 expression kinetics in peripheral T cells of DP-BB and Lewis rats. *Transplant. Proc.*

55. Kampinga, J., H. Groen, F. Klatter, B. Meedendorp, R. Aspinall, R. Roser & P. Nieuwenhuis. 1992. Postthymic T cell development in rats: an update. *Biochem. Soc. Trans. 20*: 191.

56. Coutinho, G. C., Delassus, S., Kourilsky, Bandeira, A. Coutinho, A. 1994. developmental shift in the patterns of interleukin production in early post-natal life. *Eur. J. Immunol. 24*: 1858.

57. Stutman, O. 1986. Postthymic T-cell development. *Immunol. Rev. 91*: 159.

58. Mackay, C. R. 1993. Homing of naive, memory and effector lymphocytes. *Current Opinion Immunol. 5*: 423.

59. Smith, H., I.-M. Chen, R. Kubo & K. S. K. Tung. 1989. Neonatal Thymectomy Results in a Repertoire Enriched in T Cells Deleted in Adult Thymus. *Science 245*: 749.

60. Smith, H., Sakamoto, Y., Kasai, K., Tung, K.S.K. 1991. Effector and regulatory cells in autoimmune oophoritis elicited by neonatal thymectomy. *J. Immunol. 147*: 2928.

61. Penhale, W. J., Farmer, A., Irvine, W.J. 1975. Thyreoiditis in T cell-depleted rats. Influence of strain, radiation dose, adjuvants and antilymphocyte serum. *Clin. Exp. Immunol. 21*: 362.

62. Penhale W.J., S., P.A., Huxtable, C.R., sutherlsand, R.J., Pethick, D.W. 1990. Induction of diabetes in PVG/c strain rats by manipulation of the immune system. *autoimmunity 7*: 169.

63. Bonomo, A., Kehn, P.J., Shevach, E.M. 1995. Post-thymectomy autoimmunity: abnormal T-cell homeostasis. *Immunol Today 16*: 61.

64. Sakaguchi, S., Fukuma, K., Kuribayashi, K., Masuda, T. 1985. Organ-specific autoimmune diseases induced in mice by elimination of T cell subset. *J. Exp. Med. 161*: 72.

65. Gleeson, P., Toh, B-H., van Driel, I.R. 1996. Organ-specific autoimmunity induced by lymphpenia. *Immunol Rev. 149*: 95.

66. Kojima, A., Tanaka-Kojima, Y., Sakakura, T., Nishizuka, Y. 1976. Spontaneous development of autoimmune thyreoiditis in neonatally thymectomized mice. *Lab. Investigation 34*: 550.

67. Kojima, A., Prehn, R. 1981. Genetic susceptibility to post thymectomy autoimmune disease in mice. *Immunogenetics 14*: 15.

68. Taguchi, O. & Y. Nishizuka. 1980. Autoimmune oophoritis in thymectomized mice: T cell requirement in adoptive cell transfer. *Cl. Exp. Immunol. 42*: 324.

69. Jacob, H. J., A. Pettersson, D. Wilson, Y. Mao, Å. Lernmark & E. S. Lander. 1992. Genetic dissection of autoimmune type I diabetes in the BB rat. *Nature Genetics 2*: 56.

70. Hotnum, L., Jackerot, M., Markholst, H. 1995. The rat T cell lymphopenia resistance gene (Lyp) maps between D4Mit6 and Npy on RNO4. *Mamm. Genome 6*: 371.

71. Markholst, H., S. Eastman, D. Wilson, B. E. Andreasen & A. Lernmark. 1991. Diabetes segregates as a single locus in crosses between inbred BB rats prone or resistant to diabetes. *J Exp Med 174*: 297.

72. Klöting, I., Vogt, L., Serikawa, T. 1995. Locus on chromosome 18 cosegregates with diabetes in the BB/OK subline. *Diabete & Metabolism (Paris) 21*: 338.

73. Thiele, H. G., F. Koch, F. Haag & W. Wurst. 1989. Evidence for normal thymic export of lymphocytes and an intact RT6a gene in RT6 deficient diabetes prone BB-rats. *Thymus 14*: 137.

74. Crisa, L., P. Sarkar, D. J. Waite, F. Haag, F. Koch-Nolte, T. V. Rajan, J. P. Mordes, E. S. Handler, H.-G. Thiele, A. A. Rossini & D. L. Greiner. 1993. An RT6ᵃ Gene Is Transcribed and Translated in Lymphopenic Diabetes-Prone BB Rats. *Diabetes 42*: 688.

75. Fangmann, J., R. Schwinzer, H. J. Hedrich, I. Kloting & K. Wonigeit. 1991. Diabetes-prone BB rats express the RT6 alloantigen on intestinal intraepithelial lymphocytes. *Eur J Immunol 21*: 2011.

76. Lang, F., Kastern, W. 1989. The gene for the T lymphocyte alloantigen RT6, is not linked to either diabetes or lymphopenia and is not defective in the BB rat. *Eur. J. Immunol 19*: 1785.

77. Sarkar, P., L. Crisa, U. McKeever, E. S. Handler, H.-G. Thiele, A. A. Rossini & D. L. Greiner. 1992. Diabetes prone (DP) BB/Wor rats express functional mRNA for RT6 but lack peripheral phenotypic RT6+ T lymphocytes because of a shortened T cell life span (abstract). *Am. Diabet. Assoc.*

78. Angelillo, M., D. L. Greiner, J. P. Mordes, E. S. Handler, N. Nakamura, U. McKeever & A. Rossini. 1988. Absence of RT6+ T cells in diabetes-prone biobreeding/Worcester rats is due to genetic and cell developmental defects. *J Immunol 141*: 4146.

79. Wang, J., E. Nemoto, A. Y. Kots, H. R. Kaslow & G. Dennert. 1994. Regulation of cytotoxic T cells by ecto-nicotinamide adenine dinucleotide (NAD) correlates with cell surface GPI-anchored/arginine ADP-ribosyltransferase. *J. Immunol. 153*: 4048.

80. Wang, P., S. Toyoshima & T. Osawa. 1988. Properties of a novel GTP-binding protein which is associated with soluble phosphoinosites-specific phospholipase C. *J Biochem Tokyo 103*: 137.

81. Fangmann, J., R. Schwinzer, M. Winkler & K. Wonigeit. 1990. Expression of RT6 alloantigens and the T-cell receptor on intestinal intraepithelial lymphocytes of the rat. *Transplant Proc 22*: 2543.

82. Miyasaka, M., Heron, J., Dudler, L., Cahill, R.N.P., Forni, L., Knaak, T., Truka, Z. 1983. Studies on the differentiation of T lymphocytes in sheep. I. Recognition of a T lymphocyte differentiation antigen by a monoclonal antibody T 80. *Immunology 49*: 545.

14

MOLECULAR CLONING AND CHARACTERIZATION OF A MONO(ADP-RIBOSYL)TRANSFERASE FROM HUMAN TESTIS

Isabelle Lévy,[1] André Pawlak,[1] Marie Geneviève Mattéi,[2] and Georges Guellaën[1]

[1]Unité INSERM 99
Hôpital Henri Mondor 94010
Créteil, France
[2]Unité INSERM 406
Hôpital de la Timone, 13385
Marseille, France

ABSTRACT

A human homologue of the rodent T cell mono(ADP-ribosyl)transferase RT6 mRNA was identified by a systematic partial sequencing of human testis transcripts. This messenger encodes for a precursor protein of 367 aa (MW: 41.5 kDa) which exhibits a peptide signal, consensus domains for mono(ADP-ribosyl)transferase and a C terminal part which contains three repeated motives (GEKNQKLEDH) and a region characteristic of glycophosphatidyl inositol anchored proteins. This mRNA is transcribed from a gene localized in 4q13-q21. Surprisingly, it is not expressed in human white blood cells but it exhibits a very specific testis expression in which it is likely to correspond to a new ADP-ribosyl transferase.

BACKGROUND

RT6 is a glycosylphosphatidyl inositol (GPI) anchored membrane protein (1, 2) specifically expressed at the surface of rat and mouse T lymphocytes (2, 3). The rat exhibits two alloantigens for this protein (rRT6.1 and rRT6.2) which both display a NAD-glycohydrolase activity (4, 5). Thus these proteins are able to transfer ADP-ribose, but to a still unknown physiological acceptor.

ADP-Ribosylation in Animal Tissue, edited by Haag and Koch-Nolte
Plenum Press, New York, 1997

Interestingly, in diabetes-prone Bio Breeding rats (DPBB) and NOD mice, a defect in the RT6[+] lymphocyte population correlates with an increased susceptibility to insulin-dependent diabetes mellitus (6,7). In addition, a transfusion of RT6[+] cells in DPBB rats prevented the outbreak of the disease, (8, 9). A similar defect in RT6[+] population exists in (NZWxNZB) mice which exhibit systemic lupus erythematosus (10). All these observations argue for a relationship between a decrease in RT6[+] T cell number and the susceptibility for autoimmune disease in rodents, even if the defect in RT6 expression is unlikely to be the primary event of the observed T cell lymphopenia (11–13).

Recently we characterized a large series of transcripts from human testis by partial cDNA sequencing and by comparison of these sequences with nucleic and proteic databases (14). Among those cDNA, we identified a clone (htMART) which exhibited a significant homology with the rat and mouse RT6 proteins. Since this protein might represent a marker of T cells involved in autoimmune diseases, we further characterized this clone in order to investigate whether it could be a marker in similar human disease. From the present work, it is clear that the htMART mRNA is not expressed in white blood cells but is highly specific for testis in which it is likely to correspond to a new mono(ADP-ribosyl)transferase.

RESULTS

The complete nucleic acid sequence of the htMART cDNA (Genbank accession number: U47054) revealed a 5' untranslated region of 31 bp, a 1101 bp open reading frame and a 3' untranslated region of 278 bp. The predicted amino acids sequence (367 aa; MW: 41.5 kDa) is represented in Figure 1. This sequence exhibits one potential glycosylation site (Asn^{248}).

The alignment of htMART with the rat (rRT6.1 and rRT6.2), mouse (mRT6) and rabbit (rabNART) mono(ADP-ribosyl)transferases sequences is depicted in Figure 2. This alignment reveals that htMART contains the three consensus motives (fig. 2 I, II and III) specific for enzymes catalyzing ADP-ribose transfer (15). Motif I includes a critical His^{94}; motif II is composed of 9 hydrophobic or aromatic residues; motif III contains two glutamate and several other acidic amino acids in the downstream region. Two cystein residues (Cys^{43} and Cys^{206}) are conserved among the five sequences. They might be of importance for the secondary structure of the protein.

The nucleotide alignment of htMART on itself (Fig. 3) revealed a small repeated region in the C terminal part of the mRNA. This corresponds to 3 repeated motives (GEKNQKLEDH) from G^{283} to H^{312} in its C-terminal region.

Until now, all the membranous mono ADP-ribosyl transferases have been described as glycosylphosphatidyl inositol (GPI) anchored proteins. The hydropathy plot of the protein corresponding to htMART according to Kyte and Doolitle (16) (Fig. 4) revealed a hydrophobic peptide signal from Phe^{6} to Val^{24} , and an hydrophobic C-terminal part (Leu^{344} to Leu^{367}). This result indicates that the structure of htMART protein is also compatible with a membranous GPI anchored protein (17).

The expression of the htMART mRNA was analyzed on a Northern blot of 16 different human tissues. Following hybridization of htMART cDNA, an abundant 1.8 kb messenger was detected only in testis under stringent conditions (Fig. 5). Under low hybridization conditions, related messengers were detected in skeletal muscle and heart.

This gene was localized in the human genome. Out of the 150 metaphase cells examined following *in situ* hybridization, 223 silver grains were associated with chromo-

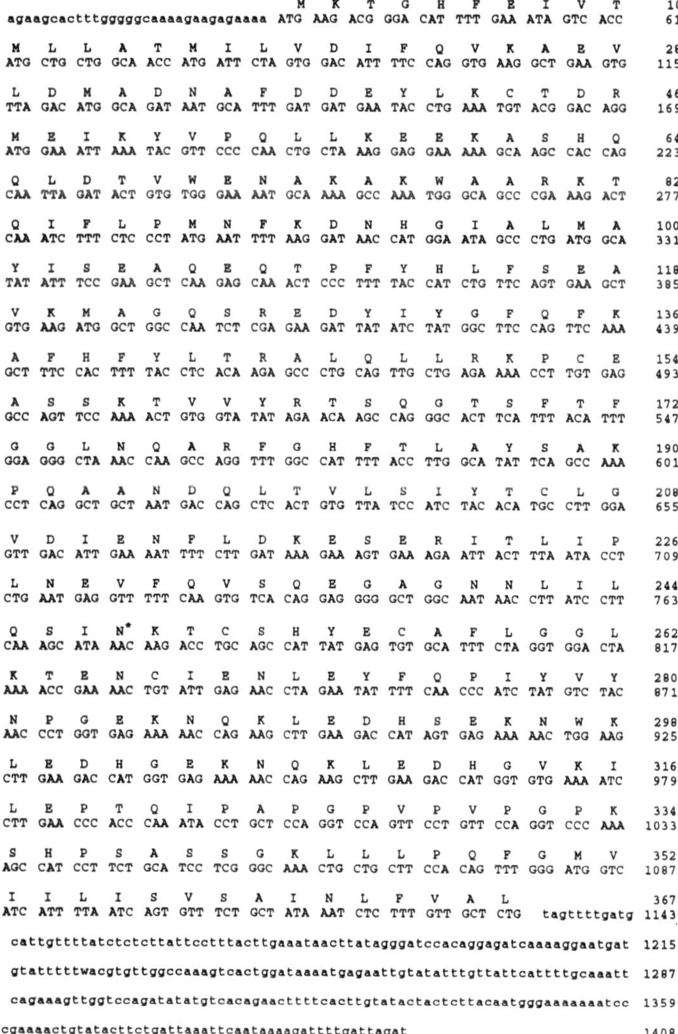

Figure 1. Predicted amino acid sequence of htMART. Asparagine[250] (*) is the only possible N-glycosylation site in the human sequence.

somes and 81 of these (36.3%) were located on Chromosome 4. The distribution of grains on this chromosome was not random: 63/81 (77.7%) of them mapped to the q13- q21 region of Chromosome 4 long arm (data not shown). These results allowed us to map the htMART gene to the 4q13-q21 bands of the human genome.

DISCUSSION

The predicted htMART amino acid sequence has several common features with known mono ADP-ribosyl transferases. Firstly, the htMART sequence exhibits around 35% identities and 60 % of similarities with the whole length of the rodent and rabbit mono-ADP ribsoyl transferase amino acid sequences. This includes a conserved region, framed by the signal peptide and consensus region I of ADP-ribosyl transferase; it corre-

```
                    10        20                  30        40
htMART    MKTGHFEIVTMLLATMILVDIFQV-------------KAEVLDMADNAFDDEYLKCTDRME
rRT6.1    ----MPSNICKFFLTWWLIQQVTG----------LTGPLMLDTAPNAFDDQYEGCVNKME
rRT6.2    ----MPSNICKFFLTWWLIQQVTG----------LTGPLMLDTAPNAFDDQYEGCVNKME
mRT6      ----MPSNNFKFFLTWWLTQQVTG----------LAVPFMLDMAPNAFDDQYEGCVEDME
rabNART   --MWVPAVANLLLLSLGLLEAIQAQSHLVTRRDLFSQETPLDMAPASFDDQYVGCAAAMT
                     -    •                           •• • -••••• • •

           50        60        70        80                  90
htMART    IKYVPQLLKEEKASHQQLDTVWENAKAKWAAR---------KTQI --LPMNFKDNHGIA
rRT6.1    EK-APLLLKEDFNKSEKLKVAWEEAKKRWNN---------IKPSMS--YPKGFNDFHGTA
rRT6.2    EK-APLLLQEDFNMNAKLKVAWEEAKKRWNN---------IKPSRS--YPKGFNDFHGTA
mRT6      KK-APQLLQEDFNMNEELKLEWEKAEIKWKE---------IKNCMS--YPAGFHDFHGTA
rabNART   AA-LPHLNLTEFQVNKVYADGWALASSQWRERSAWGPEWGLSTTRLPPPPAGFRDEHGVA
                •  •  -      •  •  -•                      •  -• • •• •
                                                                I

          100       110       120       130       140       150
htMART    LMAYISEAQEQTPFYHLFSEAVKMAGQSREDYIYGFQFKAFHFYLTRALQLLRKPCEASS
rRT6.1    LVAYTGS------IGVDFNRAVREFKEN----PGQFHYKAFHYYLTRALQLLS-NG---D
rRT6.2    LVAYTGS------IAVDFNRAVREFKEN----PGQFHYKAFHYYLTRALQLLS-NG---D
mRT6      LVAYTGN------IHRSLNEATREFKIN----PGNFHYKAFHYYLTRALQLLS-DQ---G
rabNART   LLAYTANS----PLHKEFNAAVRQAGRSRAHYLQHFSFKTLHFLLTEALQLLGRDQRMPR
          •-••          • -          •-•- •-  •• •••••
                                                 II

          160       170       180       190       200       210
htMART    KTVVYRTSQGTSFTFGGLNQ-ARFGHFTLAYSAKPQAA-----NDQLTVLSIYTCLGVDI
rRT6.1    CHSVYRGTK-TRFHYTGAGS-VRFGQFTSSSLSKTVAQSPEFFSDDGTLFIIKTCLGVYI
rRT6.2    CHSVYRGTK-TRFHYTGAGS-VRFGQFTSSSLSKKVAQSQEFFSDHGTLFIIKTCLGVYI
mRT6      CRSVYRGTN-VRFRYTGKGS-VRFGHFASSSLNRSVATSSPFFNGQGTLFIIKTCLGAHI
rabNART   CRQVFRGVHGLRFRPAGPGTTVRLGGFASASLKNVAAQ--QFGED--TFFGIWTCLGVPI
          •-•       •      •     •   •   -•          •  • •••• •

          220       230       240       250       260
htMART    ENFLDKESERITLIPLNEVFQVSQEGAGN---NLILQSINKTCSHYECAFLGGLKTENCI
rRT6.1    KEFSFYPDQEEVLIPGYEVYQKVRTQGYN---EIFLDSPKRKKSNYNCLYS---------
rRT6.2    KEFSFRPDQEEVLIPGYEVYQKVRTCGYN---EIFLDSPKRKKSNYNCLYS---------
mRT6      KHCSYYTHEEEVLIPGYEVFHKVKTQSVERYIQISLDSPKRKKSNFNCFYSG--------
rabNART   QGYSFFPGEEEVLIPPFETFQVINASRPAQGPARIYLKALGKRSSYNCEYIK--------
          -    •••  •  - --
          III
```

Figure 2. Alignment of htMART with rat rRT6.1 (18), rRT6.2 (1), mouse mRT6 (19) and rabbit (rabNART)(20). (•) and (-) indicates identical or similar residues among the four sequences. The roman numeral indicate the three consensus regions for mono(ADP-ribosyl)transferase activity.

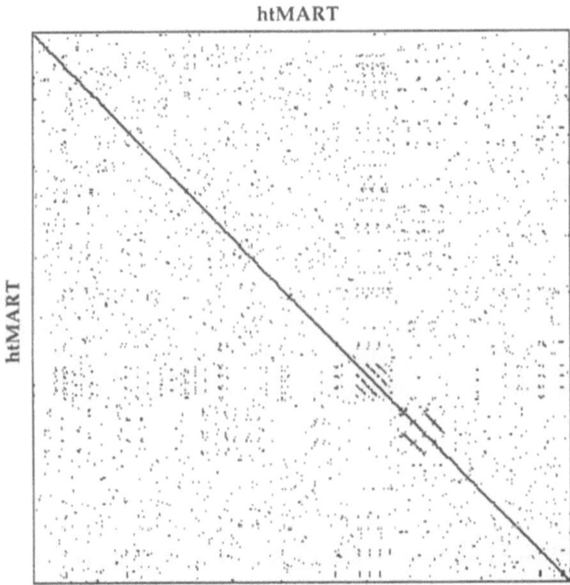

Figure 3. Self alignment of htMART. This alignment was done using the DIAGON program and a window of 7 nucleotides.

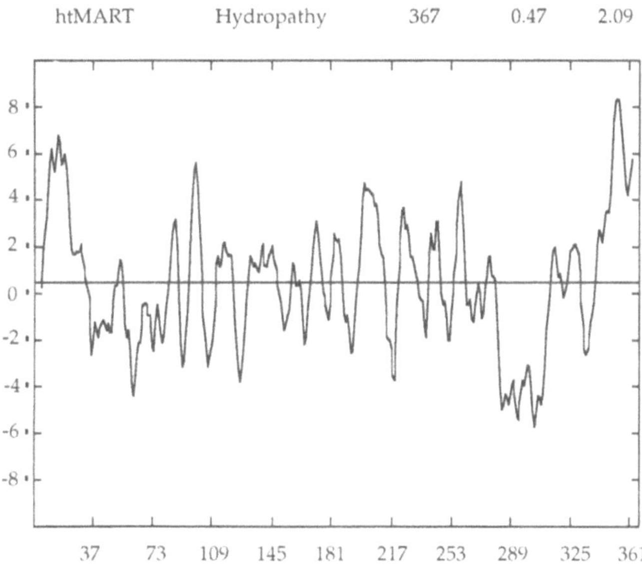

Figure 4. Hydropathy plot of htMART. This plot was determined according to Kyte and Doolitle (16).

sponds to the initial hit indicating a homology between htMART and the rRT6 sequences. Secondly, htMART has the potential structure of a membranous GPI anchored protein (19) as observed for other mono ADP-ribosyl transferases (1,2). Based on the alignment of htMART with the rodent sequences, the htMART signal peptide would include the aa 1 to 24 and the cleavage site of the C-terminal tail would be inside the Ser doublet 341–342. Thus the mature htMART would be 320 amino acids long. Thirdly, this sequence exhibits

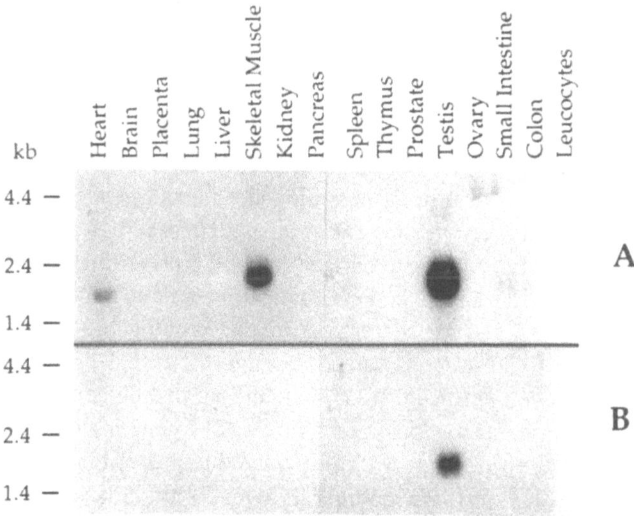

Figure 5. Northern blot from 16 different human tissues, hybridized with htMART cDNA and washed under low (A) or high (B) stringency conditions, and exposed for 1 night. Exposure for A and B were 1 and 3 nights respectively. The size of the markers is indicated on the left hand side.

the three consensus regions specific for mono ADP-ribosyl transferases. Region I contains the critical His[94] which plays a role in hydrogen bonding. Region II contains a series of hydrophobic residues which are important for the positioning of the nicotinamide and adenine moieties. Region III has two specific features: i) 2 Glu residues (Glu[218] and Glu[220]) are present at a slightly shifted position as compared to rat (18), mouse RT6 (19) and rabbit ADP-ribosyl transferase (20) (i.e. at position 222 of the human sequence); ii) 26 acidic residues in the region upstream the hydrophobic C-terminal end. Altogether, these observations argue for htMART as being a new human ADP-ribosyl transferase. Only its expression will determine whether this protein exhibits such an enzymatic activity, as it has been done recently for rRT6.1 and rRT6.2 (21). Fourthly, htMART contains 6 Cys residues, of which two (Cys[42] and Cys[206]) are at the same position as in the rat, mouse and rabbit sequences. The three others are in the downstream sequence, excluding the hydrophobic C-terminal part. Thus htMART has conserved the potential to form at least two intrachain bonds as described for the rRT6.2 (1). The htMART mRNA is longer than the rat, mouse and rabbit mono ADP-ribosyl transferase messages. This additional stretch contains a motif (GEKNQKLEDH) repeated three times, with the exception of 2 amino acids changes in the central motif. Screening of sequences and motif databases with this sequence did not provide any useful information on its role in the protein function. These repeats are also present in the genomic DNA, thus indicating that they do not result from a cloning artefact (Koch-Nolte and Thiele, personal communication).

Using the htMART cDNA as probe, we could establish that htMART transcript is specific for human testis and is unlikely to be a marker of T-lymphocytes. Even under low stringency washing conditions, htMART cDNA did not detect any signal in human leukocytes mRNA. Other groups have analyzed the RT6 expression in rat and mouse tissues, and have looked for cross reactive messengers in different human tissues. They clearly established that: i) RT6 is exclusively expressed in T lymphocyte and absent from rat testis (23); ii) a human genomic probe, having strong similarities with the rodent sequences (23), was unable to detect any transcript in the same human tissues as the ones analyzed in this study.

Altogether these data indicate that htMART and the rodent sequences belong to the mono-ADP-ribosyl transferase family. However, their sites of expression are clearly different and neither the RT6 specific for rodent T cell, nor the htMART expressed in human testis have their respective counterparts in the other species.

Initial attempts have been made in order to identify a human gene susceptible to encode for a rat RT6 homologue (23). Using a probe, derived from the conserved consensus regions I and II (see Fig. 1), these authors cloned a human genomic sequence which exhibited around 80% identities with rat and mouse RT6, whereas we observed only 34 % identities of this sequence with the predicted htMART protein. In addition this genomic sequence has three in frame stop codons, indicating that it corresponds to a non transcribed pseudogene (23). During the course of this cloning procedure, no genomic sequences related to htMART were isolated, although the initial primers were chosen in a very conserved region among species (23). In addition, this human RT6 pseudogene is located in 11q13 (24), whereas we localized the htMART gene in 4q13-q21 (25). Therefore it is clear that this pseudogene is more closely related to rodent RT6 than to htMART.

Thus it is clear that, at the genomic DNA level, the htMART and the rodent RT6 gene are not related. Although these genes might have a common ancestor, the rodent RT6 gene family is now represented by a pseudogene in human, whereas the htMART gene which has related sequence in monkey, dog and cow, does not have a rodent counterpart (25). Thus, both at the RNA and the DNA level, there is no more cross hybridization among those genes between human and rat. These observations demonstrate that only the partial sequencing associated with sequence comparison was able to establish a link among those genes.

In conclusion, htMART is not expressed in human lymphocyte and it is likely to represent a new mono ADP-ribosyl transferase specific from human testis. Until now, only one report in the literature has described a possible mono ADP-ribosylation of testis histone H2B and H3 under the control of the gonadotropin-testosterone system (26). Thus, htMART is of potential interest for the investigation of a new signal transduction pathway involving ADP-ribosylation in testis.

REFERENCES

1. Koch, F., Haag, F., Kashan, A. and Thiele, H.G. Primary structure of rat RT6.2, a non glycosylated phophatidylinositol-linked surface marker of postthymic T cells. (1990) *Proc. Nat. Acad. Sci. 87*, 964–967.
2. Koch, F., Thiele, H.G. and Low, M.G. Release of the rat T cell alloantigen RT-6.2 from cell membranes by phosphatidylinositol-specific phospholipase C. (1986) *J. Exp. Med., 164*, 1338–1343.
3. Zolkiewska, A., Nightingale, M.S. and Moss, J. Molecular characterization of NAD:arginine ADP-ribosyltransferase from rabbit skeletal muscle. (1992) *Proc. Nat. Acad. Sci., 89*, 11352–11356.
4. Haag, F., Andresen, V., Karsten, S., Koch-Nolte, F. and Thiele H.G. Both allelic form of the rat T cell differentiation marker RT6 display nicotinamide adenine dinucleotide (NAD)-glycohydrolase activity, yet only RT6.2 is capable of automodification upon incubation with NAD. (1995) *Eur. J. Immunol., 25*, 2355–2361.
5. Takada, T., Iida, K. and Moss, J. Expression of NAD glycohydrolase activity by rat mammary adenocarcinoma cells transformed with rat T cell alloantigen RT6.2. (1994) *J. Biol. Chem. 269*, 9420–9423.
6. Greiner, D.L., Handler, E.S., Nakano, K., Mordes, J.P. and Rossini, A.A. Absence of the RT6 T cell subset in diabetes-prone BB/W rats. (1986) *J. Immunol., 136*, 148–151.
7. Prochazka, M., Gaskins, H.R., Leiter, E.H., Koch-Nolte, F., Haag, F. and Thiele, H.G. Chromosomal localisation, DNA polymorphism, and expression of RT6, the mouse homologue of rat T-lymphocyte differentiation marker RT6. (1991) *Immunogenetics, 33*, 152–156
8. Burstein, D., Mordes, J.P., Greiner, D.L., Stein, D., Nakamura, N., Handler, E.S. and Rossini, A.A. (1989) Prevention of diabetes in BB/Wor rat by single transfusion of spleen cells. Parameters that affect degree of protection. *Diabetes, 38*, 24–30
9. Fowell, D. and Masson, D. (1993) Evidence that the T cell repertoire of normal rats contains cells with the potential to cause diabetes. Characterization of the CD4+ T cell subset that inhibits this autoimmune potential. J. Exp. Med. 177, 627–636.
10. Koch-Nolte, F., Klein, J., Hollman, C., Kühl, M., Haag, F., Gaskins, H.R., Leiter, E. and Thiele H.G. Defects in the structure and expression of the genes for the T cell marker RT6 in NZW and (NSBxNZW)F1 mice. (1994) *Intern. Immunol. 7*, 883–890.
11. Fangman, J., Schwinzer, R., Hedrich, H.J., Klöting, I. and Wonigeit, K. Diabetes-prone BB rat express the RT6 alloantigen on intestinal intraepithelial lymphocytes. (1991) *Eur. J. Immunol., 21*, 2011–2013.
12. Crisà, L., Sarkar, P., Waite, D.J., Haag, F., Koch-Nolte, F., Rajan, T.V., Mordes, J.P., Handler, E.S., Thiele, H.G., Rossini, A.A. and Greiner, D.L. An RT6 gene is transcribed and translated in lymphopenic diabetes-prone BB rats. (1993) *Diabetes, 42*, 688–695.
13. Crisa, L., Greiner, D.L., Mordes, J.P., MacDonald, R.G., Handler, E.S., Czech, M.P. and Rossini, A.A. Biochemical studies of RT6 alloantigens in BB/Wor and normal rats. (1990) *Diabetes, 39*, 1279–1288.
14. Pawlak, A, Toussaint, C., Lévy, I., Bulle, F., Poyard, M., Barouki, R. and Guellaen, G. Characterization of a large population of mRNAs from human testis (1995) *Genomics, 26*, 151–158
15. Takada, T., Iida, K. and Moss, J. Conservation of a common motif in enzymes catalyzing ADP-ribose transfer. Identification of domains in mammalian transferases. *J. Biol. Chem., 270*, 541–544.
16. Kyte A.M. and Doolitle, F. A simple method displaying the hydropathic character of a protein (1982) *J. Mol. Biol. 157*, 105–132.
17. Ferguson, M.A.J. and Williams, A.F. Cell-surface anchoring of proteins via glycosyl-phosphatidylinositol structures *Ann. Rev. Biochem.* (1988) *57*, 285–320
18. Haag, F., Koch, F. and Thiele, H.G. Nucleotide and deduced amino acid sequence of the rat T-cell alloantigen RT6.1. (1990) *Nucl. Ac. Res. 18*, 1047.

19. Koch, F., Haag, F. and Thiele, H.G. Nucleotide and deduced amino acid sequence of the mouse homologue of the rat T-cell differentiation marker RT6. (1990) *Nucl. Ac. Res.*, *18*, 3636.

20. Zolkiewska, A., Nightingale, M.S. and Moss, J. Molecular characterization NAD:arginine ADP-ribosyl-transferase from rabbit skeletal muscle. (1992) *Proc. Nat. Acad. Sci. 89*, 11352–11356.

21. Koch-Nolte, F., Petersen, D., Balasubramanian, S., Haag, F., Kahlke, D., Willer, T., Kastelein, R., Bazan, F. and Thiele, H.G.(1996) Mouse T-cell membrane proteins Rt6–1 and Rt6–2 are arginine/protein mono(AD-Pribosyl)transferases and share secondary structure motifs with ADP-ribosylating bacterial toxins. (1996) *J. Biol. Chem.*, 271, 7686–7693.

22. Koch, F., Haag, F., Kashan, A. and Thiele, H.G. Construction of a rat T-cell hybridoma cDNA expression library and isolation of a cDNA clone for the rat T-cell differentiation marker RT6.2.(1989) *Immunology*, *67*, 344–350.

23. Haag, F., Koch-Nolte, F., Külh, M., Lorenzen, S. and Thiele, H.G. Premature stop codons inactivate the RT6 genes of the human and chimpanze species. (1994) *J. Mol. Biol. 243*, 537–546.

24. Koch-Nolte, F., Haag, F., Kühl, M., van Heyningen, V., Hoovers, J., Grzeschik, K.H., Singh, S. and Thiele, H.G. Assignment fo the human RT6 gene to 11q13 by PCR screening of somatic cell hybrids and in situ hybridization. (1993) *Genomics*, *18*, 404–406.

25. Lévy, I., Wu, Y.Q., Roeckel, N., Bulle, F., Pawlak, A., Siegrist, S., Mattéi, M.G. and Guellaën, G.(1996) Human testis specifically expresses a homologue of the rodent T lymphocytes RT6 mRNA.(1996) *FEBS letters* , *382*, 276–280.

26. Kurokawa, T., Fujimura, Y., Takahashi, K., Chono, E., and Ishibashi, S. (1988) Reduction of mono ADP-ribosylation of histones in rat testis by gonadotropin-testosterone system. *Biochem. Biophys. Res. Comm. 215*, 808–813.

15

MOLECULAR CLONING AND CHARACTERIZATION OF LYMPHOCYTE AND MUSCLE ADP-RIBOSYLTRANSFERASES

Ian J. Okazaki, Hyun-Ju Kim, and Joel Moss

Pulmonary-Critical Care Medicine Branch
National Heart, Lung, and Blood Institute
National Institutes of Health
Bethesda, Maryland 20892

1. ABSTRACT

Mono-ADP-ribosylation, catalyzed by ADP-ribosyltransferases, is a posttranslational modification of proteins in which the ADP-ribose moiety of NAD is transferred to an acceptor protein(arginine). Several of the bacterial toxin ADP-ribosyltransferases have been well characterized in their ability to alter cellular metabolism. It has been postulated that these bacterial toxins mimic the actions of transferases from mammalian cells. We have cloned and characterized ADP-ribosyltransferases from rabbit and human skeletal muscle, and mouse lymphocytes. The muscle transferases are glycosylphosphatidylinositol (GPI)-anchored proteins that are conserved among species. Two distinct transferases, termed Yac-1 and Yac-2 were cloned from mouse lymphoma (Yac-1) cells. The Yac-1 transferase, like the muscle enzymes, is a GPI-linked exoenzyme. The Yac-2 transferase, on the other hand, is membrane-associated but appears not to be GPI-linked. In contrast to Yac-1, the Yac-2 enzyme had significant NAD glycohydrolase activity and may preferentially hydrolyze NAD.

The bacterial toxin ADP-ribosyltransferases contain three noncontiguous regions of sequence similarity, which are involved in formation of the catalytic site. Alignment of the deduced amino acid sequences of the mammalian transferases and the rodent RT6 enzymes, along with results from site-directed mutagenesis of the muscle enzyme, are consistent with the notion of a common mechanism of NAD binding and catalysis among ADP-ribosyltransferases.

2. INTRODUCTION

Mono-ADP-ribosylation, catalyzed by ADP-ribosyltransferases, involves transfer of the ADP-ribose moiety of NAD to an acceptor protein. Several bacterial toxin ADP-ribosyltransferases have been shown to modify different amino acids in target proteins leading to alterations in cellular metabolism. Examples of ADP-ribosylation catalyzed by bacterial toxin transferases include cholera toxin and the related heat-labile enterotoxin of *Escherichia coli* that modify a specific arginine in $G_{s\alpha}$, the α-subunit of the stimulatory guanine nucleotide-binding (G) protein, which results in the stimulation of adenylyl cyclase[1]. These toxins also modify free arginine and simple guanidino compounds. Diphtheria toxin[2] and *Pseudomonas aeruginosa* exotoxin A[3] ADP-ribosylate a modified histidine in elongation factor 2 leading to the inhibition of protein synthesis and cell death.

An NAD:arginine ADP-ribosyltransferase activity, similar to that of cholera toxin, has been detected in numerous animal tissues[4]. Several vertebrate transferases of the type have been cloned and characterized including those from rabbit[5] and human[6] skeletal muscle, chicken heterophils[7] and erythroblasts[8], and mouse lymphoma cells[9,10]. The glycosylphosphatidylinositol (GPI)-anchored skeletal muscle transferases are conserved across species[6]. It was shown that the muscle enzyme in C2C12 mouse myoblasts modified integrin $\alpha7$[11]. In chicken heterophils, two closely related transferases were detected in heterophil granules, which appear to modify *in vitro*, p33[12], a granule protein related to the myeloid inhibitor membrane protein (*mim-1*), and non-muscle actin[13], with resulting inhibition of actin polymerization. Incubation of mouse cytotoxic T lymphocytes (CTL) with NAD resulted in the ADP-ribosylation of membrane proteins and inhibition of CTL proliferation and cytotoxicity[14]. Two ADP-ribosyltransferases from mouse T cell lymphoma (Yac-1) cells, termed Yac-1 and Yac-2, which were subsequently cloned and characterized[9,10], are the focus of this paper. We also present evidence for the existence of regions of amino acid sequence similarity among several of the bacterial toxins as well as the cloned mammalian ADP-ribosyltransferases, which are thought to form, in part, the catalytic cleft involved in NAD binding and ADP-ribose transfer.

3. RESULTS AND DISCUSSION

3.1. Cloning of the Mammalian ADP-Ribosyltransferases

A family of vertebrate ADP-ribosyltransferases have been purified that, like cholera toxin, modify arginine or other simple guanidino compounds. Of the mammalian ADP-ribosyltransferases, those from rabbit and human skeletal muscle were first to be cloned[5,6]. The nucleotide and deduced amino acid sequences of the human skeletal muscle enzyme were 80.8 and 81.3% identical, respectively, to those of the rabbit. The muscle transferases are glycosylphosphatidylinositol (GPI)-anchored and conserved across species[6] (see accompanying chapter, "The $\alpha7$ integrin-chain as a target protein for cell surface mono-ADP-ribosylation in muscle cells").

GPI-anchored transferases were also found in some murine T cell lymphoma and hybridoma cells[15], and in mouse cytotoxic T lymphocytes (CTL)[14]. Incubation of CTL in the presence of NAD resulted in the ADP-ribosylation of membrane proteins, inhibition of CTL proliferation, and, to a lesser extent, cytotoxicity. The suppressive effects of NAD on CTL were prevented by treatment of the cells with phosphatidylinositol-specific phospholipase C (PI-PLC), which releases most GPI-linked proteins from the cell surface[16], before incubation with NAD, consistent with the conclusion that a GPI-linked ADP-ribosyltransferase was responsible for modulating CTL function[14]. In addition, ADP-ribosylation of a 40-kDa CTL membrane protein (p40) resulted in the inhibition of p56[lck], a tyrosine kinase that exists in a complex with the 40-kDa protein[17]. Release of the membrane-bound transferase following treatment with PI-PLC prevented the NAD-induced suppression of kinase activity[17]. The relationship between the inhibition of p56[lck], following ADP-ribosylation of p40, and the inhibition of CTL proliferation has not been established. These data are consistent with a role for ADP-ribosylation in lymphocyte function.

Two lymphocyte ADP-ribosyltransferases, termed Yac-1[9] and Yac-2[10], were cloned from mouse lymphoma (Yac-1) cells. The deduced amino acid sequence of the Yac-1 transferase was 77 and 75% identical, to those from the human and rabbit skeletal muscle enzymes, respectively; that of the Yac-2 transferase was only 28 and 30% identical. Nucleotide and deduced amino acid sequences of the Yac-1 and Yac-2 transferases were 52 and 32% identical, respectively. Further, the Yac-1 and Yac-2 enzymes were only ~30% identical to the rodent RT6 family of proteins which are NAD glycohydrolases and, under certain conditions, NAD:arginine ADP-ribosyltransferases[18–21].

Despite the limited overall sequence identity among the mammalian transferases and RT6 proteins, these enzymes possess regions of sequence similarity that have been noted in several of the bacterial toxin ADP-ribosyltransferases[22]. In the bacterial toxin transferases, these sequences form, in part, the active site cleft for NAD binding and catalysis (Fig. 1). The H-R region contains a nucleophilic histidine or arginine capable of hydrogen bond formation with NAD. Based on sequence alignment, crystal structure, photoaffinity labeling, and site-directed mutagenesis, His21 of diphtheria toxin (DT)[23–25], His440 of *Pseudomonas aeruginosa* exotoxin A (ETA)[26,27], Arg7 of cholera toxin (CT) and the related heat-labile enterotoxin of *E. coli* (LT)[28–30], Arg9 of pertussis toxin (PT)[31–33] and perhaps Arg164 of the chicken heterophil transferases[7,21] are the H-R region amino acids responsible for NAD binding. In the bacterial toxins, the hydrophobic region comprises closely spaced aromatic and hydrophobic amino acids that form a pocket for the adenine and nicotinamide rings of NAD. Conservation of hydrophobic amino acids that interact with the aromatic rings of NAD are evident in the alignment of the hydrophobic region of DT and ETA; Trp50, Phe53, Tyr54 and Tyr65 of DT correspond to Trp466, Phe469,

A			**B**		
RMT	177	VFRGV	RMT	231	GYSFFPGEEEVLIP
Yac-1	172	·y···	Yac1	226	······E·······
Yac-2	159	·····	Yac2	213	aL·V··E·R·····
Rt6-1	144	·y··T	Rt6-1	200	HC·yytH·······
RT6.1	144	·y··T	RT6.1	200	Ef··y·Dq······
RT6.2	144	·y··T	RT6.2	200	Ef··R·Dq······
DT	19	Syr·T	DT	136	EG·SSV----·YiNN
ETA	438	Gyr·T	ETA	544	·PEEg·RL·Tilg
CT	5	ly·aD	CT	105	---·PH·Y·q··SAL

Figure 1. Alignments of consensus regions of bacterial toxins mammalian ADP-ribosyltransferases, and rodent RT6 proteins. A. The H-R region contains a nucleophilic histidine or arginine. B. The acidic region contains the active-site glutamate. DT, diphtheria toxin; ETA, *Pseudomonasaeruginosa* exotoxin A; PT, pertussis; CT, cholera toxin; RMT, rabbit skeletal muscle transferase; Yac-1, Yac-1 ADP-ribosyltransferase; Yac-2, Yac-2 ADP-ribosyltransferase;RT6.1, rat RT6.1 alloantigen; RT6.2, rat RT6.2 alloantigen; Rt6–1, mouse Rt6 locus 1 protein. Sequences are in the single letter code with the position of the first amino acid following the name of the protein. ● indicates amino acid identity. Lower case letter indicates conservative differences from the rabbit muscle transferase. — indicates gap inserted to optimize alignment.

Tyr470 and Tyr481 of ETA[26]. The aromatic-rich segment of PT encompasses amino acids 82–98, based on structural alignment of toxins[22]. The acidic region contains the catalytic glutamate that facilitates the nucleophilic attack of the N-glycosidic bond of NAD. Glu148 of DT[34–36], Glu112 of CT and LT[29,37–39], Glu553 of ETA[40,41] and Glu129 of PT[42,43] were identified by photoaffinity labeling, site-directed mutagenesis and three-dimensional studies as the strictly conserved, active-site glutamate.

Alignment of the deduced amino acid sequence of the rabbit muscle transferase with those of several of the bacterial toxin transferases along with data from site-directed mutagenesis and computer modeling demonstrated that Arg174 and Glu238 with Glu240 are the critical amino acids in the H-R and acidic region, respectively[21,44]. Likewise, alignment of the Yac-1 and Yac-2 transferase sequences with those of the muscle enzyme and bacterial toxins revealed apparent conservation of critical amino acids thought to be involved in catalysis. As shown in Fig. 1, Arg174 of Yac-1 and Arg161 of Yac-2 appear to be the conserved H-R region arginines; Glu223 and Glu235 of Yac-1 and Glu220 and Glu222 of Yac-2 are the apparent active site glutamates. Although Yac-1 contains a glutamate at position 234 and Yac-2 contains an arginine at position 221, site-directed mutagenesis of Glu239 in the corresponding position in the rabbit muscle transferase demonstrated that this amino acid was not crucial for activity[44].

The lymphocyte differentiation antigen CD38, catalyzes the formation and hydrolysis of cyclic ADP-ribose leading to overall conversion of NAD to ADP-ribose and nicotinamide[45]. In addition, soluble CD38 expressed in a baculovirus system ADP-ribosylated lysozyme, Il-2 and myoglobin by a nonenzymatic mechanism in which CD38-generated ADP-ribose became attached to a cysteine residue via a thioglycosidic bond[46]. The skeletal muscle and lymphocyte transferases, in contrast, are arginine-specific ADP-ribosyltransferases.

3.2. Characterization of the Lymphocyte ADP-Ribosyltransferases

On Northern analysis, a labeled Yac-1 cDNA probe hybridized with a major 1.6-kb band and a minor 6.0-kb band in poly(A)$^+$ RNA from mouse cardiac and skeletal muscle consistent with the significant amino acid sequence identity of the Yac-1 and muscle transferases[9]. There was also faint hybridization with a 1.6-kb band in poly(A)$^+$ RNA from mouse lung, possibly due the presence of a related lung transferase. There was no hybridization with mRNA from the spleen, despite the fact that the Yac-1 transferase was cloned from lymphoid cells. There was, however, very weak hybridization with poly(A)$^+$ RNA from Yac-1 cells consistent with a low abundance of Yac-1 transferase mRNA in these cells and a scarcity or absence of transferase mRNA in splenic lymphocytes[9]. In contrast, a labeled Yac-2 cDNA probe hybridized with 1.6- and 2.0-kb bands from mouse testis and weakly with 1.6-kb bands from mouse cardiac and skeletal muscle[10]. There was no hybridization of mouse Rt6 locus 1 (Rt6–1) cDNA with poly(A)$^+$ RNA from Yac-1 cells, which distinguishes the Yac-1 and Yac-2 transferases from the rodent RT6 proteins.

To compare enzyme activities and localization, rat mammary adenocarcinoma (NMU) cells were transformed with the coding region cDNAs of the Yac-1 and Yac-2 ADP-ribosyltransferases[9,10]. In NMU cells transformed with the Yac-1 transferase cDNA, enzyme activity (ADP-ribosylagmatine formation) was detected in the soluble and membrane fractions, similar to that found in native Yac-1 cells. Total transferase activity in transformed cells, however, was ~400-times that in Yac-1 cells. In Yac-2-transformed NMU cells, transferase activity was membrane-associated but, in contrast to cells expressing the Yac-1 enzyme, demonstrated significant basal NAD glycohydrolase activity.

When cells expressing the Yac-1 transferase were treated with 0.1 U of PI-PLC, ~25% of enzyme activity was released into the medium whereas Yac-2 transferase activity was not released following incubation with as much as 1.0 U of PI-PLC[10]. Hydropathy profiles of the Yac-1 and Yac-2 transferases (Fig. 2) revealed the presence of hydrophobic amino termini. The Yac-1, but not the Yac-2 enzyme, however, contained a hydrophobic carboxy terminal signal peptide characteristic of GPI-linked proteins.

Yac-1 and Yac-2 transferases were expressed as GST-fusion proteins to determine kinetic constants[10]. The K_m values for NAD with 20 mM agmatine as ADP-ribose acceptor were 118 ± 17 and 142 ± 13 µM for Yac-1 and Yac-2, respectively (mean ± S.D., n=4). Values for agmatine in the presence of 100 µM NAD were 9.4 ± 1.7 and 15 ± 4.9 mM, respectively. Values of V_{max} in the presence of 20 mM agmatine for Yac-1 and Yac-2 were 19 ± 5 and 8 ± 3 pmol·min⁻¹·µg⁻¹, respectively.

4. SUMMARY

Cloning of the muscle and lymphocyte ADP-ribosyltransferases has assisted in characterization of these enzymes. The GPI-anchored skeletal muscle transferase modifies the adhesion molecule integrin α7 on the surface of myocytes which has been proposed to alter myogenesis[11]. The Yac-1 enzyme has properties similar to those of the ADP-ribosyltransferase found in CTL which modifies surface proteins leading to inhibition of CTL proliferation. The Yac-2 protein is an ADP-ribosyltransferase with significant NAD glycohydrolase activity expressed in mouse muscle and testis.

All mammalian ADP-ribosyltransferases thus far cloned contain regions of amino acid sequence similarity that can be aligned with those of several bacterial toxin transferases and the rodent RT6 family of proteins, consistent with the hypothesis that these enzymes possess a common mechanism of catalysis.

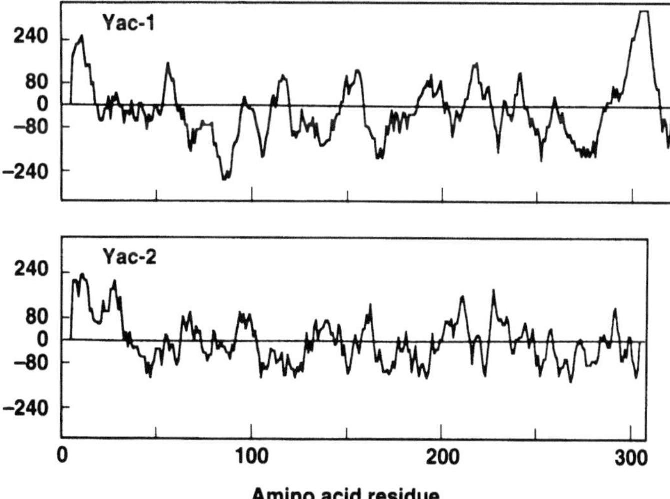

Figure 2. Hydropathy profile of Yac-1 and Yac-2 ADP-ribosyltransferases. The Kyte-Doolittle hydrophobicity profiles (GeneWorks, IntelliGenetics Inc.) of the deduced amino acid sequences of the Yac-1 and Yac-2 ADP-ribosyltransferases are shown. Numbering of amino acids is at the bottom.

The inhibition of CTL proliferation and cytotoxicity following ADP-ribosylation of membrane proteins is consistent with the hypothesis that ADP-ribosyltransferases are involved in immune regulation; the Yac-1 and Yac-2 transferases as well as other transferases and NADases may play a role in this process.

6. REFERENCES

1. Moss, J. & M. Vaughan. 1988. ADP-ribosylation of guanyl nucleotide-binding proteins by bacterial toxins. *Adv. Enzymol. 61*: 303–379.
2. Collier, R. J. 1990. Diphtheria toxin: Structure and function of a cytocidal protein. In: *ADP-ribosylating toxins and G proteins: Insights into signal transduction.* J. Moss, and M. Vaughan (eds). American Society of Microbiology, Washington, D.C., p. 3–19.
3. Wick, M. J., & B. H. Iglewski. 1990. *Pseudomonas aeruginosa* exotoxin A. In: *ADP-ribosylating toxins and G proteins: Insights into signal transduction.* J. Moss, and M. Vaughan (eds). American Society of Microbiology, Washington, D.C., p. 31–43.
4. Okazaki, I. J., & J. Moss. 1996. Mono-ADP-ribosylation: A reversible posttranslational modification of proteins. *Advances in Pharmacology.* T. J. August, M. W. Anders, F. Murad, & J. T. Coyle (eds). Academic Press, San Diego, CA. *35*: 247–280.
5. Zolkiewska, A., M. S. Nightingale, & J. Moss. 1992. Molecular characterization of NAD:arginine ADP-ribosyltransferase from rabbit skeletal muscle. *Proc. Natl. Acad. Sci. 89*: 11352–11356.
6. Okazaki, I. J., A. Zolkiewska, M. S. Nightingale, & J. Moss. 1994. Immunological and structural conservation of mammalian skeletal muscle glycosylphosphatidylinositol-linked ADP-ribosyltransferases. *Biochem. 33*: 12828–12836.
7. Tsuchiya, M., N. Hara, K. Yamada, H. Osago, & M. Shimoyama. 1994. Cloning and expression of cDNA for arginine-specific ADP-ribosyltransferase from chicken bone marrow cells. *J. Biol. Chem. 269*: 27451–27457.
8. Davis, T., & S. Shall. 1995. Sequence of a chicken erythroblast mono(ADP-ribosyl)transferase-encoding gene and its upstream region. *Gene 164*: 371–372.
9. Okazaki, I. J., H-. J. Kim, N. G. McElvaney, E. Lesma, & J. Moss. 1996. Molecular characterization of a glycosylphosphatidylinositol-linked ADP-ribosyltransferase from lymphocytes. *Blood*, in press.
10. Okazaki, I. J., H-. J. Kim, & J. Moss. 1996. Cloning and characterization of a novel cell surface lymphocyte NAD:arginine ADP-ribosyltransferase. *J. Biol. Chem.*, submitted.
11. Zolkiewska, A., & J. Moss. 1993. Integrin α7 as substrate for a glycosylphosphatidylinositol-anchored ADP-ribosyltransferase on the surface of skeletal muscle cells. *J. Biol. Chem. 268*: 25273–25276.
12. Mishima, K., M. Terashima, S. Obara, K. Yamada, K. Imai, & M. Shimoyama. 1991. Arginine-specific ADP-ribosyltransferase and its acceptor protein p33 in chicken polymorphonuclear cells: Co-localization in the cell granules, partial characterization, and *in situ* mono(ADP-ribosyl)ation. *J. Biochem. 110*: 388–394.
13. Terashima, M., K. Mishima, K. Yamada, M. Tsuchiya, T. Wakutani, & M. Shimoyama. 1992. ADP-ribosylation of actins by arginine-specific ADP-ribosyltransferase purified from chicken heterophils. *Eur. J. Biochem. 204*: 305–311.
14. Wang, J., E. Nemoto, A. Y. Kots, H. R. Kaslow, & G. Dennert. 1994. Regulation of cytotoxic T cells by ecto-nicotinamide adeninne dinucleotide (NAD) correlates with cell surface GPI-anchored/arginine ADP-ribosyltransferase. *J. Immunol. 153*: 4048–4058.
15. Soman, G., A. Haregewoin, R. C. Hom, & R. W. Finberg. 1991. Guanidine group-specific ADP-ribosyltransferase in murine cells. *Biochem. Biophys. Res. Commun. 176*: 301–308.
16. Ferguson, M. A. J., & A. F. Williams. 1988. Cell surface anchoring of proteins via glycosylphosphatidylinositol structures. *Ann. Rev. Biochem. 57*: 285–320.
17. Wang, J., E. Nemoto, & G. Dennert. 1996. Regulation of CTL by ecto-nicotinamide adenine dinucleotide (NAD) involves ADP-ribosylation of a p56[lck]-associated protein. *J. Immunol. 156*: 2819–2827.
18. Takada, T., K. Iida, & J. Moss. 1994. Expression of NAD glycohydrolase activity by rat mammary adenocarcinoma cells transformed with rat T cell alloantigen RT6.2. *J. Biol. Chem. 269*: 9420–9423.
19. Haag, F., V. Andresen, S. Karsten, F. Koch-Nolte, & Thiele, H-. G. 1995. Both allelic forms of the rat T cell differentiation marker RT6 display nicotinamide adenine dinucleotide (NAD)-glycohydrolase activity, yet only RT6.2 is capable of automodification upon incubation with NAD. *Eur. J. Immunol. 25*: 2355–2361.

20. Maehama, T., H. Nishina, S. Hoshino, Y. Kanaho, & T. Katada. 1995. NAD⁺-dependent ADP-ribosylation of T lymphocyte alloantigen RT6.1 reversibly proceeding in intact rat lymphocytes. *J. Biol. Chem. 270*: 22747–22751.

21. Koch-Nolte, F., D. Petersen, S. Balasubramanian, F. Haag, D. Kahlke, T. Willer, R. Kastelein, F. Bazan, & H-. G. Thiele. 1996. Mouse T cell membrane proteins *Rt6–1* and *Rt6–2* are arginine/protein mono(ADP-ribosyl)transferases and share secondary structure motifs with ADP-ribosylating bacterial toxins. *J. Biol. Chem. 271*: 7686–7693.

22. Rappuoli, R., & M. Pizza. 1991. Structure and evolutionary aspects of ADP-ribosylating toxins. In: *Sourcebook of Bacterial Proteins Toxins*. J. E. Alouf, J. H. Freer (eds). Academic Press Ltd., San Diego, CA, p. 1–21.

23. Papini, E., G. Schiavo, D. Sandona, R. Rappuoli, & C. Montecucco. 1989. Histidine 21 is at the NAD⁺ binding site of diphtheria toxin. *J. Biol. Chem. 264*: 12385–12388.

24. Johnson, V. G., & P. Nicholls. 1994. Histidine-21 does not play a major role in diphtheria toxin catalysis. *J. Biol. Chem. 269*: 4349–4354.

25. Blanke, S. R., K. Huang, B. A. Wilson, E. Papini, A. Covacci, & R. J. Collier. 1994. Active-site mutation of diphtheria toxin catalytic domain: Role of histidine-21 in nicotinamide adenine dinucleotide binding and ADP-ribosylation of elongation factor 2. *Biochem. 33*: 5155–5161.

26. Carroll, S. F., & R. J. Collier. 1988. Amino acid sequence homology between the enzymic domains of diphtheria toxin and *Pseudomonas aeruginosa* exotoxin A. *Mol. Microbiol. 2*: 293–296.

27. Han, X. Y., & D. R. Galloway. 1995. Active site mutations of *Pseudomonas aeruginosa* exotoxin A. Analysis of the His⁴⁴⁰ residue. *J. Biol. Chem. 270*: 679–684.

28. Burnette, W. N., V. L. Mar, B. W. Platler, J. D. Schlotterbeck, M. D. McGinley, K. S. Stoney, M. F. Rohde, & H. R. Kaslow. 1991. Site-specific mutagenesis of the catalytic subunit of cholera toxin: Substitution of lysine for arginine 7 causes loss of activity. *Infect. Immun. 59*: 4266–4270.

29. Pizza, M., M. Domenighini, W. Hol, V. Giannelli, M. R. Fontana, M. M. Giuliani, C. Magagnoli, S. Peppoloni, R. Manetti, & R. Rappuoli. 1994. Probing the structure-activity relationship of *Escherichia coli* LT-A by site-directed mutagenesis. *Mol. Microbiol. 14*: 51–60.

30. Sixma, T. K., K. H. Kalk, B. A. M. van Zanten, Z. Dauter, J. Kingma, B. Witholt, & W. G. J. Hol. 1993. Refined structure of *Escherichia coli* heat-labile enterotoxin, a close relative of cholera toxin. *J. Mol. Biol. 230*: 890–918.

31. Stein, P. E., A. Boodhoo, G. D. Armstrong, S. A. Cockle, M. H. Klein, & R. J. Read. 1994. The crystal structure of pertussis toxin. *Structure 2*: 45–57.

32. Burnette, W. N., W. Cieplak, V. L. Mar, K. T. Kaljot, H. Sato, & J. M. Keith. 1988. Pertussis toxin S1 mutant with reduced enzyme activity and a conserved protective epitope. *Science 242*: 72–74.

33. Kaslow, H. R., J. D. Schlotterbeck, V. L. Mar, & N. W. Burnette. 1989. Alkylation of cysteine 41, but not cysteine 200 decreases the ADP-ribosyltransferase activity of the S1 subunit of pertussis toxin. *J. Biol. Chem. 264*: 6386–6390.

34. Domenighini, M., C. Magagnoli, M. Pizza, & R. Rappuoli. 1994. Common features of the NAD-binding and catalytic site of ADP-ribosylating toxins. *Mol. Microbiol. 14*: 41–50.

35. Carroll, S. F., J. A. McCloskey, P. F. Crain, N. J. Oppenheimer, T. M. Marschner, & R. J. Collier. 1985. Photoaffinity labeling of diphtheria toxin fragment A with NAD: Structure of the photoproduct at position 148. *Proc. Natl. Acad. Sci. 82*: 7237–7241.

36. Tweten, R. K., J. T. Barbieri, & R. J. Collier. Diphtheria toxin. Effect of substituting aspartic acid for glutamic acid 148 on ADP-ribosyltransferase activity. *J. Biol. Chem. 260*: 10392–10394.

37. Sixma, T. K., S. E. Pronk, K. H. Kalk, E. S. Wartna, B A. M. van Zanten, B. Witholt, & W. G. J. Hol. 1991. Crystal structure of cholera toxin-related heat-labile enterotoxin from *E. coli. Nature 351*: 371–377.

38. Tsuji, T. T. Inoue, A. Miyama, K. Okamoto, T. Honda, & T. Miwatani. 1990. A single amino acid substitution in the A subunit of *Escherichia coli* enterotoxin results in loss of its toxic activity. *J. Biol. Chem. 265*: 22520–22525.

39. Lobet, Y., C. W. Cluff, & W. Cieplak, Jr. 1991. Effect of site-directed mutagenic alterations on ADP-ribosyltransferase activity of the A subunit of *Escherichia coli* heat-labile enterotoxin. *Infect. Immun. 59*: 2870–2879.

40. Carroll, S. F., & R. J. Collier. 1987. Active site of *Pseudomonas aeruginosa* exotoxin A. Glutamic acid 553 is photolabeled by NAD and shows functional homology with glutamic acid 148 of diphtheria toxin. *J. Biol. Chem. 262*: 8707–8711.

41. Douglas, C. M., & R. J. Collier. 1987. Exotoxin A of *Pseudomonas aeruginosa*: Substitution of glutamic acid-553 with aspartic acid drastically reduces toxicity and enzymic activity. *Infect. Immun 169*: 4967–4971.

42. Barbieri, J. T., L. M. Mende-Mueller, R. Rappuoli, & R. J. Collier. 1989. Photolabeling of Glu-129 of the S1 subunit of pertussis toxin with NAD. *Infect. Immun. 57*: 3549–3554.

43. Pizza, M., A. Bartoloni, A. Prugnola, S. Silvestri, & R. Rappuoli. 1988. Subunit S1 of pertussis toxin: Mapping of the regions essential for ADP-ribosyltransferase activity. *Proc. Natl. Acad. Sci. 85*: 7521–7525.

44. Takada, T., K. Iida, & J. Moss. 1995. Conservation of a common motif in enzymes catalyzing ADP-ribose transfer. *J. Biol. Chem. 270*: 541–544.

45. Howard, M. C., J. C. Grimaldi, J. F. Bazan, F. E. Lund, L. Santos-Argumedo, R. M. E. Parkhouse, T. F. Walseth, & H. C. Lee. 1993. Formation and hydrolysis of cyclic ADP-ribose catalyzed by lymphocyte antigen CD38. *Science 262*: 1056–1059.

46. Grimaldi, J. C., S. Balasubramanian, N. H. Kabra, A. Shanafelt, J. F., Bazan, G. Zurawski, & M. C. Howard. 1995. CD38-mediated ribosylation of proteins. *J. Immunol. 155*: 811–817.

MOLECULAR CLONING AND CHARACTERIZATION OF ARGININE-SPECIFIC ADP-RIBOSYLTRANSFERASES FROM CHICKEN BONE MARROW CELLS

Makoto Shimoyama, Mikako Tsuchiya, Nobumasa Hara, Kazuo Yamada, and Harumi Osago

Department of Biochemistry
Shimane Medical University
Izumo 693, Japan

ABSTRACT

Among a number of tissues and peripheral blood cells in chicken, leukocytes, bone marrow cells, liver and spleen showed high ADP-ribosyltransferase activity, with leukocytes having the highest. Density gradient centrifugation of the leukocytes revealed that the leukocyte ADP-ribosyltransferase originates in the polymorphonuclear cells, so called heterophils. Subcellular distribution of the cells showed the localization of the enzyme in the granule fraction. Based on the obtained amino acid sequences of arginine-specific ADP-ribosyltransferase purified from chicken peripheral heterophils, two arginine-specific ADP-ribosyltransferase cDNAs (designated AT1 and AT2) were obtained from chicken bone marrow cells. Each cDNA encodes a different peptide of 312 amino acid residues. Homology of the deduced amino acid sequences between AT1 and AT2 was 78.3%. Arginine-specific ADP-ribosyltransferase activity was detected in culture medium of COS 7 cells transiently transfected with AT1 cDNA, while activity from the cells transfected with AT2 cDNA was found in both culture medium and cell lysate. AT1 transferase required 2-mercaptoethanol (MSH) for the activity and in the presence of NaCl, the activity was inhibited while the AT2 enzyme was activated by either agent. Highly conserved regions were observed among the deduced amino acid sequences of AT1, AT2, chicken erythroblast and rabbit and human skeletal muscle ADP-ribosyltransferases, and rodent T-cell surface antigen RT6. Two forms of the transferase with much the same properties as AT1 and AT2 proteins, regarding the effect of NaCl and MSH, were detected in bone marrow cells. Based on these results it seems that AT1 and AT2 cDNAs encode the two forms of arginine-specific ADP-ribosyltransferase detected in chicken bone marrow cells.

ADP-Ribosylation in Animal Tissue, edited by Haag and Koch-Nolte
Plenum Press, New York, 1997

A

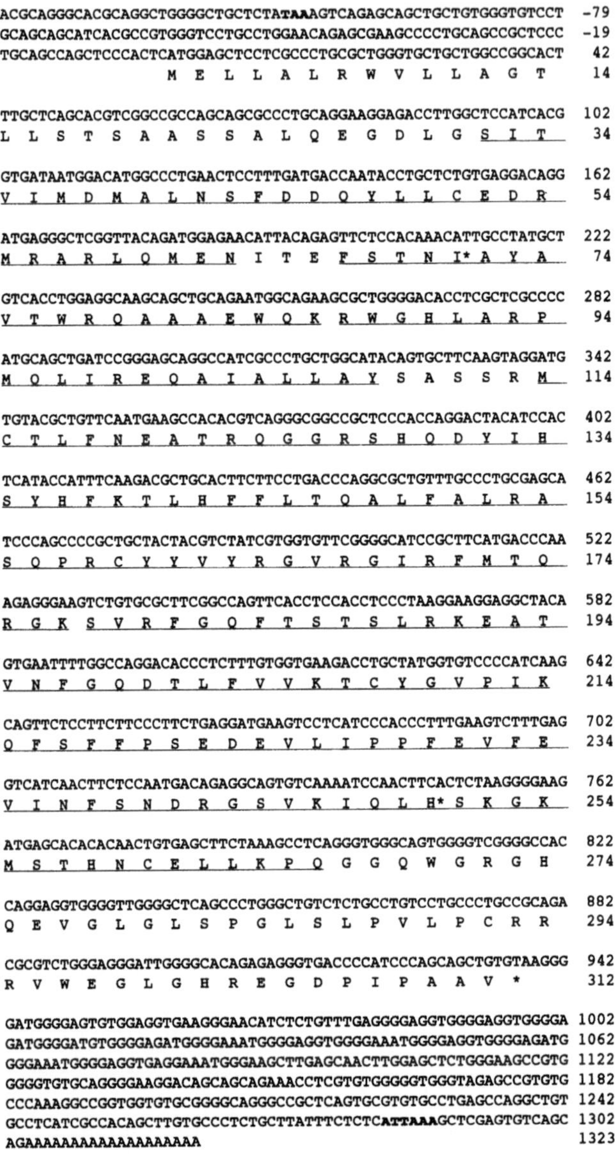

Figure 1. Nucleotide and deduced amino acid sequence of (A) AT1 and (B) AT2. Amino acid sequences are indicated below the nucleotide sequence. Nucleotides and amino acids are numbered from the beginning of the open reading frames. In-frame stop codons and polyadenylation signals are indicated in bold letters. Amino acid sequences obtained from the purified heterophil ADP-ribosyltransferase are underlined. I* and H* indicate the difference between the DNA sequence information and peptide sequence obtained from the purified protein (8).

INTRODUCTION

It has been proposed that ADP-ribosylation of proteins, resulting from the transfer of the ADPR moiety of NAD$^+$ to an amino acid in a protein acceptor, is critical to the regulation of cellular activities.

B

```
CAGGGCACGCAGGCTGGGGCTGCTCTATAAAGTCAGAGCAGCTGCTGTGGGTGTCCTGCA  -76
GCAGCATCACGCCGTGGGTCCTGCCTGGAACAGAGCGAAGCCCCTGCAGCCGCTCCCTGC  -16
AGCCAGCTCCCACTCATGGAGCTCCTCGCCCTGCGCTGGGTGCTGCTGGCCGGCACCTTG   45
                M   E   L   L   A   L   R   W   V   L   L   A   G   T   L    15

CTCAGCACGTCGGCCGCCAGCAGCGCCCTGCAGGAAGGAGACCTTGGCTCCATCACGGTG  105
L   S   T   S   A   A   S   S   A   L   Q   E   G   D   L   G   S   I   T   V   35

ATAATGGACATGGCCCCAAACTCCTTTGATGACCAATATGTGGGCTGCGCGCACGTGATG  165
I   M   D   M   A   P   N   S   F   D   D   Q   Y   V   G   C   A   H   V   M   55

TGGGCTAATCTACAGAAGCTGAAGTGTACTGAGTTCGCCAGAAATTATGCCTATGCTGTG  225
W   A   N   L   Q   K   L   K   C   T   E   F   A   R   N   Y   A   Y   A   V   75

GGTTGGAGGAAAGCAGCTGCAGAATGGCAGAAGCGCTGGGGATACCTCGCTCACCCCATG  285
G   W   R   K   A   A   A   E   W   Q   K   R   W   G   Y   L   A   H   P   M   95

CAGCTGAGACCAGAGCAGGCCATTGCCCTGCTGGCATACAGTGCTGCAAGCAACCTGTAC  345
Q   L   R   P   E   Q   A   I   A   L   L   A   Y   S   A   A   S   N   L   Y   115

CAGCAGTTCAATGCAGCCACACGTCAGGGTGGATGCTCCCACCAGTACTACGTCCACTTT  405
Q   Q   F   N   A   A   T   R   Q   G   G   C   S   H   Q   Y   Y   V   H   F   135

TACCACTTCAAGACGCTGCACTTCCTCCTGACCCAGGCGCTGTTTGCCCTGCGAGCATCC  465
Y   H   F   K   T   L   H   F   L   L   T   Q   A   L   F   A   L   R   A   S   155

CAGCCCCGCTGCTACTACGTCTATCGTGGTGTTCGGGGCATCCGCTTCATGACCCAAAGA  525
Q   P   R   C   Y   Y   V   Y   R   G   V   R   G   I   R   F   M   T   Q   R   175

GGGAAGTCTGTGCGCTTCGGCCAGTTCACCTCCACCTCCCTAAGGAAGGATGTCGCAGTG  585
G   K   S   V   R   F   G   Q   F   T   S   T   S   L   R   K   D   V   A   V   195

AATTTTGGCCAGGACACCTTCTTTGTGGTGAAGACCTGCTATGGTGTCCCCATCAAGCAG  645
N   F   G   Q   D   T   F   F   V   V   K   T   C   Y   G   V   P   I   K   Q   215

TTCTCCTTCTACCCTTCTGAGGATGAAGTCCTCATCCCACCCTTTGAAGTCTTTGAGGTC  705
F   S   F   Y   P   S   E   D   E   V   L   I   P   P   F   E   V   F   E   V   235

ACCAACTTCTGCACTGGCAACGGTAGAATTCAAATCTATCTCCGCTCTAAGGGGAAGATG  765
T   N   F   C   T   G   N   G   R   I   Q   I   Y   L   R   S   K   G   K   M   255

AGCAGACACAACTGTGAGCTTCTAAAGCCTCGGGGTGGGCAGTGGGGTCGGGGCCACCAG  825
S   R   H   N   C   E   L   L   K   P   R   G   G   Q   W   G   R   G   H   Q   275

GAGGTGGGGTTGGGGCTCAGCCCTGGGCTGGCTCTGCCCGTCCTGCCCTGCTCCAACTGC  885
E   V   G   L   G   L   S   P   G   L   A   L   P   V   L   P   C   S   N   C   295

AGCTGTTGGGGATCAGGGCACAGAGCAGGTGACCCCATCCCAGCAGCTGTGTAAGGGGAT  945
S   C   W   G   S   G   H   R   A   G   D   P   I   P   A   A   V   *       312

GGGGAGTGTGGAGGTGAAGGGAACATCTCTGTTTGTGGGGAGGTGGGGATGTGGGGAGGT 1005
GGGGAGATGGGGAGGTGGGGAGGTGGGGAAATGGGAAGCTTGAGCAACGTGGAGCTCTGT 1065
GAAGCTGTGGGGGTGTGCAGGGGAAGCGCAGCAGCAGAAACCTCGTGTGGGGGTGGGTAG 1125
AGCCGTGTGCCCAAAGGCCGGTGGTGTGCGGGGCAGGGCCGCTCAGTGCGTGTGCCTGAG 1185
CCAGGCTGTGCCTCATCGCCACAGCTTGTGCCCTCTGCTTATTTCTCTCATTAAAGCTCG 1245
AGTGTCAGCAGAAAAAAAAAAAAAAAAAAAAAA                            1277
```

Figure 1. (*Continued*).

Peterson et al (1) reported the purification and characterization of a 38-kDa membrane-associated arginine-specific ADP-ribosyltransferase from rabbit skeletal muscle. Recent works including molecular cloning revealed that the muscle enzyme is anchored in the membrane through a glycosylphosphatidylinositol tail (2, 3). In addition to the membrane transferase, soluble arginine-specific ADP-ribosyltransferases have been purified from chicken polymorphonuclear cell granules to homogeneity, and kinetic, physical and regulatory properties (4–7) and molecular cloning and characterization of the transferase described (8). To elucidate the physiological role of ADP-ribosylation, precise molecular knowledge of the transferase is notable.

RESULTS AND DISCUSSION

Tissue Distribution of Arginine-Specific ADP-Ribosyltransferase in the Chicken

Among a number of tissues and peripheral blood cells tested, liver, spleen, bone marrow cells, lung and peripheral leukocytes showed high arginine-specific ADP-ribosyltransferase activity, with leukocytes having the highest (Table I). The leukocytes were further fractionated by Ficoll centrifugation and it was observed that the leukocyte ADP-ribosyltransferase originates in the polymorphonuclear leukocyte, so called heterophils. To clarify the subcellular distribution of the enzyme in heterophils, isolated heterophils were suspended in 0.25 M sucrose and the preparation was pressurized with nitrogen gas. Nuclei and unbroken cells were pelleted by centrifugation of the cavitate and the postnuclear fraction was fractionated on Percoll gradients (9). The major peak of ADP-ribosyltransferase migrated with the granule enzymes, β-glucuronidase and lysozyme and there was an unequivocal dissociation from alkaline phosphatase and DNA (4). Thus heterophil ADP-ribosyltransferase was apparently localized in the granules. The transferase was purified to homogeneity through sodium chloride extract and CM-cellulose, gel filtration, Phenyl 5-PW and mono S column chromatographies. Purification of the transferase was achieved 219-fold with a yield of 14.4% (4). The enzyme preparation was apparently homogeneous, as determined by SDS-PAGE analysis. The molecular mass of heterophil transferase was 27.5 kDa (see Fig. 3).

The enzyme was electrophoresed and transferred to a PVDF membrane and the N-terminal sequence was determined. Further, the peptides digested with proteases were purified by reverse-phased HPLC and sequenced. By aligning the obtained sequences, 222 amino acid residues in six combined peptides were determined.

Table 1. Tissue distribution of ADP-ribosyltransferase in the chicken

Tissue	Specific activity (nmol/mg/h)
Leukocytes	372
Spleen	39.7
Bone marrow cells	28.7
Lung	10.5
Liver	4.83
Heart	1.53
Skeletal muscle	1.48
Erythrocytes	0.066

The values are means of two separate experiments. The enzyme assays were carried out with casein as the exogenous acceptor, and 0.1 mM benzamide, an inhibitor of poly(ADP-ribosyl)polymerase, was added to the reaction mixture (4).

Molecular Cloning and Characterization of Arginine-Specific ADP-Ribosyltransferases from Chicken Bone Marrow Cells

Based on the obtained amino acid sequences of heterophil ADP-ribosyltransferase, a pair of oligonucleotide primer was prepared. Using cDNA from chicken bone marrow cells, PCR was done with these primers and about a 500-bp fragment was obtained. Direct sequencing of the DNA fragment revealed that the deduced amino acid sequence encoded by the fragment contained peptide sequences obtained from the purified ADP-ribosyltransferase. This indicates that the fragment is part of ADP-ribosyltransferase cDNA.

Using the fragment as a probe, a cDNA library of chicken bone marrow cells was screened and two related but not identical clones, AT1 and AT2, were isolated. Both clones contained a possible stop codon, a polyadenylation signal, and poly(A) tail but lacked the portion encoding the N-terminal part, including the start methionine of the transferase. To obtain the lacking portion, 5'-RACEs were carried out, and the complete open reading frames for the two clones were obtained. The composite nucleotide sequences of a cDNA clone named AT1 has an open reading frame of 939 bp surrounded by 139 bp of a 5'-noncoding sequence, including an in-frame stop codon, and 387 bp of a 3'-noncoding sequence containing a poly(A) tail preceded by a polyadenylation signal AT-TAAA (Fig. 1). The open reading frame predicts a protein of 312 amino acids with a molecular mass of 35,318 Da. The deduced amino acid sequence includes all the peptide sequences derived from the heterophil transferase. The N-terminal sequence of the purified ADP-ribosyltransferase identified by Edman degradation starts from residue 32 of the deduced amino acid sequence of AT1. The deduced amino acid sequence of AT1 N-terminus having a hydrophobic sequence may serve as a leader sequence because amino acid 1 to 22 satisfies the rules for a signal sequence (10). This assumption is consistent with our observations that heterophil transferase is located in the cell granules (4) and that the purified transferase lacks the hydrophobic N-terminus. The other cDNA, AT2 has a similar structure to AT1 but the 3'-untranslated region contains 7 repeats of octanucleotide GGGGAXXT, whereas AT1 has thirteen repeats. Interestingly, between AT1 and AT2

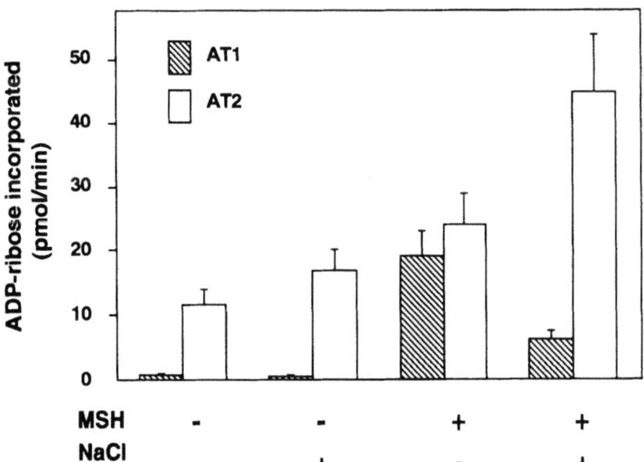

Figure 2. Effects of MSH and/or NaCl on the transferase activities in culture medium of AT1 and AT2 cDNA-transfected COS 7 cells. Concentrated culture medium of COS 7 cells transfected with pSVAT1 or pSVAT2 was incubated with [adenylate-^{32}P]NAD$^+$ and casein in the presence or absence of 5 mM MSH or 200 mM NaCl (8).

Table 2. ADP-ribosyltransferase activity of transfected COS 7 cells

	ADP-ribosyltransferase activity			
	Cell lysate		Medium	
Plasmid	nmol/min/plate (mg protein/plate)			
pSVsport	ND	(2.20 ± 0.35)	ND	(0.73 ± 0.13)
pSVAT1	0.08 ±0.01	(1.98 ±0.17)	3.51 ± 0.32	(0.68 ± 0.16)
pSVAT2	5.09 ±0.85	(2.58 ±0.86)	4.62 ± 0.82	(0.85 ± 0.15)

ADP-ribosyltransferase activities in cell lysates or culture medium of cDNA-transfected cells were assayed by measuring the formation of ADP-ribosylarginine. Data are means of values ± standard deviation from three plates in the same transfection experiment (8).

cDNAs, nucleotide homologies of the coding region were 89.4% while much higher homologies as 98 and 100% were observed in 5'- and 3'-uncoding sequences, respectively.

To determine if the isolated cDNAs encode catalytically active arginine-specific ADP-ribosyltransferases, the coding regions of AT1 and AT2 were subcloned in a mammalian expression vector pSVsport, and were transfected to COS 7 cells with a cationic lipid mixture. Three days after transfection, cells and culture medium were harvested and activity of the transferase was assayed by measuring the formation of ADP-ribosylarginine. The transferase activity was observed mainly in culture medium of pSVAT1 transfected cells while the lysate of pSVAT2-transfected cells also contained the activity (Table II). This activity was not detected in control cells transfected with pSVsport. Furthermore, α-anomer of ADP-ribosylarginine was the predominant form at the early phase of the reaction, which means that the enzyme catalyzes α-anomer formation (11) and AT2-enzyme also showed the same stereospecificity.

Chicken heterophil transferase requires a reducing agent such as mercaptoethanol (MSH) or dithiothreitol for the activity to transfer ADPR of NAD^+ to a model acceptor casein and the activity is inhibited by salts (4). Thus, the effects of these compounds on the AT1 and AT2 enzyme activities were tested, respectively. As expected, MSH was an absolute requirement for the activity of AT1 transferase and maximal activity was obtained with 5 mM MSH (Fig. 2). Further, the addition of 200 mM NaCl decreased the activity to 31% of the maximal activity. On the other hand, the AT2 enzyme did not require MSH,

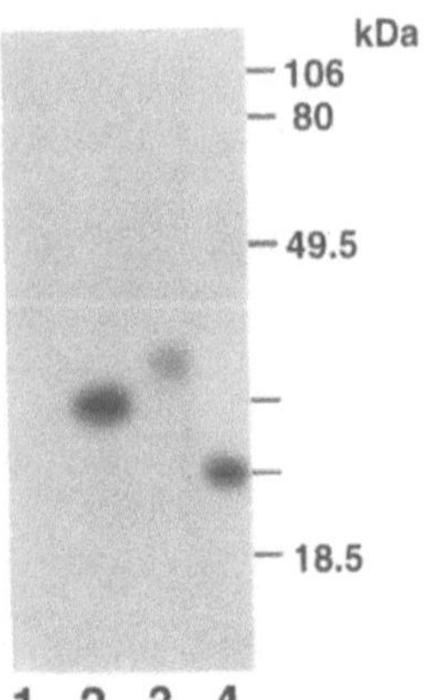

Figure 3. Detection of arginine-specific ADP-ribosyltransferase activity of cDNA-transfected COS 7 cells. Native heterophil transferase (120 ng) (lane 4) or concentrated culture medium of COS 7 cells transfected with pSVsport (100 μg) (lane 1), pSVAT1 (100 μg) (lane 2) or pSVAT2 (14 μg) (lane 3), were subjected to the zymographic *in situ* gel assay (8).

but the activity was increased about 2-fold with 5 mM MSH. The addition of 200 mM NaCl further stimulated the AT2 activity to 1.8-fold of that with MSH only.

To determine the size of arginine-specific ADP-ribosyltransferase expressed in AT1 or AT2 cDNA-transfected cells, the concentrated culture medium of the cells was fractionated by SDS/PAGE and a zymographic *in situ* gel assay was carried out. Heterophil enzyme was also subjected to gel assay. As shown in Figure 3, molecular masses of AT1 and AT2 transferases were 32 and 34 kDa, respectively, while heterophil transferase was 27.5 kDa. The molecular mass of recombinant AT1 protein was larger than that of the native transferase. We presume that COS 7 cells lack a specific peptidase to cleave the C-terminal peptide fragment.

To detect the mRNA of ADP-ribosyltransferase in several tissues and cells of chicken, total RNAs from chicken bone marrow cells, peripheral heterophils, spleen, and liver and human HL-60 cells were subjected to Northern blot analysis with ^{32}P-labeled AT1 cDNA coding region as a probe. A band corresponding 1.5-kb with chicken bone marrow cells but not with other preparations was detected.

The similarity of deduced amino acid sequences around active site glutamic acid of mammalian skeletal muscle ADP-ribosyltransferases was observed in chicken bone marrow cell and erythroblast ADP-ribosyltransferase and rodent T-cell antigen RT6 (2, 8, 12, 13). These sequences were particularly conserved in muscle and chicken ADP-ribosyltransferases. The rabbit and human sequences are EEEVLIPP and the chicken sequences are EDEVLIPP.

Evidence for Two Forms of ADP-Ribosyltransferase in Chicken Bone Marrow Cells

To determine if bone marrow cells contain the two enzymes corresponding to AT1 and AT2 proteins, the cells were homogenized and centrifuged. The supernatant was subjected to CM-cellulose and eluted with 0.5 M NaCl. CM-passed and -eluted fractions contained ADP-ribosyltransferase activity. As expected, these enzymes showed different properties regarding the effects of MSH and NaCl. CM-cellulose-eluted enzyme was inhibited by sodium chloride and activated by MSH while the CM-cellulose-passed enzyme was activated with either agent. In the presence of both reagents, the two enzymes showed a different activity, low and high as was seen in the case of recombinant AT1 and AT2 proteins (Fig. 3). From these results, we concluded that the CM-cellulose eluted enzyme has much the same properties as the heterophil enzyme and AT1 cDNA protein and that properties of the CM-cellulose-passed enzyme resembles that of the recombinant AT2 protein. Amino acid sequencing of the CM-cellulose-passed enzyme may show whether this enzyme is the same as the protein expressed with AT2 cDNA.

REFERENCES

1. Peterson, J. E., Jacquiline, S.-A. L. & Graves, D. J. 1990. Purification and partial characterization of arginine-specific ADP-ribosyltransferase from skeletal muscle microsomal membranes. *J. Biol. Chem.* 265: 17602–17609.
2. Zolkiewska, A., Nightingale, M. S. & Moss, J. 1992. Molecular characterization of NAD:arginine ADP-ribosyltransferase from rabbit skeletal muscle. *Proc. Natl. Acad. Sci. USA.* 89: 11352–11356.
3. Okazaki, I. J., Zolkiewska, A., Nightingale, M. S. & Moss, J. 1994. Immunological and structural conservation of mammalian skeletal muscle glycosylphosphatidylinositol-linked ADP-ribosyltransferase. *Biochemistry.* 33: 12828–12836.

4. Mishima, K., Terashima, M., Obara, S., Yamada, K., Imai, K. & Shimoyama, M. 1991. Arginine-specific ADP-ribosyltransferase and its acceptor protein p33 in chicken polymorphonuclear cells: co-localization in the cell granules, partial characterization, and *In Situ* mono(ADP-ribosyl)ation. *J. Biochem. 110*: 388–394.

5. Terashima, M., Mishima, K., Yamada, K., Tsuchiya, M., Wakutani, T. & Shimoyama, M. 1992. ADP-ribosylation of acins by arginine-specific ADP-ribosyltransferase purified from chicken heterophils. *Eur. J. Biochem. 204*: 305–311.

6. Terashima, M., Yamamori, C. & Shimoyama, M. 1995. ADP-ribosylation of Arg28 and Arg206 on the actin molecule by chicken arginine-specific ADP-ribosyltransferase. *Eur. J. Biochem. 231*: 242–249.

7. Yamada, K., Tsuchiya, M., Nishikori, Y. & Shimoyama, M. 1994. Automodification of arginine-specific ADP-ribosyltransferase purified from chicken peripheral heterophils and alteration of the transferase ativity. *Arch. Biochem. Biophys. 308*: 31–36.

8. Tsuchiya, M., Hara, N., Yamada, K., Osago, H. & Shimoyama, M. 1994. Cloning and expression of cDNA for arginine-specific ADP-ribosyltransferase from chicken bone marrow cells. *J. Biol. Chem. 269*: 27451–27457.

9. Borregaard, N., Heiple, J. M., Simons, E. R. & Clark, R. A. 1983. Subcellular localization of the b-cytochrome component of the human neutrophil microbicidal oxidase. *J. Cell Biol. 97*: 52–61.

10. von Heijne, G. 1985. Signal sequences. *J. Mol. biol. 184*: 99–105.

11. Tsuchiya, M., Tanigawa, Y., Mishima, K. & Shimoyama, M. 1986. Determination of ADP-ribosylarginine anomers by reverse-phase high-performance liquid chromatography. *Anal. Biochem. 157*: 381–384.

12. Davis, T. & Shall, S. 1995. Sequence of a chicken erythroblast mono(ADP-ribosyl)transferase-encoding gene and its upstream region. *Gene. 164*: 371–372.

13. Koch, F., Haag, F., Kashan, A. & Thiele, H.-G. 1990. Primary structure of rat RT6.2, a nonglycosylated phosphatidylinositol-linked surface marker of postthymic T cells. *Proc. Natl. Acad. Sci. USA. 87*: 964–967.

17

PURIFICATION, CHARACTERISATION, AND MOLECULAR CLONING OF A CHICKEN ERYTHROBLAST MONO(ADP-RIBOSYL)TRANSFERASE

Terence Davis, Jamal S. M. Sabir, Manoochehr Tavassoli, and Sydney Shall

Cell and Molecular Biology Laboratory
School of Biological Sciences
University of Sussex
Brighton, Sussex, BN1 9QG
England

ABSTRACT

We have purified an arginine-specific mono(ADP-ribosyl)transferase from chicken erythrocytes. The purified transferase was free from poly (ADP-ribose) polymerase activity. The molecular weight of the purified enzyme was estimated to be 27.5 kDa by gel filtration through Sephadex G-75 in a non-denaturing solvent. Activity gel experiments indicate that the active enzyme has an apparent molecular weight in SDS gels of about 28 kDa. The optimum pH of the reaction is about 8.0. The K_m value for NAD^+ of the purified enzyme is about 130 μM. Small molecular weight inhibitors of poly (ADP-ribose) polymerase have no significant effect on the mono ADP-ribosyl transferase enzyme activity. A number of inhibitors of the arginine-specific mono(ADP-ribosyl)transferase activity have been identified. Among the more effective inhibitors are 1,4 naphthoquinone, 5,8-dihydroxy-1,4-naphthoquinone, 4-amino-1-naphthol and 1,2-naphthoquinone.

We have also cloned a mono(ADP-ribosyl)transferase from chicken erythroblasts. This gene has been expressed in *E. coli* and ADP-ribosylation activity has been demonstrated using histones as substrate. The activity is shown to be arginine-specific by the use of poly-L-arginine as substrate. Use of a specific inhibitor has shown that this enzyme is indeed a mono(ADP-ribosyl)transferase and not a NAD glycohydrolase activity. The sequence of this gene is very similar to several other mono(ADP-ribosyl)transferase genes. There are thus at least three different chicken mono(ADP-ribosyl)transferase genes in the blood system alone; this suggests that there is a quite large family of mono(ADP-ribosyl)transferase genes in animals. We have also isolated the promoter region of this chicken gene and are able to identify several standard motifs in this promoter.

ADP-Ribosylation in Animal Tissue, edited by Haag and Koch-Nolte
Plenum Press, New York, 1997

1. INTRODUCTION

Mono-(ADP-ribosyl)ation is a post-translational modification of proteins involving the transfer of an ADP moiety from NAD^+ to specific amino acids in specific proteins by enzymes termed mono(ADP-ribosyl)transferases. This type of protein modification is the basis of a number of bacterial toxins such as the cholera, diphtheria, pertussis and some of the clostridial toxins [1,2]. In almost all these cases, the acceptor proteins are so-called G-proteins. The bacterial toxin genes have been cloned but they show rather low levels of sequence homology, except for a few amino acids at the active site [3]. In addition, it has been demonstrated that there are endogenous mono-(ADP-ribosyl) transfereases in all organisms examined, and that they seem to play a role in regulating cellular process [1,2]. The endogenous substrates are generally unknown, but are thought to include proteins involved in signal transduction. The first eukaryotic mono(ADP-ribosyl)transferase gene to be cloned was a rabbit skeletal muscle, arginine-specific enzyme [4].

We describe here the purification and characterisation of an arginine-specific mono(ADP-ribosyl)transferase from chicken erythrocytes. We also describe the molecular cloning and expression of a cDNA for mono(ADP-ribosyl)transferase from chicken erythroblasts. We also describe the cloning of the promoter region of this gene from chicken DNA. Finally, we compare these sequences with other sequences from related genes. This is part of a program to elucidate the physiological function of these apparently ubiquitous enzymes.

2. MATERIALS AND METHODS

i. *Isolation of cDNA sequence.* An AEV-transformed chicken erythroblast cDNA library in λgt10 was screened with a probe made from nucleotides 51 to 300 and 684 to 800 of the coding sequence of the rabbit mono(ADP-ribosyl)transferase sequence very generously supplied by Dr. Joel Moss (NIH, Bethesda, USA) [4]. Two identical clones were isolated, called λ411 and λ412.

ii. *Isolation of genomic sequence.* The 5' part of the clone λ411 was used to screen a chicken genomic library in λGem11 (Promega, Madison, WI, USA), and the clone λG114 was isolated.

iii. *DNA sequencing.* The cDNA insert from the λ411 clone (clone p411) was sequenced in an Applied Biosciences Model 373A automatic sequencer. In addition, a 1.8 Kb *Sst* I fragment (clone p1812) of the genomic clone was also sequenced. Both strands were sequenced in both directions.

iv. *Sub-cloning the cDNA into an expression vector.* The chicken cDNA containing the 3' 286 amino acids of the chicken erythroblast gene was cut with the enzymes *Nsi* I and *Eco* RI and ligated into linear pUC 9 which had been cut by *Pst* I and *Eco* RI. The resulting plasmid called pNE1 has a promoter which is inducible by IPTG, and a coding sequence consisting of the first 9 amino acids of the *E. coli* β-galactosidase gene fused to 279 amino acids of the chicken mono(ADP-ribosyl)transferase gene.

v. *Expression of the chicken cDNA in E.coli.* Plasmids pNE1 and pUC9 as a control, were transfected into *E. coli* strain JM109 and liquid cultures in YT medium were grown at 37°C until an A_{600} of 1.0 was achieved. One ml of this culture was diluted with 4.0 ml of YT medium containing 40 mM IPTG and the culture was incubated at 37°C and samples were removed at 1 hr, 2 hr and at 4 hr for enzyme assay.

vi. *Assay for mono(ADP-ribosyl)transferase enzyme activity.* The bacterial samples were microfuged and the pellets were resuspended in 200 µl of 50 mM Tris-HCl, pH 8.0, containing 5 mg/ml lysozyme and a mix of several protein inhibitors. This was incubated at 4°C for 5 min. Then 25 µl of 250 mM EDTA, pH 8.0, was added and the sample was incubated at 37°C for 15 min. After microcentrifugation, the supernatants were used for enzyme assay.

The enzyme extract (50 µl) was added to 50 µl of a solution containing;100 mM potassium phosphate buffer, pH 7.0, 600 µg/ml of histone F2b (Sigma), 20 µM NAD^+ and 200 µCi/ml [adenylate-^{32}P] NAD (500 Ci/mMol). The mixture was incubated for 30 min at 30°C. the reactions were stopped by spotting the samples onto 3MM filter papers and immersing them in 10% (w/v) TCA for 10 min. The filters were washed 3 times in 10% (w/v) TCA, followed by once in acetone for 10 minutes. The filters were air-dried and were placed in scintillation vials in 10 ml of water and counted for 10 min in a Cerenkoc Counter. For comparison, four sets of assays were done; 1. control pUC9, 2. pNE1 - uninduced cells, 3. pNE1 - induced cells, 4. pNE1 - induced cells in the presence of the transferase inhibitor 200 µM 1,4-naphthaquinoline. The assays in this case were incubated for three hr. In some reactions, 150 µg/ml of poly-L-arginine was used as substrate instead of histone.

3. RESULTS

A mono(ADP-ribosyl)transferase activity was purified from chicken erythrocytes in a 7 step procedure. In all the chromatography steps only one peak of enzyme activity able to mono-ADP-ribosylate poly-L-arginine was observed, except in the very last step of the purification. A mono Q column chromatography separated three peaks of enzyme activity. This may compare with the four different arginine-specific enzyme activities which have been identified in turkey erythrocytes [16–18].

The optimum pH of the purified protein was determined to be around pH 8.0. There is a fairly sharp pH optimum, with an asymmetric, not bell-shaped, pH dependence. The estimated K_m of the enzyme was 130 ± 3.22 µM NAD^+. This K_m value is much higher than the 7 µM found with the turkey erythrocyte enzyme [19], or than the 36 uM for the turkey erythrocyte transferase B [16,17] or the 15 µM for the turkey erythrocyte transferase C [16], but only slightly higher than the 70 µM of the hen liver nuclear enzyme [20].

Activity gel analysis revealed an enzyme activity against histones and poly-L-arginine that migrated with a molecular weight of about 29 kDa. Analysis of the product of the enzyme reaction revealed that snake venom phosphodiesterase produced only AMP, and therefore we conclude that no contamination with poly (ADP-ribose) polymerase was present.

In a search for enzyme inhibitors, we discovered that 1,4-naphthoquinone has an apparent K_i of 4.9 µM. Three other inhibitors with K_i values in the region of 30–50 µM are 5,8-dihydroxy-1,4-naphthoquinone, 4-amino-1-naphthol and 1,2 naphthoquinone.

Using the rabbit skeletal muscle, mono(ADP-ribosyl)transferase gene, very generously given to us by Joel Moss (NIH, Bethesda, USA), we isolated an homologous sequence from a chicken erythroblast cDNA library. Then we used this cDNA sequence to isolate a genomic DNA sequence which contains both the cDNA sequence and the putative promoter sequence of this gene. The combined DNA sequences are shown in Figure 1.

The genomic sequence has a 300 amino acid open reading frame coding for a putative protein of 35.5 kDa; followed by a stop codon and a polyadenylation signal. The intron shown in the figure has the consensus splice acceptor and donor sites. In order to

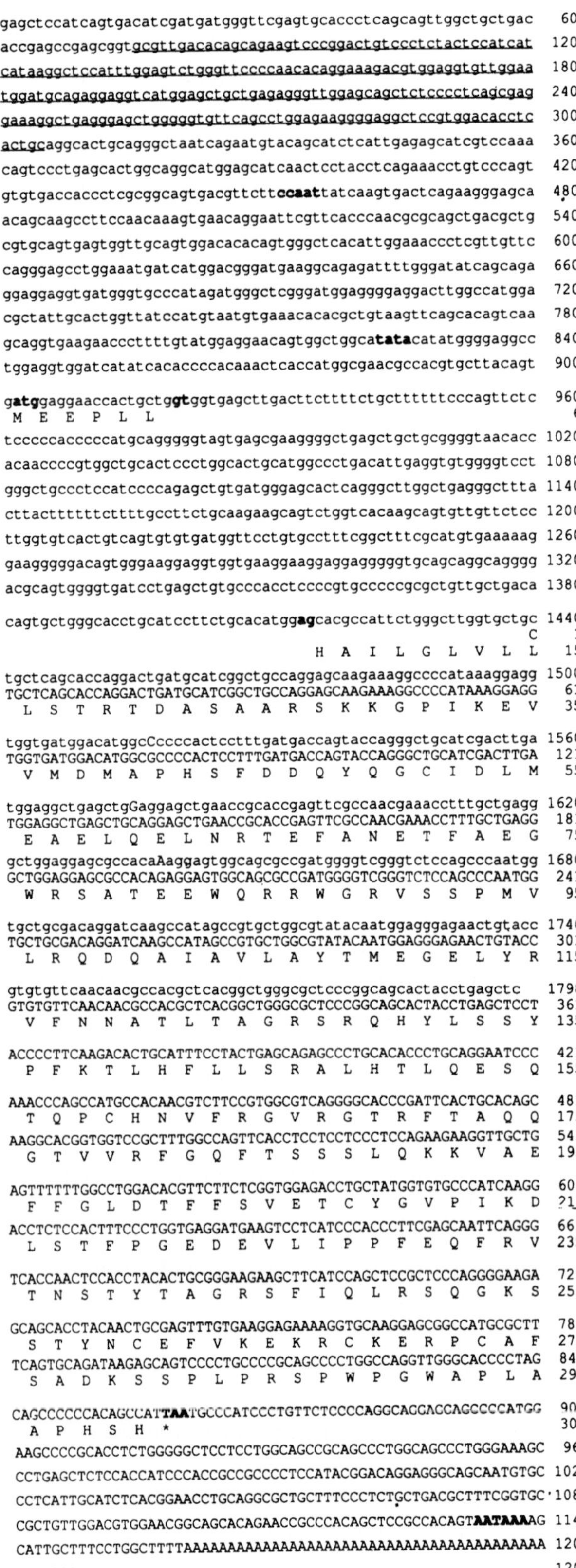

```
gagctccatcagtgacatcgatgatgggttcgagtgcaccctcagcagttggctgctgac            60
accgagccgagcggtgcgttgacacagcagaagtcccggactgtccctctactccatcat           120
cataaggctccatttggagtctgggttccccaacacaggaaagacgtggaggtgttggaa           180
tggatgcagaggaggtcatggagctgctgagagggttggagcagctctcccctcagcgag           240
gaaaggctgagggagctggggggtgttcagcctggagaagggggaggctccgtggacacctc         300
actgcaggcactgcagggctaatcagaatgtacagcatctcattgagagcatcgtccaaa           360
cagtccctgagcactggcaggcatggagcatcaactcctacctcagaaacctgtcccagt           420
gtgtgaccaccctcgcggcagtgacgttcttccaattatcaagtgactcagaagggagca           480
acagcaagccttccaacaaagtgaacaggaattcgttcacccaacgcgcagctgacgctg           540
cgtgcagtgagtggttgcagtggacacacagtgggctcacattggaaaccctcgttgttc           600
cagggagcctggaaatgatcatggacgggatgaaggcagagattttgggatatcagcaga          660
ggaggaggtgatgggtgcccatagatgggctcgggatggagggg aggacttggccatgga          720
cgctattgcactggttatccatgtaatgtgaaacacgctgtaagttcagcacagtcaa            780
gcaggtgaagaacccttttgtatggaggaacagtggctggcatatacatatggggaggcc          840
tggaggtggatcatatcacaccccacaaactcaccatggcgaacgccacgtgcttacagt          900

gatggaggaaccactgctggtggtgagcttgacttctttctgctttttcccagttctc            960
 M   E   E   P   L   L                                                   6
tcccccacccccatgcaggggggtagtgagcgaagggg ctgagctgctgcggggtaacacc       1020
acaacccgtggctgcactccctggcactgcatggccctgacattgaggtgtggggtcct          1080
gggctgccctccatccccagagctgtgatgggagcactcagggcttggctgagggcttta         1140
cttacttttttcttttgccttctgcaagaagcagtctggtcacaagcagtgttgttctcc         1200
ttggtgtcactgtcagtgtgtgatggttcctgtgcctttcggctttcgcatgtgaaaaag         1260
gaaggggggacagtgggaaggaggtggtgaaggaaggaggaggggggtgcagcaggcagggg        1320
acgcagtggggtgatcctgagctgtgccccacctccccgtgcccccgcgctgttgctgaca        1380

cagtgctgggcacctgcatccttctgcacatggagcacgccattctgggcttggtgctgc        1440
                                            C                         1
                          H   A   I   L   G   L   V   L   L            15
```

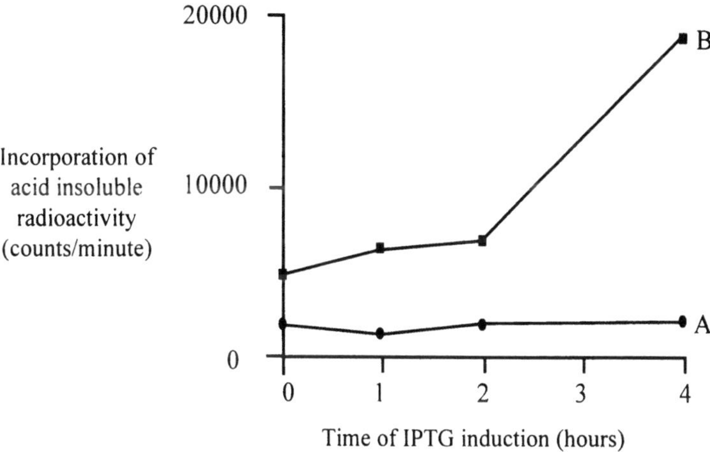

Figure 2. Induction of mono(ADP-ribosyl)transferase activity with IPTG. The bacteria were incubated with 40 mM IPTG for the indicated times and samples were taken for assay of mono(ADP-ribosyl)transferase activity in a 30 min. assay. (A) = control culture of pUC9 cells; (B) = bacterial cells containing the plasmid pNE1.

prove that this sequence does indeed represent a mono(ADP-ribosyl)transferase gene, we decided to clone this gene into a bacterial expression vector and determine whether the protein which was produced showed mono(ADP-ribosyl)transferase enzyme activity.

The induction of enzyme activity in pNE1 which contains a plasmid with the cDNA of the putative enzyme inserted, (Fig. 2) increases rapidly after a lag period of 2 to 3 hr, compared with with pUC9 control. The higher activity found with pNE1 at zero time of induction, compared to the control, presumably reflects the basal level of expression from the pUC9 promoter. The level of enzyme activity with the pUC9 extract is not zero; this may be explained by the presence of a mono(ADP-ribosyl)transferase enzyme activity, which has been reported to be present in *E. coli* [5].

The mono(ADP-ribosyl)transferase enzyme activity in the bacterial extracts increases with time of assay (Fig. 3). The apparent enzyme activity in extracts from the un-induced pNE1 and pUC9 bacteria, also increase with time, consistent with the conclusions reached from the data shown in Fig.2. Adding a specific mono(ADP-ribosyl)transferase enzyme inhibitor results is a decrease in the apparent enzyme activity.

The enzyme activity which we demonstrate seems to be arginine-specific because when poly-L-arginine is used as a substrate instead of histone, there is a much larger incorporation of radio-activity into acid-insoluble material (Fig. 4).

In addition to the coding region of this protein, we have sequenced 900 bp upstream from the Met start codon into the promoter region (Fig. 1). This sequence has paired CCAAT (nt 452) and TATA (nt(819) sequence motifs. There is also a sequence at nucleotides 76 to 305 which shows about 80% identity to a chicken repeat element called CR-1. The function of these repeat elements is unknown, but they have been shown to occur frequently at the 5' ends of many genes in the chicken genome [6]. Four other sequences at nucleotides 210, 256,697 and 824 resemble Sp-1 binding motifs found in the 5' region of the gene encoding the rat poly (ADP-ribose) polymerase gene [7] and the mouse *p12* gene [8]. Two of these four sequences are in the region that shows homology to the chicken CR-1 repeats. There are also two possible AP-1 sites at nucleotides 463 and 616; the first is identical to the AP-1 consensus sequence, while the second has two mis-matches [9].

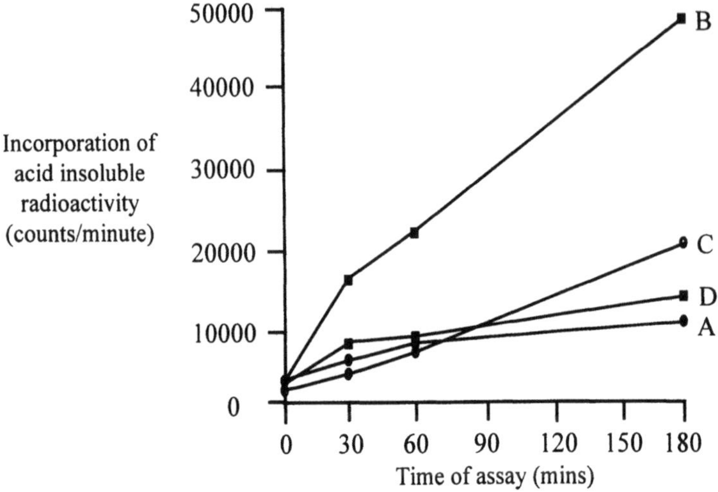

Figure 3. Mono(ADP-ribosyl)transferase enzyme activity in bacterial extracts. (A) = control culture of bacteria carrying only pUC9; (B) = induced bacteria carrying the plasmid pNE1, which is a chicken cDNA gene for an arginine-specific enzyme [expressing the cDNA]; (C) = uninduced pNE1 bacteria; (D) = induced pNE1 bacteria in the presence of 20 μM 1,4 naphthoquinoline.

4. DISCUSSION

Using the rabbit, skeletal muscle cDNA as a probe, we have isolated an homologous sequence from a chicken erythroblast library. The chicken sequence is highly homologous to the rabbit sequence (47% amino acid identity). This gene was then expressed in *E. coli* and the anticipated enzyme activity, with histone as substrate was demonstrated. Consequently, it is clear that we have indeed isolated the the chicken erythroblast homologue of

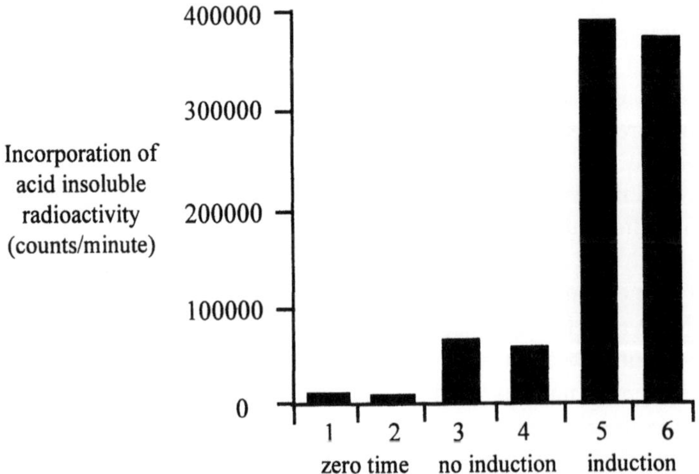

Figure 4. Enzyme assay with poly-arginine as the substrate. Each experiment was done twice. Columns 1,2 are zero time with cells containing the expression plasmid pNE1 and induced with IPTG. Columns 3,4 are uninduced cells containing pNE1. Columns 5,6 are cells induced with IPTG and containing the expression plasmid pNE1. The assay was for 1 hour with 100 μg/ml of poly-L-arginine as substrate.

Table 1. Percentage amino acid identity between members of the
mono(ADP-ribosyl)transferase family

SAMPLE[a]	CHAT1	CHAT2	RMAT	HMAT	RNRT6	MMRT6
CEAT	49.5	52.0	47.2	47.0	35.6	32.7
CHAT1		80.1	41.9	43.1	35.0	31.7
CHAT2			44.0	45.1	35.0	34.5
RMAT				81.1	38.2	38.4
HMAT					38.2	38.4
RNRT6						71.0

[a]CEAT = chicken erythroblast, CHAT1 = chicken heterophil 1, CHAT2 = chicken
heterophil 2, RMAT = rabbit skeletal muscle, HMAT = human skeletal muscle,
RNRT6 = rat lymphocyte, MMRT6 = mouse lymphocyte.

the rabbit skeletal muscle gene. The activity was shown to be arginine-specific by using
poly-arginine as a substrate. Use of a specific inhibitor shows that the activity is truly that
of a mono-(ADP-ribosyl)-transferase and not due to related NAD-glycohydrolase activity.

The gene we have cloned encodes a protein that is highly homologous (approximately 50% identity) to six other mono(ADP-ribosyl)transferases; namely two genes from
chicken heterophils [10], one from human [11] and rabbit [4] skeletal muscle and one each
from rat and mouse T cells [12–14]. The degree of amino acid identity between all these
genes is shown in Table 1. The extent of amino acid identity ranges from 30% to 80%.
There seems to be more homology for the same gene between species, than there is within
a single species. This report is the first time that different mono(ADP-ribosyl)transferases
have been cloned from different tissues of the same species. This indicates that there is a
very large family of mono(ADP-ribosyl)transferases.

There is considerable homology spread throughout the length of these proteins (Fig.
2). In particular, there is almost perfect conservation around the active site. Indeed, comparison of this region not only among the mono(ADP-ribosyl)transferases, but including
also the bacterial and phage enzymes and the poly (ADP-ribose) polymerase enzymes reveals some very striking conservations (Fig. 3). The active site glutamate is universally
conserved. There are almost always an hydrophobic amino acid on the carboxyl side of
the active site glutamate. The two amino acids on the amino side of the glutamate are
characteristic of the specific group in the classification; the polymerizing enzymes all
have a tyrosine and asparagine or aspartate in these postions; whereas the mono enzymes
mostly have glutamine or glutamate.

The termini of the proteins show significantly less homology. The skeletal muscle
and the T cell enzymes are glycosylphosphatidylinositol (GPI)-anchored proteins on the
external surface of the cells, and have hydrophobic termini characteristic of these proteins,
whereas the chicken protein is a cytoplasmic red cell enzyme [11]. A major difference between the skeletal enzyme and the red cell enzyme, is that the amino acids required for the
GPI anchor are absent in the red cell enzyme. However, a Kyte-Doolittle hydropathy plot
shows that the central sequences of the two proteins are remarkably similar (not shown),
which suggests that the 3D structures might be very similar.

Two other proteins that have extensive homology are the RT6 allo-antigens (Fig.
1–3). These are GPI-anchored proteins in rat and mouse T cells [10]. They show 32%
identity/43% similarity with the rabbit and chicken proteins; the active site is 100% con-

Table 2. Comparison of the active site regions of the mono (ADP-ribosyl) transferases, bacterial toxins and poly (ADP-ribose) polymerases

MONO		ACTIVE SITE REGION				
HUMAN	E	E	E	V	L	I
RABBIT	E	E	E	V	L	I
CHICKEN HET1	E	D	E	V	L	I
CHICKEN HET2	E	D	E	V	L	I
CHICKEN ERYTH	E	D	E	V	L	I
RAT RT6.1	Q	E	E	V	L	I
RAT RT6.2	Q	E	E	V	L	I
MOUSE RT6	E	E	E	V	L	I

TOXINS						
C.BOT 3N	Q	L	E	V	L	L
C.BOT 3P	Q	L	E	M	L	L
C.LIM C3	Q	L	E	V	L	L
S.AUREUS	Q	Q	E	V	L	L
B.CEREUS	Q	Y	E	L	L	L
P.AERU ETA	R	L	E	T	I	L
E.COLI LT	E	Q	E	V	S	A
V.CHOL CT	E	Q	E	V	S	A
V.FISH HALOVIBRIN	Q	N	E	L	R	L
B.PERT	Q	S	E	Y	L	A
C.DIPH	S	V	E	Y	I	N
A.LIP DRAT	E	G	E	Y	L	V
R.RUB DRAT	E	G	E	Y	L	V
T4 PHAGE	E	M	E	V	I	L

POLY						
HUMAN	Y	N	E	Y	I	V
RAT	Y	N	E	Y	I	V
MOUSE	Y	N	E	Y	I	V
BOVINE	Y	N	E	Y	I	V
CHICKEN	Y	N	E	Y	I	V
XENOPUS	Y	N	E	Y	I	V
DROS	Y	N	E	Y	I	V
SPR	Y	N	E	F	I	I
THAL	Y	N	E	Y	I	V
CEL	Y	D	E	Y	V	M
CONCENSUS	E	Ac	E	Hy	Hy	Hy
	Q	Am		Y		A
	Y	Hy				

(Ac = acid, Am = amide, Hy = hydrophobic)

These regions are the binding sites for the NAD molecule. Mutation of the 3rd amino acid E reduces transferase activity by 200-1000 fold, but does not reduce the k_m for NAD or the rate of glycohydrolase activity. Therefore, this E is essential for the actual transfer of ADP-ribose to the target amino acid. Presumably the ADP-ribose covalently attaches to this E.

served and the hydropathy plots are similar to the transferases (data not shown). Very recently, these proteins have been clearly shown to be mono(ADP-ribosyl)transferase enzymes as well [21]. Finally, there are 4 cysteines which are present in all 8 proteins, which adds support to the proposal that the 3D structures of these proteins might be very similar.

In addition to the cDNA, we also cloned and sequenced 900 bp upstream from the Met start codon of the promoter region of the chicken erythroblast enzyme. This promoter has the features of an household gene.

5. ACKNOWLEDGMENTS

We would like especially to thank Dr. Joel Moss (NIH, Bethesda, MD, USA) for very generously providing his clone of the skeletal muscle mono(ADP-ribosyl)transferase gene without which this work would not have been possible. This work was supported by the British Cancer Research Campaign.

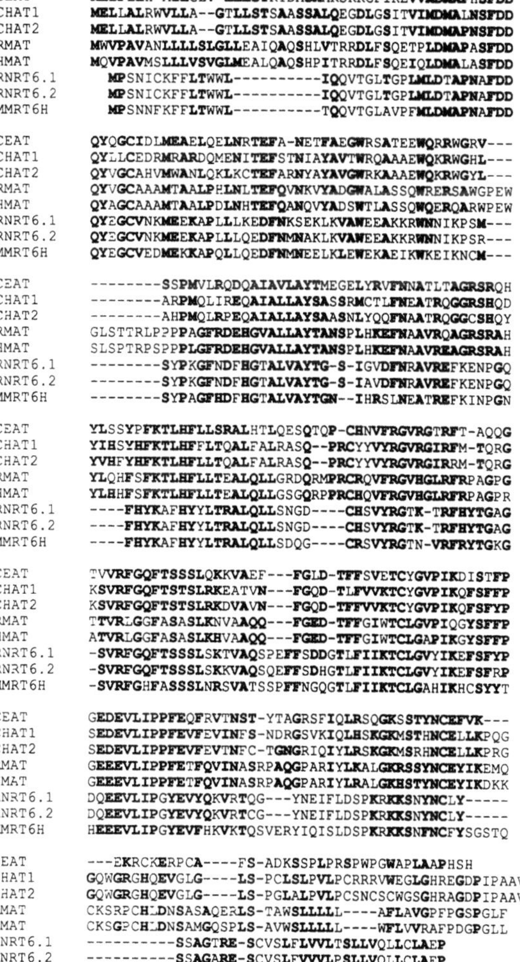

Figure 5. Comparison of the protein sequences of members of the mono(ADP-ribosyl)transferase family. Amino acids conserved or identical in at least a majority of the sequences are in bold type. At least four sequences have to be common before they are included. The rules for conservation that were used are; A-G, L-I=V-M, R=K=H, D=E, N=Q, F=Y=W, S=T, C; CEAT = chicken erythroblast, CHAT1, CHAT2 = chicken heterophil enzymes 1 and 2, RMAT = rabbit skeletal muscle, HMAT = human skeletal muscle, RNRT6.1, RNRT6.2 = rat lymphocyte (two alleles of the same gene), MMRT6H = mouse lymphocyte.

6. REFERENCES

1. Lowery, R. G. and Ludden, P. W. 1990, Endogenous ADP ribosylation in procaryotes. In *ADP-ribosylating toxins and G- proteins: insights into signal transduction.* (Moss, J. & Vaughan, M. eds.) Amer. Soc Microbiol., Washington, pp 458–477.
2. Williamson, K. C. and Moss, J. 1990. Mono(ADP-ribosyl)transferases and ADP-ribosyl arginine hydrolases: a mono-ADP-ribosylation cycle in animal cells. In *ADP-ribosylating toxins and G- proteins: insights into signal transduction.* (Moss, J. & Vaughan, M. eds.) Amer. Soc Microbiol., Washington, pp 493–510.
3. Takada, T., Iida, K. and Moss, J. 1994. Expression of NAD glycohydrolase activity by rat mammary adenocarcinoma cells transformed with T cell allo-antigen RT6.2. *J. Biol. Chem.* **269:** 9420- 9423.
4. Zolkiewska, A., Nightingale, M. S. and Moss, J. 1992. Molecular characterization of NAD:arginine ADP-ribosyl transferase from rabbit skeletal muscle. *Proc. Acad. Sci. USA* **89:** 11352–11356.
5. Yamamoto, T., Tamura, T. and Yokota, T. 1984 Primary structure of heat labile enterotoxin produced by *Escherichia coli* pathogenic for humans. J. Biol. Chem. **259,** 5037- 5044.
6. Stumph, W. E., Hodgson, C. P., Tsai, M-J. and O'Malley, B. W. 1984.Genomic structure and possible retroviral origin of the chicken CR1 repetitive DNA sequence family. *Proc. Natl. Acad. Sci. USA.* **81:** 6667- 6671.
7. Potvin, F., Roy, R.J., Poirier, G.G., and Guerin, S. L. 1993. The US-1 element from the gene encoding rat poly (ADP-ribose) polymerase binds the transcription factor Sp1. *Eur. J. Biochem.* **215:** 73–80.

8. Guerin, S. L., Pothier, F., Robidoux, S., Gosselin, P. and Parker, M. G. 1990. Identification of a DNA binding site for the transcription factor-GC2 in the promoter region of the p12-gene and repression of its positive activity by upstream regulatory activities. J. Biol. Chem. **265:** 22035- 22043.

9. Lee, W., Mitchell, P.J., and Tjian, R. (1987) Purified transcription factor AP-1 interacts with TPA-inducible enhancer elements. Cell **49**, 741–752.

10. Tsuchiya, M., Hara, N., Yamada, K., Asago, H. and Shimoyama, M. 1994. Cloning and expression of cDNA for arginine-specific (ADP-ribosyl) transferase from chicken bone marrow cells. *J. Biol. Chem.. 269:* 27451- 27457.

11. Okazaki, I. J., Zolkiewska, A., Nightingale, M. S. and Moss, J. 1994. Immunological and structural conservation of mammalian skeletal muscle glycosylphosphatidylinositol-linked ADP-ribosyl transferases. *Biochemistry 33:* 12828–12836.

12. Haag, F., Koch, F. and Thiele H-G. 1990. Nucleotide and deduced amino acid sequence of the rat T cell alloantigen RT6.1. Nucleic Acids Res. **18,** 1047.

13. Koch, F., Haag, F. and Thiele, H-G. 1990. Nucleotide and deduced amino acid sequence of the mouse homologue of the rat T cell differentiation marker RT6. Nucleic Acids Res. **18,** 3686.

14. Koch, F., Haag, F., Kashan, A. and Thiele, H-G. 1990. Primary structure of rat RT6.2, a nonglycosylated, phosphatidylinositol-linked surface marker of postthymic T cells. Proc. Natl. Acad. Sci. USA **87,** 964–967.

15. Sabir, J., Tavassoli, M. and Shall, S. 1992. in ADP-ribosylation reactions (Poirier, G. P. & Moreau, P. eds.) Springer-Verlag, New-York. pp 397- 401.

16. Moss, J., Stanley, S. J. and Watkins, P. A. 1980. Isolation and properties of an NAD- and guanidine-dependent ADP-ribosyl transferase from turkey erythrocytes. *J. Biol. Chem.* **255:** 5838–5840.

17. West, R. E. and Moss, J. 1986. Amino acid-specific ADP-ribosylation: Specific NAD:arginine mono-ADP-ribosyltransferase associated with turkey erythrocyte nuclei and plasma membranes. Biochemistry **25,** 8057–8062.

18. Yost, D. A. and Moss, J. 1983. Amino acid-specific ADP-ribosylation. Evidence for two distinct NAD:Arginine ADP-ribosyl transferases in turkey erythrocytes. *J. Biol. Chem. 258:* 4926–4929.

19. Osborne, J. C., Stanley, S. J. and Moss, J. 1985. Kinetic mechanism of two NAD:arginine ADP-ribosyltransferases: The soluble salt-stimulated transferase from turkey-erythrocytes and choleragin, a toxin from *Vibrio cholera.* Biochemistry. **24,** 5235- 5240.

20. Tanigawa, Y., Tsuchiya, M., Imai, Y. and Shimoyama, M. 1984. (ADP- ribosyl) transferase from hen liver nuclei. *J. Biol. Chem.. 259;* 2022–2029.

21. Koch-Nolte, F., Petersen, D., Balasubramanian, S., Haag, F., Kahlke, D., Willeer, T., Kastelein, R., Bazan, F. and Thiele, H-G. 1996. Mouse T cell membrane proteins Rt6–1 and Rt6–2 are arginine/protein mono (ADP- ribosyl) transferases and share secondary structure motifs with ADP- ribosylating bacterial toxins. *J. Biol. Chem. 271:* 7686–7693.

18

MOLECULAR CHARACTERISATION OF A FUNGAL MONO(ADP-RIBOSYL)TRANSFERASE

Martha Deveze-Alvarez, Jesús García-Soto, and Guadalupe Martínez-Cadena

Instituto de Investigación en Biología Experimental
Facultad de Química
Universidad de Guanajuato, Apdo. postal 187
Guanajuato, Gto., 36000 México

ABSTRACT

A soluble arginine-specific mono(ADP-ribosyl)transferase was detected in dormant spores of *Phycomyces blakesleeanus*. Soluble proteins incubated with [^{32}P]NAD revealed, after a two dimensional electrophoretic separation, three major ADP-ribosylated substrates with molecular weights of 38, 37, and 36 kDa and pI values of 6.9, 8.1 and 4.6, respectively. The addition of $MgCl_2$ stimulated the (ADP-ribosyl)transferase activity. This enzymatic activity was stimulated by 250 µM NO-releasing agent sodium nitroprusside and inhibited with 8 mM benzamide, 0.4 mM *meta*-IodoBenzylGuanidine (MIBG), and 0.5 mM novobiocin. The three ADP-ribosylation inhibitors affected the germination of *Phycomyces* spores. The concentrations necessary to inhibit 50% of the spore germination of *Phycomyces* were 0.05 mM, 0.2 mM, and 8 mM for novobiocin, MIBG, and benzamide, respectively. All the above inhibitors affected the germination process to the same extent, that is, they inhibited the tube protuberation, leaving the spores as swollen cells. These data suggest that ADP-ribosylation may be involved in the germination process of *Phycomyces*, particularly in germ-tube formation.

INTRODUCTION

Mono-ADP-ribosylation is a post-translational modification of proteins by the enzymatic transfer of ADP-ribose from NAD resulting in alteration of the functional properties of the respective proteins. Both, soluble and particulate mono(ADP-ribosyl)transferases have been identified in different eukaryotic cells and tissues. They have been purified from turkey red blood cells, hen and frog liver, and rabbit skeletal muscle. All of these enzymes catalyze the transfer of ADP-ribose from NAD to arginine [1]. Also, a cysteine-specific (ADP-ribosyl)transferase has been purified from human erythrocytes [2].

ADP-Ribosylation in Animal Tissue, edited by Haag and Koch-Nolte
Plenum Press, New York, 1997

Phycomyces blakesleeanus is a zygomycete fungus whose spores present an endogenous dormancy, that is, they need an activation treatment to germinate. The germination process involves a morphological transition from elliptical spore to a spherical cell with subsequent tube formation and development into mycelial cells in a suitable culture medium. Immediately after the activation treatment, the cytoplasmic cAMP levels are transiently elevated, suggesting that this second messenger might be the trigger of spore germination [for review, see 3]. Mutants that do not germinate show no increase in cAMP [4, 5], strongly supporting the hypothesis that this nucleotide is a critical agent in breaking *Phycomyces* spore dormancy. Other rapid metabolic changes include the increase in the concentration of fructose 2,6-bisphosphate, and in the enzymatic activities of trehalase and glycerol-3-phosphatase. These enzymatic activities are regulated by phosphorylation, probably through a cAMP dependent-protein kinase [3, 6, 7].

We are interested in the elucidation of the biochemical events underlying spore germination in the zygomycete fungus *Phycomyces blakesleeanus*. In this study we characterised the mono(ADP-ribosyl)transferase(s) from dormant spores of *Phycomyces*.

METHODS

Strains

The wild type strain of *P. blakesleeanus* NRRL 1555 (-) was used throughout this work.

Media and Growth Conditions

The fungus strain was maintained and propagated in YPG medium [8]. Sporangiospores were produced on YPG solidified with 2% (w/v) agar from cultures incubated at 24°C for 6 days under diffused light.

Preparation of Cell-Free Extracts

Cell-free extracts from dormant spores were obtained basically as described by Carrillo-Rayas *et al.*[9].

Assay of ADP-Ribosylation

ADP-ribosylation was carried out in a reaction mixture (40 µl) composed of: 40 mM MES (pH 6.0), 1–10 µM [^{32}P]NAD (3–5 µCi/assay) and 75–100 µg protein (as indicated in Results). When indicated, 10 mM MgCl$_2$, 1 mM ATP, and 1 mM GTP were added separately or combined. The reaction was terminated by adding 100 µl 50% (w/v) trichloroacetic acid (TCA). Precipitates were collected into GF/B Whatman filters, washed, dried, and counted for radioactivity in a Beckman LS 7800 spectrophotometer. To visualize endogenous ADP-ribosylated proteins, assays were done as described before, but the reaction was terminated by the addition of SDS-PAGE solubilization buffer (2X). Samples were boiled for 3 min and subjected to 10% SDS-PAGE electrophoresis [10]. After electrophoresis the gels were stained with Coomassie blue, dried, and exposed (7–10 days) to X-OMAT films for autoradiography. For the two dimensional gel electrophoresis, the reaction was terminated by the addition of lysis buffer (9.8 M urea, 2% NP-40, 2% (v/v) ampholytes pH 7–9, 100 mM DTT) and separated by IEF-SDS-PAGE as described by Bravo [11]. When (ADP-ribosyl)transferase activity was assayed with 75 mM agmatine as exogenous substrate, after 90 minutes of incubation at 30 °C samples were diluted 5-fold with water. [^{14}C]ADP-ribosyl agmatine was separated from [^{14}C]NAD on a Partisil 10 SAX column (4.6 x 250 mm) used with a mobile phase of 25 mM potassium phosphate buffer (pH 4.7) with a flow rate of 1 ml/min [12]. The diluted fraction (250 μl) was injected onto the HPLC column and fractions of 0.5 ml were collected and counted for radioactivity as stated before. Under these experimental conditions [^{14}C]ADP-ribosylated agmatine presented a retention time of 5 min, while [^{14}C]NAD displayed a retention time of 8 min.

RESULTS AND DISCUSSION

Extracts of dormant spores of *Phycomyces blakesleeanus* were separated by centrifugation in a sucrose continuous gradient. Collected fractions were incubated with [^{32}P]NAD to detect (ADP-ribosyl)transferase activity. Fig. 1 shows that only one band of ADP-ribosylated endogenous protein is present in that sucrose gradient fraction which corresponds to soluble proteins. Also, when albumin was used as exogenous substrate in all the gradient fractions, no [^{32}P]ADP-ribosylated albumin was observed in those fractions corresponding to membrane proteins (data not shown), discarding the presence of a

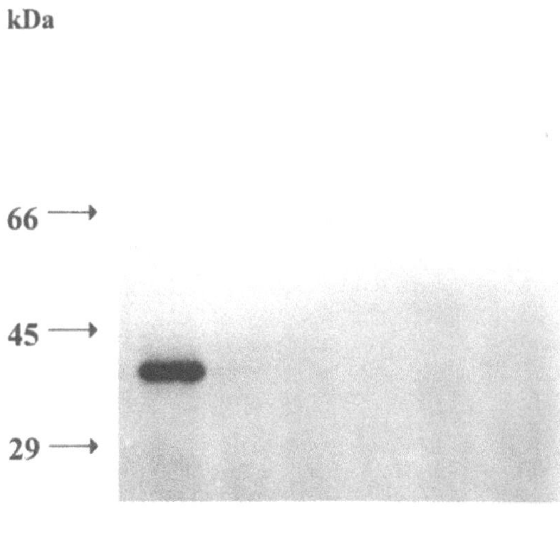

Figure 1. Distribution of (ADP-ribosyl)transferase activity in a sucrose continuous gradient. Dormant spores (5 ml of 10^{10}) were washed and resuspended in 50 mM Tris-HCl buffer (pH 7.0) containing 10% (v/v) glycerol, 5 mM EGTA, 5 mM EDTA, and 5 μg antipain per ml and mixed with an equal volume of glass beads (0.45–0.50 mm diameter) and broken in a Braun model MSK cell homogeniser for 180 s while cooling with a stream of CO_2. The crude extract was layered on top of a sucrose continuous gradient (20 - 60%) and centrifuged at 22,400 rpm on a SW28 rotor for 20 h. Afterwards, the gradient was fractionated in 1-ml aliquots, and protein concentration and density were measured in each of them. Different fractions were mixed and (ADP-ribosyl)transferase assayed using [^{32}P]NAD as a substrate. Proteins were separated by SDS-PAGE followed by radioautography. Lanes are: gradient fractions 1–10, F1; fractions 11–20, F2; fractions 21–25, F3; fractions 26–35, F4, fractions 36–42, F5; gradient residue, F6.

particulate (ADP-ribosyl)transferase. Therefore, we used a 100,000 x g supernant of the extracts of the dormant spores to characterize this soluble enzymatic activity. Using casein as exogenous substrate, this activity augmented linearly up to 200 µg of protein and 90 minutes of incubation of the assay, after which no change of activity was observed. This activity showed an optimum pH at 6.5. The enzymatic activity was lost after incubation at 95°C for 5 min, therefore discarding the addition of ADP-ribose adducts to the protein. As mentioned before, a broad [^{32}P]ADP-ribosylated band with molecular weight of 37 kDa was observed by SDS-PAGE and autoradiography as a very good substrate for this enzyme. This radiolabeled band was resolved by isoelectrofocusing gel separation in three proteins of 38, 37, and 36 kDa with pI values of 6.9, 8.1, and 4.6, respectively (Fig. 2). Meyer *et* al. [13] have reported different chemical treatments to determine the amino acid bound to the ADP-ribose. Thus, we treated the [^{32}P]ADP-ribosylated proteins with hydroxylamine to determine arginine modification, and with $HgCl_2$ to see cysteine modification by the *Phycomyces* enzyme. No significant decrease in radioactivity was observed in the presence of 1 mM $HgCl_2$. The radioactivity was partially released from the endogenous [^{32}P]ADP-ribosylated substrates after 30 min in the presence of 3 M NH_2OH (pH 7.0). Additionally, this mono(ADP-ribosyl)transferase used as an ADP-ribose acceptor a compound containing a guanidino moiety such as agmatine. These results indicate that the enzyme is a NAD:arginine-specific (ADP-ribosyl)transferase.

It has been reported that Mg^{2+} stimulates the enzymatic activity of (ADP-ribosyl)transferases detected in different cellular types [14]. We investigated the (ADP-ribo-

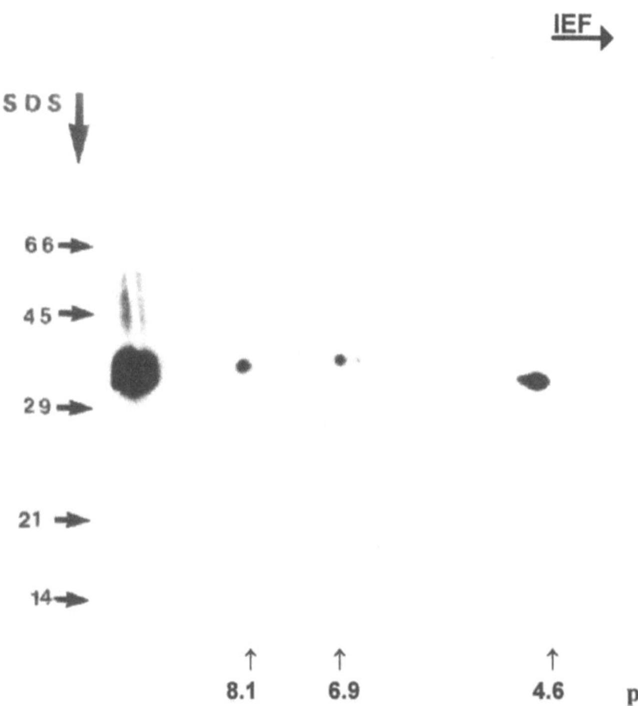

Figure 2. Two dimensional electrophoresis of [^{32}P]ADP-ribosylated soluble proteins. The cytosol (100 µg) from dormant spores was incubated under ADP-ribosylation conditions as described in Methods. Afterwards proteins were separated by two-dimensional gel electrophoresis followed by radioautography. This same result was obtained in three other experiments.

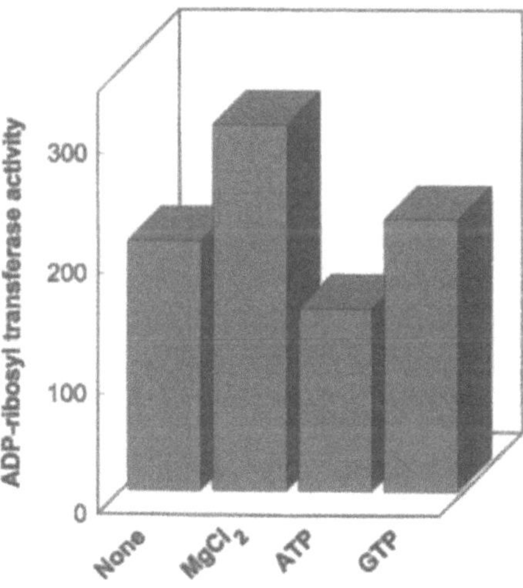

Figure 3. Effect of MgCl₂ and nucleotides on (ADP-ribosyl)transferase activity. The cytosol (100 µg of protein) was incubated under ADP-ribosylation conditions in the presence of casein (100 µg) as exogenous substrate and 10 mM MgCl₂, 1 mM GTP or 1 mM ATP were added as indicated. After 1 h of incubation at 30 °C, samples were pre-cipitated with 50% TCA and processed as described in Methods. The activity of (ADP-ribosyl)transferase is expressed as pmol of ADP-ribose incorporated/h/mg.

syl)transferase activity using casein as exogenous substrate in the absence or presence of MgCl₂, GTP, or ATP. Fig. 3 shows that only the addition of MgCl₂ to the assay stimulated the transferase activity, GTP had no effect and a slight decrease in the enzymatic activity can be observed in the presence of ATP. Also, we investigated the ADP-ribosylation of the endogenous proteins in the absence or presence of MgCl₂ alone or in combination with GTP or ATP. Two additional [^{32}P]ADP-ribosylated proteins with molecular weights of 55 and 57 kDa, respectively, were observed when the assay was done in the absence of Mg^{2+} and in the presence of GTP or ATP. On the other hand, the addition of Mg^{2+} together with either or both nucleotides eliminated the ADP-ribosylation of the 57 and 55 kDa bands, but intensified the radiolabeling in the band of 37 kDa.

The compounds sodium nitroprusside [15] and benzamide [16, 17] have been de-scribed as positive and negative effectors of ADP-ribosylation, respectively. We investi-gated the effect of sodium nitroprusside, a NO-releasing agent, and benzamide on the [^{32}P]ADP-ribosylation of the 37 kDa protein; for this, we incubated the cytosol under ADP-ribosylation conditions in the presence of different concentrations of these com-pounds. The densitometric analysis of the radiolabelled bands was performed and in Fig.4(A) it can be observed that the intensity of the band increased with the sodium nitro-prusside concentration, revealing a 300% stimulation with 250 µM sodium nitroprusside. The intensity of the band decreased as a function of the benzamide concentration, showing a 50% inhibition with 8 mM benzamide.

Also, the effect of two other strong mono-ADP-ribosylation inhibitors, MIBG and novobiocin, was studied. Fig.4(B) shows the densitometric analysis of the 37 kDa protein band. It can be observed a decrease of the absorbance of the 37 kDa band by incubation of the extracts with either MIBG or novobiocin indicating a 25% inhibition with 0.4 mM MIBG and a 36% inhibition with 0.5 mM novobiocin. These data are a little higher than

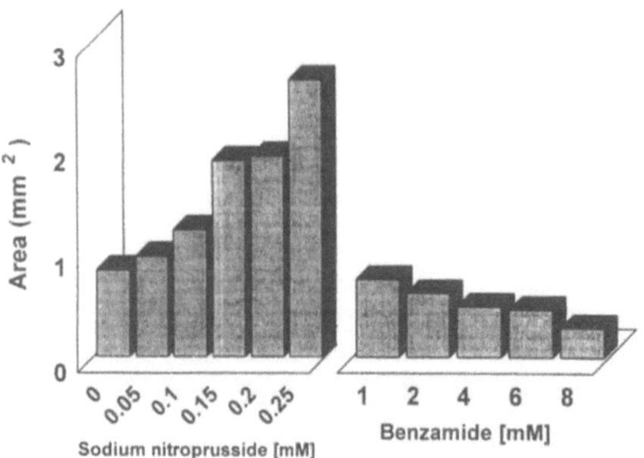

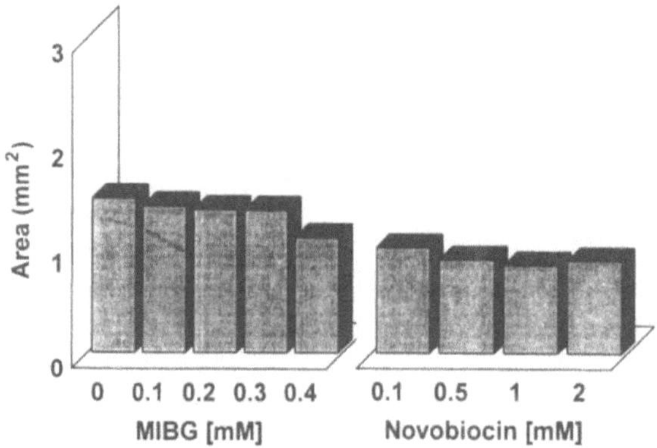

Figure 4. Effect of nitroprusside, benzamide, MIBG and novobiocin on ADP-(ribosyl)transferase activity. The cytosol was incubated under ADP-ribosylation conditions and where indicated different concentrations of sodium nitroprusside, benzamide, MIBG or novobiocin were added to the reaction mixture. After 1 h of incubation proteins were separated by 10% SDS-PAGE followed by autoradiography. The densitometric analysis of the 37 kDa band of all the lanes is presented.

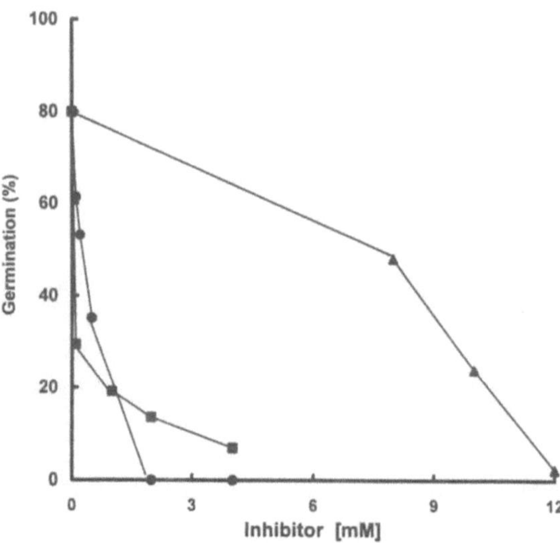

Figure 5. Effect of different inhibitors on the germination of the spores of *Phycomyces*. Spores (5 x 10^5 per ml) were inoculated in flasks containing liquid Sutter medium, 30 mM ammonium acetate and different concentrations of benzamide (▲), MIBG (●) or novobiocin (■). After 20 h of growth, germination was followed microscopically. 200 cells were counted.

those reported for the mono(ADP-ribosyl)transferase activity (IC_{50} = 0.1 mM for MIBG and 0.28 mM for novobiocin) from turkey erythrocyte membranes [17, 18].

Since we observed an inhibitory effect on the (ADP-ribosyl)transferase activity of spore extracts, we investigated the effect of these drugs on the germination process of *Phycomyces*. Fig.5 shows that the concentrations necessary to inhibit 50% of the spore germination of *Phycomyces* were 0.05 mM, 0.2 mM, and 8 mM for novobiocin, MIBG, and benzamide, respectively. All the above inhibitors affected the germination process to the same extent, that is, they inhibited the tube protuberation, leaving the spores as swollen cells. These data suggest that ADP-ribosylation may be involved in the germination process of *Phycomyces,* particularly in germ-tube formation.

The results presented here indicate the presence of a soluble NAD:arginine (ADP-ribosyl)transferase in dormant spores of the fungus *Phycomyces blakesleeanus* which could be involved in the germination process of the spores, particularly in germ-tube extrusion, and of five major soluble endogenous substrates of 57, 55, 38, 37, and 36 kDa, respectively.

ACKNOWLEDGMENTS

M.A.D.A. is a predoctoral student with a scholarship of CONACyT. This work was supported by the Consejo Nacional de Ciencia y Tecnología of México (Grants N305 and N3032) and the Third World Academy of Sciences (Grant BC90–111).

REFERENCES

1. Williamson, K.C., & J. Moss. 1990. Mono-ADP-ribosyltransferases and ADP-ribosylarginine hydrolases: a mono-ADP ribosylation cycle in animal cells. In *ADP-ribosylating toxins and G proteins: Insights into signal transduction.* pp 493–510. Edited by J. Moss & M. Vaughan. American Society for Microbiology. Washington, D.C.

2. Jacobson, M.K., P.T. Loflin, N. Aboul-Ela, M. Minmuang, J. Moss, & E.L. Jacobson 1990. Modification of plasma membrane protein cysteine residues by ADP-ribose *in vivo. J. Biol. Chem. 265*: 10825–10828.

3. Van Laere, A. J., J. A. Van Assche, & B. Furch. 1987. The sporangiospore: dormancy and germination. In *Phycomyces.* pp. 247–279. Edited by E. Cerdá-Olmedo & Lipson. Cold Spring Harbor Laboratory, Cold Spring Harbor, NY.

4. Van Laere, A. J. & F. Rivero. 1986. Properties of a germination mutant of *Phycomyces blakesleeanus. Arch. Microbiol. 145*: 290–294.

5. Rivero, F. & Cerdá-Olmedo, E. 1987. Spore activation by acetate, propionate and heat in *Phycomyces* mutants. *Mol. Gen. Genet. 209*: 149–153.

6. Van Laere, A.J. & P. Hendrix. 1983. Cyclic AMP-dependent *in vitro* activation of trehalase from dormant *Phycomyces blakesleeanus* spores. *J. Gen. Microbiol. 129*: 3287–3290.

7. Van Mulders, R.M. & A. J. Van Laere. 1984. Cyclic AMP, trehalase and germination of *Phycomyces blakesleeanus* spores. *J. Gen. Microbiol. 130*: 541–547.

8. Bartnicki-Garcia S. & W. J. Nickerson. 1962. Nutrition, growth, and morphogenesis of *Mucor rouxii. J. Bacteriol. 84*: 841–858.

9. Carrillo-Rayas, M.T., J. Garcia-Soto,. & G. Martínez-Cadena. 1988. 12-O-tetradecanoyl phorbol-13-acetate interferes with germination of *Phycomyces blakesleeanus* sporangiospores. *FEBS Lett. 238*: 441–444.

10. Laemmli, U.K. 1970. Cleavage of structural proteins during the assembly of the head of bacteriophage T4. *Nature 227*: 680–685.

11. Bravo, R. 1984. Two dimensional gel electrophoresis: a Guide for the beginner. In *Two-dimensional gel electrophoresis of proteins.* pp. 3- 36. Edited by J.E. Celis & R. Bravo. Academic Press Inc. NY.

12. Moss, J., A. Yost, & S. J. Stanley. 1983. Amino acid-specific ADP-ribosylation. *J. Biol. Chem. 258*: 6466–6470.

13. Meyer, T., R. Koch, W. Fanick, & H. Hilz. 1988. ADP-ribosyl proteins formed by pertussis toxin are specifically cleaved by mercury ions. *Biol. Chem. Hoppe-Seyler 369*: 579–583.

14. Soman, G., S. J. Mickelson, C. F. Louis, & D. J. Graves. 1984. NAD:guanidino group specific mono ADP-ribosyltransferase activity in skeletal muscle. *Biochem. Biophys. Res. Commun. 120*: 973–980.

15. Brüne, B. & E. G. Lapetina. 1989. Activation of cytosolic ADP-ribosyltransferase by nitric oxide-generating agents. *J. Biol. Chem. 264:* 8455–8458.

16. Tanigawa, Y., M. Tsuchiya, Y. Imai, & M. Shimoyama. 1984. ADP-ribosyltransferase from hen liver nuclei. *J. Biol. Chem. 259*: 2022–2029.

17. Banasik, M., H. Komura, M. Shimoyama,. & K. Ueda. 1992. Specific inhibitors of poly(ADP-ribose) synthetase and mono(ADP-ribosyl)transferase. *J. Biol. Chem. 267*: 1569–1575.

18. Smets, L.A., C. Loesberg, M. Janssen, & H. Van Rooij. 1990. Intracellular inhibition of mono(ADP-ribosylation) by *meta*-iodobenzylguanidine: specificity, intracellular concentration and effects on glucocorticoid-mediated cell lysis. *Biochim. Biophys. Acta 1054*: 49–55.

19

USE OF THE EST DATABASE RESOURCE TO IDENTIFY AND CLONE NOVEL MONO(ADP-RIBOSYL)TRANSFERASE GENE FAMILY MEMBERS

Rickmer Braren,[1] Kathrin Firner,[1] Sriram Balasubramanian,[2] Fernando Bazan,[2] Heinz-Günter Thiele,[1] Friedrich Haag,[1] and Friedrich Koch-Nolte[1]

[1]Department of Immunology
University Hospital
D-20246 Hamburg, Germany
[2]DNAX Research Institute of Molecular and Cellular Biology
Palo Alto, California 94304

ABSTRACT

We searched the database of expressed sequence tags (dbEST) for relatives of the known human and murine mono(ADP-ribosyl)transferases (mADPRT), poly(ADP-ribosyl)polymerases (PARP), ADP-ribosyl cyclases, and ADP-ribosylarginine hydrolases (ARH). By May 31, 1996, all of the known enzymes except for RT6 were represented in dbEST by exact sequence matches from mouse and/or human tissues. Several ESTs show significant sequence similarity but not identity to known mADPRTs. We isolated, cloned, and sequenced the corresponding genes. Our results show that seven human ESTs stem from a novel gene, provisionally designated *LART*, which is specifically expressed in *l*ymphatic tissues. Five human ESTs stem from a novel gene, here designated *TART1*, which is specifically expressed in *t*estis. This gene is also represented by a single mouse EST. One other mouse EST stems from a distinct gene, here designated *TART2*, which is also expressed in *t*estis. These genes have similar exon/intron structures. The predicted LART and TART1 gene products contain hydrophobic N- and C-terminal signal peptides characteristic for GPI-anchored surface proteins, TART2 lacks the GPI-anchor signal peptide. The predicted native proteins show 28–42% sequence identity to one another. They each contain four cysteine residues that probably form conserved disulfide bonds. They each also contain a conserved glutamic acid residue within the proposed active site motif. LART and TART1 show interesting deviations from the surrounding consensus sequence.

ADP-Ribosylation in Animal Tissue, edited by Haag and Koch-Nolte
Plenum Press, New York, 1997

BACKGROUND

The systematic sequencing of cDNA fragments has been coined the turning point in genome research (1). Within the last 6 months, the number of these so called expressed sequence tags (ESTs) in the publicly accessible database (dbEST) has more than doubled. At the time of the Hamburg workshop, dbEST contained more than 500.000 entries. Most entries derive from human and mouse cDNA libraries of the IMAGE consortium (1). It has been estimated that 78% of the positionally cloned human genes had an exact sequence match in the dbEST on Feb. 1, 1996 (2). We report here the use of this powerful database resource for identifying transcripts of known genes encoding mADPRTs and related enzymes. Moreover we have applied the information obtained for cloning three novel mADPRT gene family members.

RESULTS AND DISCUSSION

We screened the database of expressed sequence tags (dbEST) for sequences related to known mADPRTs and related enzymes using the BLAST program (3). The results are summarized in Figure 1. The single known mammalian PARP (4), the two known mammalian ADPR-cyclases CD38 and BST-1 (5), and the single known mammalian ARH (6) all are represented in dbEST by exact sequence matches in human and/or mouse ESTs (Fig. 1A). No further human or mouse ESTs with significant sequence identity to these genes were detected[*]. Of the known mADPRTs, only the muscle transferase (7, 8) (here designated MART), is represented by exact sequence match ESTs. Neither of the two T cell mADPRTs, Rt6–1 and Rt6–2 (9, 10), is yet represented in dbEST. However, dbEST does contain many ESTs with significant sequence similarity but not identity to known mADPRTs (Fig. 1B).

With oligonucleotides derived from the mADPRT-related ESTs we isolated, cloned, and sequenced the novel genes from which these ESTs derive. Our results show that seven human ESTs derive from a novel gene which is specifically expressed in lymphatic tissues (11). We tentatively designate this gene *LART*. All of these ESTs are from the same fetal liver/spleen library of the IMAGE consortium. The expression pattern of *LART* as judged by Northern blot analysis is very similar to that observed for RT6 in rodents (10). Thus, it is conceivable that the LART gene product compensates for the universal loss of RT6 in the human species by the premature stop codons contained in its gene (12) (see also Thiele et al., this volume).

Five human ESTs represent transcripts from a novel gene, designated *TART1*, which is specifically expressed in testis (11). One of these ESTs, indeed, was isolated at IN-SERM, Paris, from a testis cDNA library (13, 14) and evidently corresponds to the gene described by Guellaen et al. in this volume. *TART1* is also represented by a single mouse EST. One further mouse EST stems from a distinct gene, designated *TART2*, which is also expressed in testis. *TART2* evidently corresponds to the *YAC2* gene described by Moss et al. in this volume.

Like the other vertebrate mADPRTs cloned to date, LART, TART1, and TART2 also contain hydrophobic N-terminal signal peptides. This leads to the prediction that they

---

[*] dbEST contains one other cyclase-related EST (accession number N20756), which is derived from a novel gene of the parasite worm *Schistosoma mansonii*.

A)

family	gene	human	mouse	other
mADPRT	RT6-1	0	0	0
	RT6-2	0	0	0
	MART	1	9	0
	LART	7	0	0
	TART-1	5	1	0
	TART-2	0	1	0
	CHAT-1	0	0	0
	CHAT-2	0	0	0
	CEAT	0	0	0
PARP	PARP	16	7	0
ADPR cyclase	CD38	6	0	0
	BST-1	4	0	0
	other	-	-	1*
ARH	ARH	0	3	0

B)

gene	accession number	date entered	species	EST project
LART	T70606	15.3.95	human	IMAGE
	T70872	15.3.95	human	IMAGE
	R07880	5.4.95	human	IMAGE
	N70349	14.3.96	human	IMAGE
	N76036	29.3.96	human	IMAGE
	W04280	22.4.96	human	IMAGE
	W04281	22.4.96	human	IMAGE
MART	C03716	22.2.96	human	TOKYO
	W08722	25.4.96	mouse	IMAGE
	W18805	2.5.96	mouse	IMAGE
	W34749	13.5.96	mouse	IMAGE
	W20908	20.5.96 *	mouse	IMAGE
	W41414	20.5.96 *	mouse	IMAGE
	W40714	20.5.96 *	mouse	IMAGE
	W41430	20.5.96 *	mouse	IMAGE
	W42131	21.5.96 *	mouse	IMAGE
	W12573	21.5.96 *	mouse	IMAGE
TART1	Z24839	29.7.93	human	EMBL
	R35364	2.5.95	human	IMAGE
	H12146	26.6.95	human	IMAGE
	T19112	8.8.95	human	PARIS
	L49677	1.1.96	human	TIGR
	W36909	15.5.96	mouse	IMAGE
TART2	W12489	21.5.96 *	mouse	IMAGE

Figure 1. Entries matching vertebrate mono(ADP-ribosyl)transferases, poly(ADP-ribosyl)polymerase, ADP-ribo-sylprotein hydrolase, and ADP-ribosylcyclases in dbEST. A) Number of exact sequence match ESTs in dbEST. (*) CD38/BST-1-related EST from *Schistosoma mansonii*. B) Entry date and source of mADPRT-related ESTs in dbEST. Note that 7 (*) of the 23 mADPRT-related ESTs were deposited in dbEST while their genes were being discussed at the Hamburg workshop.

1	2	3	4	5	6	7	8	9	10	11	12	13		
■	83.0	39.4	40.5	39.8	31.1	33.7	38.6	36.9	42.0	49.6	50.8	50.8	1	mMart
	■	37.9	37.9	37.9	28.8	31.8	36.7	35.5	39.8	48.1	49.2	49.6	2	hMART
		■	71.2	73.9	36.0	39.8	36.7	34.1	39.8	39.8	41.3	39.4	3	rRT6.2
			■	79.2	32.2	37.9	34.5	33.5	39.4	38.6	40.9	37.5	4	mRt6-1
				■	34.1	37.9	36.0	34.2	41.7	39.8	41.7	39.0	5	mRt6-2
					■	71.2	29.9	28.5	39.0	33.7	33.3	34.1	6	mTart1
						■	29.9	28.7	41.7	37.5	36.4	37.1	7	hTART1
							■	60.5	37.1	41.3	40.5	39.0	8	hLART
								■	36.9	40.0	38.1	37.3	9	mLart
									■	39.4	42.8	42.4	10	mTart2
										■	81.4	59.5	11	cHAT1
											■	62.5	12	cHAT2
												■	13	cEAT

Figure 2. Percent amino acid sequence identities of mammalian and avian mADPRTs. Sequences for the predicted native polypeptides (i.e. without N- and C-terminal signal peptides) were aligned and the percent amino acid sequence similarities were calculated with the DNAstar software on a Macintosh PC. m = mouse, h = human, r = rat, c = chicken.

are processed to the extracellular environment. LART and TART1, like RT6 and MART, also contain a C-terminal signal peptide characteristic for GPI-anchored membrane proteins. In contrast, TART2 lacks a GPI-anchor signal sequence and, in this respect, resembles the three mADPRTs cloned so far from the chicken (see Shimoyama et al., this volume, and Shall et al., this volume). Nevertheless, TART2 shows a lower degree of sequence identity to the chicken mADPRTs than the GPI-anchored MART (Fig. 2).

Like the other known vertebrate mADPRTs, LART, TART1 and TART2 each contain four cysteine residues that probably form conserved disulfide bonds. Their predicted secondary structure motifs are very similar to those of Rt6 (11, 15) and are compatible with a fold akin to that of the bacterial toxins (16) and to the catalytic domain of PARP (17). They also contain a glutamic acid residue in the proposed active site crevice. LART and TART1 do show interesting deviations from the surrounding consensus sequence (Fig. 3). LART, remarkably, contains two basic residues instead of the usual acidic resi-

Group	Sequence	Name
ART1	F F P E E E E V L I P P F E T F Q V	mMart
	F F P G E E E V L I P P F E T F Q V	hMART
ART2	F R P D Q E E V L I P G Y E V Y Q K	rRT6.2
	Y Y T H E E E V L I P G Y E V F H K	mRt6-1
	S F P R E E E V L I P G Y E V Y H K	mRt6-2
ART3	N K E D D S V L I P L S E V F Q V	mTart1
	D K E S E R I T L I P L N E V F Q V	hTART1
ART4	D I S L K K E V L V P P Y E L F E V	mLart
	Y F S L K K E V L I P P Y E L F K V	hLART
ART5	V F P E E R E V L I P P H E V F L V	mTart2
ART6	F F P S E D E V L I P P F E V F E V	CHAT1
	F Y P S E D E V L I P P F E V F E V	CHAT2
ART7	T F P G E D E V L I P P F E Q F R V	chEBAT

Figure 3. Variations of a theme around the proposed active site glutamic acid residue (*) in vertebrate mADPRTs.

dues directly upstream of the predicted catalytic glutamic acid (KKE). In TART1, the characteristic cluster of acidic residues appears to be shifted upstream by two residues. It is possible that these sequence differences translate into differences in enzymatic activities, e.g. differences in amino acid specificity.

We have mapped the chromosomal localization and determined the exon/intron structures of the genes for RT6, MART, LART, and TART1 (11, 18, 19). While the genes map to different chromosomal positions, their structures are quite similar: the signal peptides are encoded by separate exons while the predicted native polypeptide is encoded by a single, unusually large exon. The similarities in structure and sequence suggest that these genes are derived from a common ancestor. Indeed, we have cloned and sequenced fragments of the species-homologues of *RT6*, *MART*, *LART*, and *TART1* from a number of mammalian species including monkey, rabbit, mouse, pig and sheep (F. K.-N. et al., in preparation). The presence of homologues in different mammalian species and the relatively low degree of sequence conservation between different genes (see Fig. 2) suggest that the duplication events generating the respective ancestral genes of RT6, MART, LART, and TART predate the mammalian radiation. Whether the origin of these genes predates the divergence of mammals and birds, i.e. whether RT6-, MART-, LART- and TART-homologues exist also in birds, remains to be determined.

Have most or all mammalian genes for mADPRTs and related enzymes already been identified? Our results indicate that the mADPRT and ARH gene families are rather small compared, for example, to those of protein kinases and phosphatases. However, considering that neither Rt6–1 nor Rt6–2 are yet represented in dbEST and further, that expression of mADPRT genes appears to be tightly restricted in quantity as well as in tissue-specifity, it is quite possible that mouse and human genomes contain additional family members that have not yet appeared in dbEST. Moreover, considering the great flexibility in amino acid sequences observed in procaryotic mADPRTs (see also Bazan and Koch-Nolte, this volume), it is also conceivable that other mammalian mADPRT subfamilies exist with amino acid sequence similarities to known mADPRTs below the limit of detectability with conventional database search programs. In any case, the rapidly expanding EST and full genome databases have already proven to be and promise to remain rewarding resources for hunters of genes encoding mADPRTs and related enzymes.

ACKNOWLEDGMENT

This work was supported by grant No310 from the Deutsche Forschungsgemeinschaft to F. K.-N.

REFERENCES

1. Boguski, M. 1995. The turning point in genome research. *TIBS 20*: 295–296.
2. Bassett, D. E., M. S. Boguski & P. Hieter. 1996. Yeast genes and human disease. *Nature 379*: 589–590.
3. Altschul, S. F., W. Gish, W. Miller, E. W. Myers & D. J. Lipman. 1990. Basic local alignment search tool. *J. Mol. Biol. 215*: 403–410.
4. DeMurcia, G. & J. MénissierDeMurcia. 1994. Poly(ADP-ribose)polymerase: a molecular nick-sensor. *Trends Biochem. Sci. 19*: 172–176.
5. Itoh, M., K. Ishihara, H. tomizawa, H. Tanaka, Y. Kobune, J. Ishikawa, T. Kaisho & T. Hirano. 1994. Molecular cloning of murine BST-1 having homology with CD38 and Plysia ADP-ribosyl cyclase. *Biochem. Biophys. Res. Commun. 203*: 1309–1317.

6. Moss, J., S. J. Stanley, M. S. Nightingale, J. J. Murtagh, L. Monaco, K. Mishima, H. C. Chen, K. C. Williamson & S. C. Tsai. 1992. Molecular and immunological characterization of ADP-ribosylarginine hydrolases. *J. Biol. Chem. 267*: 10481–8.

7. Zolkiewska, A., M. S. Nightingale & J. Moss. 1992. Molecular characterization of NAD:arginine ADP-ribosyltransferase from rabbit skeletal muscle. *Proc. Natl. Acad. Sci. USA 89*: 11352–11356.

8. Koch-Nolte, F., M. Kühl, F. Haag, M. Cetkovich-Cvrlje, E. H. Leiter & H. G. Thiele. 1996. Assignment of the human and mouse genes for muscle ecto mono ADP-ribosyltransferase to a conserved linkage group on human Chromosme 11p15 and mouse Chromosome 7. *Genomics 36*: 215–216.

9. Koch, F., F. Haag & H. G. Thiele. 1990. Nucleotide and deduced amino acid sequence for the mouse homologue of the rat T-cell differentiation marker RT6. *Nucleic Acids Res. 18*: 3636.

10. Hollmann, C., F. Haag, M. Schlott, A. Damaske, H. Bertuleit, M. Matthes, M. Kühl, H. G. Thiele & F. Koch-Nolte. 1996. Molecular characterization of mouse T-cell ecto-ADP-ribosyltransferase Rt6: cloning of a second functional gene and identification of the Rt6 gene products. *Mol. Immunol. 33*: 807–817.

11. Koch-Nolte, F., F. Haag, R. Braren, M. Kühl, J. Hoovers, S. Balasubramanian, F. Bazan & H. G. Thiele. 1996. Two novel human members of an emerging mammalian gene family related to mono-ADP-ribosylating bacterial toxins. *Genomics* in press.

12. Haag, F., F. Koch-Nolte, M. Kühl, S. Lorenzen & H. G. Thiele. 1994. Premature stop codons inactivate the *RT6* genes of the human and chimpanzee species. *J. Mol. Biol. 243*: 537–546.

13. Pawlak, A., C. Toussaint, I. Levy, F. Bulle, M. Poyard, R. Barouki & G. Guellaen. 1995. Characterization of a large population of mRNAs from human testis. *Genomics 26*: 151–8.

14. Lévy, I., Y. Q. Wu, N. Roeckel, F. Bulle, A. Pawlak, S. Siegrist, M. G. Mattéi & G. Guellaen. 1996. Human testis specifically expresses a homologue of the rodent T lymphocytes RT6 mRNA. *FEBS Lett. 382*: 276–280.

15. Koch-Nolte, F., D. Petersen, S. Balasubramanian, F. Haag, D. Kahlke, T. Willer, R. Kastelein, F. Bazan & H. G. Thiele. 1996. Mouse T cell membrane proteins Rt6–1 and Rt6–2 are arginine/protein mono ADP-ribosyltransferases and share secondary structure motifs with ADP-ribosylating bacterial toxins. *J. Biol. Chem. 271*: 7686–7693.

16. Domenighini, M., C. Magagnoli, M. Pizza & R. Rappuoli. 1994. Common features of the NAD-binding and catalytic site of ADP-ribosylating toxins. *Mol. Microbiol. 14*: 41–50.

17. Ruf, A., J. M. DeMurcia, G. M. DeMurcia & G. E. Schulz. 1996. Structure of the catalytic fragment of poly(ADP-ribose)polymerase from chicken. *Proc. Natl. Acad. Sci. USA 93*: 7481–7485.

18. Haag, F., G. Kuhlenbäumer, F. Koch-Nolte, E. Wingender & H.-G. Thiele. 1996. Structure of the gene encoding the rat T cell ecto-ADP-ribosyltransferase RT6. *J. Immunol. 157*: 2022–2030.

19. Koch-Nolte, F., M. Kühl, F. Haag, M. Cetkovich-Cvrlje, E. H. Leiter & H. G. Thiele. 1996. Assignment of the human and mouse genes for muscle ecto mono ADP-ribosyltransferase to a conserved linkage group on human Chromosme 11p15 and mouse Chromosome 7. *Genomics 36*: 215–216.

20

MOUSE RT6 LOCUS 1 AND RAT RT6.2 ARE NAD$^+$

Arginine ADP-Ribosyltransferases with Auto-ADP-Ribosylation Activity

Rita Bortell,[1] Mark Rigby,[1] Linda Stevens,[2] Joel Moss,[2] Toshihiro Kanaitsuka,[1] John Mordes,[1] Dale Greiner,[1] and Aldo Rossini[1]

[1]Diabetes Division
University of Massachusetts Medical School
373 Plantation St., #218
Worcester, Massachusetts 01605
[2]Pulmonary-Critical Care Medicine Branch
National Heart, Lung, and Blood Institute
National Institutes of Health
Bethesda, Maryland 20892–1590

ABSTRACT

We report that rat RT6.2 and recombinant mouse Rt6 locus 1 proteins possess auto-ADP-ribosyltransferase activity and that Rt6, but not RT6, catalyzes the ADP-ribosylation of exogenous histones. Based on NH_2OH sensitivity, it appeared that the ADP-ribose was attached to arginine residues on proteins. We also observed that the NAD$^+$ concentration in culture medium correlates inversely with the proliferation of rat RT6$^+$ T cells. The data suggest that lymphocyte surface ADP-ribosyltransferases could be involved in signaling and immunoregulatory processes.

BACKGROUND

RT6 is expressed post-thymically on T lymphocytes in the rat. Two allotypes, RT6.1 and RT6.2, of 25–35 kDa and 24–26 kDa, respectively, are known (1). Two mouse homologues of the rat RT6 gene, Rt6 locus 1 and Rt6 locus 2, have been identified (2). Both mouse genes are transcribed (2,3), but expression of Rt6 protein has not been documented.

ADP-Ribosylation in Animal Tissue, edited by Haag and Koch-Nolte
Plenum Press, New York, 1997

T cells expressing RT6 in the rat subserve an immunoregulatory function (4). The mechanism is unknown, but RT6.2 is now known to have nicotinamide adenine dinucleotide (NAD^+) glycohydrolase activity (5,6). NADases catalyze the hydrolysis of NAD^+ to nicotinamide and adenosine diphosphoribose (ADPR); some also possess ADP-ribosyltransferase activity and covalently link ADPR to acceptors (7). Some ADP-ribosyltransferases like cholera and pertussis toxins can modulate cellular function.

RESULTS

1. Recombinant Rat RT6.2 and Mouse Rt6 Locus 1 Have ADP-Ribosyltransferase Activity

Recombinant rat RT6.2 (BV/RT6.2) and mouse Rt6 locus 1 (BV/Rt6) baculoviruses (BV) were produced by cotransfection of Sf9 cells with baculovirus and a transfer plasmid. ADP-ribosyltransferase assays were performed on culture supernatants of Sf9 cells infected with BV/RT6.2, BV/Rt6, or wild-type BV. In assays containing [^{32}P-adenylate]NAD but not histone, ^{32}P-labeled proteins were observed at ≈26 kDa in the recombinant rat RT6.2 lane and at ≈31 kDa in the recombinant mouse Rt6 lane (Fig. 1). Immunoblotting confirmed the identity of labeled bands in Figure 1 as RT6.2 and Rt6 (data not shown). In the presence of histone and [^{32}P-adenylate]NAD, mouse Rt6 but not rat RT6.2 catalyzed the incorporation of [^{32}P] into histone (Fig. 1). No bands were detected with supernatants from BV-infected Sf9 cells incubated with histone and [^{32}P-adenylate]NAD. The absence of inhibition by the addition of free ADPR to assays in the presence of [^{32}P-adenylate]NAD and histone excluded any free [^{32}P]ADP-ribose interme-

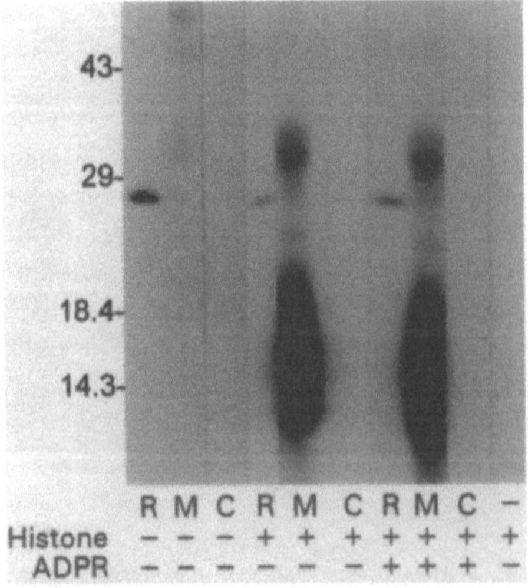

Figure 1. ADP-ribosyltransferase activities of rat RT6.2 and mouse Rt6. Supernatants from cultures of Sf9 cells infected with BV/RT6.2 (15 µl, R) BV/Rt6 locus 1 (25 µl, M), or wild type baculovirus (25 µl, C) were assayed in the presence or absence of histone VII-S and ADPR. Proteins were separated in 12% gels that were exposed to X-ray film for 2 hr.

diate; the data suggest that radiolabeling of the proteins was catalyzed by an ADP-ribosyl-transferase.

2. ADP-Ribose-Protein Linkage Analysis

ADP-ribose-protein linkages were deduced from chemical stability analyses. ^{32}P-labeled proteins, prepared as in Fig. 1, were incubated with $HgCl_2$, HCl, NH_2OH, or NaCl. The ADP-ribose-protein bond was cleaved by NH_2OH with significantly higher loss of label at 2 hr than at 20 min (data not shown). These results are consistent with ADP-ribose-arginine linkages.

3. Immunoprecipitated Rat RT6.2 and Intact Rat T Cells Catalyze Auto-ADP-Ribosylation

To establish that the auto-ADP-ribosylation activity of recombinant RT6.2 was also catalyzed by RT6 expressed on T cells, RT6.2 was immunoprecipitated from cell extracts. Cell-derived RT6.2 was immunoprecipitated and then incubated *in vitro* with [32P-adenylate]NAD[+], electrophoresed, and transferred to nitrocellulose. Autoradiography demonstrated ^{32}P-labeled bands at ≈26 kDa and immunoblotting identified them as RT6 (data not shown). Because RT6 is expressed principally on the surface of T cells, we also incubated T lymphocytes in the presence of [32P-adenylate]NAD[+]. Figure 2A shows that cell lysates yielded a predominant ^{32}P-labeled band at ≈26 kDa that was confirmed by immunoblotting to be RT6.2 (Fig. 2B).

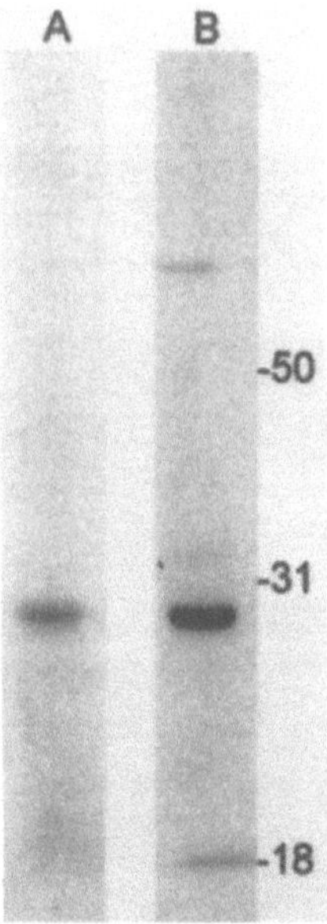

Figure 2. ADP-ribosyltransferase activity and immunoblot analysis of intact RT6.2[+] WF T cells. Cells were incubated with [32P-adenylate]NAD[+] and then detergent extracted. After separation in 10% gels, proteins were transferred to nitrocellulose and exposed to x-ray film. After autoradiography (Lane A), the same blot was analyzed by immunoblotting (Lane B) using the 1126 rabbit polyclonal anti-RT6 peptide antiserum.

4. The Proliferative Response of Rat T Cells Is Inhibited by Incubation with Ecto-NAD$^+$

To determine if NAD$^+$ could affect lymphocyte activation, rat T cells incubated with NAD$^+$ were activated with mitogens and assayed for proliferation. NAD$^+$ inhibited in a dose-dependent manner the proliferative response to three mitogens (Fig. 3).

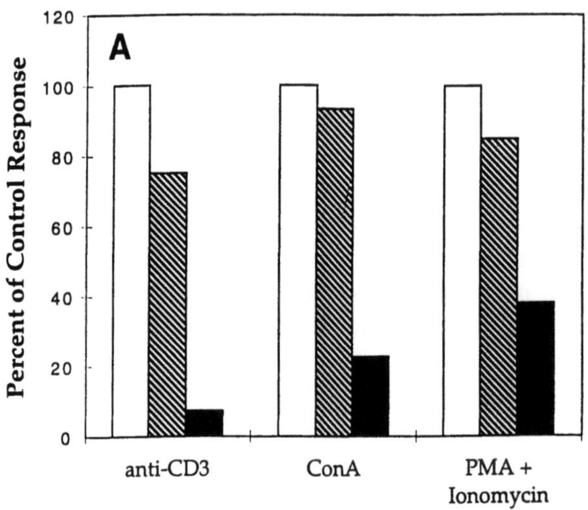

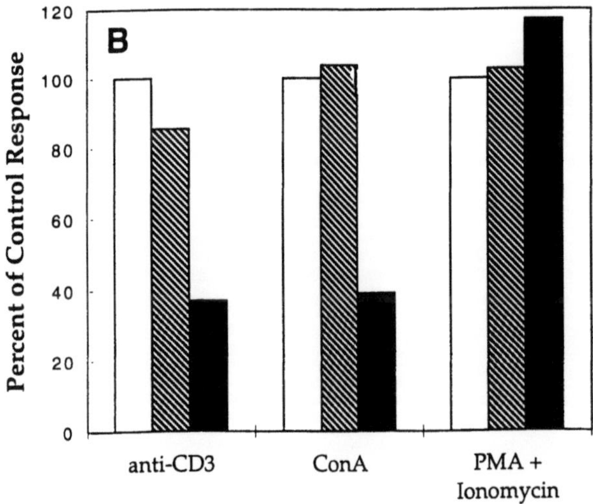

Figure 3. Inhibition of WF rat T cell proliferation by ecto-NAD$^+$. Cells were stimulated with anti-CD3 mAb, concanavalin A (con A), or phorbol myristate acetate (PMA) plus ionomycin. Stimulated cells were cultured in the absence (open bars) or presence of 20 μM (shaded bars) or 200 μM (solid bars) NAD$^+$. [^{3}H]Thymidine incorporation (% of cells cultured in the absence of NAD$^+$) was determined after 40 (Panel A) or 64 hr (Panel B). Data shown for each condition are averages of triplicate cultures from one of four experiments.

DISCUSSION

These data confirm and extend our report that recombinant rat RT6.2 is an NAD^+ glycohydrolase (5). They show that RT6.2 isolated from rat T cells, recombinant rat RT6.2, and recombinant mouse Rt6 locus 1 proteins all possess auto-ADP-ribosyltransferase activity. Recombinant mouse Rt6, but not rat RT6.2, also catalyzed the ADP-ribosylation of histone. ADP-ribose-protein bonds produced by both mouse Rt6 and rat RT6.2 were labile to NH_2OH, with a time course consistent with ADP-ribosylation of the guanidino group of arginine, and were stable to HCl, $HgCl_2$, and NaCl, excluding ADP-ribosylation of cysteine, serine/threonine, or glutamate.

The presumptive substrate of cell surface RT6.2 is extracellular NAD^+ but little is known about the biological effects of ecto-NAD^+. In the mouse, cytotoxic T lymphocyte proliferation and killing were inhibited *in vitro* by ecto-NAD^+ in a dose-dependent manner (8). In this study, we demonstrated that ecto-NAD^+ inhibits rat T cell proliferation in a dose-dependent manner. The data are consistent with a possible linkage between the catalytic activity of RT6.2 and the proliferative capability of cells that express it.

Autoimmunity in the BB rat is modulated by $RT6^+$ T cells (4), and mouse autoimmunity is associated with abnormal expression of the Rt6 gene (2,3). Our findings that RT6.2 and Rt6 locus 1 proteins have ADP-ribosyltransferase activity, that they can be auto-ADP-ribosylated, and that ecto-NAD^+ inhibits rat T cell proliferation, suggest that these molecules could play important roles in immunoregulation.

REFERENCES

1. Koch, F., A. Kashan, and H.-G. Thiele. 1988. The rat T-cell differentiation marker RT6.1 is more polymorphic than its alloantigenic counterpart RT6.2. *Immunology* 65:259–265.
2. Prochazka, M., H. R. Gaskins, E. H. Leiter, F. Koch-Nolte, F. Haag, and H.-G. Thiele. 1991. Chromosomal localization, DNA polymorphism, and expression of *Rt-6*, the mouse homologue of rat T-lymphocyte differentiation marker *RT6*. *Immunogenetics* 33:152–156.
3. Koch-Nolte, F., J. Klein, C. Hollmann, M. Kühl, F. Haag, H. R. Gaskins, E. Leiter, and H.-G. Thiele. 1995. Defects in the structure and expression of the genes for the T cell marker *Rt6* in NZW and (NZB x NZW)F$_1$ mice. *Int. Immunol.* 7:883–890.
4. Crisá, L., J. P. Mordes, and A. A. Rossini. 1992. Autoimmune diabetes mellitus in the BB rat. *Diabetes/Metab. Rev.* 8:9–37.
5. Takada, T., K. Iida, and J. Moss. 1994. Expression of NAD glycohydrolase activity by rat mammary adenocarcinoma cells transformed with rat T cell alloantigen RT6.2. *J. Biol. Chem.* 269:9420–9423.
6. Haag, F., V. Andresen, S. Karsten, F. Koch-Nolte, and H.-G. Thiele. 1995. Both allelic forms of the rat T cell differentiation marker RT6 display nicotinamide adenine dinucleotide (NAD)-glycohydrolase activity, yet only RT6.2 is capable of automodification upon incubation with NAD. *Eur. J. Immunol.* 25:2355–2361.
7. Zolkiewska, A., I. J. Okazaki, and J. Moss. 1994. Vertebrate mono-ADP-ribosyltransferases. [Review]. *Mol. Cell. Biochem.* 138:107–112.
8. Wang, J., E. Nemoto, A. Y. Kots, H. R. Kaslow, and G. Dennert. 1994. Regulation of cytotoxic T cells by ecto-nicotinamide adenine dinucleotide (NAD) correlates with cell surface GPI-anchored/arginine ADP-ribosyltransferase. *J. Immunol.* 153:4048–4058.

21

EXPRESSION AND COMPARATIVE ANALYSIS OF RECOMBINANT RAT AND MOUSE RT6 T CELL MONO(ADP-RIBOSYL)TRANSFERASES IN *E. COLI*

Stefan Karsten,[2] Jens Schröder,[2] Cristina Da Silva,[1] Dominik Kahlke,[2] Heinz-Günter Thiele,[2] Friedrich Koch-Nolte,[2] and Friedrich Haag[2]

[1]Department of Physiological Chemistry
Hamburg University
[2]Department of Immunology
University Hospital Hamburg-Eppendorf
Martinistrasse 52, D-20246
Hamburg, Germany

ABSTRACT

Recombinant RT6 proteins of rat and mouse were analyzed for NAD-metabolizing, i.e. mono(ADP-ribosyl)transferase, NAD-glycohydrolase (NADase) and ADP-ribosyl cyclase activities. The results reveal surprising intra- as well as inter-species differences in enzyme activities. While mouse Rt6 proteins were found to be strong arginine-specific transferases, but comparatively weak NADases, the opposite held true for rat RT6, for which transferase activity could only be detected in the form of arginine-specific auto-ADP-ribosylation, displayed by RT6.2 but not by RT6.1. NADase activity of rat RT6 was not accompanied by production of cyclic ADPR (cADPR). Rat RT6 gained potent arginine-specific transferase activity by exchange of a single amino acid for the corresponding residue of the mouse proteins.

INTRODUCTION

RT6 denotes a system of T-cell specific GPI-linked membrane proteins, encoded by two highly polymorphic alleles (RT6.1 and RT6.2) of a single gene locus in the rat, and by two tandemly linked genes (Rt6–1 and Rt6–2) in the mouse [1,2]. Originally characterized as T-cell differentiation antigens in the rat, they have recently been found to belong to the

ADP-Ribosylation in Animal Tissue, edited by Haag and Koch-Nolte
Plenum Press, New York, 1997

family of mono(ADP-ribosyl)transferases [3], and to possess NAD⁺-metabolizing enzyme activity [4–7].

Aside from the nuclear poly(ADPribose)polymerase (PARP), two structurally distinct groups of NAD⁺-metabolizing enzymes have been characterized on the molecular level to date, mono(ADP-ribosyl)transferases (mADPRTs) and ADP-ribosyl cyclases. The relationship between the enzymatic activities mediated by each of the groups is not yet wholly clear, and the observation of NAD-glycohydrolase (NADase) activity in conjunction with members of both groups indicates that some overlap may exist [5, 8–10]. Members of both families (RT6 and CD38) have been found on the cell surfaces of lymphocytes, and their enzymatic activities have been implicated in the modulation of T-cell activation [11, 12]. The precise mechanisms of action have not been established. They are thought to involve the modification of target proteins involved in signal transduction in one case [13], and the production of intracellular cyclic ADP-ribose as a second messenger in the other [14].

As a basis for a more detailed investigation of their enzymatic activity, recombinant rat and mouse RT6 were produced in an *E. coli*-based expression system.

RESULTS

Arginine-Specific Mono(ADP-Ribosyl)Transferase Activity

An affinity peptide (*FLAG/His*-tag, Kodak/IBI), recognized by monoclonal antibody M2, was incorporated into the C-terminus of recombinant RT6 proteins (RT6.FH). This provided a convenient assay for transferase activity, since the light chain of the M2 antibody (M2L) proved to be a substrate for ADP-ribosylation by mouse Rt6.FH (Fig. 1A, lanes 1 and 2). In contrast, rat RT6.FH did not modify M2L (lanes 3 and 4). Instead, rat RT6.2, but not RT6.1 modified itself (lane 3 vs. lane 4). All modification reactions were shown to be arginine-specific by analysis of the chemical stability of the ADPribose-pro-

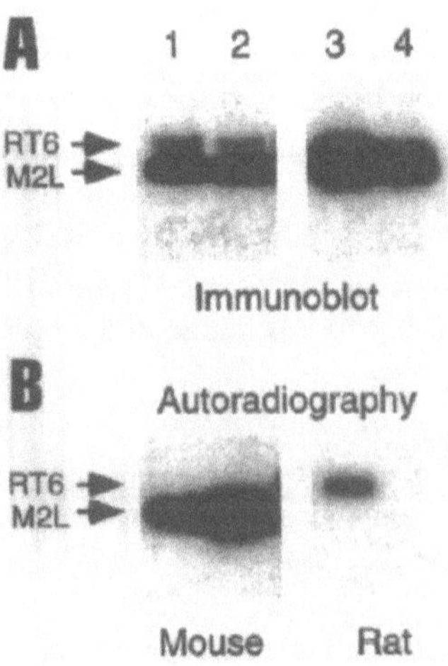

Figure 1. Analysis of arginine-specific mono(ADP-ribosyl)transferase activity. Recombinant mouse Rt6–1.FH and Rt6–2.FH (lanes 1 and 2), as well as rat RT6.2.FH and RT6.1.FH (lanes 3 and 4) were immobilized on M2-sepharose and incubated with [³²P]-NAD⁺ as described [7]. A. Immunoblot showing relative amounts of RT6 and the M2 light chain (M2L). B. Autoradiography showing incorporation of radiolabel into proteins.

tein bond [15] (not shown). Corresponding results were obtained using arginine-rich histones as a substrate (data not shown).

NAD-Glycohydrolase and ADPR-Cyclase Activities

NADase activity of recombinant RT6 proteins was investigated by thin-layer chromatography [5]. In this assay, rat RT6 showed much stronger NADase activity than either of the mouse Rt6 proteins (Fig. 2A, lane 4 vs. lanes 2 and 3). Rat RT6.1 and RT6.2 allotypes displayed comparable levels of activity (data not shown). We investigated a possible ADPR cyclase activity of rat RT6 proteins by a sensitive HPLC-based method [16]. Cyclic ADPR could not be detected at any time as a reaction product in supernatants of immobilized recombinant RT6 incubated with NAD$^+$ (Fig. 2B). To rule out that this could be due to rapid hydrolysis of cADPR to ADPR, an assay was performed using nicotinamide guanosine dinucleotide (NGD) instead of NAD [17]. This substrate is readily converted to cyclic GDPR (cGDPR) by ADPR cyclases, but, unlike cADPR, is not further hydrolyzed to GDPR. In these experiments, as before, cyclase activity of rat RT6 proteins could not be demonstrated (not shown).

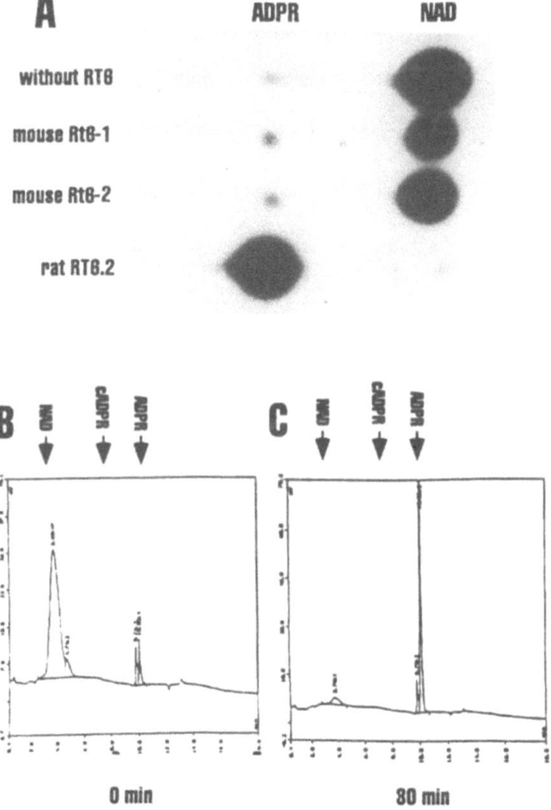

Figure 2. Analysis of NADase (A) and ADPR cyclase (B) activities. A. RT6.FH proteins were purified and incubated with [^{32}P]-NAD$^+$ as in Figure 1. The reaction supernatants were analyzed by thin layer chromatography as described [5]. B. Purified recombinant rat RT6.2.FH was incubated with NAD$^+$, and the supernatant was analyzed by HPLC at the indicated times as described [16].

The C-Terminal Region Is Crucial for Determining Enzyme Activity

To further investigate the molecular basis for the differences in enzymatic behavior between rat and mouse RT6, chimeric proteins combining the N-terminal region of rat with the C-terminal region of mouse RT6, and *vice versa*, were constructed. Analysis of relative transferase and NADase activities showed that the C-terminal portion was decisive in determining enzyme activity. Chimeric proteins containing the C-terminal region of mouse Rt6 displayed strong mADPRT and weak NADase activities, while the reverse held true for those enzymes with a C-terminus derived from the rat sequence (not shown).

Exchange of a Single Amino Acid Residue between Rat and Mouse RT6 Leads to a Switch in Enzymatic Phenotype

Both rat alloantigens differ from mouse Rt6 and other arginine-specific transferases two residues N-terminal of the proposed 'catalytic' glutamic acid E189 (Fig. 3C, see also Bredehorst et al., this volume) where they display glutamine (Q) in place of glutamic acid

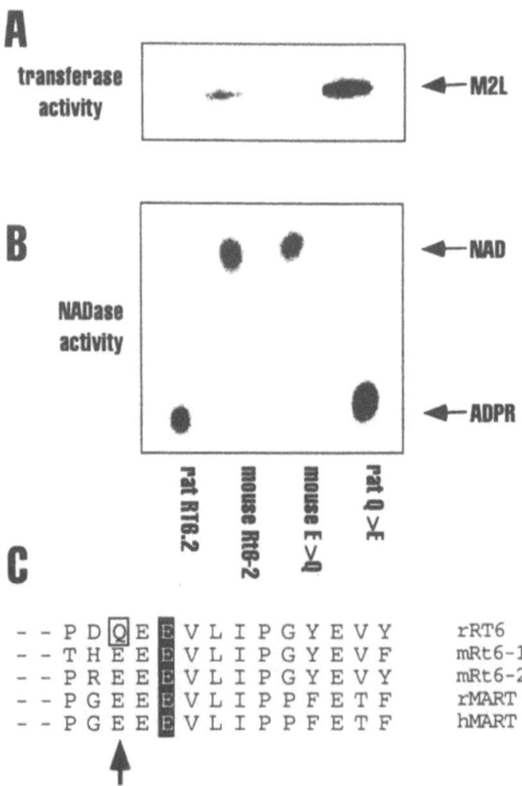

Figure 3. Mutation of a single amino acid in rat RT6 confers arginine-specific transferase activity. Mono(ADP-ribosyl)transferase (A) and NADase (B) activities of wild-type and mutant RT6 proteins were analyzed as in Fig. 1. C. The predicted 'catalytic' glutamic acid residue (E189) of eukaryotic mono(ADP-ribosyl)transferases (see Bredehorst et al., this volume) is highlighted in black. The glutamine (Q187) that distinguishes rat RT6 from known arginine-specific transferases is boxed.

(E). The mutation of Q187 to E in rat RT6 led to a dramatic gain of transferase activity (Fig. 3A), while not affecting NADase activity (Fig. 3B). Conversely, exchange of E187 to Q in mouse Rt6 abolished transferase activity, but did not lead to a gain in NADase activity.

DISCUSSION

The RT6 proteins of rat and mouse were expressed in enzymatically active form in the *E. coli* periplasm, and assayed for NAD-metabolizing, i.e. mono(ADP-ribosyl)transferase, NADase and ADP-ribosyl cyclase activities. The results reveal surprising intra- as well as inter-species differences in enzyme activities.

The Rt6 proteins of the mouse display strong arginine-specific transferase, but comparatively weak NADase activity. In contrast, rat RT6 proteins displayed only weak transferase, but strong NADase activity. In the assays employed in these experiments, transferase activity was detected only in the form of arginine-specific auto-ADP-ribosylation, displayed by the allotype RT6.2 but not by RT6.1.

The finding that cADPR is not an intermediate in the NADase activity associated with rat RT6 supports the concept that mono(ADP-ribosyl)transferases and ADPR cyclases represent different enzyme families with distinct structures and distinct activities, that exert their effects on T cells via distinct mechanisms.

It is important to emphasize that despite the differences in enzymatic activity beween RT6 proteins of different species, similar effects, i.e. inhibition of proliferation, have been reported to be mediated by extracellular NAD^+ in both rat and mouse T-cells [11,18]. The finding that mutation of a single amino acid residue (Q187) of rat RT6 leads to gain of arginine-specific transferase activity identifies this residue as a critical one for the catalytic process. A similar Q/E polymorphism is also found between pertussis and cholera toxins at the corresponding position [7]. These two enzymes differ with respect to their amino acid specificity, although they belong to a common subfamily of prokaryotic mono(ADP-ribosyl)transferases, and modify a common class of target proteins (the alpha subunit of heterotrimeric G-proteins) [19]. In this context it is important to emphasize that the ADP-ribosylation assays employed in this study were primarily directed at the detection of arginine-specific transferase activity. It is therefore conceivable that rat and mouse RT6 differ in their amino acid specificity, possibly reflecting species-specific mutations in their respective target proteins.

ACKNOWLEDGMENTS

This work was supported by grant Ha 2369 from the Deutsche Forschungsgemeinschaft DFG to FH and HGT, as well as by a stipend from the Sandoz Foundation for Therapeutic Research to FH.

REFERENCES

1. Koch, F., F. Haag, A. Kashan & H. G. Thiele. 1990. Primary structure of rat RT6.2, a nonglycosylated phosphatidylinositol-linked surface marker of postthymic T cells. *Proc Natl Acad Sci USA 87*: 964.

2. Prochazka, M., H. R. Gaskins, E. H. Leiter, F. Koch-Nolte, F. Haag & H. G. Thiele. 1991. Chromosomal localization, DNA polymorphism, and expression of Rt-6, the mouse homologue of rat T-lymphocyte differentiation marker RT6. *Immunogenetics 33*: 152.

3. Zolkiewska, A., M. S. Nightingale & J. Moss. 1992. Molecular characterization of NAD:arginine ADP-ribosyltransferase from rabbit skeletal muscle. *Proc. Natl. Acad. Sci. USA 89*: 11352.

4. Takada, T., K. Iida & J. Moss. 1994. Expression of NAD glycohydrolase activity by rat mammary adenocarcinoma cells transformed with rat T cell alloantigen RT6.2. *J Biol Chem 269*: 9420.

5. Haag, F., V. Andresen, S. Karsten, F. Koch-Nolte & H.-G. Thiele. 1995. Both allelic forms of the rat T cell differentiation marker RT6 display nicotinamide adenine dinucleotide (NAD)-glycohydrolase activity, yet only RT6.2 is capable of automodification upon incubation with NAD. *Eur. J. Immunol. 25*: 2355.

6. Maehama, T., H. Nishina, S. Hoshino, Y. Kanaho & T. Katada. 1995. NAD+-dependent ADP-ribosylation of T lymphocyte alloantigen RT6.1 reversibly proceeding in intact rat lymphocytes. *J. Biol. Chem. 270*: 22747.

7. Koch-Nolte, F., D. Petersen, S. Balasubramanian, F. Haag, D. Kahlke, T. Willer, R. Kastelein, F. Bazan & H.-G. Thiele. 1996. Mouse T cell membrane proteins Rt6–1 and Rt6–2 are arginine/protein mono(ADPribosyl)transferases and share secondary structure motifs with ADP-ribosylating bacterial toxins. *J Biol Chem 271*: 7686.

8. Gelman, L., P. Deterre, H. Gouy, L. Boumsell, P. Debre & G. Bismuth. 1993. The lymphocyte surface antigen CD38 acts as a nicotinamide adenine dinucleotide glycohydrolase in human T lymphocytes. *Eur J Immunol 23*: 3361.

9. Zocchi, E., L. Franco, L. Guida, U. Benatti, A. Bargellesi, F. Malavasi, H. C. Lee & F. A. De. 1993. A single protein immunologically identified as CD38 displays NAD+ glycohydrolase, ADP-ribosyl cyclase and cyclic ADP-ribose hydrolase activities at the outer surface of human erythrocytes. *Biochem Biophys Res Commun 196*: 1459.

10. Grimaldi, J. C., S. Balasubramanian, N. H. Kabra, A. Shanafelt, J. F. Bazan, G. Zurawski & M. C. Howard. 1995. CD38-mediated ribosylation of proteins. *J. Immunol. 155*: 811.

11. Wang, J., E. Nemoto, A. Y. Kots, H. R. Kaslow & G. Dennert. 1994. Regulation of cytotoxic T cells by ecto-nicotinamide adenine dinucleotide (NAD) correlates with cell surface GPI-anchored/arginine ADP-ribosyltransferase. *J. Immunol. 153*: 4048.

12. Lund, F., N. Solvason, J. C. Grimaldi, R. M. E. Parkhouse & M. Howard. 1995. Murine CD38: an immunoregulatory ectoenzyme. *Immunol Today 16*: 469.

13. Wang, J., E. Nemoto & G. Dennert. 1996. Regulation of CTL by ecto-nicotinamide adenine dinucleotide (NAD) involves ADP-ribosylation of a $p56^{lck}$-associated protein. *J. Immunol. 156*: 2819.

14. Howard, M., J. C. Grimaldi, J. F. Bazan, F. E. Lund, A. L. Santos, R. M. Parkhouse, T. F. Walseth & H. C. Lee. 1993. Formation and hydrolysis of cyclic ADP-ribose catalyzed by lymphocyte antigen CD38. *Science 262*: 1056.

15. Cervantes-Laurean, D., D. E. Minter, E. L. Jacobsen & M. K. Jacobsen. 1993. Protein glycation by ADP-ribose: studies of model conjugates. *Biochemistry 32*: 1528.

16. Guse, A. H., C. P. da Silva, F. Emmrich, G. A. Ashamu, B. V. Potter & G. W. Mayr. 1995. Characterization of cyclic adenosine diphosphate-ribose-induced Ca2+ release in T lymphocyte cell lines. *J Immunol 155*: 3353.

17. Graeff, R. M., T. F. Walseth, K. Fryxell, W. D. Branton & H. C. Lee. 1994. Enzymatic synthesis and characterizations of cyclic GDP-ribose. A procedure for distinguishing enzymes with ADP-ribosyl cyclase activity. *J Biol Chem 269*: 30260.

18. Rigby, M. R., R. Bortell, L. A. Stevens, J. Moss, T. Kanaitsuka, H. Shigeta, J. P. Mordes, D. L. Greiner & A. A. Rossini. 1996. Rat RT6.2 and mouse Rt6 locus 1 are NAD^+:arginine ADP ribosyltransferases with auto-ADP ribosylation activity. *J. Immunol. 156*: 4259.

19. Moss, J. & M. Vaughan. 1990. *ADP-ribosylating toxins and G proteins: Insights into signal transduction.* Washington DC: American Society for Microbiology.

22

MOLECULAR CHARACTERIZATION OF RAT T LYMPHOCYTE ALLOANTIGEN RT6.1 AS AN ADP-RIBOSYLTRANSFERASE

Tomohiko Maehama and Toshiaki Katada

Department of Physiological Chemistry
Faculty of Pharmaceutical Sciences
University of Tokyo
Hongo, Tokyo 113, Japan

ABSTRACT

A family of glycosylphosphatidylinositol-linked ADP-ribosyltransferases, of which cDNAs were cloned from various mammalian cells, possess a common Glu-rich motif (EEEVLIP) near their carboxyl termini. Although the first Glu in the common motif is replaced by Gln (Q^{207}EEVLIP) in rat T lymphocyte alloantigens RT6.1 and RT6.2, the two RT6s appear to have ADP-ribosyltransferase activity. To investigate the significance of the Glu-rich motif in the enzyme activity, we produced a mutant RT6.1, in which Gln^{207} was replaced by Glu (Q207E), together with wild-type RT6s, in *Escherichia coli*. The recombinant RT6.1 and RT6.2 displayed extremely low auto-ADP-ribosylation, though the latter modification was somewhat higher than the former one. In contrast, much higher the auto-modification was observed for Q207E mutant. Moreover, the mutant could effectively ADP-ribosylate agmatine as a substrate. Thus, the single amino acid mutation of RT6.1 caused remarkable increase in its ADP-ribosyltransferase activity, indicating that the Glu-rich motif near the carboxy terminus plays an important role in the enzyme activity.

BACKGROUND

ADP-ribosylation is one of the post-translational modifications of proteins, in which the ADP-ribose moiety of NAD^+ is transferred to specific amino acid residues of the target proteins. Recently, several endogenous ADP-ribosyltransferases have been identified and characterized in eukaryotic cells. In particular, NAD^+:arginine ADP-ribosyltransferases which are anchored in the cell surface via glycosylphosphatidylinositol (GPI) linkage have

ADP-Ribosylation in Animal Tissue, edited by Haag and Koch-Nolte
Plenum Press, New York, 1997

rat RT6.1 and RT6.2	207	**Q**EEVLIP
Q207E mutant of RT6.1	207	**E**EEVLIP
rabbit muscle ART	238	EEEVLIP
human muscle ART	238	EEEVLIP
chicken bone marrow ART	222	EDEVLIP
chicken erythroblast ART	208	EDEVLIP

Figure 1. The amino acid sequences of a Glu-rich motif in NAD$^+$:arginine ADP-ribosyltransferases and the amino acid residue mutated in the present study. Position of the first amino acid is indicated by the number following the protein name. *ART*, ADP-ribosyltransferase.

been well characterized (1–3). They possess a conserved Glu-rich motif of EE(or D)EVLIP near their carboxy termini (see Fig. 1).

We have recently reported that rat T lymphocyte alloantigen RT6.1 displayed auto-ADP-ribosylation in intact lymphocyte (4). Haag *et al.* (5) also reported that RT6.2 exhibited auto-ADP-ribosylation in intact cells. However, the first Glu in the above Glu-rich motif is replaced by Gln (Q^{207}EEVLIP) in RT6.1 and RT6.2. In the present study, we produced recombinant proteins of RT6.1 and its mutant (Q207E) to explore the significance of the Glu-rich motif in the enzyme activity. Moreover, a recombinant RT6.2 was also produced to investigate the enzymatic difference between two allelic forms.

RESULTS AND DISCUSSION

The recombinant RT6.1, RT6.2 and a mutant RT6.1 (Q207E), fused with maltose-binding protein (MBP) were expressed in *E. coli* and then affinity-purified by means of amylose column chromatography. Then, we investigated ADP-ribosyltransferase activity of the three recombinant RT6s by measuring their auto-ADP-ribosylation, and the results are shown in Fig. 2*A*. The auto-modification could be observed in RT6.2. Although auto-ADP-ribosylation of RT6.1 was not apparent under the present conditions, it became obvious with the longer exposure time of the gel to an imaging plate in accordance with our

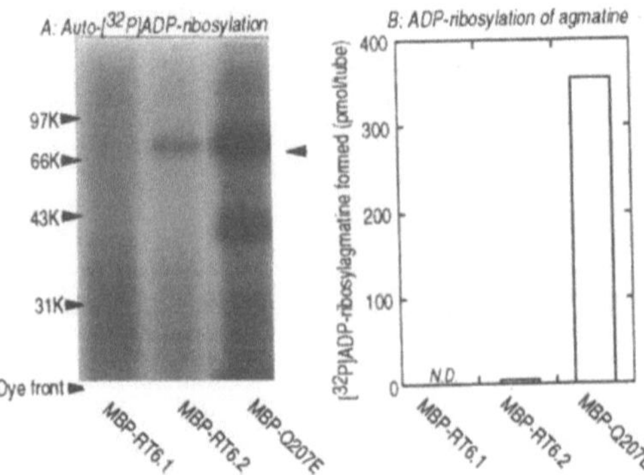

Figure 2. Stimulation of ADP-ribosyltransferase activity by the single amino acid mutation (Q207E). *A*. After the incubation with [^{32}P]NAD$^+$, recombinant RT6s (1 μg each) were separated by SDS-PAGE, then the gel was subjected to autoradiography. *B*. Recombinant RT6s (1 μg each) were incubated with [^{32}P]NAD$^+$ and agmatine, then [^{32}P]ADP-ribosylagmatine produced was measured. *N.D.*, not detected.

previous paper (4). In contrast to the wild-type RT6s, auto-ADP-ribosylation of Q207E mutant occurred at a considerable extent.

Since arginine derivatives could serve as substrates for the previously identified GPI-linked ADP-ribosyltransferases, we next investigated whether agmatine was ADP-ribosylated by the recombinant RT6s (Fig. 2*B*). RT6.1 and RT6.2 did not significantly catalyze the ADP-ribosylation of agmatine. In contrast, Q207E mutant effectively ADP-ribosylated agmatine. This correlated with the finding that auto-ADP-ribosylation was enhanced by Q207E mutation. These results indicate that the single amino acid mutation of RT6.1 causes remarkable increase in its ADP-ribosyltransferase activity. In contrast, there was no significant difference among three recombinant RT6s in their NAD$^+$ glycohydrolase activities (data not shown).

Haag *et al.* (5) reported that RT6.2 did, but RT6.1 did not, display auto-ADP-ribosylation in T lymphocytes. Our present results obtained with the recombinant RT6s also showed that ADP-ribosyltransferase activity of RT6.2 was slightly higher than that of RT6.1. However, the significant ADP-ribosyltransferase activity was only observed for Q207E mutant which contained the common Glu-rich motif. Recently, Takada *et al.* (6) reported the significance of the Glu-rich motif in rabbit muscle ADP-ribosyltransferase activity; the enzyme activity was inhibited by the single amino acid mutation of the first or the third Glu in the motif. Thus, the significance of the Glu-rich motif near carboxy termini seems to be general properties of GPI-linked ADP-ribosyltransferases. The present Q207E mutant might be a very useful tool for studying the function of RT6s in T lymphocytes, due to its enhanced ADP-ribosyltransferase activity.

REFERENCES

1. Zolkiewska, A., M. S. Nightingale, & J. Moss. 1992. Molecular characterization of NAD:arginine ADP-ribosyltransferase from rabbit skeletal muscle. *Proc. Natl. Acad. Sci. USA 89*: 11352–11356.
2. Okazaki, I.J., A. Zolkiewska, M. S. Nightingale, & J. Moss. 1994. Immunological and structural conservation of mammalian skeletal muscle glycosylphosphatidylinositol-linked ADP-ribosyltransferases. *Biochemistry 33*: 12828–12836.
3. Tsuchiya, M., N. Hara, K. Yamada, H. Osago, & M. Shimoyama. 1994. Cloning and expression of cDNA for arginine-specific ADP-ribosyltransferase from chicken bone marrow cells. *J. Biol. Chem. 269*: 27451–27457.
4. Maehama, T., H. Nishina, S. Hoshino, Y. Kanaho, & T. Katada. 1995. NAD$^+$-dependent ADP-ribosylation of T lymphocyte alloantigen RT6.1 reversibly proceeding in intact rat lymphocytes. *J. Biol. Chem. 270*: 22747–22751.
5. Haag, F., V. Andresen, S. Karsten, F. Koch-Nolte, and H.-G. Thiele. 1995. Both allelic forms of the rat T cell differentiation marker RT6 display nicotinamide adenine dinucleotide (NAD)-glycohydrolase activity, yet only RT6.2 is capable of automodification upon incubation with NAD. *Eur. J. Immunol. 25*: 2355–2361.
6. Takada, T., K. Iida, and J. Moss. 1995. Conservation of a common motif in enzymes catalyzing ADP-ribose transfer. *J. Biol. Chem. 270*: 541–544.

23

USING SECONDARY STRUCTURE PREDICTIONS AND SITE-DIRECTED MUTAGENESIS TO IDENTIFY AND PROBE THE ROLE OF POTENTIAL ACTIVE SITE MOTIFS IN THE RT6 MONO(ADP-RIBOSYL)TRANSFERASES

Kay Bredehorst, Karsten Wursthorn, Heinz-Günter Thiele, Friedrich Haag, and Friedrich Koch-Nolte

Department of Immunology
University Hospital
D-20246 Hamburg, Germany

ABSTRACT

The RT6 T cell mono(ADP-ribosyl)transferases are expressed as GPI-anchored membrane proteins by mature T lymphocytes. We performed secondary structure prediction analyses of RT6 with a profile based neural network system based on multiple alignments of RT6 with other vertebrate mono(ADP-ribosyl)transferases (mADPRTs). The results reveal a linear order of predicted ßsheets/αhelix in RT6 that are quite similar to those in the catalytic subunit of the four known crystal structures of mono-ADP-ribosylating bacterial toxins. Recognizable amino acid similarities occur throughout the region of predicted structural homology to the bacterial toxins. Three residues which have been shown to be important for catalysis in bacterial toxins (e.g. R^9, S^{52} and E^{129} in pertussis toxin) occur in a similar context also in RT6 (R^{126}, S^{147} and E^{189}). We have mutated these residues in RT6 by site-directed mutagenesis. The RT6 mutants exhibit remarkably similar alterations in enzymatic phenotype as those reported for mutations of the proposed analagous residues in bacterial toxins. These results support the hypothesis that eu- and procaryotic mADPRTs share a common fold and have a common ancestry.

ADP-Ribosylation in Animal Tissue, edited by Haag and Koch-Nolte
Plenum Press, New York, 1997

BACKGROUND

The catalytic subunit of the four known crystal structures of ADP-ribosylating bacterial toxins have a very similar Pacman-like fold despite only very little amino acid sequence similarity (Fig. 1) (1–5). Fortysix amino acids in two ß sheets and an α helix form the presumptive active site crevice. Three amino acids which are highlighted in the schematic diagram of *E. coli* heat labile enterotoxin (LT) in Figure 1 are important for enzyme activity for most of these toxins. The glutamic acid residue in ß6 has been implicated as a key catalytic residue by photoaffinity labeling with NAD^+ and by site directed mutagenesis (6–9). The arginine residue in ß1 and the serine residue in ß3 are linked by a hydrogen bond in LT and pertussis toxin and have been proposed to play a role in maintaining the conformation of the active site (5) (these residues are histidine and tyrosine in pseudomonas exotoxin A, diphtheria toxin, and the chicken poly(ADP-ribose)polymerase) (1, 2, 10). Site directed mutagenesis studies indicate that these residues also are essential for enzyme activity of the toxins (11–13). A powerful profile based neural network system for the prediction of secondary structures in proteins, PHDsec, has been developed by Ross, Sanders and coworkers at the EMBL in Heidelberg (14, 15). In cases where several related amino acid sequences are known, this program allows highly accurate predictions to be made on the basis of multiple amino acid sequence alignments. We report here the application of this program and site directed mutagenesis to identify residues in the RT6 T cell mADPRTs that may be analogues of catalytic site residues in the bacterial toxins.

RESULTS

We have performed secondary structure prediction analyses for RT6 and other vertebrate mADPRTs using the email server for the PHDsec program at the EMBL. The results

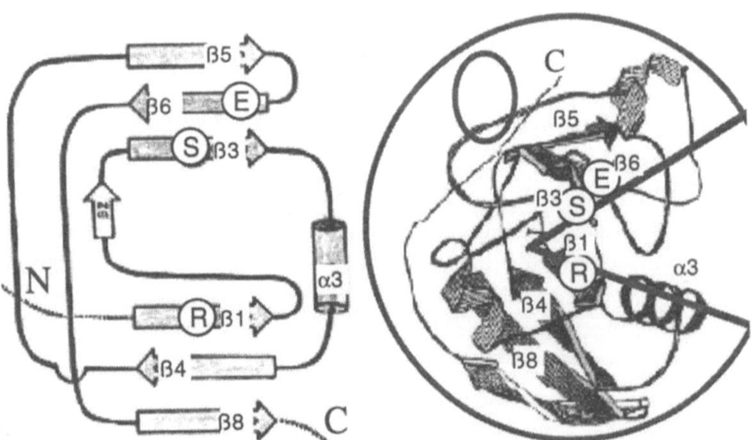

Figure 1. The four crystal structures of ADP-ribosylating bacterial toxins have a very similar Pacman-like fold despite only very little amino acid sequence similarity. Two ß sheets form the upper and lower jaws of the active site crevice. Three amino acids which are highlighted in these schematic diagrams of LT (R in ß1, S in ß3, and E in ß6) have been shown to be important for enzyme activity for most of the toxins. Secondary structure predictions indicate that vertebrate mADPRTs adopt a similar fold. The nomenclature used is that for *E. coli* heat labile enterotoxin (3).

indicate that the N-terminal portion of vertebrate mADPRTs is dominated by helices and loops and the C-terminal portion by ß sheets (16). The latter contains a linear order of predicted ß sheets and α helix that is remarkably similar to that found in the crystal structures of bacterial toxins (1–5).

Recognizable amino acid sequence similarities occur throughout the region of predicted structural homology to the bacterial toxins. The degree of sequence similarity is highest in the secondary structure units that line the active site crevice in the bacterial toxins (Fig. 2). Note in particular that three highly conserved amino acid residues occur in a similar context of predicted secondary structure units in the eucaryotic proteins as in the bacterial toxins.

We have tested the prediction that these residues (R126, S147 and E189 in ß1, ß3-α3, and ß6, respectively) are important for catalytic activity of mouse Rt6–2 by site directed mutagenesis using soluble, C-terminally tagged Rt6 expressed in the *E. coli* periplasm. While wild type Rt6–2 efficiently ADP-ribosylates the light chain of the M2 antibody (Fig. 3, lane 1), none of the mutants show any detectable transferase activity. Even conservative substitutions of E189 to aspartic acid (lane 3) or of R126 to histidine or lysine are not tolerated (lanes 8 and 9). Similar results were obtained for the NAD-glyco-

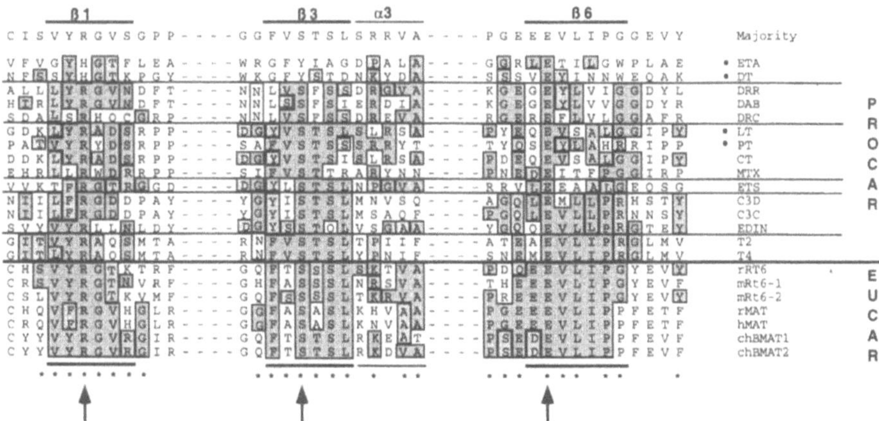

Figure 2. Alignment of presumptive active site residues in Rt6 and other mono(ADP-ribosyl)transferases. Multiple sequence alignment of vertebrate mADPRTs was performed with the MaxHom program and the resulting alignment was used as input for secondary structure prediction analyses with the PHDsec program (14, 15). Nomenclature of conserved secondary structure units is as in Fig. 1. The four bacterial toxins with known three dimensional structures are marked by • on the right. Distinct subfamilies are separated by horizontal lines. Residues conserved in at least three distinct subfamilies are boxed. The presumptive catalytic arginine, serine, and glutamic acid residues are marked by arrows on the bottom. Residues in the eucaryotic enzymes which occur also in at least three of the six subfamilies of bacterial enzymes are marked by a * on the bottom. ETA: Pseudomonas exotoxin A; DT: diphtheria toxin; DRR: *Rhodospirillum rubrum* dinitrogenase reductase ADP-ribosyltransferase; DAB: *Azosporillium brasiliense* ADP-ribosyltransferase; DRC *Rhodospirillum capsulatus* ADP-ribosyltransferase; LT: *E. coli* heat labile enterotoxin; PT: pertussis toxin; CT: cholera toxin; MTX: *Bacillus sphaericus* mosquitocidal toxin; ETS: pseudomonas extoxin S; C3C and C3D: *Clostridium botulinum* type C and type D phage exoenyzmes C3; EDIN: epidermal cell differentiation inhibitor from *Staphylococcus aureus*; T2 and T4: gpALT ADP ribosyltransferase from *E. coli* bacteriophages T2 and T4; rRT6: rat T cell marker RT6, mRt6–1 and mRt6–2: mouse T cell markers Rt6–1 and Rt6–2; rMAT and hMAT: rabbit and human skeletal muscle ADP-ribosyltransferases; chBMAT1 and chBMA2: chicken bone marrow ADP-ribosyltransferases 1 and 2. Reprinted from Ref. 16 (F. Koch-Nolte et al.,*The Journal of Biological Chemistry Vol. 271*: 7686–7693, 1996.) with kind permission from The American Society for Biochemistry and Molecular Immunology, Bethesda, MD.

hydrolase and automodification activities of rat RT6 mutants. All R126 and E189 mutants of rat RT6 show >50 fold reduction of NAD-glycohydrolase and automodification activities, some S147 mutants retain residual activity.

DISCUSSION

Our RT6 mutants exhibit remarkably similar alterations in enzymatic phenotype as mutants of bacterial toxins, where conservative substitutions of E in ß6 and R in ß1 abolish enzyme activity, but some S mutants in ß3-α3 retain residual acitivity. These results support the hypothesis that eu- and procaryotic mADPRTs share a common fold and perhaps a common ancestry. Note that the results of our secondary structure predictions agree only partially with alignments between mammalian and bacterial mADPRTs made by Takada et al. (17). Our results agree fully with alignments that were recently reported independently by Domenighini and Rappuoli (18).

It is of note that the predicted catalytic domain of the vertebrate mADPRTs is encompassed entirely in the ß-sheet rich C-terminal half of the vertebrate proteins. The sig-

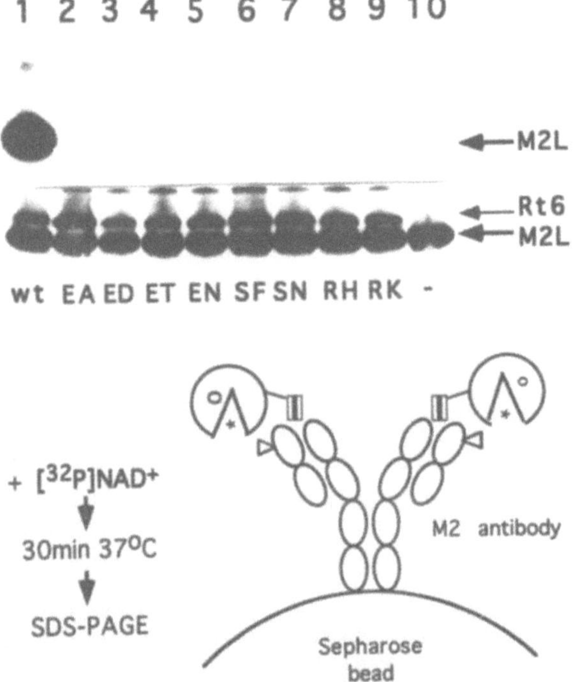

Figure 3. Comparative analysis of the transferase activity of wild-type and mutant mouse Rt6–2. Residues E189, S147, or R126 in mouse Rt6 were mutated by site directed mutagenesis (mutated residues are indicated on the bottom using the single letter code). Wild-type and mutant Rt6 were produced in an *E. coli* periplasma expression system and, as illustrated schematically on the bottom, were precipitated from perpiplasma lysates with the M2 antibody directed against the engineered C-terminal FLAG tag. Immunoprecipitates were incubated with radiolabeled NAD⁺ for 30 minutes and then analyzed by SDS-PAGE and Western-blotting. The immunostain of this blot in the middle shows that similar amounts of wild-type and mutant Rt6 proteins were precipitated by the M2 antibody, the light chain of which appears just below the RT6 band. The autoradiograph of this blot on top shows covalent modification of the light chain of the M2 antibody by ADP-ribose only in lane 1 with wild-type Rt6. Lane 1: wild-type Rt6, lanes 2–5: mutants of E189, lanes 6 & 7: mutants of S147, lanes 8 & 9: mutants of R126, lane 10: control precipitate from mock-transformed *E. coli*.

nificance of the additional, mainly helical section in the N-terminal half of the eucaryotic proteins is presently unknown and open for speculation. Interesting possibilities include roles in ligand binding or translocation across the cell membrane.

ACKNOWLEDGMENT

This work was supported by grants Ko1144/1–2 and No310/1–1 from the Deutsche Forschungsgemeinschaft to F.K-N.

REFERENCES

1. Allured, V. S., R. J. Collier, S. F. Carrol & D. B. McKay. 1985. Structure of exotoxin A of Pseudomonas aeruginosa at 3.0 Angstrom resolution. *Proc. Natl. Acad. Sci. USA 83*: 1320–1324.

2. Choe, S., M. J. Bennett, G. Fujii, P. M. Curmi, K. A. Kantardjieff, R. J. Collier & D. Eisenberg. 1992. The crystal structure of diphtheria toxin. *Nature 357*: 216–222.

3. Sixma, T. K., S. E. Pronk, K. H. Kalk, E. S. Wartna, B. vanZanten, B. Witholt & W. G. Hol. 1991. Crystal structure of a cholera toxin-related heat-labile enterotoxin from *E. coli. Nature 351*: 371–377.

4. Stein, P., A. Boodhoo, G. D. Armstrong, S. A. Cockle, M. H. Klein & R. J. Read. 1994. The crystal structure of pertussis toxin. *Structure. Curr. Biol. 2*: 45–57.

5. Domenighini, M., C. Magagnoli, M. Pizza & R. Rappuoli. 1994. Common features of the NAD-binding and catalytic site of ADP-ribosylating toxins. *Mol. Microbiol. 14*: 41–50.

6. Carrol, S. F. & R. J. Collier. 1984. NAD+ binding site of diphtheria toxin: identification of a residue within the nicotinamide subsite by photochemical modification with NAD. *Proc. Natl. Acad. Sci. USA 81*: 3307–3311.

7. Barbieri, J. T., L. M. Mende-Meuller, R. Rappuoli & R. J. Collier. 1989. Photolabeling of Glu129 of the s-1 subunit of pertussis toxin with NAD. *Infect. Immunol. 57*: 3549–3554.

8. Tweten, R. K., J. T. Barbieri & R. J. Collier. 1985. Diphtheria toxin. Effect of substituting aspartic acid for glutamic acid 148 on ADP-ribosyltransferase activity. *J. Biol. Chem. 260*: 10392–10394.

9. Antoine, R., A. Tallett, S. vanHeyningen & C. Locht. 1993. Evidence for a catalytic role of glutamic acid 129 in the NAD-glycohydrolase activity of the pertussis toxin S1 subunit. *J. Biol. Chem. 268*: 24149–24155.

10. Ruf, A., J. M. DeMurcia, G. M. DeMurcia & G. E. Schulz. 1996. Structure of the catalytic fragment of poly(ADP-ribose)polymerase from chicken. *Proc. Natl. Acad. Sci. USA 93*: 7481–7485.

11. Burnette, W. N., W. Cieplak, V. L. Mar, K. T. Kaljot, H. Sato & J. M. Keith. 1988. Pertussis toxin S1 mutant with reduced enzyme activity and a conserved protective epitope. *Science 242*: 72–74.

12. Papini, E., G. Schiavo, D. Sandona, R. Rappuoli & C. Montecucco. 1989. Histindine 21 is at the NAD+ binding site of diphtheria toxin. *J. Biol. Cem. 264*: 12385–12388.

13. Blanke, S. R., K. Huang, B. A. Wilson, E. Papini, A. Covacci & R. J. Collier. 1994. Active site mutations of the diphtheria toxin catalytic domain: role of histidine-21 in nicotinamide adenine dinucleotide binding and ADP-ribosylation of elongation factor 2. *Biochemistry 33*: 5155–5161.

14. Rost, B. & C. Sander. 1993. Improved prediction of protein secondary structure by use of sequence profiles and neural networks. *Proc. Natl. Acad. Sci. USA 90*: 7558–7562.

15. Sander, C. & R. Schneider. 1994. The HSSP database of protein structure-sequence alignments. *Nucleic Acids Res. 22*: 3597–3599.

16. Koch-Nolte, F., D. Petersen, S. Balasubramanian, F. Haag, D. Kahlke, T. Willer, R. Kastelein, F. Bazan & H. G. Thiele. 1996. Mouse T cell membrane proteins Rt6–1 and Rt6–2 are arginine/protein mono ADP-ribosyltransferases and share secondary structure motifs with ADP-ribosylating bacterial toxins. *J. Biol. Chem. 271*: 7686–7693.

17. Takada, T., K. Iida & J. Moss. 1995. Conservation of a common motif in enzymes catalyzing ADP-ribose transfer. Identification of domains in mammalian transferases. *J. Biol. Chem. 270*: 541–544.

18. Domenighini, M. & R. Rappuoli. 1996. Three conserved consensus sequences identify the NAD-binding site of ADP-ribosylating enzymes, expressed by eukaryotes, bacteria and T-even phages. *Mol. Microbiol. 21*: 667-674.

24

REGULATION OF CYTOTOXIC T CELL FUNCTIONS BY A GPI-ANCHORED ECTO-ADP-RIBOSYLTRANSFERASE

Jin Wang, Eiji Nemoto, and Gunther Dennert

USC/Norris Comprehensive Cancer Center
University of Southern California
School of Medicine
Los Angeles, California 90033

ABSTRACT

Protein mono-(ADP-ribosyl)transferases (ADPRTs) catalyze transfer of the ADP-ribose moiety from nicotinamide adenine dinucleotide (NAD) to specific amino acids. We recently described presence of an enzyme with this activity on cytotoxic T cells (CTL). Incubation of CTL with micromolar concentrations of NAD causes inhibition of cell proliferation and cytolytic activity. ADPRT can be released by bacterial phosphoinositol specific phospholipase C, indicating that it is a glycosylphosphatidylinositol (GPI) anchored exo-enzyme. Enzymatic release of ADPRT results in inability of NAD to modulate CTL function. Expression of ADPRT was found to be regulated, in quiescent CTL ADPRT is expressed at significant levels, however, upon TCR crosslinking it is rapidly released by an anchor hydrolyzing mechanism. This results in relative insensitivity to the inhibitory action of NAD. The question how ADPRT regulates T cell functions was investigated by incubating CTL with radioactively labeled NAD which causes modification of several proteins, pointing to potential candidates in these regulatory processes. We found that the protein tyrosine kinase $p56^{lck}$ but not $p59^{fyn}$ exists in a digitonin resistant complex with a 40 kD protein, which in its ADP-ribosylated form suppresses $p56^{lck}$ kinase activity. ADP-ribosylation of this protein is mediated by the arginine specific protein mono-ADPRT, presumably utilizing ecto-NAD as substrate. Release of the ADPRT by GPI-specific phospholipase C results in failure of ecto-NAD to downmodulate $p56^{lck}$ kinase activity. Concomitant to suppression of the kinase by ecto-NAD, CD8 mediated transmembrane signaling is found to be inhibited, whereas transmembrane signaling *via* CD3 is only slightly affected.

ADP-Ribosylation in Animal Tissue, edited by Haag and Koch-Nolte
Plenum Press, New York, 1997

BACKGROUND

The regulation of lymphocytes responding to antigenic stimuli has been an area of intense interest in immunology. Many components are involved forming intricate cascades designed to eliminate invading microorganisms. Central to these responses is the recognition of antigens by specific receptors which serve to trigger a myriad of intracellular processes. Over the last few years it has become clear that although antigen specific receptors play an obligatory role in immune responses, signals received through coreceptors may determine the precise type of the response, be it stimulatory or inhibitory. Molecules on the lymphocyte surface, involved in transducing or modulating these second signals are therefore of much interest in studies of lymphocyte regulation. We recently found a glycosylphosphatidylinositol (GPI) anchored enzyme on the surface of cytotoxic T cells (CTL) that utilizes extracellular NAD to transfer ADP-ribose groups to arginine residues of cell surface proteins (1). In as much as this observation was unexpected, expression of the enzyme was intriguing in view of the well known fact that bacterial toxins like cholera and pertussis toxin have similar activity and strongly impair cell functions (2). The observation did raise several questions, one of which is where might the substrate of this enzyme i.e., nicotinamide adenine dinucleotide (NAD) be coming from? NAD is an intracellular metabolite whose extracellular origin and function is far from clear. While this remains to be answered, it is well established that enzymes that utilize NAD are quite common on lymphoid cells. For example cell surface marker CD38 (3,4,5) had been described as an enzyme that utilizes NAD to form cyclic ADP-ribose, a potent second messenger (4,6) which may mobilize Ca^{2+} independent of IP3. Another group of enzymes are NADases which have been described on many cells, including lymphocytes and some of which are GPI-anchored (7,8). The function of NADases is not known but their ubiquitous distribution suggests that for reasons unknown, extracellular concentrations of NAD are quite low because of the action of the NADases. A prediction therefore is that incubation of lymphocytes with low concentrations of NAD should, for the case that these enzyme do indeed fulfill a function, lead to detectable effects. We therefore set out to examine what effects NAD may exert on lymphocyte functions and whether these effects are associated with expression of ADP-ribosyltransferase on the cell surface of affected cells.

RESULTS

Cytotoxic T Cells Express an Arginine-Specific Protein Mono-ADP-Ribosyltransferase on Their Cell Surface

The observation that an ADP-ribosyltransferase (ADPRT) is expressed on lymphocytes was originally made with murine cytotoxic T cells (CTL) which had been generated in an allogeneic mixed lymphocyte reaction. When these cells were harvested 10 days after culture initiation (1), incubated with [^{32}P]-NAD and their membranes assayed by SDS-PAGE followed by autoradiography, strong labeling of various proteins was seen (Fig. 1A). This labeling was found to be due to mono-ADP-ribosylation of arginine residues by virtue of release of ADP-ribose from proteins by hydroxylamine (1). Evidence that labeling is mediated by a resident cell surface ADPRT was provided by two kinds of experiments. In one, isolated CTL membranes were separated on SDS-PAGE, followed by incubation of the gels with arginine rich histone and [^{32}P]-NAD (1). Subsequent autora-

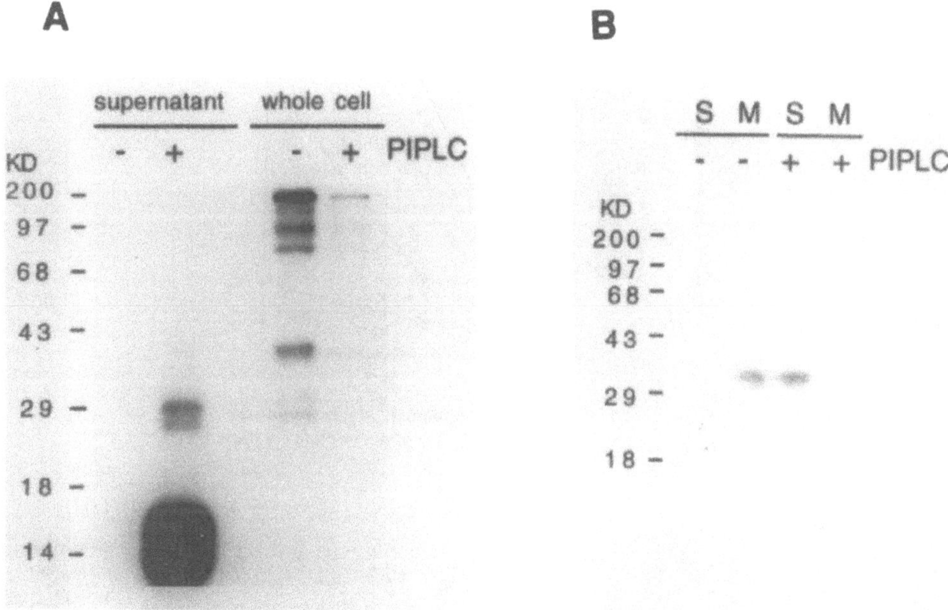

Figure 1. PIPLC releases ADP-ribosyltransferase from intact CTL. **A:** CTL were incubated in BSS either with (+) or without (-) 5 U/ml PIPLC for 1 hr at 37°C. Cells were collected, labeled with [^{32}P]-NAD and crude cell lysates analyzed by SDS-PAGE (1). To tissue culture supernatants histone and [^{32}P]-NAD were added using the conditions for ADP-ribosylation reactions of histone (1), then analyzed by SDS-PAGE. **B:** CTL particulate fraction or corresponding amounts of concentrated culture supernatant from cells treated with PIPLC were separated on SDS-PAGE. ADP-ribosyltransferase activity was detected by zymographic assay using histone and [^{32}P]-NAD as substrates.

diography revealed that an entity with an apparent MW of 35 kD, able to label histone is present in the gel bed (Fig. 1B). That the enzyme responsible for labeling resides on the outside of CTL was likely, as NAD is not able to penetrate cells. We speculated that the enzyme might be GPI-anchored because a recently cloned enzyme with similar properties, expressed on skeletal muscle, is attached to the cell membrane in this way (9). Indeed incubation of CTL with bacterial phosphoinositol specific phospholipase C (PIPLC) releases enzyme able to label histone and visualized in the zymographic assay (Fig. 1B). However, the enzyme is also released by PIPLC from whole cells where cell extracts were tested for ADPRT activity by the agmatine assay (Fig. 2A).

Ecto-NAD Has Suppressive Effects on CTL Functions Requiring Presence of the ADPRT

Demonstration of a protein mono-ADPRT on CTL and its ability to label cell surface proteins does not in itself show that this enzyme has a function. To provide evidence for this, CTL were incubated with NAD and then assayed for various responses. Figure 2B shows that preincubation of CTL for only 3 hr with 1, 10 or 100 μM NAD exerts significant inhibitory effects on the proliferative response upon culture with stimulator cells on days 2 or 3 after restimulation. To examine whether other functions are similarly affected, the ability of NAD treated CTL to lyse targets was examined. Results in Fig. 2C show that preincubation of CTL but not targets with NAD leads to a significant inhibition of cell

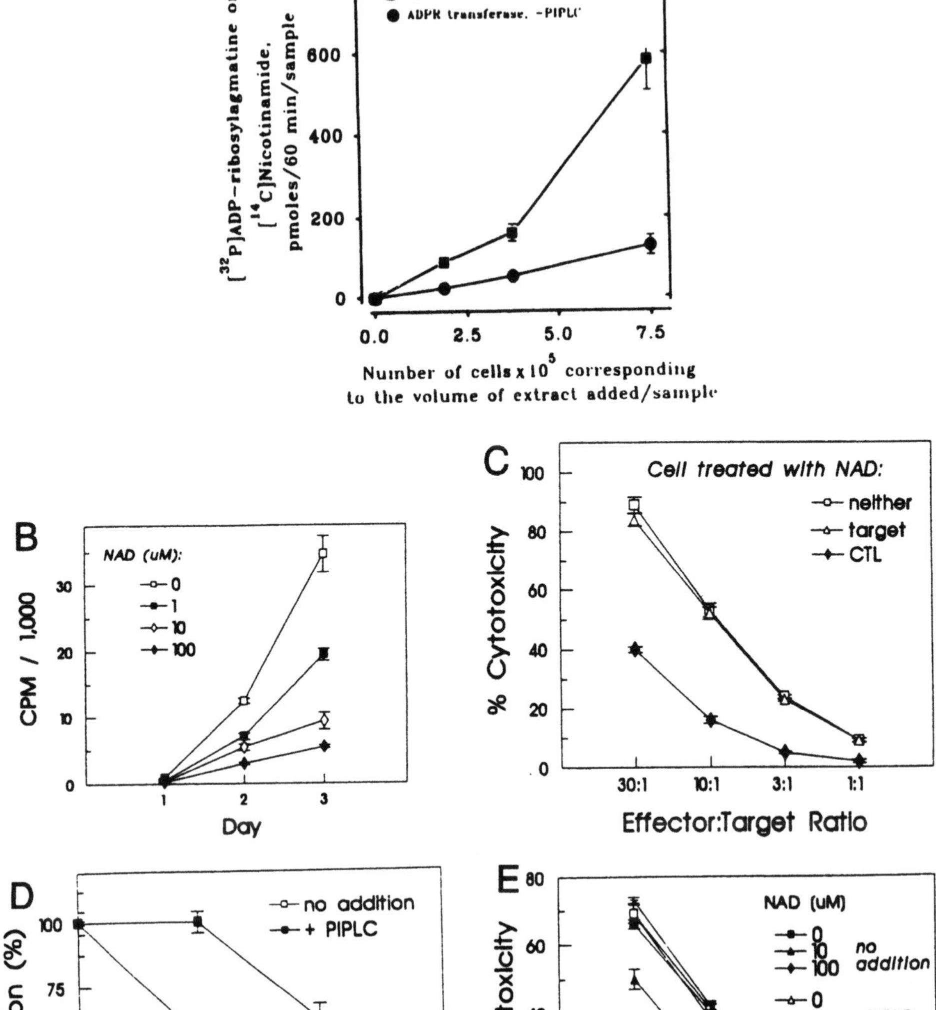

Figure 2. **A:** PIPLC releases ADP-ribosylagmatine-forming activity from CTL. CTL were incubated with or without 5 U/ml PIPLC in BSS for 1 hr at 37°C and ADP-ribosyltransferase activity was determined in the cell culture medium (1). **B:** NAD inhibits CTL proliferation. CTL were treated with NAD for 3 hr, washed and stimulated with irradiated BALB/c cells for three days in the absence of NAD. Cell proliferation on day three is plotted (1). **C:** NAD inhibits CTL activity. P815 targets or CTL were incubated for 3 hr with 100 μM NAD at 37°C. After washing targets were labeled with ^{51}Cr, mixed with effector cells at the ratios shown and cytotoxicity determined 4 hr later. **D:** Treatment of CTL with PIPLC reduces the effect of ecto-NAD. CTL were incubated with or without PIPLC as in Fig. 1 followed by treatment with NAD as indicated, for 3 hr at 37°C. Cells were washed and cultured for two days with irradiated BALB/c stimulator cells to assay the proliferative response. Cell growth was calculated as % growth from [^{3}H]Tdr incorporation on day two into cells not treated with NAD (1). **E:** CTL were incubated or not with PIPLC, treated with the indicated concentrations of NAD for 3 hr, then assayed for cytotoxic activity on P815 targets in a 4 hr assay.

mediated cytotoxicity. It therefore appears that incubation of CTL with relatively low concentrations of NAD exerts sustained effects on CTL cytotoxic activity and proliferation.

The modulation of CTL functions by NAD, while predicted from the presence of the protein mono-ADPRT on the cell surface of CTL, does not in itself show that it is this enzyme that is responsible for the effects of NAD, other enzymes could be involved. There is no easy approach to prove that it is indeed the identified ADPRT which is responsible for the observed effects. Various NAD derivatives and breakdown products, among those nicotinamide, ADP-ribose, cyclic ADP-ribose, 5'-AMP, adenosine and NMN were not able to mimic the inhibitory effect of NAD, making unlikely the action of CD38 and NADases (1). On the other hand incubation of CTL with PIPLC,which causes release of GPI-anchored proteins, induces relative insensitivity to the NAD mediated inhibitory effects. Figures2D,E show that PIPLC-reated CTL sustain decreased inhibition by incubation with NAD in the proliferative and cytotoxic response, consistent with the notion that the NAD effect is due to a GPI-anchored cell surface component, presumably the ADPRT. The correlation between expression of ADPRT on CTL and failure of NAD derivatives to mimic the NAD effect therefore provides presumptive evidence for the action of this enzyme in NAD-ediated modulation of their function.

Expression of ADPRT on the CTL Surface Is Regulated by an Active Release Mechanism

One prediction from the effects ecto-NAD has on CTL function is that if the identified ADPRT is indeed responsible for these effects, its activity should be regulated. An indication that this may be the case was that cell membranes from CTL, restimulated *in vitro* with anti-CD3, were found to lack detectable enzyme activity (10). This lack of enzyme activity could be due to either removal of the enzyme from the cell surface or its *in situ* inactivation. To examine whether enzyme activity is released from the cell surface upon stimulation *via* the TCR, CTL were incubated on anti-CD3 coated tissue culture plates for various lengths of time and culture supernatants assayed for ADPRT activity using the agmatine assay. Figure 3A shows that ADPRT activity appears very rapidly in the tissue culture medium, reaching plateau values by about 60 min. Concomitant to the appearance of enzyme activity in the tissue culture medium, labeling of cells incubated with $[^{32}P]$-NAD (10) decreases dramatically, indicating that loss of enzyme from the cell surface results in failure of $[^{32}P]$-NAD to label membrane associated proteins. Also note that cell viability is not affected by release of the enzyme.

Release of ADPRT from the CTL surface is not only induced by stimulation *via* the TCR but also by other stimuli including incubation with PMA and IL-2, the kinetics and efficiency of this release being similar to that induced by anti-CD3 (10). These results suggest that cell activation in general,and not just a signal through the TCR,induces the release mechanism. The question as to what the precise mechanism of release might be was explored. We excluded an active secretory process,as stimulation of CTL in the presence of monensin and brefeldin A does not interfere with release of the enzyme (10). It is likely that ADPRT is directly cleaved from the cell surface by an enzymatic process, but it is not known whether a phospholipase or protease are involved (10).

Ecto-NAD Appears to Modulate Transmission of Second Signals

Given the observation that incubation of CTL with NAD leads to ADP-ribosylation of various cell surface proteins concomitant to inhibition of CTL functions, the question arises as to what the regulatory pathway may be that leads to inhibition of CTL functions. Cell activation is initiated at the cell surface by engagement of receptors followed by transmission of signals into the cell. Considering the localization of the ADPRT on the cell surface and the inability of NAD to penetrate cell membranes (11), it is most likely that action of the ADPRT is on the level of receptor engagement and signal transmission. Therefore, transmembrane signal transmission initiated by receptor crosslinking was assayed in NAD-reated CTL (12). When CTL are treated with NAD, then incubated with stimulator cells and the generation of inositolphosphates assayed, it is seen that the generation of IP1, IP2 and IP3 is significantly inhibited (Fig. 3B). This suggests that signal transmission initiated by receptor cross-linking is affected by ecto-NAD. To examine which receptors may be involved, CTL were treated with NAD, then reacted with either anti-CD3 or anti-CD8 and a cross-linking second antibody. Results in Fig. 3C show that inositol phosphate generation induced by anti-CD3 is not affected in NAD-retreated cells. In contrast, the generation of inositol phosphates induced by cross-linking of CD8 is significantly reduced (Fig. 3D). These results are consistent with the notion that ecto-NAD regulates signaling through the CD8 coreceptor while having little effect on signaling through CD3 (12). The observed inhibitory effect of NAD on stimulator cell-induced generation of inositol phosphates therefore may be due primarily to inhibition of CD8 coreceptor signaling, rather than inhibition of TCR signaling (12).

Ecto-NAD Has Direct Inhibitory Effects on the Activity of the Protein Kinase p56^{lck}

The observation that signaling through CD8 is inhibited by preincubation of cells with NAD suggests that either the CD8 molecule itself or a molecule associated with CD8 is affected. Experiments to examine whether CD8 is ADP-ribosylated did not however indicate that this is the case (12). Therefore the possibility was examined that a CD8-associated molecule may be involved. It is well documented that the *src*-related protein tyrosine kinase p56^{lck} can be associated with CD8, and that cross-linking of CD8 may activate this enzyme (13,14). To examine whether this pathway of transmembrane signaling is affected, NAD-reated cells were reacted with anti-CD8 and a cross-linking second antibody, lysed in 1% digitonin buffer, and p56^{lck} precipitates were assayed for kinase activity. Results in Fig. 4A show that within 5 min of CD8 cross-linking, auto-phosphorylation of p56^{lck} is strongly stimulated, a response that is completely inhibited in NAD-preincubated cells. It therefore appears that in NAD-treated CTL anti-CD8-mediated activation of p56^{lck}

Figure 3. ADPRT is released from CTL upon CD3 crosslinking and PMA stimulation (10). **A:** CTL were cultured on anti-CD3-coated plates in RPMI 1640 10% FCS for the indicated times. ADPRT activity was determined by agmatine assay (1). CTL were also labeled with [^{32}P]-NAD and whole cell lysates separated by SDS-PAGE. Autoradiographs were scanned and integrated values from each band plotted. Cell viabilities were determined by trypan blue exclusion. Tissue culture supernatants from CTL treated with 5 U/ml PIPLC in RPMI 1640 10% FCS for 1 hr were used as positive control (10). **B:** Activation of CTL is inhibited by NAD (12). CTL labeled with [^{3}H]-inositol were treated with NAD for 3 hr, followed by incubation with P815 target cells (at a ratio of 1:2) at 37°C for 5 min. Cells were lysed and assayed for inositol phosphates (12) in triplicates. **C:** CTL labeled with [^{3}H]-inositol were treated with NAD for 3 hr, followed by incubation in complete medium for 5 min at 37°C containing 1:100 final dilution of anti-CD3 (500 AA2) ascites. Cells were harvested for inositol phosphate assay as in B. **D:** CTL labeled with [^{3}H]-inositol were treated with NAD for 3 hr then incubated with a 1:10 dilution of anti-CD8 (AD4 [15] supernatant) on ice for 30 min. After washing cells were reacted with 10 µg/ml goat anti-mouse IgG+IgM followed by incubation at 37°C for 5 min. Cells were then harvested for inositol phosphate assay.

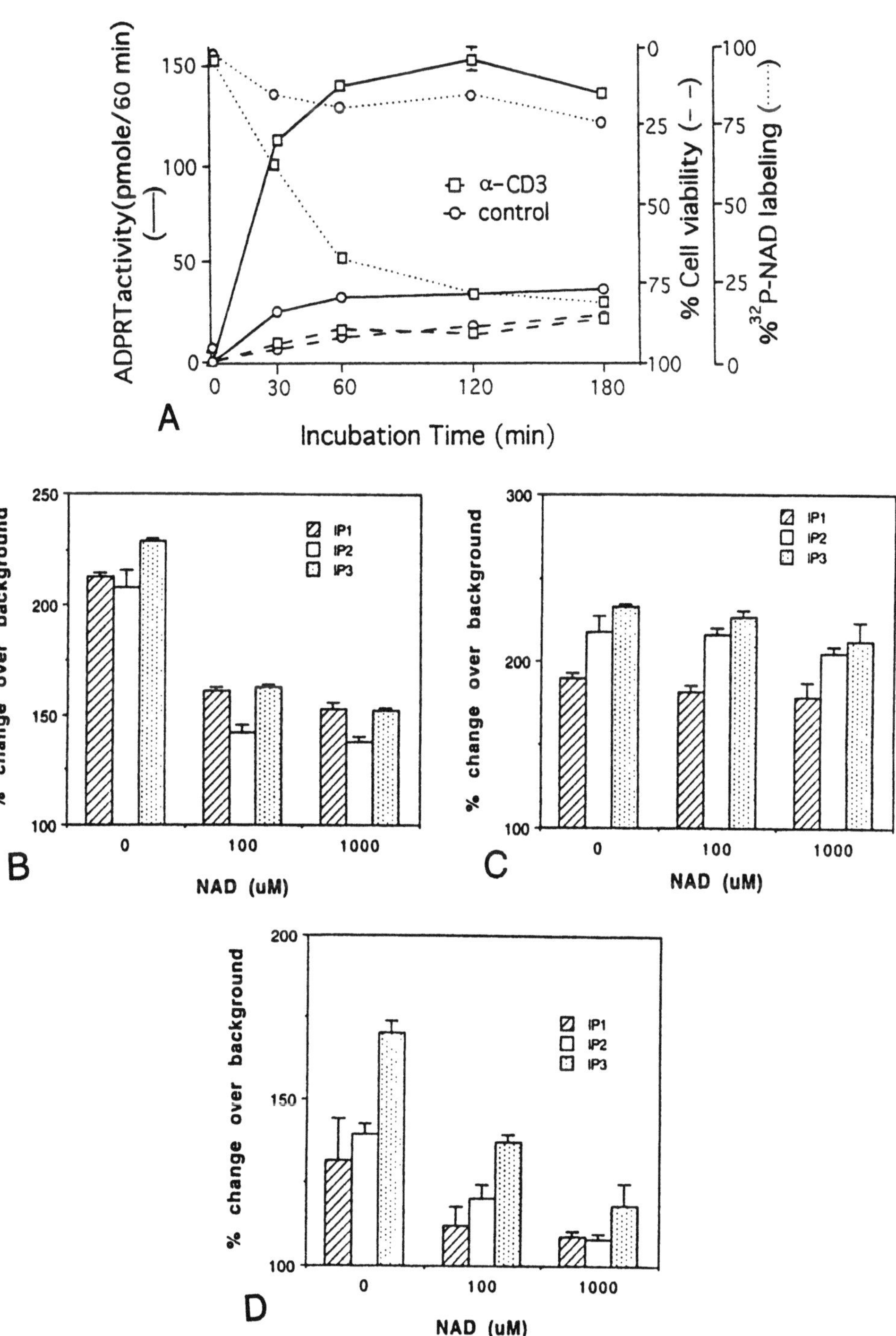

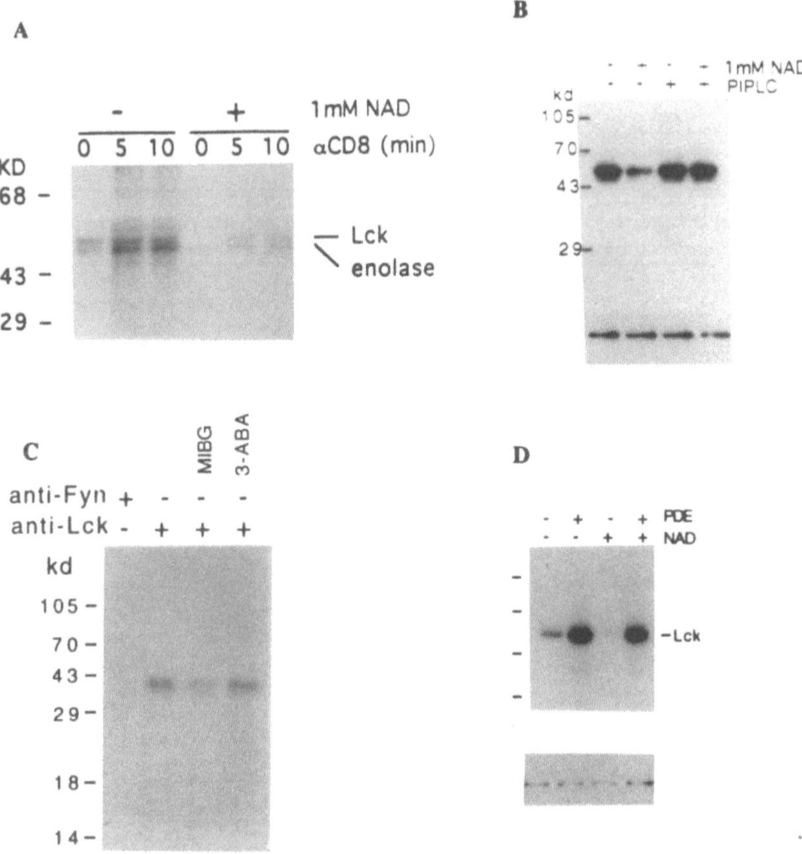

Figure 4. NAD inhibits p56^{lck} but not p59^{fyn} kinase activity. **A:** CTL were treated with 1 mM NAD for 3 hr, followed by incubation with 1:10 anti-CD8 (AD4 [15]) on ice for 30 min. Cells were incubated in complete RPMI 1640 medium containing 10 μg/ml anti-mouse IgG+IgM at 37°C for 0, 5, and 10 min, lysed in digitonin buffer and p56^{lck} precipitates assayed for autophosphorylation and enolase activity (protein equivalent to 10^7 cells per lane). **B:** Treatment of intact CTL with PIPLC abrogates NAD mediated suppression of p56^{lck} activity. CTL (10^7 cells per lane) were treated or not with PIPLC then incubated with 1 mM NAD for 3 hr at 37°C. Digitonin cell lysates were precipitated with anti-p56^{lck} and tested for autophosphorylation activity. Western blots for p56^{lck} are shown in the lower panel. **C:** CTL membranes were labeled with [^{32}P]-NAD in the presence or absence of 1 mM MIBG or 10 mM 3-ABA as indicated. Digitonin lysates were prepared, precipitated with anti-p59^{fyn} or anti-p56^{lck} and separated on SDS-PAGE. **D:** Treatment of p56^{lck} immunoprecipitates with PDE increases kinase activity. CTL membranes were first treated or not treated with 1 mM NAD as indicated. Next digitonin lysates were prepared from these membranes and p56^{lck} immunoprecipitated. Immunoprecipitates were treated with PDE and autophosphorylation activity of p56^{lck} was determined. Quantities of p56^{lck} assayed by Western blot are shown in the lower panel.

does not take place, raising questions as to the underlying mechanism of this effect. To examine this, we first assessed whether the observed inhibitory effects of NAD require presence of ADPRT on the cell surface. CTL were treated with PIPLC, membranes prepared, incubated with NAD and immunoprecipitates tested for p56^{lck} autophosphorylation activity. Results in Fig. 4B show that while preincubation with NAD inhibits kinase autophosphorylation activity, pretreatment with PIPLC before incubation with NAD, completely abrogates the inhibitory effect of NAD. These results are consistent with the notion that the GPI-anchored ADPRT plays a role in NAD mediated regulation of p56[lck] activity.

One mechanism by which ADPRT could regulate p56[lck] is by ADP-ribosylation, another is that an ADP-ribosylated protein associates with the kinase and modulates its activity. To find out, CTL membranes were labeled with [32P]-NAD, and digitonin lysates were precipitated with anti-kinase antibodies. Results in Fig. 4C show that anti-p56[lck] but not anti-p59[fyn] precipitates a labeled 40 kD protein (p40). Labeling of this protein is inhibited by the presence of meta-iodobenzyl guanidine (MIBG) but less by 3-aminobenzamide (3-ABA) in support of the notion that the arginine specific mono(ADP-ribosyl)transferase (15) is responsible for the labeling. The possibility that p40 is in fact the ADPRT, bound to the kinase,was examined by assaying the ability of p56[lck] precipitates to ADP-ribosylate agmatine. No activity was demonstrable, suggesting that p40 is not the ADPRT. Again it was shown that removal of the ADPRT from the cell surface by PIPLC leads to failure of [32P]-NAD to label p40 (12).

Association of an ADP-ribosylated protein (p40) with p56[lck] points to a regulatory function of this protein in kinase activity. Thus, it is possible that removal of ADP-ribose groups from p40 changes p56[lck] activity. To examine this p56[lck] immunoprecipitates containing p40 were treated with snake venom phosphodiesterase (PDE) then tested for autophosphorylation activity. Figure 4D shows that incubation of cell membranes with NAD causes complete loss of autophosphorylation activity in p56[lck] immunoprecipitates. In contrast membranes treated with PDE show a dramatic increase in p56[lck] autophosphorylation activity. When membranes were first treated with NAD followed by incubation with PDE, kinase activity was restored. These results support the notion that presence of the ADP-ribose groups on p40 exerts a suppressive effect on the activity of p56[lck].

The observation that p40 in its ADP-ribosylated form inhibits kinase activity raises questions as to the precise localization of this protein relative to the ADPRT which is on the cell surface and p56[lck] which is bound to the inner leaflet of the cytoplasmic membrane. Given this configuration one would postulate that the p40 protein should be labeled by [32P]-NAD in intact cells. Results showed that labeled p40 can be immunoprecipitated with anti-p56[lck] regardless of whether cell membranes or whole cells were incubated with [32P]-NAD (12). This suggests that p40 is accessible to ecto-NAD as well as the ADPRT. Moreover treatment of either cell membranes or intact cells with PDE removes label from p40, consistent with the notion that the ADP-ribose groups are accessible to macromolecules, i.e. the PDE on the outside of the cell (12).

DISCUSSION

The expression on various cell types of enzymes which utilize NAD has been an unresolved mystery for many years. The ubiquitous presence of NADases was thought to enable uptake of NAD into the cell by virtue of the fact that nicotinamide can penetrate cell walls and then be converted into NAD inside the cell (11). This mechanism does not however address the question how NAD, which is presumably generated inside cells, reaches the extracellular space. Expression of enzymes synthesizing cyclic ADP-ribose from NAD or able to attach ADP-ribose to proteins points to pathways in which NAD may have functions in cell regulation. The observation then that an arginine-specific ADPRT is expressed on murine CTL opened the possibility to explore the regulatory action ecto-NAD might have on lymphocyte functions.

We found that µM concentrations of NAD exert profound effects on CTL (1). Thus, cell proliferation induced by stimulator cells is severely impaired as well as cytolytic activity on target cells. These effects do not require the continuous presence of NAD, as pre-incubation and subsequent removal lead to similar results. These observations suggest that ecto-NAD induces more than transient effects, a finding that is consistent with action of an ADPRT whose role is to ADP-ribosylate cell surface proteins. It is difficult however to conclusively demonstrate that the NAD-induced suppressive effect is indeed due to action of the ADPRT. The involvement of NADase or ADP-ribosyl cyclase was made unlikely by the use of reaction products of these enzymes which fail to mimic the NAD effect. However, these results constitute indirect evidence only. Another approach to show that ecto-NAD acts *via* the ADPRT was opened by the demonstration that the enzyme is anchored by a GPI-anchor in the cell surface. Thus, removal of the ADPRT by bacterial PIPLC could serve to show that NAD effects require presence of the enzyme on the cell surface. Our results verify this prediction, but this evidence is again indirect as PIPLC removes other GPI-anchored proteins.

In many instances expression of molecules that have profound effects on lymphocyte functions are strictly regulated. This has been documented for adhesion molecules, antigen receptors and cytokine receptors. Therefore, a prediction from the observation that ecto-NAD inhibits CTL function is that mechanisms should be in place that regulate its inhibitory pathway. In support we show that stimulation of CTL *via* the TCR leads to virtually quantitative release of ADPRT in active form. As a predicted consequence CTL become relatively insensitive to the inhibitory action of NAD. While at present it is not known why ADPRT is released in active form, the fact that it is released, suggests that its absence from the cell surface facilitates more efficient cell activation or function.

An important question is how the ADPRT regulates CTL function. Analysis of this question is not straightforward as incubation of cells with labeled NAD leads to ADP-ribosylation of several proteins, the identity of which is presently under investigation. But even when these proteins were to be identified it may still not be obvious which one is the most important one in cell regulation. We therefore pursued approaches in which the regulation of CTL function by NAD was scrutinized. Experiments showed that induction of inositol phosphate generation is inhibited by NAD when CTL were stimulated with stimulator cells, but not in cells in which CD3 was crosslinked. This pointed to the impairment of coreceptor signaling. In support, inositol phosphate generation induced by crosslinking of CD8 was significantly inhibited in NAD-treated cells. Support was provided for the notion that the inhibition of CD8 signaling may be due to inhibition of the protein tyrosine kinase p56^{lck}. This kinase was shown to be associated with a protein called p40 which in its ADP-ribosylated form inhibits autophosphorylation of the kinase *in vitro*. Partial removal of ADP-ribose groups by phosphodiesterase was shown to reverse the inhibitory effects on autophosphorylation. These results provide a mechanism by which the ADPRT regulates CTL function, i.e. by regulating signal transmission through CD8 and *via* inhibition of p56^{lck} activity. Caution has to be applied to this interpretation however. For once, inhibition of p56^{lck} should also exert effects on the signaling function of TCRs, which had been demonstrated in a lymphoma cell line (16). In fact incubation of CTL with NAD followed by CD3 crosslinking was found to result in some inhibition of cell activation and proliferation by the NAD treatment (12). Therefore, the failure to observe inhibition of inositol phosphate generation induced by CD3 crosslinking subsequent to NAD treatment could be due to relative insensitivity of this assay. It therefore appears that there are effects on both TCR and CD8 signaling, yet the effects on the latter are more pronounced.

Another consideration is that other cell surface molecules are also ADP-ribosylated, raising the possibility that additional molecules, perhaps involving second signals may be involved. The previous observation that CTL target binding is inhibited in NAD-treated CTL could point to this. Further experimentation is required to investigate this possibility.

Finally, given the identification of this regulatory pathway in CTL, the question arises whether other lymphocytes express ADPRT, and whether ADPRT exerts regulatory effects in these cells. Another question is, can this pathway be activated to modulate lymphocyte responses and does this open new avenues for the treatment of disease? These questions will be interesting to investigate now that the first steps in the characterization of a lymphocyte protein mono-ADPRT have been initiated.

REFERENCES

1. Wang, J., E. Nemoto, A. Y. Kots, H. R. Kaslow, and G. Dennert. 1994. Regulation of cytotoxic T cells by ecto-nicotinamide adenine dinucleotide (NAD) correlates with cell surface GPI-anchored/arginine ADP-ribosyltransferase. J. Immunol., *153*:4048.
2. Osborne, J. C., S. J. Stanley, and J. Moss. 1985. Kinetic mechanisms of two NAD:arginine ADP-ribosyltransferase: the soluble salt-stimulated transferase from turkey erythrocytes and choleragen, a toxin from *Vibrio cholerae*. Biochemistry, *24*:5235.
3. Reinherz, E. L., P. C. Kung, G. Goldstein, R. H. Levey, and S. F. Schlossman. 1980. Discrete stages of human intrathymic differentiation: analysis of normal thymocytes and leukemic lymphoblasts of T-cell lineage. Proc. Natl. Acad. Sci. USA, *77*:1588.
4. Howard, M., J. C. Grimaldi, J. F. Bazon, F. E. Lund, L. Santos-Argumedo, R. M. E. Parkhouse, T. F. Walseth, and H. C. Lee. 1993. Formation, and hydrolysis of cyclic ADP-ribose catalyzed by lymphocyte antigen CD38. Science, *262*:1056.
5. Ho, H. N., L. E. Hultin, T. Mitsuyasu, J. L. Matud, M. A. Hausner, D. Bockstoll, C. C. Chou, S. O'Rourke, J. M. G. Taylor, and J. V. Giorgi. 1993. Circulating HIV-specific CD8$^+$ cytotoxic T cells express CD38, and HLA-DR antigens. J. Immunol., *150*:3070.
6. Lee, H. C., T. F. Walseth, G. T. Bratt, R. N. Hayes, and D. L. Clapper. 1989. Structural determination of a cyclic metabolite of NAD with intracellular Ca^{2+}-mobilizing activity. J. Biol. Chem., *264*:1608.
7. Muller, H. M., C. D. Muller, and F. Schuber. 1983. NAD^+ glycohydrolasean ecto-enzyme of calf spleen cells. Biochem. J., *212*:459.
8. Han, M. K., C. Y. Yim, N. H. Au, H. R. Kim and U. H. Kim. 1994. Glycosylphosphatidylinositol-anchored NAD glycohydrolase is released from peritoneal macrophages activated by interferon γ and lipopolysaccharide. J. Leucocyte Biol., *56*:792.
9. Zolkiewska, A., M. S. Nightingale, and J. Moss. 1992. Molecular characterization of NAD:arginine ADP-ribosyltransferase from rabbit skeletal muscle. Proc. Natl. Acad. Sci. USA, *89*:11352.
10. Nemoto, E., S. Stohlman, and G. Dennert. 1996. Release of a glycosyl-phosphatidylinositol-anchored ADP-ribosyltransferase from cytotoxic T cells upon activation. J. Immunol., *156*:85.
11. Olsson, A., T. Oloffson, and R. W. Pero. 1993. Specific binding, and uptake of extracellular nicotinamide in human leukemic K-562 cells. Biochem. Pharmacol., *45*:1191.
12. Wang, J., E. Nemoto, and G. Dennert. 1996. Regulation of CTL by ecto-NAD involves ADP-ribosylation of a p56^{lck} associated protein. J. Immunol., *156*:2819.
13. Luo, K. X., and B. M. Sefton. 1990. Cross-linking of T-cell surface molecules CD4 and CD8 stimulates phosphorylation of the lck tyrosine protein kinase at the autophosphorylation site. Mol. Cell. Biol., *10*:5305.
14. Veillette, A., M. A. Bookman, E. M. Horak, and J. B. Bolen. 1988. The CD4 and CD8 T cell surface antigens are associated with the internal membrane tyrosine-protein kinase p56^{lck}. Cell, *55*:301.
15. Smets, L. A., C. Loesberg, M. Jassen, and H. van Rooig. 1990. Intracellular inhibition of mono-(ADP-ribosylation) by meta-iodobenzyl guanidine: specificity, intracellular concentration and effects on glucocorticoid mediated cell lysis. Biochem. Biophys. Acta, *1054*:49.
16. Gupta, S., A. Weiss, G. Kumar, S. Wang, and A. Nel. 1994. The T-cell antigen receptor utilizes Lck, Raf-1, and MEK-1 for activating mitogen-activated protein kinase. Evidence for the existence of a second protein kinase C-dependent pathway in an Lck-negative Jurkat cell mutant. J. Biological Chem., *269*(25):17349.

25

EFFECTS OF INHIBITORS OF ADP-RIBOSYLATION ON MACROPHAGE ACTIVATION

Cécile Le Page,[1] Catherine Pellat-Deceunynck,[2] Jean-Claude Drapier,[1] and Juana Wietzerbin[1]

[1]Unité 365 INSERM "Interférons et Cytokines"
Institut Curie
Section de Recherche
26, rue d'Ulm-75231 Paris cedex 05, France
[2]Unité 211 INSERM "Interactions Récepteurs-Ligands en Immunologie et
    Cancérologie"
Institut de Biologie-9
Quai Moncousu-44035 Nantes cedex 01, France

1. ABSTRACT

Nitric oxide (NO) is a potent mediator involved in many biological functions including macrophage cytotoxicity and non-specific immunity against parasites, bacteria and viruses. Murine macrophages possess the capacity to express the inducible NO synthase (iNOS) which is not constitutively expressed but induced at the transcriptional level by interferon gamma (IFN-γ) alone or synergistically with LPS.

We have investigated the possible role of ADP-ribosylation reactions in the signaling pathway involved in NO synthase induction, since ADP-ribosylation has been reported to be involved in the expression of certain IFN-γ and LPS-inducible genes. We found that inhibitors of ADP-ribosylation inhibited nitrite synthesis in RAW 264.7 macrophages after stimulation by IFN-γ and LPS. These ADP-ribosylation inhibitors acted by preventing NO synthase mRNA induction, without inhibiting NO synthase enzyme activity.

IRF-1, a transcription factor induced and activated by IFN-γ was recently shown to be involved in iNOS induction. We showed that inhibitors of ADP-ribosylation had no effect on IFN-γ-mediated mRNA induction of IRF-1 nor on its activation and binding to its target sequence in the iNOS gene. In addition, the inhibitors failed to impair the IFN-γ-mediated antiviral activity against VSV virus. Since induction by IFN-γ of IRF-1 and induction of the antiviral state proceed through the JAK/STAT signalling pathway, our results imply that ADP-ribosylation reactions are not involved in triggering this pathway. Although the precise mechanism requires further investigation, our results indicate that

ADP-Ribosylation in Animal Tissue, edited by Haag and Koch-Nolte
Plenum Press, New York, 1997

ADP-ribosylation is a crucial step restricted to the signalling pathway which leads to iNOS mRNA induction, as well as TNF and MHC class II induction during macrophage activation.

2. INTRODUCTION

We are interested in the study of the mechanism of action of interferon gamma (IFN-γ), a cytokine produced by activated T cells (1,2). The interaction of IFN-γ with specific cellular receptors triggers the expression of genes coding for proteins involved in the expression of its biological properties, such as antiviral activity, antibacterial and antiparasitic effects. IFN-γ is known to be a potent activator of macrophage functions. It activates macrophage cytotoxicity, TNF-α production, class II expression and nitric oxide (NO) production, which is one of the major effector molecules of macrophages activated by IFN-γ (1,3,4).

NO is produced by conversion of L-arginine to citrulline by two types of NO-synthases (4,5). Two constitutive enzymes (NOS I and NOS III), which are immediately activated by calcium influx and produce low amounts of NO, responsible for some physiological effects such as endothelial relaxation and neurotransmission, and an inducible NO synthase (iNOS or NOS II) are present in many cells, including macrophages and hepatocytes (4–6). In macrophages, this enzyme is transcriptionally induced by IFN-γ alone or synergistically with LPS or TNF-α (7,8). iNOS is responsible for the release of high amounts of NO, which results in physiological increase of macrophage cytotoxicity and non-specific immunity against parasites or bacteria but also in pathophysiological events during inflammation (4,5).

We considered the possibility of modulating the transductional events involved in iNOS induction in macrophages in order to prevent the harmful effect of large amounts of NO release. Since IFN-γ was reported to increase ADP-ribosylation in macrophages, we have studied the effect of inhibitors of ADP-ribosylation such as nicotinamide, 3-aminobenzamide or 3-methoxybenzamide on the induction of iNOS by IFN-γ and LPS.

3. RESULTS AND DISCUSSION

We have first investigated the effect of inhibitors of ADP-ribosylation on the production of nitrite (the oxidation product of NO) when RAW 264.7 macrophage cell line was stimulated with IFN-γ alone or combined with LPS. As shown in Figure 1, both nicotinamide and benzamide inhibited nitrite accumulation in the culture medium. In contrast, nicotinic acid, which does not inhibit ADP-ribosylation, did not prevent nitrite synthesis.

This inhibition of nitrite production by ADP-ribosylation inhibitors was in good correlation with a marked decrease of iNOS protein detectable by immunohistochemistry and with prevention of iNOS mRNA induction (7 and Le Page C. *et al.*, unpublished).

We and others (8,9) have shown that TNF-α is an autocrine co-factor required for iNOS induction by IFN-γ in macrophages. Therefore, we investigated whether the effect of nicotinamide was related to inhibition of TNF secretion. In fact, although nicotinamide inhibited TNF-α production, the addition of exogeneous TNF-α to RAW 264.7 macrophages stimulated in the presence of nicotinamide did not reverse the inhibitory effect on nitrite production (7). Nicotinamide also prevented the induction of MHC class II antigen

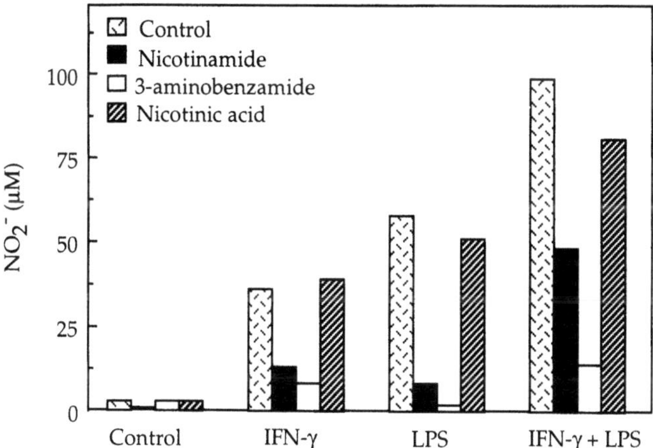

Figure 1. Effect of inhibitors of ADP-ribosylation on NO production by macrophages RAW 264.7 macrophages (5 x 10^5/ml) were cultured for 18 h in medium alone and then stimulated for 24 h with IFN-γ (200 U/ml) or LPS (20 ng/ml) alone or combined in the absence or presence of 10 mM nicotinamide, 3-aminobenzamide or nicotinic acid. Cell-free supernatants were collected and tested for their nitrite concentration using Griess reagent.

by IFN-γ (7). However, the ADP ribosylation inhibitors failed to impair the IFN-γ-mediated antiviral activity against Vesicular Stomatitis Virus (VSV), indicating that ADP ribosylation reactions are not involved in triggering the antiviral machinery (Table 1).

We have analyzed the effect of ADP-ribosylation inhibitors on Interferon Responsive Factor-1 (IRF-1), a transcription factor induced and activated by IFN-γ, since this factor was recently shown to be involved in iNOS induction (10–12).

In addition, Kamijo *et al.* showed that macrophages from IRF-1 knock-out mice were unable to express iNOS after stimulation by IFN-γ alone or combined with TNF-α or LPS (13). Interestingly, iNOS induction is also impaired in IFN-γ receptor knock-out mice (14). Indeed, activation of STAT1 via IFN-γ receptor is required for IRF-1 gene induction (15,16). These observations prompted us to analyze the effect of nicotinamide on IRF-1 induction and IRF-1 activation for the binding to the target sequence present in iNOS gene. As shown in Figure 2, our results indicated that inhibitors of ADP-ribosylation had no significant effect on IFN-γ-mediated IRF-1 mRNA induction nor on the activation and subsequent binding to the IRF-1 protein to its target sequence.

Table 1. Inhibitors of ADP-ribosylation did not impair interferon-g-mediated induction of the antiviral state

	Interferon titer* (U/ml)	
Treatment	Exp.1	Exp. 2
None	4,000	250
Nicotinamide 1 mM	4,000	250
Nicotinamide 10 mM	4,000	250
3-methoxybenzamide 10 mM	4,000	250

*Interferon titer: The reciprocal of the dilution which protects 50 % of murine L cells from the pathogenic effect of Vesicular Stomatitis Virus (VSV)

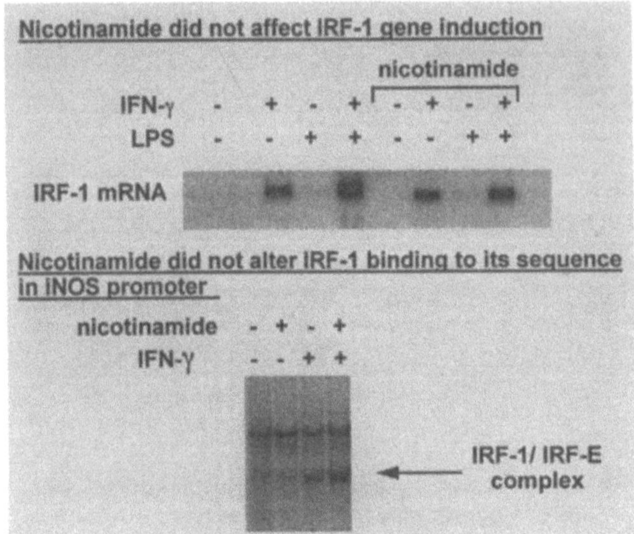

Figure 2. Effect of inhibitors of ADP-ribosylation on interferon responsive factor-1 (IRF-1) *Top panel*: RAW 264.7 macrophages (10 x 10^6) were cultured in medium only or with 200 U/ml IFN-γ, 20 ng/ml LPS or both, in the presence or absence of 10 mM nicotinamide. After 1 h incubation, total RNA was extracted and examinated by Northern blot analysis for IRF-1 and actin (not shown) mRNA expression. No significant modifications or IRF-1 mRNA levels were observed when quantification versus actin mRNA levels was determined by densitometric scanning. *Bottom panel*: Nuclear extracts prepared using 20 x 10^6 cells stimulated in the indicated conditions were submitted to electrophoretic mobility shift assay using a [^{32}P] radiolabeled purified oligonucleotide that corresponded to the IRF-1 binding sequence in the murine iNOS promoter (12).

Taking together, the results reported here showed that nicotinamide and benzamide prevent the induction of NO synthase activity in RAW 264.7 macrophages by IFN-γ (see scheme in Figure 3). Inhibition of endogenous production of TNF-α by stimulated macrophages does not seem to be primarily responsible for the effect of ADP-ribosylation inhibitors on iNOS induction. The effect of nicotinamide is due to prevention of induction of iNOS mRNA and not to the inhibition of the enzyme activity.

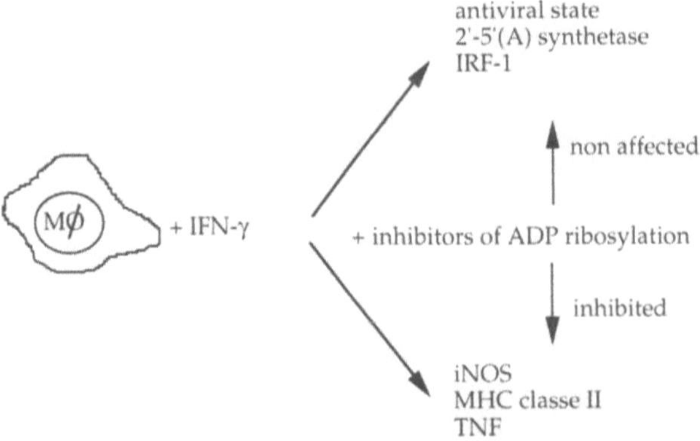

Figure 3. Effect of inhibitors of ADP-ribosylation on interferon-γ signaling pathways.

The inhibition of ADP-ribosylation failed to impair the classical IFN-γ-mediated JAK/STAT signaling pathway(s) (induction of antiviral state, 2′-5′ (A) synthetase, induction and activation of IRF-1, etc...). In contrast, our results strongly suggest that ADP-ribosylation is a crucial step in the signaling pathway which leads to iNOS, TNF-α and MHC class II induction during macrophage activation by IFN-γ.

4. ACKNOWLEDGMENTS

This work was supported by grants from INSERM and ARC (6689). The authors are grateful to Mrs Agnès Birot for expert secretarial assistance.

REFERENCES

1. Schreiber, R. D., & Celada, A. 1985. Molecular characterization of IFN-γ as a macrophage activating factor. *Lymphokine Res. 11*: 87–118.
2. Trinchieri, G., & Perrussia, B. 1985. Immune interferon: A pleiotropic lymphokine with multiple effects. *Immunol. Today 6*: 131–136 .
3. Stuehr, D. J., & Nathan, C. F. 1989. Nitric oxide: A macrophage product responsible for cytostasis and respiratory inhibition in tumor target cells. *J. Exp. Med. 169*: 153–1555.
4. Nathan, C. 1992. NO as a secretory product of mammalian cells. *FASEB J. 6*: 3051–3064.
5. Moncada, S., Palmer, R. M. J., & Higgs. E. A. 1991. Nitric oxide: Physiology, Pathophysiology and Pharmacology. *Pharmacol. Rev. 43*: 109–142.
6. Nüssler, A. K., & Billiar, T. R. 1993. Inflammation, immunoregulation and inducible nitric oxide synthase. *J. Leuk. Biol. 54*: 171–178.
7. Pellat-Deceunynck, C., Wietzerbin, J., & Drapier, J. C. 1994. Nicotinamide inhibits nitric oxide synthase mRNA induction in activated macrophages. *Biochem. J. 297*: 53–58.
8. Drapier, J. C., Wietzerbin. J., & Hibbs, Jr., J. B. 1988. IFN-γ and TNF-α induce the L-arginine-dependent cytotoxic effector mechanism in murine macrophages. *Eur. J. Immunol. 18*: 1587–1592.
9. Deng, W., Thiel, B., Tannenbaum, C. S., Hamilton, T. A., & Stuehr, D. J. 1993. Synergistic cooperation between T cell lymphokines for induction of the nitric oxide synthase gene in murine peritoneal macrophages. *J. Immunol. 151*: 322–329.
10. Lowenstein, C. J., Alley, Z. W., Raval, P., Snowman, A. M., Snyder, S., Russell, S. W., & Murphy, W. 1993. Macrophage NOS gene: Two upstream regions mediate induction by IFN-γ and LPS. *Proc. Natl. Acad. Sci. USA 90*: 9730–9734.
11. Xie, Q. W., Whisnant, R., & Nathan, C. 1993. Promoter of the mouse gene encoding calcium-independent NO synthase confers inducibility by IFN-γ and bacterial LPS. *J. Exp. Med. 177*: 1779–1784.
12. Martin, E., Nathan, C., & Xie, Q. W. 1994. Role of interferon regulatory factor-1 in induction of nitric oxide synthase. *J. Exp. Med. 180*: 977–984.
13. Kamijo, R., Harada, H., Matsuyama, T., Boslan, M., Gerecitano, T., Shapiro, D., Koh, S. I., Kimura, T., Green, S. J., Mak, T. W., Taniguchi, T., & Vilcek, J. 1994. Requirement for transcription factor IRF-1 in NOS induction in macrophages. *Science 263*: 1612–1617.
14. Kamijo, R., Shapiro, D., Le, J., Huang, S., Aguet, M., & Vilcek, J. 1993. Generation of nitric oxide and induction of major histocompatibility complex class II antigen in macrophages from mice lacking the interferon-γ receptor. *Proc. Natl. Acad. Sci. USA 90*: 6626–6630.
15. Pine, R., Canova, A., & Schindler, C. 1994. Tyrosine phosphorylated p91 binds to a single element in the ISGF-2/IRF-1 to mediate induction by IFN-α and IFN-γ, and is likely to autoregulate the p91 gene. *EMBO J. 13*: 158–164.
16. Darnell, J. E., Kerr, I. A., & Stark, G. R. 1994. JAK-STAT pathways and transcriptional activation in response to IFNs and other extracellular signaling proteins. *Science 264*: 1415–1421.

26

THE T CELL MARKER RT6 IN A RAT MODEL OF AUTOIMMUNE DIABETES

Dale L. Greiner,[1] Samir Malkani,[1] Toshihiro Kanaitsuka,[1] Rita Bortell,[1] John Doukas,[1] Mark Rigby,[1] Barbara Whalen,[1] Linda A. Stevens,[2] Joel Moss,[2] John P. Mordes,[1] and Aldo A. Rossini[1]

[1]Diabetes Division, University of Massachusetts Medical School
Biotech II, 373 Plantation Street, Suite 218
Worcester, Massachusetts 01605
[2]The Pulmonary-Critical Care Medicine Branch
National Heart, Lung, and Blood Institute
National Institutes of Health
Bethesda, Maryland 20892-1590

INTRODUCTION

The RT6 alloantigenic system of the rat was discovered in the 1970s. It was originally designated Pta, AgF, A.R.T.-2, and RTLy-2 by the laboratories involved in its characterization. Exchange of reagents demonstrated that these laboratories had identified the same system, and the official designation RT6 was assigned in 1982 (1).

Two allotypes, RT6.1 and RT6.2, with molecular weights of 25–35 kD and 24–26 kD, respectively, are known (2). RT6.1 exists in both glycosylated and non-glycosylated forms, whereas RT6.2 exists only in a non-glycosylated form (2). RT6 is a glycosyl-phosphatidylinositol (GPI)-linked membrane protein (3) expressed post-thymically on mature rat T lymphocytes (4–6). RT6 is also expressed on intestinal intraepithelial lymphocytes (IELs) (7,8), and a soluble form of the RT6 protein is found in the circulation (9). RT6 protein is not detectable on bone marrow T cell precursors, thymocytes, or non-lymphoid tissues (4,5). cDNA for both alleles has been sequenced; the deduced amino acid sequences confirm the presence of glycosylation sites on RT6.1 but not RT6.2 (10,11). The unique tissue and cell specific expression of the RT6 protein have led to investigations of the development and differentiation of RT6$^+$ cells, functions of the RT6$^+$ T cell subset, and properties of RT6 protein.

ADP-Ribosylation in Animal Tissue, edited by Haag and Koch-Nolte
Plenum Press, New York, 1997

ONTOGENY OF RT6⁺ CELLS

Developmental analyses show that RT6 protein is undetectable on peripheral T cells at birth but is detectable at low levels on splenic and lymph node T cells by postnatal days 4–5 (4–6). Maximal peripheral lymphoid expression is achieved by 2–3 months of age. RT6⁺ T cells are severely deficient in neonatally thymectomized and athymic *nu/nu* rats (5,7,8). Intrathymic injection studies demonstrate that most peripheral RT6⁺ T cells are thymus-derived (5). Expression of RT6 by peripheral T cells occurs 4–8 days after their release from the thymus, suggesting a parent-progeny relationship between RT6⁻ recent thymic emigrants and RT6⁺ peripheral T cells (4–6,12).

Thymectomy, *in vivo* depletion, and adoptive transfer experiments suggest the existence of 2 lineages of peripheral T cells in the rat: 1) RT6⁺ and pre-RT6⁺ T cells and 2) T cells that remain RT6⁻. About 50% of RT6⁻ T cells never become RT6⁺ and have been termed "true RT6⁻" T cells (6). The immune capabilities specific to each of these peripheral T cell subsets have not yet been defined.

RT6 is expressed on ≈70% of CD8⁺ and ≈50% of CD4⁺ T cells (4–6) and is also found on CD45⁺ T cells (13). It is absent from most Thy-1⁺ recent thymic migrants (12). RT6⁺ T cells mediate several helper and regulatory functions, including *in vitro* proliferation in response to allogeneic and mitogenic stimulation, participation in graft-versus-host reactivity, and suppression of mixed lymphocyte reactivity (14). Effector functions of rat T cells, in particular cytotoxic activity, are mediated by RT6⁻ T cells (14).

Expression of RT6 varies with the activation state of the cell (15). Mitogen activation of unsorted T cell populations reduces the intensity of cell surface expression of RT6 and decreases the relative percentage of RT6⁺ T cells in culture. *In vitro* studies indicate that the latter effect is due the loss of cell surface RT6 and not to a proliferative advantage of RT6⁻ T cells. In response to mitogen, *in vitro* separated populations of RT6⁺ T cells proliferate as well as RT6⁻ T cells (15), and lymphocytes from rats depleted *in vivo* of RT6⁺ T cells proliferate as well as lymphocytes from untreated control animals (16).

In contrast to RT6⁺ peripheral T cells, RT6⁺ IELs consist of both thymus-dependent and thymus-independent subsets. RT6 is highly expressed on IELs of rats that are euthymic, athymic, or congenitally deficient in peripheral T cells (7,17,18). *In situ* immunohistochemical studies have defined the ontogeny of RT6⁺ IELs (8), demonstrating that they appear in euthymic and athymic rats before TcR⁺ T cells. RT6⁺ TcR⁻ IELs are present at low levels in euthymic animals, at moderate levels in the peripherally lymphopenic rat, and at high levels in athymic rats. *In vivo* depletion of peripheral RT6⁺ T cells using a cytotoxic monoclonal antibody does not deplete RT6⁺ IELs (8). These observations suggest that RT6⁺ cells in the rat are comprised of 1) a thymus-dependent population that predominates in the peripheral lymphoid compartment and 2) a thymus-independent population that predominates in the IEL compartment. The latter includes the earliest RT6⁺ cells to appear in ontogeny and predominates in athymic *nu/nu* rats.

ROLE OF RT6⁺ REGULATORY T CELLS IN THE BB/WOR RAT MODEL OF INSULIN-DEPENDENT DIABETES MELLITUS (IDDM)

In addition to marking thymus-dependent and independent cell populations in the rat, RT6 also identifies a population of T cells with immunoregulatory capabilities. These immunoregulatory activities were first observed in the BB rat and subsequently extended to other strains and species (19,20). BB rats develop autoimmune diabetes that is clinically and immunopathologically similar to human insulin-dependent diabetes mellitus (IDDM) (21). Two inbred lines of BB rats, developed under the auspices of the National Institutes of Health, have been used in the majority of studies characterizing the function of RT6 (21,22). These are designated BB/Wor rats. Inbred diabetes-prone (DP) BB/Wor rats spontaneously develop pancreatic insulitis, selective beta cell destruction, insulin deficiency, and hyperglycemia between 60–120 days of age. The cumulative incidence of disease is >90% (22). DP-BB/Wor rats are homozygous for the *lyp* gene and exhibit severe peripheral T cell lymphopenia (20,21). Subsets circulating in abnormally low numbers in DP-BB/Wor rats include those expressing RT6 (23). Coisogenic diabetes-resistant (DR) BB/Wor rats were derived from nondiabetic DP forebears, circulate normal numbers of RT6$^+$ T cells, are free of spontaneous diabetes (22), but harbor populations of autoreactive T cells (16,24). Other colonies of DP and DR-BB rats have been established worldwide. They have separate designations and are genetically polymorphic. Some differ from the BB/Wor in immune and autoimmune characteristics including the expression of RT6 (19–21,25). Except as noted, the data discussed here were obtained from the BB/Wor rat.

We have studied the BB/Wor rat with particular reference to the role of RT6$^+$ T cells in the autoimmune process. These studies in the rat have led us to conceptualize the etiology of autoimmunity in general and IDDM in particular using the analogy of a teeter-totter (Fig. 1) (20). Our working hypothesis holds that the expression of BB rat diabetes

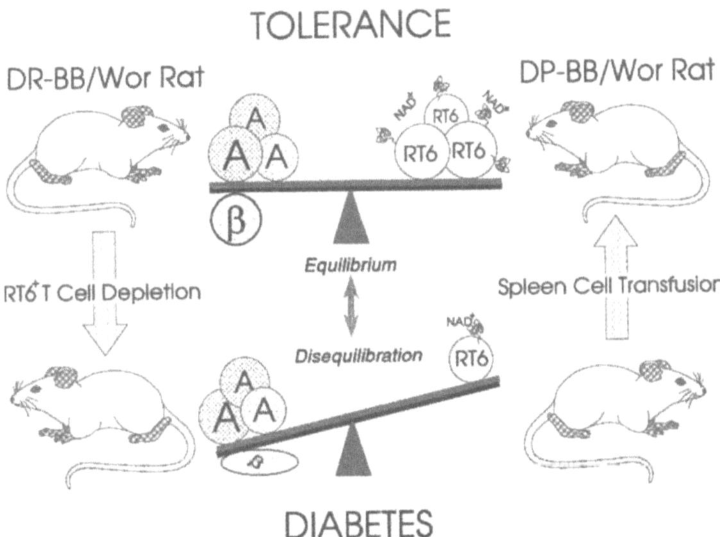

Figure 1. The role of RT6$^+$ cells in the "Balance Hypothesis" of insulin-dependent diabetes mellitus (IDDM) in the BB/Wor rat. An imbalance between autoreactive RT6$^-$ diabetogenic (A) and regulatory (RT6$^+$) T cell populations has been proposed as the basis for expression of IDDM in the BB/Wor rat. Diabetes prone (DP) BB/Wor rats spontaneously develop IDDM and are deficient in cells that express the RT6 alloantigen. Spleen cells from normal, histocompatible rats protect DP-BB/Wor rats from disease provided that RT6$^+$ T cells become engrafted. *In vivo* elimination of RT6$^+$ T cells from diabetes resistant (DR) BB/Wor rats induces autoimmune disease. The figure indicates schematically that RT6 is a glycosyl-phosphatidylinositol (GPI)-linked membrane protein with nicotinamide adenine dinucleotide (NAD$^+$) glycohydrolase activity.

depends on the relative balance that exists between RT6⁻ autoreactive cells and RT6⁺ regulatory cells that should normally prevent pancreatic beta cell destruction.

The bottom half of Figure 1 depicts the absence of self-tolerance, beta cell destruction, and the development of IDDM. On the right is the spontaneously diabetic DP-BB/Wor rat. The presence of autoreactive cell populations is inferred from the observation that spleen cells from such animals adoptively transfer IDDM (20,21). The absence of a regulatory cell population in these lymphopenic animals was then inferred from experiments showing that lymphocyte transfusions prevent IDDM provided that RT6⁺ donor T cells become engrafted (21). This is depicted schematically in the top right of Figure 1 as the restoration of a protective equilibrium.

Additional support for the hypothesis is provided by the DR-BB/Wor rat, depicted on the left. Although free of spontaneous IDDM, adoptive transfer experiments proved that they harbor autoreactive cells capable of recognizing and destroying normal beta cells (16,24). We hypothesized that the DR-BB/Wor rat is protected from diabetes by the presence of an RT6⁺ regulatory population. This hypothesis was confirmed by demonstrating that >50% of young DR-BB rats become diabetic if they are depleted *in vivo* of RT6⁺ T cells (16). This effect was readily observed in conventionally housed DR-BB/Wor rats that have been exposed to common rat viruses, whereas viral antibody free (VAF) DR rats required co-administration of an immune system activator such as polyinosinic-polycytidylic acid (poly I:C) (21).

More data supporting the balance hypothesis have come from adoptive transfer studies using MHC-compatible athymic recipients given autoreactive cells from RT6-depleted DR-BB/Wor rats (24,26). It was shown that the admixture of RT6⁺ lymph node T cells from intact, nondiabetic DR rats prevented the adoptive transfer of IDDM by the autoreactive T cells.

The balance hypothesis would predict that a deficiency of RT6⁺ regulatory cells in the absence of a robust population of RT6⁻ diabetogenic effector cells should not lead to disease. This prediction was tested and confirmed by depletion of RT6⁺ T cells in several non-BB rat strains; the procedure failed to induce IDDM in PVGc, PVGu, YOS, and LEW MHC congenic rats (24,27).

The immunoregulatory capability of RT6⁺ T cells is a key element in the balance hypothesis of IDDM, but it is important to stress that factors in addition to RT6⁺ T cells affect that balance. Not all inbred BB rats develop IDDM; about 10% of lymphopenic DP-BB/Wor rats do not become diabetic despite a deficiency of RT6⁺ T cells (22). In addition, the diabetes resistant DR-BB/Ed rat is lymphopenic and deficient in RT6⁺ T cells, but free of spontaneous IDDM (25). The DR-BB/Ed rat may be deficient in both regulatory and beta cell autoreactive T cell populations (25).

To summarize, an equilibrium between RT6⁻ autoreactive cells and RT6⁺ regulatory cells in large measure determines the expression of autoimmunity in BB/Wor rats and may represent a critical determinant of immune regulation in all rat strains. Perhaps for this reason the cellular activities, enzymology, and molecular biology of RT6 have been studied extensively.

MECHANISMS BY WHICH RT6⁺ T CELLS MAY MODULATE IMMUNE FUNCTION

Modulation of Th1 and Th2 Cells

The precise phenotype of the regulatory CD4⁺ RT6⁺ T cell that is capable of preventing IDDM in the BB/Wor rat is not known (26), but available data suggest that a

CD4$^+$ RT6$^+$ CD45$^-$ phenotype is likely (28). Those data were generated in studies using irradiation and thymectomy to induce diabetes in PVG-RT1^u rats. These animals were protected from hyperglycemia by the injection of syngeneic cells bearing this mature T cell phenotype. The functional activity of CD4$^+$ T cells is known to correlate with the expression of CD45 (29).

Cells expressing the CD4$^+$ RT6$^+$ CD45$^-$ phenotype express high levels of the mRNA encoding interleukin 4 (IL-4) (28), raising the possibility that RT6$^+$ T cells regulate autoimmunity in BB rats by modulating the balance between Th1 and Th2 type cells. This concept has been supported by the observation that CD4$^+$RT6$^+$ T cells prevent IDDM in DP-BB/Wor rats (30) and that adoptive transfer of IDDM can be prevented by CD4$^+$RT6$^+$ T cells (26) that are CD45$^-$ (BW, unpublished observations). This concept has also been supported by cytokine mRNA analyses in the BB/Wor rat. Reverse transcriptase-polymerase chain reaction (RT-PCR) was used to measure mRNA encoding the type 1 cytokines interferon gamma (IFN-γ) and IL-12 and the type 2 cytokines IL-4 and IL-10. In both DP-BB/Wor and RT6-depleted DR-BB/Wor rats, IFN-γ and IL-12 mRNAs were found in islets before and during disease onset, whereas mRNAs encoding IL-4 and IL-10 mRNA were present at low abundance in infiltrating T cells (31). These results suggest that Th1 lymphocytes predominate over Th2 lymphocytes in insulitis lesions induced by the depletion of RT6$^+$ regulatory T cells. The hypothesis that RT6$^+$ cells modulate the balance between Th1 and Th2 type cells is attractive because it predicts that peripheral immunoregulation may occur in a non-antigen-specific manner. Nonetheless, the hypothesis does require the presence of RT6$^+$ T cells at the site of inflammation to secrete the cytokines required to modulate effector cell function. How T cells expressing RT6 differentiate to become Th2 type cells upon activation is unknown, but the process may involve signals that are generated via the RT6 protein itself.

RT6 Mediated Signal Transduction

We have hypothesized that signals transduced across the cell membrane by the RT6 molecule modulate RT6$^+$ cells, altering their immune capabilities or function. In partial support of this hypothesis, we have recently demonstrated that crosslinking RT6 can modulate IL-2 receptor expression via cytoplasmic signal transduction (unpublished observations). We observed that binding of allele specific rat anti-RT6 monoclonal antibodies, followed by crosslinking with rabbit anti-rat Ig in the presence of 10 ng/ml PMA, induced cell surface expression of IL-2R on >50% of rat T cells. These activated T cells did not proliferate spontaneously, but could be driven to proliferate by adding exogenous IL-2. To delineate possible signaling pathways mediating this activity, we used immunoprecipitation, *in vitro* kinase assays, and PAGE to show that RT6 is associated with protein tyrosine kinase activity and 5 or more phosphoproteins. Immunoprecipitation of RT6-associated proteins with antibodies of known specificity identified two of the phosphoproteins as the *src* kinases p56*lck* and p60*fyn*. Because RT6 lacks membrane spanning and cytoplasmic domains, we hypothesize that RT6 may transmit its signal cytoplasmically via these molecules.

RT6 Enzymatic Activities

An additional mechanism by which RT6 protein could affect immune regulation is via its recently described enzymatic activities. The seminal observation was that rat RT6.2 is a nicotinamide adenine dinucleotide (NAD$^+$) glycohydrolase (32). Subsequently, RT6.2

protein was found to possess auto-ADP-ribosyltransferase activity, although it does not appear to catalyze the ADP-ribosylation of exogenous acceptors such as histones (33,34). The ADP-ribosyl-protein bonds in auto-ADP-ribosylated rat RT6.2 were stable to $HgCl_2$ and HCl, but labile to NH_2OH, consistent with an ADP-ribosyl-arginine linkage. ADP-ribosyltransferases are known to modulate cellular function. For example, several bacterial enzymes, including pertussis and cholera toxins, catalyze ADP-ribosylation and alter the activity of guanine nucleotide-binding proteins (35). To determine if the enzymatic activities of rat RT6.2 protein could affect the function of rat RT6.2$^+$ T cells, the effect of substrate availability on lymphocyte proliferation was examined. An inverse correlation was observed between NAD^+ concentration in the medium and the ability of rat T cells to respond to anti-CD3, concanavalin A, and phorbol myristate acetate plus ionomycin (34).

Neither the biochemical mechanisms by which NAD^+ alters lymphocyte proliferation nor the associated cytokine secretion profile of RT6$^+$ T cells activated in the presence of mitogen and NAD^+ are known. It is plausible to speculate, however, that NAD^+-catabolizing surface proteins like RT6 mediate signaling and immunoregulatory processes. Such speculation is particularly intriguing because RT6 is known to circulate in a soluble form (9) and NADase activity has been detected in the serum of rats (RB, unpublished observations). Soluble RT6 with enzymatic activity could represent another mechanism by which RT6 regulates immune activity in a non-antigen-specific fashion at sites of inflammation remote from the RT6-expressing cell itself.

CONCLUSION

Studies of RT6$^+$ T cells and the RT6 protein as they relate to modulating immune function and autoimmunity in the BB/Wor rat continue to enhance our understanding of autoimmunity. Regulatory cell populations have now been identified in the nonobese diabetic (NOD) mouse model of IDDM (36), and mouse homologues of rat RT6, designated Rt6, have been identified (37,38). Although mice have not yet been shown to synthesize Rt6 protein, mRNA levels of Rt6 appear to vary inversely with the expression of autoimmunity in both the NOD mouse (38) and the NZB mouse model of lupus erythematosus (39). Mouse Rt6 proteins, like their rat counterparts, have been found to possess NAD^+ glycohydrolase activity, ADP-ribosyltransferase activity, or both (34,40).

A human RT6 gene has been described but appears not to be functional (41). Many human cell types do exhibit both NAD^+ glycohydrolase and ADP-ribosyltransferase activities, however, and mRNA encoding a homologue of RT6 that could possess ADP-ribosyltransferase activity has recently been described in human testis (42). It will be of interest to determine if these activities are involved in the regulation of human autoimmunity.

REFERENCES

1. Lubaroff, D. M., G. Butcher, C. DeWitt, T. J. Gill,III, E. Günther, J. Howard, and K. Wonigeit. 1983. Fourth international workshop on alloantigenic systems in the rat. *Transplant. Proc.* 15:1683

2. Koch, F., A. Kashan, and H.-G. Thiele. 1988. The rat T-cell differentiation marker RT6.1 is more polymorphic than its alloantigenic counterpart RT6.2. *Immunology* 65:259–265.

3. Koch, F., H. G. Thiele, and M. G. Low. 1986. Release of the rat T cell alloantigen RT-6.2 from cell membranes by phosphatidylinositol-specific phospholipase C. *J. Exp. Med.* 164:1338–1343.

4. Thiele, H.-G., F. Koch, and A. Kashan. 1987. Postnatal distribution profiles of Thy-1$^+$ and RT6$^+$ cells in peripheral lymph nodes of DA rats. *Transplant. Proc.* 19:3157–3160.

5. Mojcik, C. F., D. L. Greiner, E. S. Medlock, K. L. Komschlies, and I. Goldschneider. 1988. Characterization of RT6 bearing rat lymphocytes. I. Ontogeny of the RT6$^+$ subset. *Cell. Immunol.* 114:336–346.

6. Mojcik, C. F., D. L. Greiner, and I. Goldschneider. 1991. Characterization of RT6-bearing lymphocytes. II. Developmental relationships of RT6$^-$ and RT6$^+$ T cells. *Develop. Immunol.* 1:191–201.

7. Fangmann, J., R. Schwinzer, and K. Wonigeit. 1991. Unusual phenotype of intestinal intraepithelial lymphocytes in the rat: Predominance of T cell receptor α/β^+/CD2$^-$ cells and high expression of the RT6 alloantigen. *Eur. J. Immunol.* 21:753–760.

8. Waite, D. J., M. C. Appel, E. S. Handler, J. P. Mordes, A. A. Rossini, and D. L. Greiner. 1996. Ontogeny and immunohistochemical localization of thymus dependent and thymus independent RT6$^+$ cells in the rat. *Am. J. Pathol.* (In Press)

9. Waite, D. J., E. S. Handler, J. P. Mordes, A. A. Rossini, and D. L. Greiner. 1993. The RT6 rat lymphocyte alloantigen circulates in soluble form. *Cell. Immunol.* 152:82–95.

10. Koch, F., F. Haag, A. Kashan, and H.-G. Thiele. 1990. Primary structure of rat RT6.2, a nonglycosylated phosphatidylinositol-linked surface marker of postthymic T cells. *Proc. Natl. Acad. Sci. USA* 87:964–967.

11. Haag, F., F. Koch, and H.-G. Thiele. 1990. Nucleotide and deduced amino acid sequence of the rat T-cell alloantigen RT6.1. *Nucleic Acids Res.* 18:1047

12. Hosseinzadeh, H. and I. Goldschneider. 1993. Recent thymic emigrants in the rat express a unique antigenic phenotype and undergo post-thymic maturation in peripheral lymphoid tissues. *J. Immunol.* 150:1670–1679.

13. Groen, H., F. A. Klatter, N. H. C. Brons, A. S. Wubbena, P. Nieuwenhuis, and J. Kampinga. 1995. High-frequency, but reduced absolute numbers of recent thymic migrants among peripheral blood T lymphocytes in diabetes- prone BB rats. *Cell. Immunol.* 163:113–119.

14. Greiner, D. L., J. P. Mordes, M. Angelillo, E. S. Handler, C. F. Mojcik, N. Nakamura, and A. A. Rossini. 1988. Role of regulatory RT6$^+$ T-cells in the pathogenesis of diabetes mellitus in BB/Wor rats. In Frontiers in diabetes research: Lessons from animal diabetes II. E. Shafrir and A. E. Renold, editors. John Libbey, London. 58–67.

15. Hunt, H. D. and D. M. Lubaroff. 1992. Identification of functional T cell subsets and surface antigen changes during activation as they relate to RT6. *Cell. Immunol.* 143:194–211.

16. Greiner, D. L., J. P. Mordes, E. S. Handler, M. Angelillo, N. Nakamura, and A. A. Rossini. 1987. Depletion of RT6.1$^+$ T lymphocytes induces diabetes in resistant BioBreeding/Worcester (BB/W) rats. *J. Exp. Med.* 166:461–475.

17. Fangmann, J., R. Schwinzer, H.-J. Hedrich, I. Klöting, and K. Wonigeit. 1991. Diabetes-prone BB rats express the RT6 alloantigen on intestinal intraepithelial lymphocytes. *Eur. J. Immunol.* 21:2011–2015.

18. Fangmann, J., R. Schwinzer, H.-J. Hedrich, and K. Wonigeit. 1993. Demonstration of RT6 expression on a CD3$^-$ population of intestinal intraepithelial lymphocytes of athymic nude rats. *Transplant. Proc.* 25:2789–2790.

19. Rossini, A. A., J. P. Mordes, E. S. Handler, and D. L. Greiner. 1995. Human autoimmune diabetes mellitus: Lessons from BB rats and NOD mice—*Caveat emptor. Clin. Immunol. Immunopathol.* 74:2–9.

20. Mordes, J., R. Bortell, J. Doukas, M. R. Rigby, B. J. Whalen, D. Zipris, D. L. Greiner, and A. A. Rossini. 1996. The BB/Wor rat and the balance hypothesis of autoimmunity. *Diabetes/Metab. Rev.* (In Press)

21. Crisá, L., J. P. Mordes, and A. A. Rossini. 1992. Autoimmune diabetes mellitus in the BB rat. *Diabetes/Metab. Rev.* 8:9–37.

22. Guberski, D. L. 1994. Diabetes-prone and diabetes-resistant BB rats: Animal models of spontaneous and virally induced diabetes mellitus, lymphocytic thyroiditis, and collagen-induced arthritis. *ILAR News* 35:29–37.

23. Greiner, D. L., E. S. Handler, K. Nakano, J. P. Mordes, and A. A. Rossini. 1986. Absence of the RT-6 T cell subset in diabetes-prone BB/W rats. *J. Immunol.* 136:148–151.

24. McKeever, U., J. P. Mordes, D. L. Greiner, M. C. Appel, J. Rozing, E. S. Handler, and A. A. Rossini. 1990. Adoptive transfer of autoimmune diabetes and thyroiditis to athymic rats. *Proc. Natl. Acad. Sci. USA* 87:7718–7722.

25. Joseph, S., A. G. Diamond, W. Smith, J. D. Baird, and G. W. Butcher. 1993. BB-DR/Edinburgh: A lymphopenic, non-diabetic subline of BB rats. *Immunology* 78:318–328.

26. Whalen, B. J., D. L. Greiner, J. P. Mordes, and A. A. Rossini. 1994. Adoptive transfer of autoimmune diabetes mellitus to athymic rats: Synergy of CD4$^+$ and CD8$^+$ T cells and prevention by RT6$^+$ T cells. *J. Autoimmun.* 7:819–831.

27. Ellerman, K. and A. Like. 1995. Regulatory T cells in rat autoimmune diabetes are strain dependent. *Diabetes* 44 (Suppl. 1):138A (Abstr.)

28. Fowell, D. and D. Mason. 1993. Evidence that the T cell repertoire of normal rats contains cells with the potential to cause diabetes. Characterization of the CD4$^+$ T cell subset that inhibits this autoimmune potential. *J. Exp. Med.* 177:627–636.

29. Bell, E. B. 1992. Function of CD4 T cell subsets in vivo: expression of CD45R isoforms. [Review]. *Seminars in Immunology* 4:43–50.

30. Mordes, J. P., D. L. Gallina, E. S. Handler, D. L. Greiner, N. Nakamura, A. Pelletier, and A. A. Rossini. 1987. Transfusions enriched for W3/25$^+$ helper/inducer T lymphocytes prevent spontaneous diabetes in the BB/W rat. *Diabetologia* 30:22–26.

31. Zipris, D., D. L. Greiner, S. Malkani, B. J. Whalen, J. P. Mordes, and A. A. Rossini. 1996. Cytokine gene expression in islets and thyroids of BB rats: Interferon gamma and IL-12 p40 mRNA increase with age in both diabetic and insulin treated nondiabetic BB rats. *J. Immunol.* 156:1315–1321.

32. Takada, T., K. Iida, and J. Moss. 1994. Expression of NAD glycohydrolase activity by rat mammary adenocarcinoma cells transformed with rat T cell alloantigen RT6.2. *J. Biol. Chem.* 269:9420–9423.

33. Haag, F., V. Andresen, S. Karsten, F. Koch-Nolte, and H.-G. Thiele. 1995. Both allelic forms of the rat T cell differentiation marker RT6 display nicotinamide adenine dinucleotide (NAD)-glycohydrolase activity, yet only RT6.2 is capable of automodification upon incubation with NAD. *Eur. J. Immunol.* 25:2355–2361.

34. Rigby, M. R., R. Bortell, L. A. Stevens, J. Moss, T. Kanaitsuka, H. Shigeta, J. P. Mordes, D. L. Greiner, and A. A. Rossini. 1996. Rat RT6.2 and mouse Rt6 locus 1 are NAD$^+$:arginine ADP-ribosyltransferases with auto-ADP-ribosylation activity. *J. Immunol.* (In Press)

35. Okazaki, I. J. and J. Moss. 1994. Common structure of the catalytic sites of mammalian and bacterial toxin ADP-ribosyltransferases. *Mol. Cell. Biochem.* 138:177–181.

36. Rashba, E. J., E.-P. Reich, C. A. Janeway, and R. S. Sherwin. 1993. Type 1 diabetes mellitus: an imbalance between effector and regulatory T cells? *Acta Diabetol.* 30:61–69.

37. Koch, F., F. Haag, and H.-G. Thiele. 1990. Nucleotide and deduced amino acid sequence for the mouse homologue of the rat T-cell differentiation marker RT6. *Nucleic Acids Res.* 18:3636

38. Prochazka, M., H. R. Gaskins, E. H. Leiter, F. Koch-Nolte, F. Haag, and H.-G. Thiele. 1991. Chromosomal localization, DNA polymorphism, and expression of *Rt-6*, the mouse homologue of rat T-lymphocyte differentiation marker *RT6*. *Immunogenetics* 33:152–156.

39. Koch-Nolte, F., J. Klein, C. Hollmann, M. Kühl, F. Haag, H. R. Gaskins, E. Leiter, and H.-G. Thiele. 1995. Defects in the structure and expression of the genes for the T cell marker *Rt6* in NZW and (NZB x NZW)F$_1$ mice. *Int. Immunol.* 7:883–890.

40. Koch-Nolte, F., D. Petersen, S. Balasubramanian, F. Haag, D. Kahlke, T. Willer, R. Kastelein, F. Bazan, and H. G. Thiele. 1996. Mouse T cell membrane proteins Rt6–1 and Rt6–2 are arginine protein mono(AD-Pribosyl)transferases and share secondary structure motifs with ADP-ribosylating bacterial toxins. *J. Biol. Chem.* 271:7686–7693.

41. Haag, F., F. Koch-Nolte, M. Kühl, S. Lorenzen, and H.-G. Thiele. 1994. Premature stop codons inactivate the RT6 genes of the human and chimpanzee species. *J. Mol. Biol.* 243:537–546.

42. Lévy, I., Y. Q. Wu, N. Roeckel, F. Bulle, A. Pawlak, S. Siegrist, M. G. Mattéi, and G. Guellaën. 1996. Human testis specifically expresses a homologue of the rodent T lymphocytes RT6 mRNA. *FEBS Lett.* 382:276–280.

27

MONO(ADP-RIBOSYL)TRANSFERASE GENES AND DIABETES IN NOD MICE

Is There a Relationship?

Marina Cetkovic-Cvrlje, Sung-Don Yang, and Edward H. Leiter

The Jackson Laboratory
Bar Harbor, Maine 04609

1. SUMMARY

The answer to the question posed by the title (is there a relationship between aberrant *Art* gene expression and IDDM pathogenesis in NOD mice?) remains elusive. Conclusions are currently based almost entirely upon analysis of mRNA transcript levels rather than on T-cell- specific mono-ADP ribosylation activities. Our unpublished data, as well as data published in abstract form by Dr. L. Chatenoud and colleagues (48) indicate that gene transcription is not impaired in splenic leukocytes of older NOD mice, including those with spontaneous IDDM development. Based upon the limited data showing that there may be reduced expression of *Art* gene products in the earliest T cell immigrants from the NOD thymus, one would have to surmise that If there is a reglatory defect, it may be in allowing single positive thymic T cells to emigrate before they are fully mature. Therefore, development of anti-*Art* monoclonal antibody together with further studies regarding functions of mono(ADP-ribosyl)transferase in immunoregulation of different subpopulation of T-cells, may finally resolve the role that altered mono(ADP-ribosyl)transferase activties play in the pathogenesis of IDDM in NOD mice.

2. THE NOD MOUSE: DEFECTS IN INNATE AND ADAPTIVE IMMUNITY UNDERLIE DIABETOGENESIS

Nonobese Diabetic (NOD) is an inbred mouse strain that spontaneously develops autoimmune, T cell-mediated insulin dependent diabetes mellitus (IDDM). Initiation of the defining histopathologic lesion, insulitis, requires an interaction between CD4[+] and CD8[+] T cells. Numerous reviews describing IDDM immunopathogenesis in NOD mice have appeared recently (1–5). The genome of this strain has been extensively characterized and specific chromosomal regions have been shown to segregate with IDDM sus-

ADP-Ribosylation in Animal Tissue, edited by Haag and Koch-Nolte
Plenum Press, New York, 1997

ceptibility [putatively termed *Idd* (Insulin dependent diabetes) loci until their molecular bases are discovered]. At least 40% of the relative risk is contributed by the unique MHC haplotype ($H2^{g7}$). Development of overt IDDM is a polygenic threshold liability phenomenon (6), requiring a complex interaction of $H2^{g7}$ with numerous other MHC-unlinked genes. IDDM etiopathogenesis in the NOD mouse is frequently contrasted with that in the BB-DP (diabetes-prone) rat (5- 7). The BB rat model and the role of the rat T cell alloantigen RT6 in diabetogenesis is discussed in detail elsewhere in this volume. The most striking immunophenotypic difference between the two rodent models is the phenomenon of T-lymphoaccumulation in peripheral lymphoid organs of NOD mice, compared to T-lymphopenia in peripheral lymphoid organs of BB-DP rats. A reciprocal imbalance in natural killer (NK) cell function is observed, with NK activity virtually absent in NOD mice, and unusually high in BB-DR rats. Despite these differences in peripheral T cell numbers, T cells have been shown to be essential mediators of beta cell destruction in both models. Another feature shared by both the BB-DP rat and the NOD mouse is the ability of bone marrow to transfer IDDM into irradiated F1 hybrids (5). Perhaps more comparable to the spontaneous IDDM developing in the NOD mouse is that induced by RT6.1 depletion of the BB/Wor-DR (diabetes-resistant) substrain, which is not T-lymphopenic prior to RT6 depletion (8). Yet even IDDM induced by RT6.1 antibody-mediated immunomodulation in these nominally diabetes-resistant BB/Wor-DR rats can be distinguished from IDDM etiopathogenesis in NOD. The NOD strain's genetic susceptibility to IDDM can only be discerned in a specific pathogen-free (SPF) environment. Immunostimulation engendered by immune responses to any of a variety of microbial pathogens suppresses diabetogenesis in NOD mice. This is apparently also the case with BB/Wor-DP rats, which exhibit an increased IDDM frequency and earlier age of onset in SPF environments (9). In contrast, efficient induction of IDDM in BB/Wor-DR rats by RT6.1 depletion requires immunomodulatory events associated either with exposure to Kilham's Rat Virus (10), or treatment with poly I:C to simulate a viral infection (9).

Despite the rather striking differences in peripheral T cell numbers distinguishing these two murine models [for review, see (5)], it should be noted that NOD mice exhibit a collection of immunodeficiencies in both innate and adaptive immunity (11). A combination of these deficiencies allows release of increased numbers of potentially islet autoreactive T cells into the periphery. Thymic defects in the ability of bone marrow derived antigen presenting cells (APC) to activate immunotolerogenic functions have been implicated (5). These APC tolerogenic defects in NOD mice are controlled by both the unique MHC haplotype ($H2^{g7}$), and non-MHC genes that limit the ability of these cells to become functionally mature. Acquisition of T cell tolerance to low abundance endocrine cell antigens may entail the spontaneous, transient expression of genes encoding these self-proteins within the thymus during the first week postpartum (12). Immature thymocytes whose T cell receptors (TCR) avidly bind self-peptides presented by MHC molecules on both haematopoietically-derived APC and thymic epithelial cells are normally stimulated to undergo apoptotic cell death. NOD thymocytes are refractile to activation of proliferation by TCR cross-linking; this, in turn, may reflect reduced secretion of IL-4 and perhaps IL-2 from T cells (13), as well as low levels of biologically active IL-1 from APC (14). The thymus of NOD mice is unusual not only in terms of altered cytoarchitecture (15), but also in that it does not undergo the involution in size typical of most other inbred strains. Figure 1 contrasts the maintenance of a large thymus in NOD/Lt female mice as a function of age. Although the strain used for comparison, NON/Lt, is a closely-related strain known to undergo an age-dependent peripheral T-lymphopenia (16), its pattern of thymic involution is typical of most inbred strains after puberty.

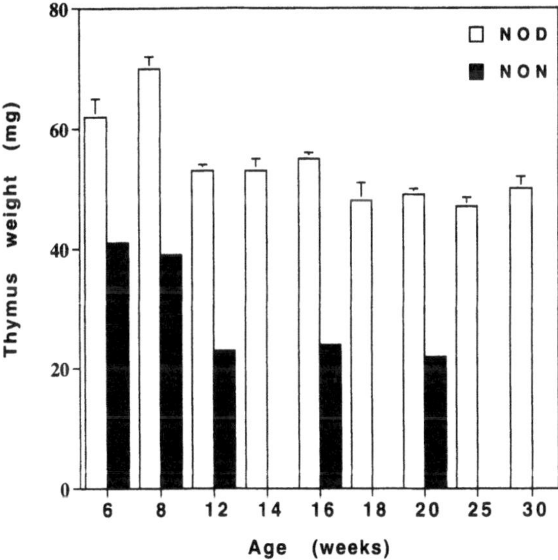

Figure 1. Delayed thymic involution in NOD/Lt compared to more typical involutions exhibited by other inbred strains (here modeled by the related NON/Lt strain, which becomes T-lymphopenic with age).

Although a preliminary study in NOD/Shi mice in Japan indicated that neonatal thymectomy reduced insulitis development in NOD mice (17), the same manipulation in SPF-NOD/Lt mice at The Jackson Laboratory failed to prevent IDDM. Absence of protection was associated with regenerating thymic residua in the mediastinal cavity as well as presence of peripheral T cells. Thymectomy of NOD mice at 3 wk of age accelerated IDDM onset (18), indicating the early colonization of peripheral lymphoid organs by autoimmune effector cells. The activation of such cells can be influenced by intrathymic exposure to islet cell (auto)antigens. Intrathymic transplantation of intact islets (19) or islet cell suspension (20) into NOD mice or BB rats (21) can prevent or retard development of IDDM.

Defects in apoptotic events in NOD thymus and in the periphery have been reported, perhaps accounting for anomalies in positive and negative selective mechanisms in the NOD mouse. APC from NOD mice apparently fail to stimulate ß-cell autoreactive T-cells to an activation state high enough to induce negative selection (5). Preliminary evidence from this laboratory indicates aberrant regulation in thymocytes and splenocytes of two genes important in triggering or preventing apoptotic cell death (increased *Bcl-2* mRNA transcripts in thymus and spleen versus reduced *Fas* transcripts in spleen compared to SWR/Bm controls). Intrathymic injection of islet cells may compensate for NOD strain-specific APC tolerogenic defects by increasing the activation state of ß-cell autoreactive T cell precursors to levels high enough to induce apoptotic cell death. NOD T-cells live longer in culture compared to T-cells of control strains and NOD splenocytes are more resistant to apoptosis induction by IL-2 withdrawal (22). These defects have been genetically linked to the region of NOD Chromosome 1 known to transmit IDDM susceptibility and containing the *Bcl2* gene whose gene product is a negative regulator of apoptosis. Another genetic susceptibility region on Chromosome 6 (*Idd6*) has recently been associated with the NOD apoptosis resistance phenotype (23).

While the "spillage" of autoreactive T-cells into the periphery may reflect incomplete negative selection as a consequence of impaired thymocyte apoptotic mechanisms, NOD mice are characterized by an inability to maintain adequate suppression of self-reactive T effectors in the periphery (14). Treatments that accelerate diabetogenesis in young prediabetic NOD mice (neonatal administration of cyclosporin, adolescent thymectomy, and cyclophosphamide adminstration) are thought to impair peripheral suppressor function (24). The potential pathogenic significance of this is well-illustrated by NOR/Lt , a recombinant congenic strain produced by outcross of NOD/Lt with C57BLKS/J. NOR/Lt mice share approximately 88.4% of their genome with NOD/Lt , including NOD alleles at numerous *Idd* susceptibility loci, including $H2^{g7}$. A major locus conferring resistance in NOR/Lt (*Idd13*) has been identified on Chromosome 2 (25). Although T-lymphoaccumulation is present in NOR/Lt as it is in NOD/Lt, leukocytic infiltrates in the pancreas rarely transit from a peri-insular, peri-vascular to an intra-islet localization. Yet autoimmune T-effectors can be generated in NOR/Lt mice , since a low percentage of older mice can be rendered diabetic by cyclophosphamide treatment (unpublished results, this laboratory). Consistent with the possibility that NOR/Lt mice maintain better peripheral inhibition of autoreactive T cells was the demonstration of normalized activation of immunoregulatory T cells in a syngeneic mixed lymphocyte reaction (SMLR) in NOR/Lt compared to NOD/Lt (26). Restoration of SMLR function by treatment of NOD mice in vivo with IL-2 or poly I:C correlated with protection from IDDM development (27).

BB-DR rats, in essence, are homologous to NOR mice in representing a MHC-identical, but diabetes-resistant control strain for BB-DP rats. Unlike BB-DP rats, BB-DR rats do not carry the *Lyp* mutation that blocks normal maturation and survival of post-thymic T cells, and hence are not T-lymphopenic. The rat T cell maturation antigen RT6 must be continuously expressed on the cell surface of BB-DR T cells for maintenance of effective regulatory control over beta cell cytotoxic T-effectors that can be generated in these diabetes-resistant rats (8). Because of the important immunoregulatory role demonstrated for a GPI-anchored cell surface mono(ADP-ribosyl)transferase expressed on the surface of rat peripheral T cells (a.k.a. RT6), it was of obvious importance to establish whether defects in the regulation of the homologous locus in the mouse were features of the pathogenic process in NOD mice.

3. MAPPING OF THE MOUSE AND HUMAN HOMOLOGS OF THE RAT RT6 GENE

The Thiele laboratory has made essential contributions to current understanding of the structure and function of the mouse homologs. Cloning of the rat RT6.2 allele by Thiele's group (28- 29) lead to a collaborative interaction with the Leiter group at The Jackson Laboratory to map the mouse homolog. This interaction began by the "impersonal" shipments of probes and PCR primers from Hamburg to Bar Harbor, and of mRNA and DNA from various strains of mice from Bar Harbor to Hamburg. However, the collaboration soon became personal when Fritz Koch-Nolte spent a summer in Bar Harbor working with Michal Prochazka and Rex Gaskins (now at NIDDK, Phoeniz, AZ, and Univ. of Illinois, Champaign-Urbana, respectively). More recently, Heinz-Günther and Gertrude Thiele also came to Bar Harbor to promote collaborative interactions among the burgeoning RT6 "club". PCR primers based upon the rat cDNA sequence were designed by the Thiele laboratory and used by them to amplify a mouse cDNA from BALB/cAnN splenic mRNA (30) . This mouse cDNA probe allowed mapping of the homologous locus in mouse to Chromosome 7 on the basis of tight linkage to the hemoglobin B (*Hbb*) locus (31). The mouse locus was initially designated *Rt6* to denote "rat T cell differentiation marker RT6 homolog". The mouse locus exhibited an unusually high level of allelic polymorphism based upon the multiplicity of restriction fragment length variants generated when DNA from various inbred strains were analyzed. Analysis of these complex patterns suggested greater genetic complexity than in the rat genome, raising the possibility of multiple related genes in tandem array or tightly linked within a short interval of Chromosome 7. This supposition was confirmed by the Thiele group's molecular identification of two tandem loci (initially designated as *Rt6–1* and *Rt6–2*). Provisional gene symbols are usually withdrawn in favor of a more functionally descriptive symbol once the precise function of a gene becomes known. A muscle-expressed mono(ADP-ribosyl)transferase (provisionally designated *Art1*) has recently been mapped very near the "*Rt6*" locus (32). Since data from the Thiele group (discussed more fully below) clearly show that "*Rt6–1*" and "*Rt6–2*" genes encode mono(ADP-ribosyl)transferases, *Art2a* will henceforth be used instead of *Rt6–1*, and *Art2b* instead of *Rt6–2*. Both *Art2a* and *Art2b* are variably expressed in an age-dependent fashion in T cells in peripheral lymphoid tissues of most inbred and wild-derived strains (33). Levels of transcription from each of the two genes can be distinguished by a RT-PCR-based-methodology (33). Sequence specific cDNAs are used to amplify cDNAs reverse transcribed from mRNA transcripts of both genes. Digestion with *Sac*I then allows the two gene products to be differentiated since a *Sac*I site is contained in *Art2a*, but not *Art2b* transcripts (33). The endonuclease-treated PCR products are separated on 1.4% agarose gels, transferred to nitrocellulose membranes, and the blot probed with a ^{32}P-labeled *Art2a*- and *Art2b* -cross-reactive cDNA probe (300 bp) generated by PCR from cloned *Art* cDNA. This probe lies 5' to the internal *Sac*I site in *Art2a* (at bp 394). This 5' probe permits visualization of the uncut *Art2b* transcript as a 600 bp product, while the *Art2a* product (digested into 350 and 250 bp fragments) is detected by the presence of the 350 bp fragment containing the 5' sequence. Sequence analysis of genomic and cDNAs showed a deletion in one of the exons of the NZW/BinJ strain's *Art2b* allele while a premature stop codon in the C57BL/6J *Art2a* allele is presumably responsible for differential expression of *Art2b* in this strain (33). Figure 2 summarizes the comparative mapping data for rats, mice, and humans. The mono(ADP-ribosyl)transferase encoding cluster of 3 genes thus far described in mice are all tightly linked on mouse Chr 7 in a region of syntenic conservation with rat Chromosome 1 containing the single RT6 locus. In the human genome, the muscle mono(ADP-ribosyl)transferase gene (*ART1)* has been separated from the non-expressed *ART2P* (pseudogene) onto thje short arm and long arm of human Chromosome 11 respectively. Two other human mono(ADP-ribosyl)transferase encoding loci have recently been identified (*ART3*, Chr 4; *ART4,* Chr. 12) (32), but mouse or rat homologs to these latter genes have not yet been identified.

4. DO *ART* GENES DEFINE THE *H1* "MINOR" HISTOCOMPATIBILITY ANTIGEN?

Figure 2 indicates that the *Art* genes map to the same position on mouse Chr 7 as the first minor histocompatibility antigen locus (*H1*) described. Minor histocompatibility loci were originally identified by T cell-mediated rejection of tissues engrafted between MHC compatible mice. Humoral responses are not elicited. Standard inbred strains of mice can be divided into 3 *Art* "haplotypes" based upon restriction fragment length polymorphisms; each haplotype is associated with one of the 3 different H1 alleles identified by transplan-

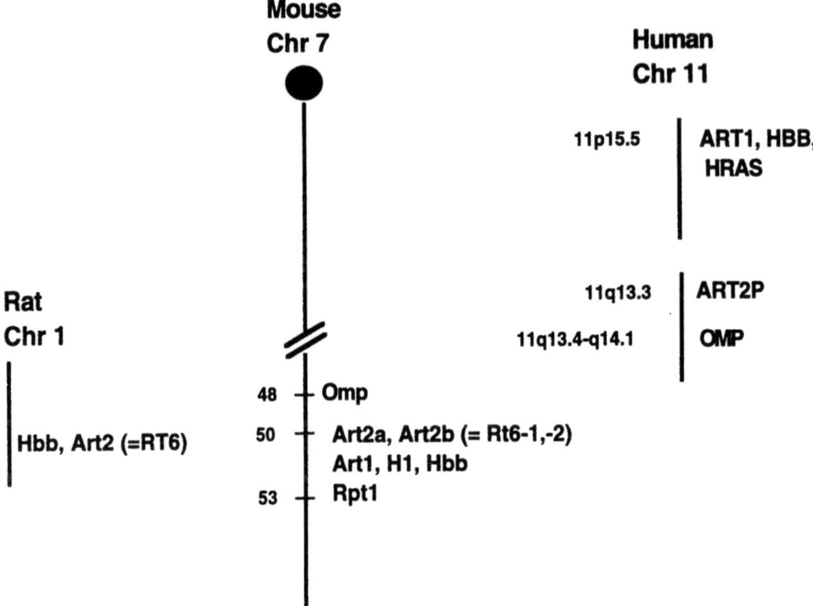

Figure 2. Comparative mapping of mono(ADP-ribosyl)transferase genes in humans, mice, and rats. (Data from Mouse Genome Database, The Jackson Laboratory).

tation analysis (34). Despite the considerable *Art* allelic polymorphisms indicated by Southern blot analysis, we and others have been unable to produce specific mouse allo-antibodies when splenic and intestinal leukocytes were used as immunogen. *H3*, one of the best characterized minor histocompatibility loci, comprises two linked loci, one stimulating class I restricted CTL, and the other class II restricted T-helper cells (35). That minor H loci can be multi-component in nature indicates that the allelic variants of the tandem *Art* genes may encode H1 antigens(34). Preliminary evidence obtained through collaboration with Dr. Derry Roopenian at The Jackson Laboratory indicates that this is not true. Co-transfection of a *Art2a* and *H2K^b* cDNAs (from C57BL/6 = H1^c) into a P815 mastocytoma failed to render them susceptible to specific lysis by a K^b restricted H1^c-specific CTL line, even though high levels of *Art2a* transcripts were detected. Conversely, an Abelson-virus transformed pre-B cell line that was killed by H1^c-specific CTL did not express either *Art2^a* or *Art2^b* transcripts. Thus, failure to transfer H1-reactivity by transfection of an *Art2a* cDNA suggests that this locus does not encode a CTL-recognized determinant. The possibility that H1 could be a product of the closely-linked *Art1* locus [the muscle-expressed mono(ADP-ribosyl)transferase] has not yet been tested.

5. BRIEF OVERVIEW OF *RT6/ART* GENE FUNCTIONS IN RAT AND MOUSE

The group of Joel Moss at NIH originally demonstrated that recombinant RT6.2 had NAD glycohydrolase activity (36). This signal discovery was confirmed and extended by the Thiele group's demonstration that native RT6.1 and RT6.2 proteins had NAD glycohydrolase activities. Whereas this latter group found that only RT6.2 exhibited NAD-de-

Table 1. Enzymatic functions associated with recombinant rat and mouse T cell-expressed mADPRT genes

Gene symbols old/new	Mono (ADP-ribosyl)transferase	NAD-glycohydrolase
Rat (1 locus with 2 alleles)		
RT6/ART2	+[1] (both alleles)	+ (both alleles)
Mouse(2 tandem loci, multiple alleles)		
Rt6-1/Art2a	+	—
Rt6-2/Art2b	+	+

[1] only auto ADP-ribosylation demonstrated. The human homolog (*ART2P*) is not expressed because of stop codons in the coding sequence.

pendent mono(ADP-ribosyl)transferase activity (37), another group succeeded in producing a bacterial recombinant rat RT6.1 fusion protein that did exhibit auto-ADP ribosylation activity (38). The Thiele group next demonstrated that Baculorvirus-produced recombinant mouse Art2a and Art2b proteins both exhibited mono(ADP-ribosyl)transferase activities (39). Whereas both recombinant proteins can ADP-ribosylate arginine residues in histone acceptor protein, only the Art2b recombinant protein exhibits intrinsic NADase activity (39). These observations, summarized in Table 1, indicate that ADP-ribosylation is the primary function of both enzyme proteins, while NAD hydrolysis appears to be a secondary catalytic function of Art2b in the absence of an ADP-ribose acceptor. The Dennert group has demonstrated that murine CTL proliferation and killing ability is inhibited by ADP-ribosylation of arginine residues in cell surface proteins (40), that CTL activation leads to release of soluble enzyme (41), and that the enzyme inhibits CTL function via inhibition of p56^{lck} kinase (42). Thus, extracellular NAD may provide the opportunity for these enzymes to mediate negative feedback inhibition of CTL activities. In this regard, it has recently been demonstrated that autoribosylation of RT6.2 molecules on rat lymphocyte cell surfaces elicited by high concentrations of extracellular NAD$^+$ inhibits ConA, anti-CD3, and PMA plus ionomycin-induced proliferation (43).

6. *ART2* GENE EXPRESSION IN NOD MICE: A ROLE IN DIABETOGENESIS?

Given the important role of T cell RT6 expression in the control of beta cell autoreactive T cells in BB-DR rats, the question of whether defects in regulation of the homologous *Art* gene cluster in NOD mice was of obvious importance. Diabetogenesis in NOD mice is the culmination of a collection of different immunodeficiencies, manifest by failure to maintain suppression of beta cell autoreactive T cells. In our initial report describing localization and expression of the mouse *Rt6/Art2* genes, we compared (by Northern blot analysis) the levels of mRNA expression in spleens of 8-wk-old NOD/Lt to IDDM-resistant NON/Lt females. Despite the significantly higher numbers of T cells in the spleen of NOD/Lt, *Art2* transcript levels were lower than in NON/Lt, which become progressively T-lymphocytopenic with age (44). This observation led to the obvious speculation that the NOD/Lt defect in ability to generate suppressor cell activity might be related to reduced Rt6/Art2 levels on a T cell basis.

Our inability to generate either polyclonal or monoclonal antibodies capable of detecting native Art2 molecules on the surface of mouse T cells has precluded a direct test of the above hypothesis. Instead, genetic approaches have been followed. Outcross of NOD to IDDM-resistant strains, followed by intercross or backcross, has permitted identification of chromosomal regions segregating for insulin dependent diabetes (*Idd*) susceptibility loci. No susceptibility locus has yet been identified at or near the *Art* gene complex on Chromosome 7. However, the finding of lower *Art2a/Art2b* transcript levels in total RNA from spleens of 5 wk-old NOD/Lt females compared to those of an MHC-identical, but IDDM-resistant control strain, NOR/Lt, suggested that early under-expression of these genes could play a role in diabetogenesis. This early difference in splenic expression of *Art* transcripts was particularly intriguing in that NOR/Lt mice exhibit a more robust SMLR, a potential indicator of ability to generate T-suppressor activity. Since MHC class I-restricted CD8+ T cells are required for initiation of IDDM in NOD mice (45- 46), reduced levels of mono-ADP-ribosylation in the earliest T cell emigrants into the periphery could permit early seeding of NOD islets by pathogenic T-effector cells. This would have to represent a rather limited period of high vulnerability, since follow-up studies have shown that NOD/Lt transcript levels increase to NOR/Lt levels as NOD mice age. High levels of transcripts for both loci can be found in spleens of diabetic NOD/Lt mice, and in islet-infiltrating leukocytes of pre-diabetic mice. Thus, the role of these gene products in keeping autoreactive NOD/Lt T cells in a quiescent state, if any, remains unclear. An urgent requirement for further progress is the need for a monoclonal antibody capable of recognizing the mouse enzyme forms on T cells.

7. GENERA DIFFERENCES AND CAVEAT EMPTOR

Dr. Aldo Rossini (Univ. of Massachusetts, Worcester) cautions users of animal models of IDDM with a "Caveat Emptor" (let the buyer beware!) (7). Although BB rats and NOD mice respectively reproduce phenotypes associated with IDDM pathogenesis in humans, there are important genera-specific distinctions that often prevent direct extrapolations of a finding in the animal models to the human situation. T cell expression of *RT6/Art* and *Thy1* genes serves as a paradigm illustration (Table 2). The Thiele group has demonstrated the presence of multiple stop codons in the human RT6 homolog that prevent transcription (47). Thy1, like RT6 , is an allotypic GPI-anchored glycoprotein expressed on cell surfaces. Its function is unknown. *THY1* is expressed on human keratinocytes, brain (probably glial cells), and fibroblasts, but not on thymocytes or on most peripheral T cells. In rats,*THY1* gene expression marks immature RT6⁻ T cells in the thymus. As thymocytes become fully mature and emigrate into the periphery, *Thy1* gene expression is silenced while RT6 transcription is activated, making the combination of Thy1⁻ and RT6⁺ staining a useful set of differentiation markers for mature peripheral T

Table 2. Genera specific differences in expression of the *RT6* versus *THY1* genes encoding GPI-anchored glycoproteins

Tissue	"RT6" expression			THY1/Thy1 expression		
	human	rat	mouse	human	rat	mouse
Thymocyte	—	—	+	—	++++	++++
Peripheral T cells	—	++++	+	+	+/-	++++
Brain, skin	—	?	?	++++	++++	++++

cells. A genera-distinct regulation is discerned in the mouse, where Thy1 is expressed on both thymocytes and peripheral T cells. Because of the lack of a serologic reagent capable of detecting cell surface *Art* gene products on mouse T cells, levels of antigen expression can only be inferred from levels of mRNA. These are very low in mouse thymocytes compared to peripheral T cells. Comparison of relative levels of *Art* gene products (both genes) to *Thy1* transcripts (both normalized to ß-actin in the same splenic RNA preparation) indicates at least 5-fold lower levels of *Art* compared to *Thy1* transcripts. These genera-specific differences in regulation of *Thy1* versus *RT6/Art* in human, rat, and mouse make it likely that different regulatory mechanisms may be employed by these three genera to regulate beta cell reactive T-effector cells

8. ACKNOWLEDGMENTS

All of the *Rt6/Art* studies initiated at The Jackson Laboratory were done with the full assistantance and collaboration with members of the Thiele laboratory, and especially Drs. Friedrich Koch-Nolte and Friedrich Haag. The model set by Heinz-Günther Thiele and his associates as to how to make scientific progress through cooperation and openess, if followed, should clarify the role of these genes in autoimmune diseases in mice in the near future. This work was supported by NIH grants DK36175, DK27722, fellowships from The Juvenile Diabetes Foundation International (to M.C.C.) and The American Diabetes Association (to S.D.Y.). Shared Resources of The Jackson Laboratory were supported by NCI Cancer Center CORE support (CA 34196).

9. REFERENCES

1. Kikutani, H., and S. Makino. 1992. The murine autoimmune diabetes model: NOD and related strains. *In* Adv. Immunol. F. J. Dixon F. J. Dixons. Academic Press, N.Y. 285.
2. Leiter, E. 1993. The nonobese diabetic mouse: a model for analyzing the interplay between heredity and environment in development of autoimmune disease. *ILAR News*. 35: 4.
3. Serreze, D. V., and E. H. Leiter. 1994. Genetic and pathogenic basis for autoimmune diabetes in NOD mice. *Current Opin. Immunol*. 6: 900.
4. Wicker, L. S., J. A. Todd, and L. _. Peterson. 1995. Genetic control of autoimmune diabetes in the NOD mouse. *Ann. Rev. Immunol*. 13: 179.
5. Serreze, D., and E. Leiter. 1995. Insulin Dependent Diabetes Mellitus (IDDM) in NOD Mice and BB Rats: Origins in Hematopoietic Stem Cell Defects and Implications for Therapy. *In* Lessons from Animal Diabetes. V. E. Shafrir E. Shafrirs. Smith-Gordon, London. 59.
6. McAleer, M. A., P. Reifsnyder, S. M. Palmer, M. Prochazka, J. M. Love, J. B. Copeman, E. E. Powell, N. R. Rodrigues, J.-B. Prins, D. V. Serreze, N. H. DeLarto, L. S. Wicker, L. B. Peterson, J. A. Todd, and E. H. Leiter. 1995. Crosses of NOD mice with the related NON strain: a polygenic model for type I diabetes. *Diabetes*. 44: 1186.
7. Rossini, A. A., E. S. Handler, J. P. Mordes, and D. L. Greiner. 1995. Human autoimmune diabetes mellitus: lessons from BB rats and NOD mice - caveat emptor. *Clin Immunol Immunopathol*. 74: 2.
8. Greiner, D. L., J. P. Mordes, E. S. Handler, M. Angelillo, N. Nakamura, and A. Rossini. 1987. Depletion of RT6.1+ T lymphocytes induces diabetes in resistant biobreeding/Worcester (BB/W) rats. *J. Exp Med*. 166: 461.
9. Like, A., D. Guberski, and L. Butler. 1991. Influence of environmental viral agents on frequency and tempo of diabetes mellitus in BB/Wor rats. *Diabetes*. 40: 259.
10. Guberski, D., V. Thomas, W. Shek, A. Like, E. Handler, A. Rossini, J. Wallace, and R. Welsh. 1991. Induction of Type 1 diabetes by Kilham's rat virus in diabetes-resistant BB/Wor rats. *Science*. 254: 1010.

11. Shultz, L. D., P. A. Schweitzer, S. W. Christianson, B. Gott, I. Birdsall-Maller, B. Tennent, S. McKenna, L. Mobraaten, T. V. Rajan, D. L. Greiner, and E. H. Leiter. 1995. Multiple defects in innate and adaptive immunological function in NOD/LtSZ-*scid* mice. *J. Immunol.* 154: 180.

12. Jolicoeur, C., D. Hanahan, and K. M. Smith. 1994. T-cell tolerance toward a transgenic ß-cell antigen and transcription of endogenous pancreatic genes in thymus. *Proc. Natl. Acad. Sci., USA.* 91: 6707.

13. Rapoport, M., D. Zipris, A. Lazarus, A. Jaramillo, D. Serreze, E. Leiter, P. Cyopick, and T. Delovitch. 1993. IL-4 reverses thymic T cell anergy and prevents the onset of diabetes in NOD mice. *J. Exp. Med.* 178: 87.

14. Serreze, D. V., and E. H. Leiter. 1988. Defective activation of T suppressor cell function in Nonobese Diabetic mice. Potential relation to cytokine deficiencies. *J. Immunol.* 140: 3801.

15. Savino, W., C. Boitard, J.-F. Bach, and M. Dardenne. 1991. Studies on the thymus in Nonobese Diabetic Mouse I. Changes in the microenvironmental compartments. *Lab. Invest.* 64: 405.

16. Leiter, E., M. Prochazka, D. L. Coleman, D. Serreze, and L. Shultz. 1986. Genetic factors predisposing to diabetes susceptibility in mice. *In* The Immunology of Diabetes Mellitus. M. Jaworsk, G. Molnar, R. Rajotte and B. Singh M. Jaworsk, G. Molnar, R. Rajotte and B. Singhs. Elsevier, Amsterdam. 29.

17. Ogawa, M., T. Maruyama, T. Hasegawa, T. Kanaya, F. Kobayashi, Y. Tochino, and H. Uda. 1985. The inhibitory effect of neonatal thymectomy on the incidence of insulitis in non-obese diabetic (NOD) mice. *Biomed. Res.* 6: 103.

18. Dardenne, M., A. Lepault, A. Bendelac, and J.-F. Bach. 1989. Acceleration of the onset of diabetes in NOD mice by thymectomy at weaning. *Eur. J. Immunol.* 19: 889.

19. Nomura, Y., E. Stein, and Y. Mullen. 1993. Prevention of overt diabetes and insulitis by intrathymic injection of syngeneic islets in newborn nonobese diabetic (NOD) mice. *Transplantation.* 56: 638.

20. Gerling, I. C., D. V. Serreze, S. W. Christianson, and E. H. Leiter. 1992. Intrathymic islet cell transplantation reduces beta cell autoimmunity and prevents diabetes in NOD/Lt mice. *Diabetes.* 41: 1672.

21. Posselt, A., A. Naji, J. Roark, J. Markmann, and C. Barker. 1991. Intrathymic islet transplantation in the spontaneously diabetic BB rat. *Ann. Surg.* 214: 363.

22. Garchon, H.-J., J.-J. Luan, L. Eloy, P. Bédossa, and J.-F. Bach. 1994. Genetic analysis of immune dysfunction in non-obese diabetic (NOD) mice: mapping of a susceptibility locus close to the *Bcl-2* gene correlates with increased resistance of NOD T cells to apoptosis induction. *Eur. J. Immunol.* 24: 380.

23. Penha-Goncalves, C., K. Leijon, L. Persson, and D. Holmberg. 1995. Type 1 diabetes and the control of dexamethazone-induced apoptosis in mice maps to the same region on chromosome 6. *Genomics.* 28: 398.

24. Bach, J.-F. 1994. Insulin-dependent diabetes mellitus as an autoimmune disease. *Endocrine Reviews.* 15: 516.

25. Serreze, D. V., M. Prochazka, P. C. Reifsnyder, M. Bridgett, and E. Leiter. 1994. Use of recombinant congenic and congenic strains of NOD mice to identify a new insulin dependent diabetes resistance gene. *J. Exp. Med.* 180: 1553.

26. Prochazka, M., D. V. Serreze, W. N. Frankel, and E. H. Leiter. 1992. NOR/Lt; MHC-matched diabetes-resistant control strain for NOD mice. *Diabetes.* 41.: 98.

27. Serreze, D. V., K. Hamaguchi, and E. H. Leiter. 1990. Immunostimulation circumvents diabetes in NOD/Lt mice. *J. Autoimmunity.* 2: 759.

28. Haag, F., F. Koch, and H. G. Thiele. 1990. Nucleotide and deduced amino acid sequence of the rat T cell alloantigen RT6.1. *Nucl. Acids Res.* 18: 1047.

29. Haag, F., F. Koch, and H. G. Thiele. 1990. The rat T cell alloantigens RT6.1 and RT6.2 differ by 12 amino acid substitutions. *Transpl. Proc.* 22: 2241.

30. Koch, F., F. Haag, and H. G. Thiele. 1990. Nucleotide and deduced amino acid sequence of the BALB/c mouse homologue of the rat T cell differentiation marker RT6. *Nucleic Acids Res.* 18: 3636.

31. Prochazka, M., H. R. Gaskins, E. H. Leiter, F. Koch-Nolte, F. Haag, and H.-G. Thiele. 1991. Chromosomal localization, DNA polymorphism, and expression of Rt-6, the mouse homologue of rat T cell lymphocyte differentiation marker RT6. *Immunogenetics.* 33: 152.

32. Koch-Nolte, F., M. Kühl, F. Haag, M. Cetkovic-Cvrlje, E. Leiter, S. Balasubramanian, F. Bazan, and H.-G. Thiele. 1996. Assignment of the human and mouse genes for muscle ecto-mono ADP-ribosyltransferase to a conserved linkage group on human Chromosome 11p15 and mouse Chromosome 7. *Genomics.* in press:

33. Koch-Nolte, F., J. Klein, C. Hollmann, M. Kuhl, F. Haag, H. R. Gaskins, E. Leiter, and H. G. Thiele. 1995. Defects in the structure and expression of the genes for the T cell marker Rt6 in NZW and (NZBxNZW)F1 mice. *Int Immunol.* 7: 883.

34. Koch-Nolte, F., M. Kuhl, C. Hollmann, F. Haag, H.-G. Thiele, M. Prochazka, and E. H. Leiter. 1995. Molecular polymorphism in the Rt6 genes of laboratory mice correlates with the allotypes of the H1 minor histocompatibility system. *Immunogenetics.* 41: 152.

35. Roopenian, D. C., A. P. Davis, G. J. Christianson, and L. E. Mobraaten. 1993. The functional basis of minor histocompatibility loci. *J-Immunol.* 151: 4595.

36. Takada, T., K. Iida, and J. Moss. 1994. Expression of NAD glycohydrolase activity by rat mammary adenocarcinoma cells transformed with rat T cell alloantigen RT6.2. *J. Biol. Chem.* 269: 940.

37. Haag, F., V. Andresen, S. Karsten, F. Koch-Nolte, and H.-G. Thiele. 1995. Both allelic forms of the rat T cell differentiation marker RT6 display nicotinamide adenine dinucleotide (NAD)-glycohydrolase activity, yet only RT6.2 is capable of automodification upon incubation with NAD. *Eur J Immunol.* 25: 2355.

38. Maehama, T., H. Nishina, S. Hoshino, Y. Kanaho, and T. Katada. 1995. NAD(+)-dependent ADP-ribosylation of T lymphocyte alloantigen RT6.1 reversibly proceeding in intact rat lymphocytes. *J Biol Chem.* 270: 22747.

39. Koch-Nolte, F., D. Petersen, S. Balasubramanian, F. Haag, D. Kahlke, T. Willer, R. Kastelein, F. Bazan, and H.-G. Thiele. 1996. Mouse T cell membrane proteins Rt6–1 and Rt6–2 are arginine/protein mono (ADP-Ribosyl) transferases and share secondary structure motifs with ADP-ribosylating bacterial toxins. *J. Biol. Chem.* 271: 7686.

40. Wang, J., E. Nemoto, A. Y. Kots, H. R. Kaslow, and G. Dennert. 1994. Regulation of cytotoxic T cells by ecto-nicotinamide adenine dinucleotide (NAD) correlates with cell surface GPI-anchored/arginine ADP-ribosyltransferase. *Journal of Immunology.* 153: 4048.

41. Nemoto, E., S. Stohlman, and G. Dennert. 1996. Release of a glycosylphosphatidylinositol-anchored ADP-Ribosyltransferase from cytotoxix T cells upon activation. *J Immunol.* 156: 85.

42. Wang, J., E. Nemoto, and G. Dennert. 1996. Regulation of CTL by ecto-nicotinamide adenine dinucleotide (NAD) involves ADP-Ribosylation of a p56[lck] -associated protein[1]. *J Immunol.* 156: 2819.

43. Rigby, M. R., R. Bortell, L. A. Stevens, J. Moss, T. Kanaitsuka, H. Shigeta, J. P. Mordes, D. L. Greiner, and A. A. Rossini. 1996. Rat RT6.2 and mouse rt6 locus 1 are NAD(+): arginine ADP ribosyltransferases with auto-ADP ribosylation activity. *J Immunol.* 156: 4259.

44. Prochazka, M., D. V. Serreze, S. M. Worthen, and E. H. Leiter. 1989. Genetic control of diabetogenesis in NOD/Lt mice: development and analysis of congenic stocks. *Diabetes.* 38: 1446.

45. Serreze, D. V., E. H. Leiter. G. J. Christianson, D. Greiner, and D. C. Roopenian. 1994. MHC class I deficient NOD-*B2m^{null}* mice are diabetes and insulitis resistant. *Diabetes.* 43: 505.

46. Wicker, L. S., E. H. Leiter, J. A. Todd, R. J. Renjilian, E. Peterson, P. A. Fischer, P. L. Podolin, M. Zijlstra, R. Jaenisch, and L. B. Peterson. 1994. ß2 microglobulin-deficient NOD mice do not develop insulitis or diabetes. *Diabetes.* 43: 500.

47. Haag, F., F. Koch-Nolte, M. Kuhl, S. Lorenzen, and H. Thiele. 1994. Premature stop codons inactivate the RT6 genes of the human and chimpanzee species. *J. Mol. Biol.* 243: 537.

48. Webb, M., P. Van Endert, F. Koch-Nolte, H.-G. Thiele, J.-F. Bach, and L. Chatenoud. 1995. Characterization of Rt6+ cells in the NOD mouse: putative candidates for active immunoregulation. *Autoimmunity.* 21: 53.

28

EXPRESSION OF THE ECTOENZYME RT6 IS NOT RESTRICTED TO RESTING PERIPHERAL T CELLS AND IS DIFFERENTLY REGULATED IN NORMAL PERIPHERAL T CELLS, INTESTINAL IEL, AND NK CELLS

Kurt Wonigeit, Astrid Dinkel, Josef Fangmann, and Hansjörg Thude

Klinik für Abdominal- und Transplantationschirurgie
Medizinische Hochschule Hannover
Hannover, Germany

1. ABSTRACT

The RT6 alloantigenic system of the rat has originally been defined on T lymphocytes of the peripheral lymphatic organs and has been considered to be selectively expressed on mature peripheral T cells. Studying NK cells and intestinal intraepithelial lymphocytes (IEL), we have now found that both cell types also express RT6 and that the expression patterns found for IEL and NK cells were markedly different from each other and also from the expression pattern previously described for T cells of the peripheral lymphatic organs. In lymph nodes, spleen, and blood both RT6$^-$ and RT6$^+$ T cells have been found and the density of RT6 expression on the positive cells has been shown to vary over a broad range. In contrast more than 98% of intestinal IEL stained for RT6 and the RT6 density was about tenfold higher than on strongly positive T cells of the peripheral lymphatic organs. Furthermore, the same high RT6 density was also found on IEL of athymic nude rats although these cells, to a large extent, lacked other T cell markers. This probably indicates that RT6 expression is an early event in the maturation of intestinal IEL which can occur already before the expression of T cell-specific membrane molecules. The conclusion that the expression of RT6 may be differently regulated in IEL and other T cell populations was further substantiated by the observation that RT6 was also present on IEL of diabetes-prone BB rats which are known to lack RT6 positive T cells in peripheral lymphatic organs. For NK cells still another pattern of RT6 expression was found. Unlike peripheral T cells and IEL, only a small subset of NK cells in blood and

ADP-Ribosylation in Animal Tissue, edited by Haag and Koch-Nolte
Plenum Press, New York, 1997

spleen expressed RT6. The percentage of RT6 positive cells was increased by in vitro stimulation of isolated NK cells with high concentrations of recombinant rat IL-2 indicating that RT6 expression may be associated with an activated state in NK cells. Taken together, these findings demonstrate that the expression of RT6 is not restricted to T cells and is differently regulated in normal peripheral T cells, intestinal IEL, and NK cells. Since it has recently been demonstrated that the RT6 gene contains two functional promoter regions with major structural disparity it is very likely that the distinct patterns of RT6 expression in different cell types reflect the differential use of the two promoters. The development of this complex control of RT6 expression in evolution may have been driven by a beneficial effect resulting from the use of the RT6 molecular function by several different lymphocyte populations.

2. INTRODUCTION

RT6 is an alloantigenic system which has been defined on T lymphocytes and has originally been called AgF, ART-2, Pta, and RT-Ly2 (1–3). It is encoded by a single gene located on chromosome 1 (4). The two alleles, $RT6^a$ and $RT6^b$, code for cell surface molecules of 25.000 to 30.000 kD attached to the lipid bilayer by a glycosylphosphatidylinositol (GPI) anchor (5,6). Recently it has been found that the sequence of RT6 is homologous to a mono(ADP-ribosyl)transferase (mADPRT) in rabbit skeletal muscle and that at least the $RT6^b$ gene product exhibits arginine-specific mADPRT activity (7–10). The products of both $RT6$ alleles are able to hydrolyse NAD^+. This observation has rendered RT6 a prototype of a new group of ectoenzymes of the lymphocyte membrane. Another member of this enzyme family is a mADPRT expressed on mouse cytotoxic T lymphocytes which is involved in regulating several lymphocyte functions including cytolytic activity, cell proliferation, and cell-cell-adhesion (11,12). The most likely mechanism underlying this modulation of lymphocyte function is an attachment of ADP-ribosyl-groups to arginine residues of other cell membrane associated proteins and thereby modification of their biological activity. In a recent review it has been delineated how this post-translational modification could translate into the regulation of various lymphocyte functions (13). Antibodies to the two allelic $RT6$ gene products have also been shown to induce proliferative responses in RT6 expressing T cells in vitro (14). Whether this effect is related to the enzymatic function is not clear since a similar effect has been reported for several GPI-anchored membrane molecules (15).

Another feature of the RT6 system which has raised considerable interest is its unique expression pattern on T cells of the peripheral lymphatic system. RT6 has been shown to be present on the majority of peripheral T cells but not on thymocytes. Therefore, RT6 expression has been regarded as representing a late process in peripheral T cell development identifying a particular maturational step (2,4,16,17). We have recently demonstrated a major exception from this pattern, by showing that intestinal intraepithelial lymphocytes (IEL) express RT6 in very high density (18). In athymic nude rats even a population of IEL was found which expressed RT6 in the absence of other T cell markers (19). These findings suggested that the expression of RT6 in IEL is differently regulated than in other peripheral T cell populations. In this article we review these findings and furthermore report on the expression of RT6 on NK cells. The analysis of RT6 on IEL and NK cells may have important implications for studies on the regulation of RT6 expression and on RT6 function.

3. MATERIALS AND METHODS

3.1. Animals

A congenic rat strain carrying the $RT6^b$ allele of the BH inbred strain on the LEW genetic background has been bred in our laboratory (3). This strain has originally ben called LEW.Ly-2.2 and was later renamed LEW.6B according to the consensus on RT6 nomenclature (1). Other strains used were LEW ($RT6^a$) and the athymic nude rat strain Han-RNU ($RT6^a$) maintained at the Medizinische Hochschule Hannover (20). The genetic traits and the origin of the different lines of diabetes-prone BB rats (DP BB/OK and DP BB/Mol) have been described elsewhere (21).

3.2. Monoclonal Antibodies

The RT6 specific mAb used in this study are directed to the serologically defined determinants RT6.1 (mAb 3G2 and P4/16) and RT6.2 (mAb Gy 1/12). The RT6.1 specificity is present on the gene product of the $RT6^a$ allele and the RT6.2 specificity on the product of the $RT6^b$ allele. Other mAb used were directed to CD4 (W3/25), CD8α (OX8), TCRβ (R73), TCRγδ (V65), CD3 (G4.18 and 1F4), and NKR-P1 (3.2.3).

3.3. Experimental Procedures

The preparation of cell suspensions and the staining procedures used have been described in detail elsewhere (18–20). Flow cytometry measurements were performed either with a FACScan or a FACStar (Becton Dickinson, Mountain View, CA). For the enrichment of NK cells in spleen cell suspensions a two step procedure was used consisting of B and T cell depletion. B cells were removed by incubation of the cell suspension on a nylon wool column. From the effluent cell suspension T cells were then depleted by panning with the TCRαβ specific mAb R73. The enriched NK cell populations were cultured in Dulbecco's medium containing sodium pyruvate (1mM), penicillin-streptomycin (100U/ml each), 2-mercaptoethanol (5×10^{-5}M), 10% fetal calf serum, and of recombinant rat IL-2 (1000U/ml). The cell density was 0.5 to 1.0×10^6/ml. Depending on the number of cells, cultures were performed in multiwell plates or small cell culture flasks.

4. RESULTS AND DISCUSSION

4.1. Comparison of RT6 Expression on T Cells of Peripheral Lymphatic Organs and on Intestinal IEL

Since the detection of the RT6 system all investigators have recognized a unique expression pattern in T cells of the various lymphatic organs (2,4,16,17). A major feature of this expression pattern is that RT6 is present on the large majority of T cells in the peripheral lymphatic organs but not on thymocytes. During ontogeny expression of RT6 is a late developmental step leading to normal numbers of RT6 positive T cells only several weeks after birth (16) and even then the density of RT6 on peripheral T cells obtained from blood, spleen, or lymph nodes remains remarkably heterogeneous (Fig. 1). In the CD4$^+$ as well as the CD8$^+$ subset a minority of cells are clearly RT6 negative, and the other cells

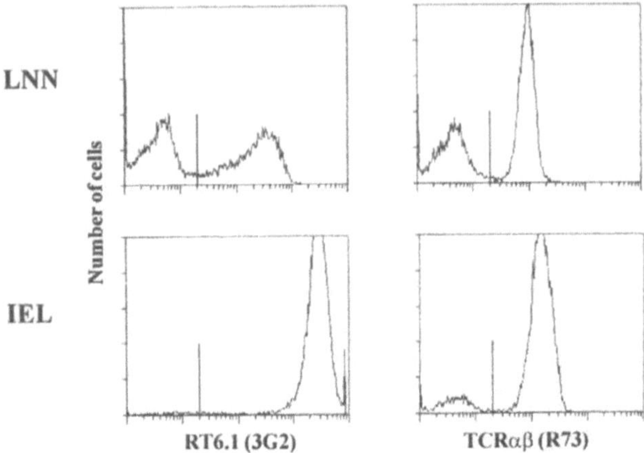

Figure 1. Expression of RT6 and αβTCR on lymph node cells (LNN) and intestinal IEL determined by flow cytometry. Single colour stainings obtained for LNN and IEL of a LEW rat expressing the RT6.1 serological specificity encoded by the *RT6ᵃ* allele are demonstrated. The RT6.1 specificity was detected with a directly labelled rat mAb (3G2), the αβTCR with a mouse mAb (R73) by an indirect staining procedure. In both cell populations the αβTCR positive cells stained with similar intensity. The variation in staining intensity within the two populations was low. By contrast, the expression of RT6 on IEL was about tenfold higher than on strongly staining LNN cells and was much more homogeneous than on LNN. The broad range of variation in RT6 expression demonstrated here for LNN is a characteristic feature of T cells in all peripheral lymphatic organs. IEL not staining for αβTCR express the γδTCR. The uniform staining pattern of more than 98% of the IEL for RT6 indicated that the RT6 expression is similar in IEL with αβTCR and γδTCR.

express RT6 in densities varying in flow cytometry measurements over two decades. This staining pattern is not only found on αβ T cells but also on γδ T cells as shown in Fig. 2 with the new γδTCR-specific antibody V65 (22).

IEL isolated from the small intestine differ from other peripheral T cell populations in that they express RT6 with very high density (about tenfold higher than strongly RT6 positive lymph node cells) and in a very uniform pattern (18). This unique expression pattern suggests that the IEL population may be particularly dependent on RT6 function. In this context it is of particular interest that mADPRTs detected in rabbit muscle and mouse cytotoxic T lymphocytes have been shown to modify the function of adhesion molecules (11,23). If RT6 has a similar function, it may play a key role in the complex interaction of IEL with the neighbouring epithelial cells and the basal membrane separating the epithelium from the lamina propria.

4.2. RT6 Expression on IEL of Diabetes-Prone BB Rats

Diabetes-prone Bio Breeding (DP BB) rats spontaneously develop an insulin-dependent diabetes mellitus closely resembling human type I diabetes. All DP BB rat strains with a high incidence of the disease suffer from a recessively inherited T cell lymphopenia. This is characterized by a marked reduction of peripheral CD4[+] T cells, an even more pronounced reduction of peripheral CD8[+] T lymphocytes, and a nearly complete failure to express CD45RC and RT6 on T cells of lymph nodes and spleen (24,25). The disturbed expression of RT6 has attracted particular attention since it could be shown that transfer of RT6[+]CD4[+] cells in young animals was able to prevent the development of diabetes in later life (26).

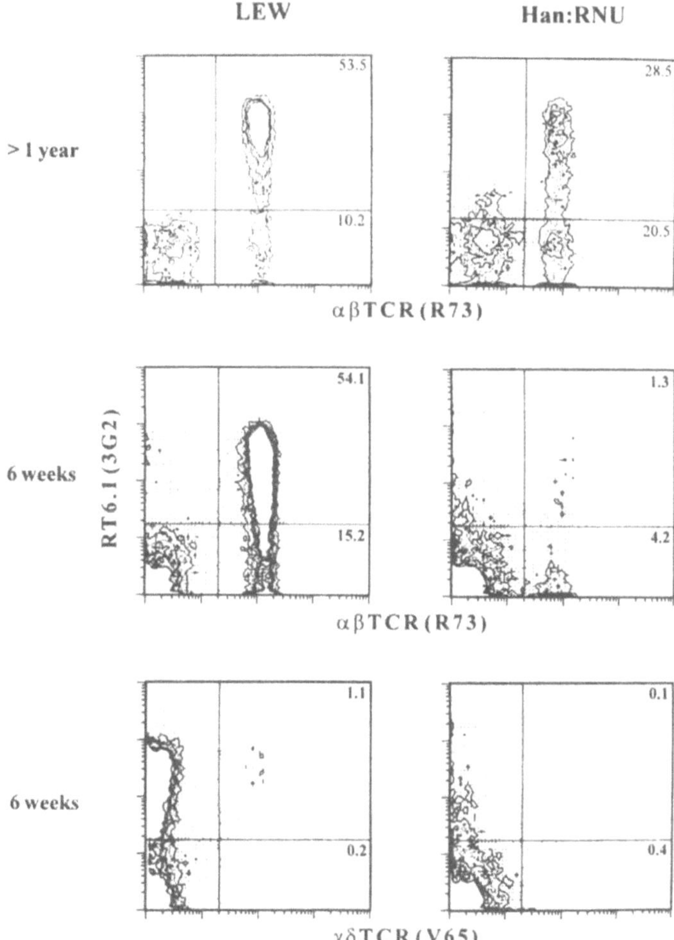

Figure 2. Comparison of RT6 expression in lymph node cells of LEW rats and athymic nude Han:RNU rats of different age. The depicted contour plots represent two colour stainings analysed by flow cytometry. Numbers given in the figure are the percentage of cells in the respective quadrant of the plot. In both the old and the young LEW rat the vast majority of αβTCR positive cells expressed RT6. In the old Han:RNU rat also a substantial number of αβTCR positive cells was present. The relative proportion of RT6 expressing cells among them, however, was clearly lower than in the age-matched LEW rat. In the young Han:RNU rat only few αβTCR positive cells were present which were mostly RT6 negative, demonstrating that T cell development in the peripheral lymphatic organs of athymic nude rats is not only delayed but also associated with disturbed maturation reflected in reduced RT6 expression. In the lower panel the RT6 expression on γδ T cells found in the lymph nodes is given for both the young euthymic and the young athymic animal. Substantial numbers of γδ T cells were present only in the euthymic rat. Their pattern of RT6 expression was similar to that of αβ T cells in the same animal. Most γδ T cells clearly expressed RT6.

The nearly complete absence of RT6 on peripheral T cells of DP BB rats raised the question of whether the failure of RT6 expression also affected the IEL population. The analysis of several DP BB strains confirmed the absence or very low expression of RT6 in T cells obtained from lymph nodes, spleen, or blood. In the isolated IEL population, however, clear-cut RT6 expression was found (21). About 50 to 60% of DP BB IEL expressed RT6 with similar density as lymph node T cells of normal rats. Unlike the IEL population of normal rats predominantly consisting of CD8$^+$ cells, the IEL population of DP BB rats comprised mostly CD4$^+$ cells. This probably indicates that the defect in T lymphocyte generation also affects the intestinal IEL. The presence of IEL with marked RT6 expression, however, made it unlikely that a defect of the *RT6* gene itself is responsible for the lymphopenia. This conclusion was supported by molecular studies and by breeding experiments demonstrating that mating of DP BB rats possessing the *RT6^a* allele with normal *RT6^b* expressing animals led to F$_1$ progeny expressing both *RT6^a* and *RT6^b* normally (27,28).

For several GPI-anchored membrane molecules it has been demonstrated that in addition to the GPI-anchored form also transmembrane forms can be expressed. Therefore, a defect in the genes responsible for generating the GPI anchor could also explain the pattern of RT6 expression in DP BB rats. In this instance the RT6 molecules on IEL of DP BB rats would be expected to represent transmembrane molecules. This was tested by examining the sensitivity of the RT6 molecules to phosphatidylinositol-specific phospholipase C (PIPLC). By treatment with this enzyme, RT6 could be removed from IEL of DP BB rats as well as from IEL and lymph node T cells of normal rats (21). In conclusion these data demonstrate that the abnormal patterns of RT6 expression are not the cause but the consequence of the lymphopenia in DP BB rats. If this is correct the finding that RT6$^+$ T cells are nearly totally absent from lymph nodes, spleen, and blood but are retained at least in part in the IEL population suggests that the regulation of RT6 expression in IEL differs from that in other T cell populations. In the peripheral lymphatic organs newly generated T cells probably die before RT6 expression is induced. This is in accordance with the view that RT6 expression in peripheral T cells reflects a late maturational step. By contrast, in IEL RT6 expression appears to be an early event. Obviously, it can be induced at least in some IEL before the lymphopenia-inducing maturational block becomes fully effective.

4.3. RT6 Expression on IEL of Athymic Nude Rats

We have previously shown that athymic nude rats can develop substantial numbers of T cells in the peripheral lymphatic organs with increasing age (20). Lymph node cells from nude rats older than 3 months are usually able to react to the T cell mitogens PHA and ConA with substantial proliferative responses and lymph node cells of animals older than 6 months can even react in allogeneic mixed lymphocyte cultures (29,30). FACS analysis of T cells from such animals revealed that these cells are also able to express RT6 (20). Interestingly, however, the relative proportion of T cells expressing RT6 is very low in young animals. Although it increases with age, it never reaches the same level as in euthymic animals (Fig. 2). When intestinal IEL of young adult nude rats (2 to 12 months of age) were analysed for T cell markers by flow cytometry only a fraction of IEL expressed CD3 and other T cell markers (Fig. 3). This fraction increased from less than 3% in two months old rats to about 50 to 70% in nude rats older than 6 months. Analysis of RT6 expression, however, revealed a similarly strong and uniform expression as in euthymic controls. Thus the predominant phenotype of IEL in young athymic nude rats is CD3$^-$RT6$^+$ (Fig. 3).

Detailed studies of the ontogeny of intestinal IEL in the mouse have demonstrated that a substantial proportion of this cell population may be generated locally from precursors directly derived from bone marrow cells (31). The maturation of these cells, however, appears to be not totally thymus-independent. Thus, it has been demonstrated that the development of a normal IEL population in the nude mouse is markedly accelerated by a thymus transplant (32). At present it is not known to what extent thymus-derived T cells contribute to the IEL population and whether thymus-derived cells or thymic factors mod-

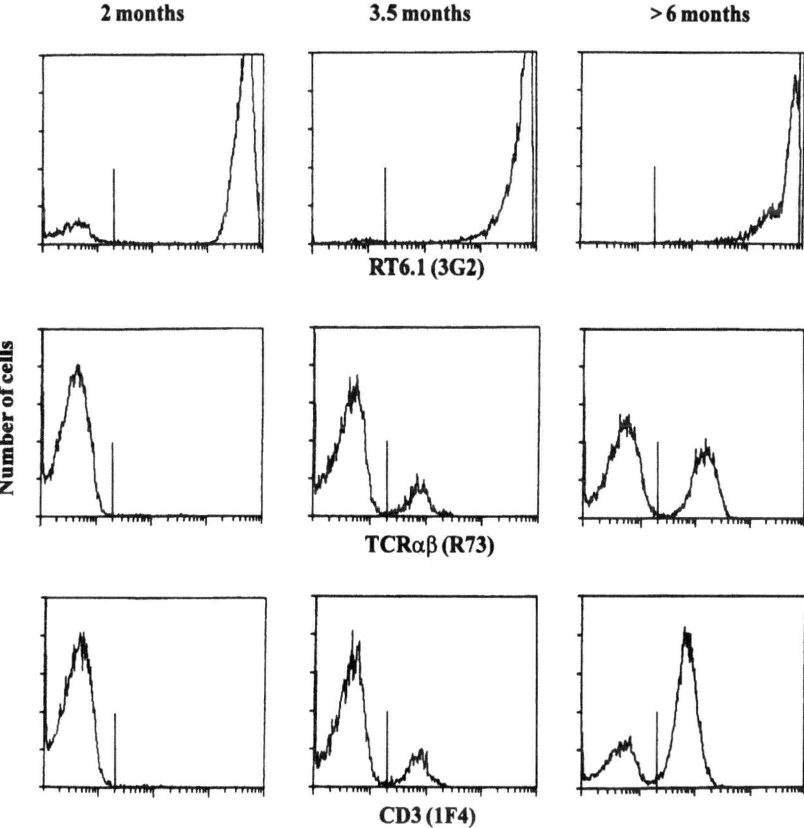

Figure 3. Age-dependent changes in the phenotype of intestinal IEL in athymic nude rats of the Han:RNU strain. Whereas in the 2 months old animal IEL did not express αβTCR or CD3, these markers became detectable in rats of higher age (3.5 months and > 6 months). This age-dependence in the expression of the TCR/CD3 complex in IEL is a characteristic feature of the Han:RNU strain and demonstrates that these animals are able to generate IEL with typical T cell features only in small numbers and with very long delay. In all three animals, however, both TCR/CD3 positive and negative IEL stained for RT6 with the same high intensity demonstrated in Fig. 1 for IEL of euthymic rats. This clearly indicated that in IEL the occurrence of RT6 on the cell membrane is not dependent on the previous expression of a functional TCR/CD3 complex.

ify the capacity for local T cell maturation in the intestine. In agreement with the results of others (33), the data presented here for nude rats demonstrate that the development of IEL with a functional TCR/CD3 complex is very much reduced and delayed pointing to a major role of the thymus in the generation of the normal IEL population in the rat. The unusual TCR/CD3 negative population most likely represents bone marrow derived precursors of the small population of TCR/CD3 positive IEL developing also in the absence of a functioning thymus. The fact that both TCR/CD3 positive and negative IEL express RT6 in very high density provides further evidence for a different mode of RT6 induction in the small intestine and in peripheral lymphatic organs.

4.4. RT6 Expression on NK Cells

Natural killer (NK) cells of the rat are characterised by the expression of the membrane molecule NKR-P1 and the absence of the CD3/TCR complex (34). Most of them express the

CD8α/α homodimer. We could now demonstrate that a subset of NK cells found in blood as well as in spleen expresses RT6 in low density (Fig. 4). This could be shown both in LEW and LEW.6B rats using the appropriate monoclonal antibodies detecting either the RT6.1 or the RT6.2 specificity. In order to study RT6 expression on NK cells in more detail, NK cells were isolated from LEW spleen and cultured for several days in medium containing high concentrations of recombinant rat IL-2. NK cell suspensions were prepared by a two step protocol consisting of B cell depletion by nylon wool column and depletion of T cells by a panning procedure with the TCR-specific antibody R73. By this means cell suspensions containing 60 to 70% NK cells were obtained. Culture of these cells with IL-2 for 48 hours led to a marked increase in the number of RT6 positive NK cells. This suggests that stimulation by IL-2 may induce RT6 expression and that RT6 positive NK cells found in the spleen and blood of normal rats may represent a population of preactivated cells. In this context it may be worthwhile to mention that a large fraction of peripheral T cells activated by mitogens or alloantigens are also able to express RT6 (35).

4.5. Allele-Specific Differences in the Expression of *RT6ᵃ* and *RT6ᵇ*: Comparison of T Cells and NK Cells

It has previously been shown that the $RT6^b$ allele is usually expressed on a larger number of T lymphocytes than the $RT6^a$ allele (36). This phenomenon has most convincingly been shown in heterozygous animals possessing both alleles. In these animals the majority of peripheral T cells is positive for RT6.1 (encoded by the $RT6^a$ allele) and RT6.2 (encoded by the $RT6^b$ allele). In addition, however, a small population of RT6.1⁻RT6.2⁺ T cells is consis-

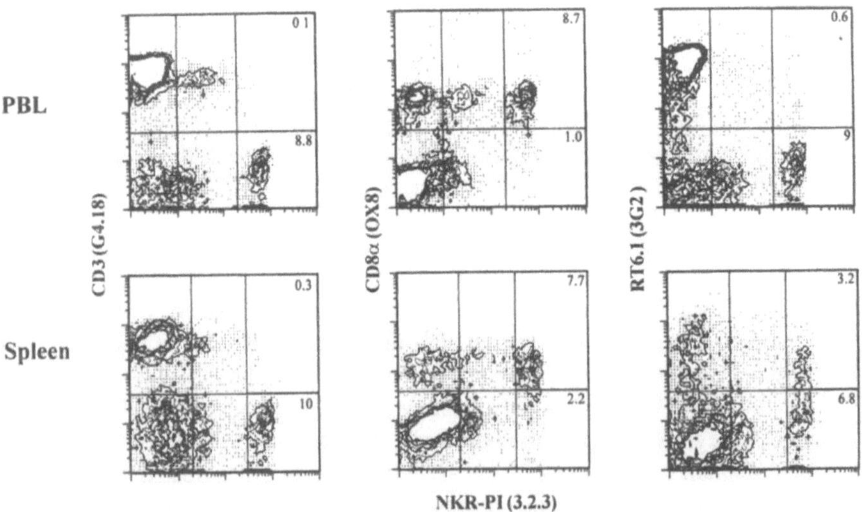

Figure 4. RT6 expression on NK cells of a LEW rat. The contour plots represent the results of two colour stainings of blood lymphocytes (PBL) and spleen cells. NK cells are characterized by expression of NKR-P1 (mAb 3.2.3) in high density and thus are found in the right hand part of all plots depicted. Both in PBL and in spleen these cells were CD3 (mAb G4.18) negative and mostly CD8α (OX8) positive. RT6 staining revealed that in PBL as well as in spleen a subset of strongly NKR-P1 positive cells also expressed RT6. The fraction of NK cells expressing RT6 was much lower in PBL than in spleen. Numbers in the figures give the percentage of cells found in the respective segment of the plot.

tently present. We have tested whether this difference in membrane expression of the two alleles is also demonstrable in NK cells. The results clearly confirmed the known asymmetry of expression in T cells but did not show it for NK cells (Fig. 5). This finding strongly supports the assumption that the expression of RT6 is differently regulated in T cells and NK cells.

4.6. Possible Mechanisms Underlying the Differential Regulation of *RT6* Expression in Different Lymphocyte Populations

Our studies on RT6 expression on T cells of the peripheral lymphatic organs, on intestinal IEL, and on NK cells have revealed a remarkable heterogeneity of expression patterns in different cell types. In T cells of the peripheral lymphoid organs the expression of RT6 is a late event in ontogeny probably depending on distinct maturational steps of T cells in this site. By contrast, IEL express RT6 uniformly in very high density. Furthermore, unlike T cells in peripheral lymphatic organs, IEL express RT6 already in the neonatal period. In athymic nude rats it can even be demonstrated on precursor cells of IEL lacking expression of the CD3/TCR complex and other T cell markers. Still another pat-

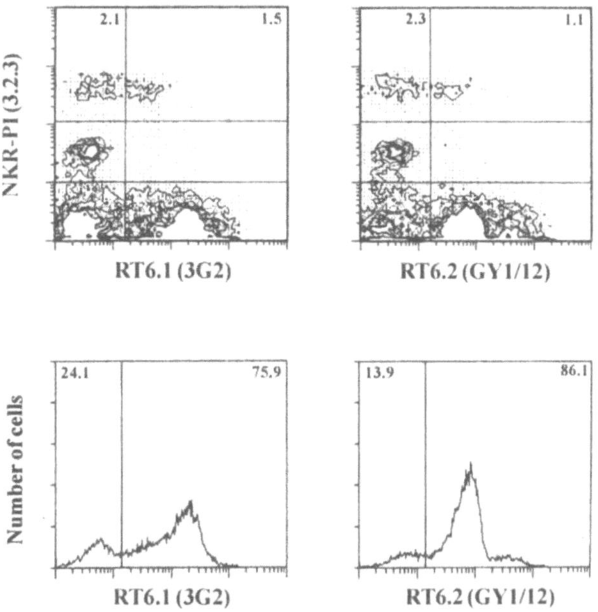

Figure 5. Comparison of the expression of the allelic specificities RT6.1 and RT6.2 on NK cells and T cells in the blood of an RT6 heterozygous rat. The upper panel shows the NK cell analysis demonstrating that the antibody directed to RT6.1 reacted with slightly more strongly NKR-P1 positive cells than the antibody to RT6.2 (1.5% versus 1.1% of all cells analysed). The lower panel depicts the analysis of T cells with an indirect staining protocol. In this case the binding of the mAb to RT6.1 or to RT6.2 was detected by a FITC-conjugated goat-anti-rat Ig antibody. By this procedure all RT6 positive cells (T cells and NK cells) and, because of Ig expression, also all B cells were stained. Unstained cells mostly represented RT6 negative T cells. With the mAb to RT6.1 (left) the number of unstained cells (24.1%) was much larger than with the mAb to RT6.2 (13.9%). This indicated that the mAb RT6.2 stained clearly more T cells than the mAb to RT6.1 confirming the known asymmetry in the expression of RT6.1 and RT6.2 in T cells. Very similar staining patterns were found in all RT6 heterozygous rats analysed.

tern is found on NK cells. In these cells RT6 expression is restricted to a subset most likely representing activated NK cells and the level of expression is rather low.

Some insight into the regulatory mechanisms underlying these highly divergent patterns of RT6 expression has come from the recent analysis of the gene structure of RT6 and in particular from the analysis of its promoter regions (37,38). These studies have shown that the *RT6* gene contains two promoter regions displaying major differences in binding sites for known transcription factors. T cells of the peripheral lymphoid organs preferentially use promoter 2 containing binding sites for *ets* and *ikaros* characteristic of regulatory regions of T cell-specific genes. With regard to the asymmetry of expression of $RT6^a$ and $RT6^b$ in T cells it is of particular interest that the promoter 2 region contains a microsatellite differing in size between the two genes. This could lead to a difference in the transcriptional activity of promoter 2 in the two RT6 alleles. If this is correct the symmetric expression of the two alleles in NK cells may indicate preferential use of promoter 1 by these cells. The low level of expression in NK cells would be in agreement with the fact that in addition to the core promoter elements only few additional activator elements have been identified in promoter 1 (38).

5. CONCLUSIONS

The two allelic genes $RT6^a$ and $RT6^b$ code for cell surface proteins functioning as ectoenzymes in rat lymphocytes. Since structurally and functionally related proteins have also been found in mouse CTL and rabbit skeletal muscle the RT6 proteins of rat and mouse most likely are members of a family of cell surface molecules with similar function. The availability of monoclonal antibodies to two allelic gene products of the rat RT6 system has allowed detailed studies of the expression of this system demonstrating the presence of RT6 proteins on most T cells of the peripheral lymphatic organs, on all intestinal IEL, and on a subset of NK cells most likely representing activated NK cells. The markedly different expression patterns in the three groups of cells suggest an unusual degree of diversity in the regulatory control of RT6 expression. This conclusion has recently been substantiated by the demonstration of a rather complex structure of the 5' end of the *RT6* gene containing two functional promoters and giving rise to multiple primary transcripts with variable 5' ends. It is likely that the development of this complex promoter structure in evolution has been driven by a beneficial effect resulting from the use of the RT6 system by several different lymphocyte populations.

6. ACKNOWLEDGMENTS

The secretarial assistance of Mrs D. Damitz is gratefully acknowledged. Furthermore we wish to thank Dr. N. Barclay (Oxford, UK), Dr. G. Butcher (Cambridge, UK), Dr. W. Chambers (Pittsburgh, USA), Dr. B. Hall (Liverpool, Australia), Dr. H.J. Hedrich (Hannover, Germany), Dr. T. Hünig (Würzburg, Germany), Dr. I. Klöting (Karlsburg, Germany), and Dr. T. Tanaka, (Sendai, Japan) for gifts of monoclonal antibodies, cell lines, and animals. The work was supported by the Deutsche Forschungsgemeinschaft (SFB 244 and SFB 265).

7. REFERENCES

1. Lubaroff, D.M., G. Butcher, C. DeWitt, T. Gill III, E. Günther, J. Howard, & K. Wonigeit. 1983. Standardized nomenclature for the rat T-cell alloantigens: Report of the Committee. *Transplant. Proc. 15*: 1683.

2. Wonigeit, K. 1979. Definition of lymphocyte antigens in rats: RT-Ly-1, RT-Ly-2, and a new MHC-linked antigen system. *Transplant. Proc. 11*: 1334–1336.

3. Wonigeit, K. 1979. Characterization of the RT-Ly-1 and RT-Ly-2 alloantigenic systems by congenic rat strains. *Transplant. Proc. 11*: 1631–1635.

4. Butcher, G.W., S. Clarke, & E.M. Tucker. 1979. Close linkage of peripheral T-lymphocyte antigen A (PtaA) to the hemoglobin variant Hbb on linkage group I of the rat. *Transplant. Proc. 11*: 1629–1630.

5. Koch, F., H.-G. Thiele, & M. Low. 1986. Release by the rat T cell alloantigen RT6.2 from cell membranes by phosphatidylinositol-specific phospholipase C. *J. Exp. Med. 164*: 1338–1343.

6. Koch, F., F. Haag, A. Kashan, & H.-G. Thiele. 1990. Primary structure of rat 6.2, a non-glycosylated phosphatidylinositol-linked surface marker of postthymic T cells. *Proc. Nat. Acad. Sci. U.S.A. 87*: 964–967.

7. Takada, T., K. Iida, & J. Moss. 1994. Expression of NAD glycohydrolase by rat mammary adenocarcinoma cells transformed with rat T cell alloantigen RT6.2. *J. Biol. Chem. 269*: 9420–9423.

8. Haag, F., V. Andresen, S. Karsten, F. Koch-Nolte, & H.G. Thiele. 1995. Both allelic forms of the rat T cell differentiation marker RT6 display nicotinamide adenine dinucleotide (NAD)-glycohydrolase activity, yet only RT6.2 is capable of automodification upon incubation with NAD. *Eur. J. Immunol. 25*: 2355–2361.

9. Maehama, T., H. Nishina, S. Hoshino, Y. Kanaho, & T. Katada. 1995. NAD$^+$-dependent ADP-ribosylation of T lymphocyte alloantigen RT6.1 reversibly proceeding in intact rat lymphocytes. *J. Biol. Chem. 270*: 22747–22751

10. Rigby, M.R., R. Bortell, L.A. Stevens, J. Moss, T. Kanaitsuka, H. Shigeta, J.P. Mordes, D.L. Greiner, & A.A. Rossini. 1996. Rat RT6.2 and mouse RT6 locus 1 are NAD$^+$: Arginine ADP ribosyltransferases with auto-ADP ribosylation activity. *J. Immunol. 156*: 4259–4265.

11. Wang, J., E. Nemoto, A.Y. Kots, H.R. Kaslow, & G. Dennert. 1994. Regulation of cytotoxic T cells by ecto-nicotinamide adenine dinucleotide (NAD) correlates with cell surface GP1-anchored/arginine ADP-ribosyltransferase. *J. Immunol. 153*: 4048–4058.

12. Wang, J. E. Nemoto, & G. Dennert. 1996. Regulation of CTL by ecto-nicotinamide adenine dinucleotide (NAD) involves ADP-ribosylation of a p56^{lck}-associated protein. *J. Immunol. 156*: 2819–2827.

13. Koch-Nolte, F., F. Haag, R. Kastelein, & F. Bazan. 1996. Uncovered: the family relationship of a T-cell-membrane protein and bacterial toxins. *Immunol. Today 17*: 402–405.

14. Wonigeit, K. Schwinzer, R. 1987. Polyclonal activation of rat T-lymphocytes by RT6 alloantisera. *Transplant. Proc. 19*: 296–299.

15. Robinson, P.J. 1991. Phosphatidylinositol membrane anchors and T-cell activation. *Immunol. Today 12*: 35–41.

16. Thiele, H.G., F. Koch, & A. Kashan. 1987. Postnatal distribution profiles of Thy-1$^+$ and RT6$^+$ cells in peripheral lymph nodes of DA rats. *Transplant. Proc. 19*: 3157–3160.

17. Mojcik, C.F., D.L. Greiner, E.S. Medlock, K.L. Komschlies, & I. Goldschneider. 1988. Characterization of RT6 bearing rat-lymphocytes. I. Ontogeny of the RT6$^-$ subset. *Cell. Immunol. 114*: 336–346.

18. Fangmann, J., R. Schwinzer, & K. Wonigeit. 1991. Unusual phenotype of intestinal intraepithelial lymphocytes in the rat: predominance of T cell receptor a/b$^+$CD2$^-$ cells and high expression of RT6 alloantigen. *Eur. J. Immunol. 21*: 753–760.

19. Fangmann, J., R. Schwinzer, H.-J. Hedrich, & K. Wonigeit. 1993. Demonstration of RT6 expression on a CD3$^-$ population of intestinal intraepithelial lymphocytes of athymic nude rats. *Transplant. Proc. 25*: 2789–2790.

20. Schwinzer, R., H.-J. Hedrich, & K. Wonigeit. 1989. T cell differentiation in athymic nude rats (rnu/rnu): demonstration of a distorted T cell subset structure by flow cytometry analysis. *Eur. J. Immunol. 19*: 1841–1847.

21. Fangmann, J., R. Schwinzer, H.-J. Hedrich, I. Klöting, & K. Wonigeit. 1991. Diabetes-prone BB rats express the RT6 alloantigen on intestinal intraepithelial lymphocytes. *Eur. J. Immunol. 21*: 2011–2015.

22. Kühnlein, P., J.-H. Park, T. Herrmann, A. Elbe, & T. Hünig. 1994. Idenfication and characterization of rat γ/δ T lymphocytes in peripheral lymphoid organs, small intestine, and skin with a monoclonal antibody to a constant determinant of the γ/δ T cell receptor. *J. Immunol. 153*: 979–986.

23. Zolkiewska, A. & J. Moss. 1993. Integrin α7 as substrate for a glycosylphosphatidylinositol-anchored ADP-ribosyltransferase on the surface of skeletal muscle cells. 1993. *J. Biol. Chem. 268*: 25273–25276.

24. Greiner, D.L., E.S. Handler, K. Nakano, J.P. Mordes, & A.A. Rossini. 1986. Absence of the RT6 T cell subset in diabetes-prone BB/W rats. *J. Immunol. 136*: 148–151.

25. Groen, H., J.M.M.M. Van der Berk, P. Nieuwenhuis, & J. Kampinga. 1989. Peripheral T cells in diabetes prone (DP) BB rats are CD45R-negative. *Thymus 14*: 145-.

26. Burstein, D., J.P. Mordes, D.L. Greiner, D. Stein, N. Nakamura, E.S. Handler, & A.A. Rossini. 1989. Prevention of diabetes in BB/Wor rat by single transfusion of spleen cells. Parameters that affect degree of protection. *Diabetes 38*: 24–30.

27. Thiele, H.-G., F. Koch, F. Haag, & W. Wurst. 1989. Evidence for normal thymic export of lymphocytes and an intact RT6a gene in RT6 deficient diabetes prone BB rats. *Thymus 14*: 137-.

28. Lang, F. & W. Kastern. 1989. The gene for the T lymphocyte alloantigen, RT6, is not linked to either diabetes or lymphopenia and is not defective in the BB rat. *Eur. J. Immunol. 19*: 1785–1789.

29. Schwinzer, R., H.J. Hedrich, & K. Wonigeit. 1987. Development of T-like cells in athymic Rowett nude rats (rnu/rnu). *Transplant. Proc. 19*: 3131–3133.

30. Hedrich, H.J., K. Wonigeit, & R. Schwinzer. In vivo alloreactivity and xenoreactivity of athymic nude rats. *Transplant. Proc. 19*: 3199–3202.

31. Guy-Grand, D., N. Cerf-Bensussan, B. Malissen, M. Malassis-Seris, C. Briottet, P. Vassali. 1991. Two gut intraepithelial CD8$^+$ lymphocyte populations with different T cell receptors: a role for the gut epithelium in T cell differentiation. *J. Exp. Med. 173*: 471–481.

32. Lefrancois, L. & S. Olson. 1994. A novel pathway of thymus-directed T lymphocyte maturation. *J. Immunol. 153*: 987–995.

33. Vaage, T.J., E. Dissen, A. Ager, I. Roberts, S. Fossum, & B. Rolstad. 1990. T cell receptor-bearing cells among rat intestinal intraepithelial lymphocytes are mainly α/β^+ and are thymus dependent. *Eur. J. Immunol. 20*: 1193–1196.

34. Chambers, W.H., N.L. Vujanovic, A.B. DeLeo, M.W. Olszowy, R.B. Herbermann, & J.C. Hiserodt. 1989. Monoclonal antibody to a triggering structure expressed on rat natural killer cells and adherent lymphokine-activated killer cells. *J. Exp. Med. 169*: 1373–1389.

35. Wonigeit, K. 1996. Expression of RT6 on activated rat T cells. In Koch-Nolte, F., and F. Haag (eds). ADP-Ribosylation in animal tissues: Structure, function, and biology of mono(ADP-Ribosyl)transferases and related enzymes. Plenum Publishing Corporation, New York in press.

36. Thiele, H.-G., F. Haag, & F. Nolte. 1993. Asymmetric expression of RT6.1 and RT6.2 alloantigens in (RT6^a x RT6^b)F$_1$ rats is due to a pretranslational mechanism. *Transplant. Proc. 25*: 2786–2788.

37. Haag, F., F. Nolte, C. Hollmann, & H.-G. Thiele. 1993. Analysis of the gene for the rat T cell alloantigen RT6: evidence for alternative splicing in the 5' region. *Transplant. Proc. 25*: 2884–2885.

38. Haag, F., G. Kuhlenbäumer, F. Koch-Nolte, E. Wingender, & H.-G. Thiele. 1996. Structure of the gene encoding the rat T cell ecto-ADP-ribosyltransferase RT6. *J. Immunol. 157*: 2022–2030.

ARGININE-SPECIFIC MONO(ADP-RIBOSYL)TRANSFERASE ACTIVITY IN HUMAN NEUTROPHIL POLYMORPHS

A Possible Link with the Assembly of Filamentous Actin and Chemotaxis

Panagiotis Kefalas, Jennifer R. Allport, Louise E. Donnelly,
Nigel B. Rendell, Stephen Murray, Graham W. Taylor, Gar Lo,
Masoud Yadollahi-Farsani, and John MacDermot

Department of Clinical Pharmacology
Royal Postgraduate Medical School
Du Cane Road
London W12 0NN, United Kingdom

ABSTRACT

Mono(ADP-ribosyl)transferase activity has been detected on the external surface of human polymorphonuclear neutrophil leucocytes (PMNs). The corresponding cDNA has been cloned and shown to be identical to that derived from human skeletal muscle. Our results suggest that mono(ADP-ribosyl)transferase is involved in the transduction pathway mediating (i) receptor-dependent re-alignment of cytoskeletal actin and (ii) chemotaxis of PMNs.

BACKGROUND AND RESULTS

Mono(ADP-Ribosyl)Transferase

Eukaryotic mono(ADP-ribosyl)transferases have been characterized, and their primary amino acid sequences deduced from cDNA clones from human (1) and rabbit skeletal muscle (2), undefined cells from chicken bone marrow (3) and chicken erythroblasts (4). In rabbit, the relevant mRNA appeared to be confined largely to skeletal muscle and heart (2), although recent Northern analysis performed in this laboratory has also revealed

ADP-Ribosylation in Animal Tissue, edited by Haag and Koch-Nolte
Plenum Press, New York, 1997

the presence of mono(ADP-ribosyl)transferase mRNA in human polymorphonuclear neu-trophil leucocytes (PMNs).

Mono(ADP-Ribosyl)Transferase in Human PMNs

Prompted by this finding we have cloned cDNA derived from the coding region (corresponding to amino acids 24–303, as numbered in reference 1) of mono(ADP-ribo-syl)transferase expressed in human PMNs. The cDNA was generated by RT-PCR amplification of human PMN mRNA with specific primers based on the reported cDNA sequence from human skeletal muscle (1). The cDNA was sequenced, and shown to be identical to that from skeletal muscle.

In preliminary experiments, mono(ADP-ribosyl)transferase activity of PMNs was found to be about 1/50th of that in skeletal or cardiac muscle. The measurement of enzyme activity involved the ADP-ribosylation of agmatine, and we had thus not excluded the possibility that free ADP-ribose might bind non-enzymatically to the primary NH_2 group of agmatine under the conditions of our assay. This reaction might then be interpreted incorrectly as indicative of mono(ADP-ribosyl)transferase activity. To address this issue, we exploited another arginine analogue, namely diethylamino(benzylidineamino)guanidine (DEA-BAG) which is an efficient substrate for the enzyme (5), but has no primary NH_2 group. DEA-BAG was synthesised, and its purity and chemical identity confirmed by HPLC and desorption electron-impact ionization mass spectrometry (5).

Incubation of approximately 10^6 intact PMNs with 1 mM DEA-BAG and 10 mM NAD^+ in a HEPES-buffered balanced salt solution yielded (ADP-ribosyl)-DEA-BAG. The product was separated from the reaction mixture by a 2-step HPLC protocol, and the identity of the product confirmed by electrospray mass spectrometry (5).

In later experiments, mono(ADP-ribosyl)transferase activity was further characterized in a more robust and simpler assay based on the ADP-ribosylation of agmatine (2). Under the conditions of the assay, enzyme activity was linear with respect to time up to 4 h ($r^2 = 0.96$, P<0.01) and cell number up to 8×10^6 cells/tube ($r^2 = 0.95$, P<0.001). The apparent K_m value for NAD^+ was determined from an NAD^+ concentration-response curve, and was 100.1 ±30.4 µM (n=4), and the V_{max} was 1.4 ±0.2 pmol ADP-ribosylagmatine/h/10^6 cells. Further details of the enzymology may be found in Donnelly *et al.*, 1996 (5).

GPI Anchor of PMN Mono(ADP-Ribosyl)Transferase

The deduced amino acid sequence of the skeletal muscle and PMN derived enzyme suggests a glycosylphosphatidylinositol (GPI) anchor, and this has been confirmed immunologically for the enzyme from human skeletal muscle (1). We have now shown that mono(ADP-ribosyl)transferase of PMNs is similarly anchored to the plasma membrane. Intact human PMNs were exposed to phosphoinositide-specific phospholipase C (PI-PLC; 1.2 units/ml), and the enzyme activity was then measured in both the cell pellet and the supernatant. After a 1 h incubation with PI-PLC, approximately 98% of the mono(ADP-ribosyl)transferase activity had been lost from the cell surface (5).

Mono(ADP-Ribosyl)ated Proteins on the Surface of Human PMNs

Intact PMNs were exposed to $[\alpha^{32}P]$-NAD^+, and labelled proteins were resolved by SDS PAGE. Labelled proteins with MWt values of 79, 67, 46, 36 and 26 kDa were identified autoradiographically, and they were shown to be mono(ADP-ribosyl)ated by hydrolysis of the adduct with snake venom phosphodiesterase to yield [^{32}P]-5'-AMP. Experiments are in progress to identify the labelled protein substrates in PMNs.

PMN Chemotaxis

PMNs are competent to migrate up a chemical gradient both *in vitro* and *in vivo*, and the receptors (*eg* for FMLP, PAF, C5a, IL-8 etc) expressed on the cell surface that bind the individual chemotaxins are all coupled to PI-PLC. The catalytic activity of this enzyme yields inositol 1,4,5-*tris*phosphate and 1,2-diacylglycerol, which mediate rises in [Ca^{2+}]$_i$ or activation of protein kinase (PK) C respectively. The roles of changes in [Ca^{2+}]$_i$ or PKC activity in chemotaxis are ambiguous however, with contradictory reports in the literature (reviewed in 6).

During chemotaxis there is continuous re-alignment of the cytoskeleton in the direction of the chemotactic gradient, contraction of actin-containing microfilaments, and adhesion or release of the cell surface to adjacent extracellular matrix proteins such as collagen, fibronectin or laminin. The process of assembly of filamentous (F) actin requires the sequential addition of monomeric or globular (G) actin to the growing cytoskeletal microfilament.

PMN Chemotaxis and Mono(ADP-Ribosyl)Transferase Activity

PMN chemotaxis was measured in a Boyden chamber by the leading front method. The cells migrated in response to 100 nM FMLP, and measurements were then made of the capacity (IC$_{50}$ values) of a panel of structurally unrelated inhibitors of mono(ADP-ribosyl)transferase activity to reduce the chemotactic response. The inhibitors were nicotinamide, novobiocin, vitamin K$_3$ and vitamin K$_1$. The capacity of DEA-BAG to inhibit chemotaxis was also measured. In addition, the IC$_{50}$ values were determined for these compounds as inhibitors of the mono(ADP-ribosyl)ation of agmatine by intact PMNs. Analysis of the data revealed a linear correlation between their IC$_{50}$ values as inhibitors of (ADP-ribosyl)-agmatine formation and as inhibitors of FMLP-dependent chemotaxis (r^2 = 0.82, P<0.01).

In subsequent studies, measurements were made of the assembly of F-actin by quantification of its capacity to bind NBD-phallacidin. Fluorescence intensity was measured by FACS. F-actin assembly was triggered with 1 μM FMLP, 1μM PAF or 4 nM C5a, and the panel of inhibitors listed above (or DEA-BAG) were compared for their capacity to reduce receptor-dependent F-actin formation. Again a linear correlation was shown between their IC$_{50}$ values as inhibitors of (ADP-ribosyl)-agmatine formation and as inhibitors of receptor-dependent chemotaxis (r^2 = 0.91, P<0.001). Further details of these studies may be found in (6).

DISCUSSION

The results presented provide unequivocal evidence for arginine-specific mono(ADP-ribosyl)transferase activity on the surface of human PMNs. Cloning studies indicate that the enzyme is identical to the mono(ADP-ribosyl)transferase of human skeletal muscle, and it is likewise attached to the plasma membrane by a GPI anchor. The kinetics of the PMN enzyme are very similar to those described for other eukaryotic mono(ADP-ribosyl)transferases, although the level of activity suggests that it is expressed at a very low level. Several unidentified proteins have been shown to be mono(ADP-ribosyl)ated on the surface of PMNs, and their further characterization is in progress.

In experiments exploiting a panel of inhibitors and a single pseudo-substrate of the enzyme, close correlations have been observed for inhibition of the ADP-ribosylation of agmatine by PMNs and for inhibition of receptor-dependent actin assembly and chemotaxis. These results prompt us to propose a role for arginine-specific mono(ADP-ribosyl)transferase in the transduction pathway mediating these responses.

We acknowledge gratefully the financial support of the Wellcome Trust (UK) and Glaxo-Wellcome (UK).

REFERENCES

1. Okazaki, I.J., A. Zolkiewska, M.S. Nightingale & J. Moss. 1994. Immunological and structural conservation of mammalian skeletal muscle glycosylphosphatidylinositol-linked ADP-ribosyltransferases. *Biochemistry 33*: 12828–12836.
2. Zolkiewska, A., M.S. Nightingale & J. Moss. 1992. Molecular characterization of NAD:arginine ADP-ribosyltransferase from rabbit skeletal muscle. *Proc. Natl. Acad. Sci. U-S-A. 89*: 11352–11356.
3. Tsuchiya, M., N. Hara, K. Yamada, H. Osago & M. Shimoyama. 1994. Cloning and expression of cDNA for arginine-specific ADP-ribosyltransferase from chicken bone marrow cells. *J. Biol. Chem. 269*: 27451–27457.
4. Davis, T., & S. Shall. 1995. Cloning of a chicken gene homologous to the rabbit mono-ADP-ribosyltransferase gene. *Biochem. Soc. Trans. 23*: 207S.
5. Donnelly, L. E., N. B. Rendell, S. Murray, J. R. Allport, G. Lo, P. Kefalas, G. W. Taylor, & J. MacDermot. 1996. Arginine-specific mono(ADP-ribosyl)transferase activity on the surface of human polymorphonuclear neutrophil leucocytes. *Biochem. J. 315*, 635–641.
6. Allport, J. R., L. E. Donnelly, B. P. Hayes, S. Murray, N. B. Rendell, K. P. Ray & J. MacDermot. 1996. Reduction by inhibitors of mono(ADP-ribosyl)transferase of chemotaxis in human neutrophil leucocytes by inhibition of the assembly of filamentous actin. *Br. J. Pharmacol.* (in press).

30

A NEWLY IDENTIFIED GLYCOSYLPHOSPHATIDYLINOSITOL-ANCHORED ARGININE-SPECIFIC ADP-RIBOSYLTRANSFERASE IN CHICKEN SPLEEN

Mikako Tsuchiya, Harumi Osago, Kazuo Yamada, and Makoto Shimoyama

Department of Biochemistry
Shimane Medical University
Izumo 693, Japan

ABSTRACT

An arginine-specific ADP-ribosyltransferase activity was detected in chicken spleen membrane fraction using a capillary electrophoresis assay and the activity was extracted by phosphatidylinositol-specific phospholipase C but not by 1 M NaCl or 1% Triton X-100. The enzyme protein was purified from chicken spleen membrane fraction to apparent homogeneity with a six-step method containing phosphatidylinositol-specific phospholipase C treatment, ammonium sulfate precipitation and conventional column chromatographies. Apparent molecular mass of the purified enzyme estimated with SDS/PAGE was 44 kDa. N-glycanase treatment of the enzyme reduced the apparent molecular size on SDS/PAGE. The enzyme was recognized by anti-cross reacting determinant antibodies. Partial amino acid sequence of the purified enzyme protein showed high homologies with primary structures of previously reported chicken arginine-specific ADP-ribosyltransferases.

INTRODUCTION

ADP-ribosyltransferases in eucaryotes were purified from turkey erythrocytes (1) and from chicken heterotrophic polymorphonuclear leukocytes (so-called heterophils) (2) in a soluble form. Based on amino acid sequences derived from the heterophil transferase, two cDNA clones AT1 and AT2 of arginine-specific ADP-ribosyltransferase were isolated from a chicken bone marrow cell cDNA library (3). The activities of two transferases were detected in culture medium of COS 7 cells transfected with AT1 or AT2 cDNA and

ADP-Ribosylation in Animal Tissue, edited by Haag and Koch-Nolte
Plenum Press, New York, 1997

these transferases had different enzymatic properties in response to 2-mercaptoethanol (MSH) or NaCl. A membrane-associated ADP-ribosyltransferase was found in mammalian cardiac and skeletal muscle tissues (4), and purified from rabbit skeletal muscle (5). The primary structures of rabbit and human skeletal muscle transferases were determined by cDNA cloning, and these transferases were shown to be glycosylphosphatidylinositol (GPI)-anchored proteins(6, 7). Recently, we developed a capillary electrophoresis method to detect ADP-ribosylarginine formation directly, for assay of arginine-specific ADP-ribosyltransferase (8). Using the method, a GPI-anchored ADP-ribosyltransferase was detected in chicken spleen membrane fraction and the transferase was purified.

RESULTS AND DISCUSSION

Chicken spleen membrane fraction was incubated with 5 mM NAD^+ and 100 mM arginine and the reaction mixture was subjected to the capillary electrophoresis assay. As shown in Fig. 1, ADP-ribosylarginine formation was detected (8). To extract the arginine-specific ADP-ribosyltransferase activity from chicken spleen membrane, the membrane was treated with 1 M NaCl, 1% Triton X-100 or phosphatidylinositol-specific phospholipase C (PI-PLC) and then centrifuged at 100,000 x g. As shown in Table 1, NaCl or Triton X-100 released less than 10% of the total ADP-ribosyltransferase activity of the membrane fraction to the 100,000 x g supernatant while PI-PLC liberated more than 80% of the activity (9). These results suggest that the transferase is associated with the membrane through the GPI-anchor. In previous work, two cDNA clones AT1 and AT2 encoding ADP-ribosyltransferases of chicken bone marrow cells were cloned (3). These transferases expressed in COS 7 cells responded differently to sulfhydryl agent and NaCl, in terms of ADP-ribosylation of a model acceptor protein, casein; AT1 transferase required thiols such as MSH for the activity and is inhibited by 200 mM NaCl in the presence of MSH while AT2 transferase did not require the thiol and the activity was increased by the addition of 200 mM NaCl. To compare the response of the spleen enzyme with those of two chicken transferases, AT1 and AT2, we examined the effect of MSH and NaCl on the activity of the transferase liberated from the spleen membrane. The activity of the trans-

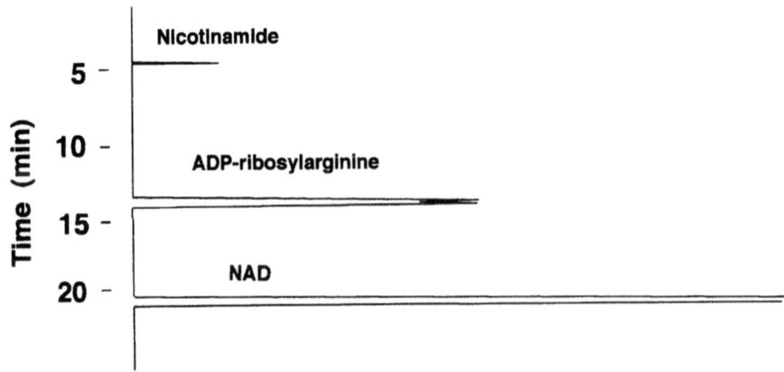

Figure 1. Electropherogram of capillary electrophoresis of ADP-ribosylarginine. Chicken spleen membrane fraction was incubated with 5 mM NAD^+ and 100 mM arginine. After incubation, the reaction solution was analyzed by capillary electrophoresis (8).

Table 1. Extraction of arginine-specific ADP-ribosyltransferase from chicken spleen membrane fraction. Total activity of the transferase in 0.5 ml of chicken spleen fraction was set at 100% (9)

Treatment	Transferase activity (%)	
	Supernatant	Precipitate
Control	1.8 ±0.2	98.4 ± 5.3
NaCl (1.0 M)	3.7 ± 0.5	89.1 ± 7.2
Triton X-100 (1.0%)	6.4 ±1.3	94.6 ±10.7
PI-PLC (10.6 munit/ml)	81.5 ±7.8	10.0 ± 2.2

ferase derived from the spleen membrane was not affected by 5 mM MSH and/or 200 mM NaCl (Fig. 2).

The enzyme protein was purified from chicken spleen membrane fraction to apparent homogeneity using PI-PLC treatment, ammonium sulfate precipitation, and sequential column chromatography on Phenyl Sepharose, DE-cellulose, Con A-Sepharose and Sephadex G-100. Apparent molecular masses of the purified enzyme estimated with SDS/PAGE were 44 kDa and 42 kDa under reducing and non-reducing conditions, respectively. These results indicate the presence of intramolecular disulfide bonds. The apparent molecular size of the transferase on SDS/PAGE was reduced by N-glycanase treatment, thereby indicating the presence of N-glycosyl modification of the enzyme protein. On Western blot analysis with anti-cross reacting determinant antibodies, the enzyme was detected. This result confirms that the enzyme is a glycosylphosphatidylinositol-anchored protein. The enzyme catalyzes the formation of the α-anomer of ADP-ribosylarginine. Partial amino acid sequence of the purified enzyme protein contained FDDQY, the sequence conserved in animal arginine-specific ADP-ribosyltransferases (3). Determination of additional amino acid sequences of the spleen transferase is required for its molecular identification.

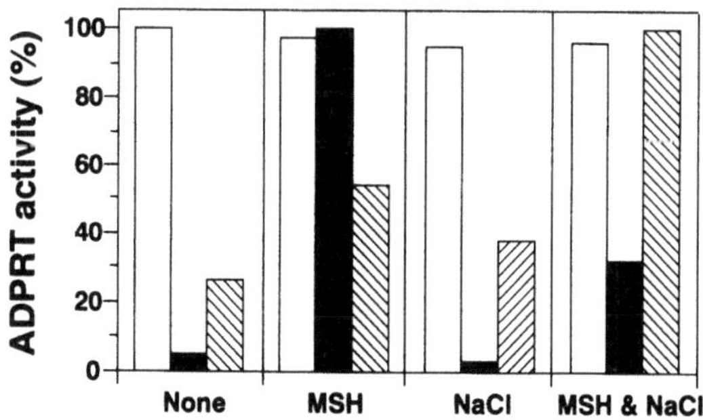

Figure 2. Effects of MSH and/or NaCl on ADP-ribosyltransferase activity released from chicken spleen membrane fraction by PI-PLC (open bars) and comparison of the effects with those on AT1 (closed bars) or AT2 (hatched bars) transferase activities. mAT; ADP-ribosyltransferase released from chicken spleen membrane fraction by PI-PLC (9). The Maximal activity of each enzyme was set at 100%.

REFERENCES

1. Moss, J., Stanley, S. J. & Watkins, P. A. 1980. Isolation and properties of an NAD- and guanidine-dependent ADP-ribosyltransferase from turkey erythrocytes. *J. Biol. Chem. 255*: 5838–5840.
2. Mishima, K., Terashima, M., Obara, S., Yamada, K., Imai, K. & Shimoyama, M. 1991. Arginine-specific ADP-ribosyltransferase and its acceptor protein p33 in chicken polymorphonuclear cells: co-localization in the cell granules, partial characterization, and *In Situ* mono(ADP-ribosyl)ation. *J. Biochem. 110*: 388–394.
3. Tsuchiya, M., Hara, N., Yamada, K., Osago, H. & Shimoyama, M. 1994. Cloning and expression of cDNA for arginine-specific ADP-ribosyltransferase from chicken bone marrow cells. *J. Biol. Chem. 269*: 27451–27457.
4. Soman, G., Tomer, K. B. & Graves, D. J. 1983. Assay of mono ADP-ribosyltransferase activity by using guanylhydrazones. *Anal. Biochem. 134*: 101–110.
5. Peterson, J. E., Jacquiline, S.-A. L. & Graves, D. J. 1990. Purification and partial characterization of arginine-specific ADP-ribosyltransferase from skeletal muscle microsomal membranes. *J. Biol. Chem. 265*: 17602–17609.
6. Zolkiewska, A. & Moss, J. 1993. Integrin a7 as substrate for a glycosylphosphatidylinositol-anchored ADP-ribosyltransferase on the surface of skeletal muscle cells. *J. Biol. Chem. 268*: 25273–25276.
7. Okazaki, I. J., Zolkiewska, A., Nightingale, M. S. & Moss, J. 1994. Immunological and structural conservation of mammalian skeletal muscle glycosylphosphatidylinositol-linked ADP-ribosyltransferase. *Biochemistry. 33*: 12828–12836.
8. Tsuchiya, M., Osago, H. & Shimoyama, M. 1995. Assay of arginine-specific adenosine-5'-diphosphate-ribosyltransferase by capillary electrophoresis. *Anal. Biochem. 224*: 486–489.
9. Tsuchiya, M., Osago, H. & Shimoyama, M. 1995. A newly identified GPI-anchored arginine-specific ADP-ribosyltransferase activity in chicken spleen. *Biochem. Biophys. Res. Commun. 214*: 760–764.

31

ROLE OF ADP-RIBOSYLATION IN ACTIVATED MONOCYTES/MACROPHAGES

Sunna Hauschildt,[1] Peter Scheipers,[2] Wolfgang Bessler,[2] Klaus Schwarz,[1] Artur Ullmer,[3] Hans-Dieter Flad,[3] and Holger Heine[3]

[1]Universität Leipzig
Institut für Zoologie
Immunbiologie
Talstr. 33, D-04103 Leipzig
[2]Universität Freiburg
Institut für Immunbiologie
Stefan-Meierstr. 8
D-79104 Freiburg
[3]Forschungszentrum Borstel
Parkallee 22, D-23845 Borstel

ABSTRACT

Stimulating monocytes/macrophages with bacterial lipopolysaccharide (LPS) results in TNF-α, IL-1, IL-6 and nitrite (NO_2^-) formation. Inhibitors of poly(ADP-ribose)polymerase inhibit release of these mediators by preventing mRNA expression indicating that ADP-ribosylation plays a crucial role in the synthesis of these mediators. Furthermore we present evidence that ADP-ribosylation is involved in modifying cellular proteins. In murine macrophages a 33 kDa cytosolic protein could be identified that in response to LPS changed its state of ADP-ribosylation, and in human monocytes we showed that the inhibitor nicotinamide prevents LPS induced phosphorylation of two cytosolic proteins of 36 kDa and 38 kDa (p36/38) LPS. Taken together these data indicate that protein modification by ADP-ribosylation may control cellular processes involved in distinct steps of monocyte/macrophage activation.

BACKGROUND

ADP-ribosylation constitutes a covalent modification by which cells regulate protein functions (1). The reaction is catalyzed by ADP-ribosyltransferases that modify a specific amino acid residue in a specific target protein. The best understood ADP-ribosyltrans-

ADP-Ribosylation in Animal Tissue, edited by Haag and Koch-Nolte
Plenum Press, New York, 1997

ferases are bacterial toxins, and lately endogeneous ADP-ribosyltransferases have been identified and characterized in eukaryotic cells (2). The endogeneous substrates for these cellular enzymes are ill defined, and their importance in the physiological regulation of cellular metabolism is poorly understood. However there is increasing evidence that ADP-ribosylation plays a regulatory role in diverse cellular processes (3, 4). To see whether this mode of regulation occurs during activation of monocytes/macrophages we studied the effect of LPS on the ADP-ribosylation of cytosolic proteins and the effect of nicotinamide, an inhibitor of ADP-ribosylation reactions, on LPS-induced mediator production and on LPS-induced posttranslational modification such as phosphorylation.

RESULTS

Stimulating murine bone-marrow derived macrophages (5×10^6/ml) for 24 h in the presence of LPS (0,1 µg/ml) results in a slight decrease in overall ADP-ribosyltransferase activity (5). To determine whether LPS influences the ADP-ribosylation state of distinct proteins, macrophages were incubated with different concentrations of LPS and after 24 h cytosolic supernatants were prepared, incubated with [^{32}P] NAD for 30 min, and proteins were seperated by SDS-PAGE as described (6). As illustrated in Fig. 1 LPS leads to the

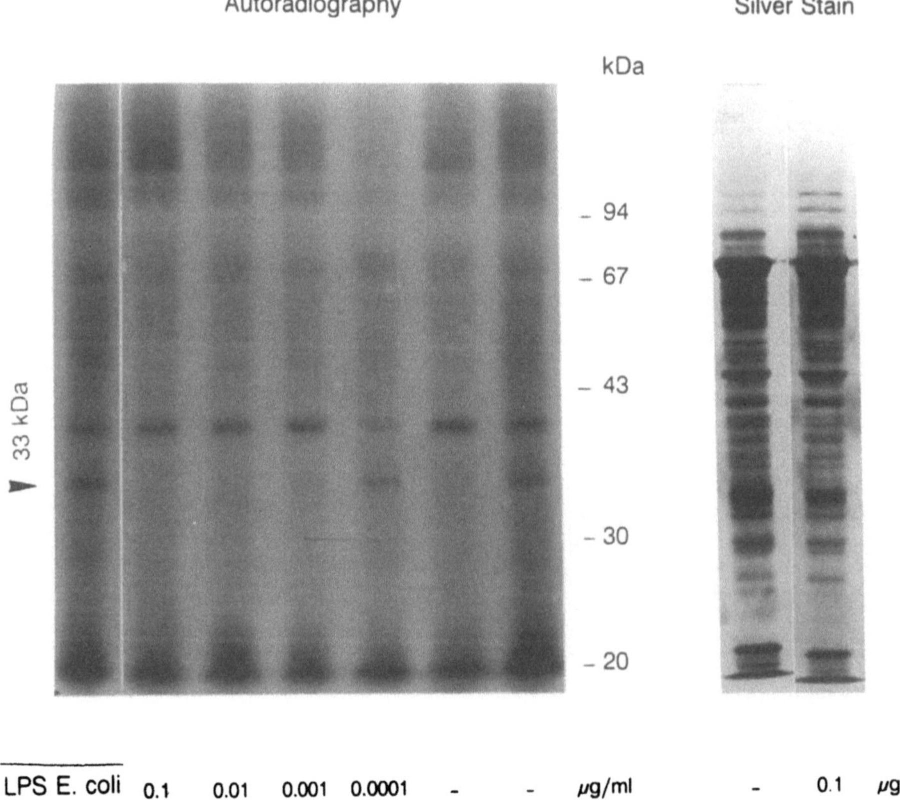

Figure 1. Dose dependency of LPS-induced altered ADP-ribosylation of p33. Macrophages (5×10^6/ml) were incubated in the presence or absence of LPS. After 24 h cytosolic supernatants were prepared and incubated with [^{32}P] NAD for 30 min. Proteins were seperated on SDS-PAGE.

disappearance of ADP-ribosylation of a protein with an apparant molecular mass of 33 kDa. Disappearance of the radioactive band was not observed until 16 h of incubation, indicating that the change in ADP-ribosylation of this protein is rather a late event possibly participating in the LPS-induced process of macrophage activation. Associated with LPS-induced activation of mouse macrophages is the secretion of NO_2^- catalyzed by the inducible NO-synthase. To see whether ADP-ribosylation is involved in NO_2^- formation, we examined inhibitors of poly(ADP-ribose)polymerase, namely nicotinamide, 3-aminobenzamide and 3-methoxybenzamide for their efficacy of preventing production of NO_2^-. All three inhibitors prevented NO_2^- release presumably by blocking biosynthetic processes leading to the synthesis of NO_2^-. Nicotinamide was also found to inhibit other LPS-induced biological responses. In human monocytes nicotinamide induced an inhibition of TNF-α and IL-6 but not IL-1 production at the mRNA and protein level (7). The inhibitor also affected the phosphate labeling of two cytosolic proteins (p36/38) which was increased after treatment with LPS (7). Increasing concentrations of nicotinamide led to a diminished phosphorylation of p36/38 (Fig. 2) indicating that ADP-ribosylation may be involved in LPS-triggered events leading to altered phosphorylation of p36/38.

DISCUSSION

We found that nicotinamide prevented LPS-induced NO_2^-, TNF-α and IL-6 formation, and that inhibition of cytokine production occurred at the mRNA level. Inhibition of NO synthase mRNA expression was described by (8). Unlike other stimuli (9, 10), whose

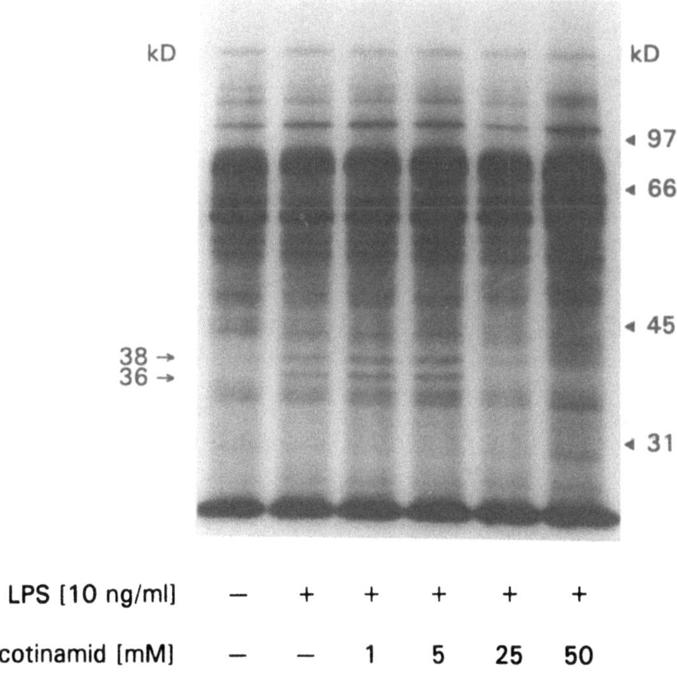

Figure 2. Effect of nicotinamide on LPS-induced altered phosphate labeling of p36/38. Before stimulating monocytes (4x10⁶/ml) with LPS (10ng/ml) for 4 h cells were preincubated with nicotinamide for 15 min. Cytosolic supernatants were prepared and incubated with [³²P] ATP for 10 min. Proteins were seperated on SDS-PAGE.

effects were shown to be blocked by the inhibitors LPS did not lead to an enhanced ADP-ribosylation in mouse macrophages. On the contrary LPS initiated the decline of ADP-ribosylation of a 33 kDa cytosolic protein, the nature of which is not known at present. Nicotinamide not only prevented release of mediators but also LPS-induced phosphorylation of two cytosolic proteins (p36/38). The high doses of nicotinamide needed to observe the effects suggest involvement of a mono(ADP-ribosyl)transferase. Inhibition of phosphorylation of p36/38 might imply that protein modification by ADP-ribosylation alters the rate of phosphorylation. Thus mono(ADP-ribosyl)transferases might modify p36/38 directly or they may act on proteins involved in phosphorylation of the two proteins. Disappearance of the phosphate labelling of p36/38 is always accompanied by a blockade of TNF-α and IL-6 production, suggesting that the two proteins may be involved in the production of the two cytokines. It is not clear whether nicotinamide blocks these reactions by virtue of its ability to inhibit endogeneous ADP-ribosyltransferases or through putative nonspecific effects. However the specificity of these inhibitors on protein phosphorylation limited to the disappearance of labelling p36/38 argues against nonspecific effects. In summary our results indicate ADP-ribosylation as a crucial protein modification involved in intracellular processes leading to activation of monocytes/macrophages.

REFERENCES

1. Hayaishi, O. & K. Ueda eds. 1982. ADP-Ribosylation Reactions: Biology and Medicine. Academic Press, New York.
2. Tanuma, S., K. Kawashima & H. Endo. 1988. Eukaryotic mono(ADP-ribosyl)transferase that ribosylates GTP-binding regulatory Gi protein. J. Biol. Chem. *263*: 5485–5489.
3. Molina Y Vedia, L.B. McDonald, B. Reep, B. Brüne, M. Di Silvio, T.R. Billiar & E.G. Lapetina. 1992. Nitric oxide-induced S-nitrosylation of glyceraldehyde-3-phosphate dehy- drogenase inhibits enzymatic activity and increases endogeneous ADP-ribosylation. J. Biol. Chem. *267*: 24929–24932.
4. Wang, J., E. Nemoto, A.Y. Kots, H. Kaslow & G. Dennert. 1994. Regulation of cytotoxic T cells by ecto-nicotinamide adenine dinucleotide (NAD) correlates with cell surface GPI anchored/arginine ADP-ribosyltransferase. J. Immunol., *153*: 4048–4058.
5. Hauschildt, S., P. Scheipers & W.G. Bessler. 1991. Inhibitors of poly(ADP-ribose)poly- merase suppress lipopolysaccharide-induced nitrite formation in macrophages. Biochem. Biophys. Res. Commun. *179*: 865–871.
6. Hauschildt, S., P. Scheipers & W.G. Bessler. 1994. Lipopolysaccharide-induced change of ADP-ribosylation of a cytosolic protein in bone marrow derived macrophages. Biochem. J.*297*: 17–20.
7. Heine, H., A.J. Ulmer, H.D. Flad & S. Hauschildt. 1995. Lipopolysaccharide-induced change of phosphorylation of two cytosolic proteins in human monocytes is prevented by inhibitors of ADP-ribosylation. J. Immunol. *155*: 4899–4908.
8. Pellat-Deceunynek, C., J. Wietzerbin & J.C. Drapier. 1994. Nicotinamide inhibits nitric oxide synthase mRNA induction in activated macrophages. Biochem. J. *297*: 53–58.
9. Agarwal, S., B.E. Drysdale & H.S. Shin. 1988. Tumor necrosis factor-mediated cytotoxicity involves ADP-ribosylation. J. Immunol. *140*: 4187–4192.
10. Schraufstatter, I.U., P.A. Hyslop, D.B. Hinshaw, R.G. Spragg, L.A. Sklar & C.G. Cochrane. 1986. Hydrogen peroxid-induced injury of cells and its prevention by inhibitors of poly(ADP-ribose)polymerase. Proc. Natl. Acad. Sci. *83*: 4908–4912.

32

EXPRESSION OF RT6, BUT NOT CD45RC, IS DISTURBED ON IMMATURE PERIPHERAL T CELLS IN THE BB RAT[*]

Herman Groen, Jennie M. Pater, Flip A. Klatter, Paul Nieuwenhuis, and Jan Rozing

Department Histology and Cell Biology
University of Groningen
Oostersingel 69/1, 9713 EZ Groningen
The Netherlands

1. INTRODUCTION

In rats, recent thymic migrants (RTM) are phenotypically characterized by the presence of Thy-1 and the lack of RT6 and CD45RC expression. Upon peripheral maturation these cells develop into Thy-1⁻ T cells that express both RT6 and CD45RC (1–3). For the bulk of RTM this transition was shown to occur through at least two subsequent intermediate stages in which T cells first express the Thy-1⁺/RT6⁺/CD45RC⁻ and later the Thy-1⁺/RT6⁺/CD45RC⁺ phenotype (1,4).

In the T lymphocytopenic Diabetes-prone BioBreeding (BB) rat, the peripheral T cell population is characterized by a severe reduction of RT6⁺ (5) and CD45RC⁺ T cells (6). Furthermore, a majority of T cells in the BB rat was found to have a reduced life span (7,8), which is likely to explain the relative overrepresentation of immature Thy-1⁺ peripheral T cells (IPT) also demonstrated in these rats (9).

Due to this particular subset distribution the question was asked whether the underrepresentation of mature Thy-1⁻ peripheral T cells (MPT) in BB rats might account for the overall underrepresentation of RT6⁺ and CD45RC⁺ T cells. However, since frequencies of RT6⁺ and CD45RC⁺ T cells were also found to be severely decreased among the few T cells that had reached the Thy-1⁻ stage of development (9), we wondered whether most peripheral T cells in BB rats might have died before the RT6⁺ stage of T cell development.

The present 3-color flow cytometry (FCM) study was conducted to assess whether lack of RT6 and CD45RC expression on T cells of the BB rat is a trait of MPT only, suggesting cell death for a majority of T cells to occur before the RT6⁺ stage of development,

---

[*] This study was supported by the Dutch Diabetes Foundation

ADP-Ribosylation in Animal Tissue, edited by Haag and Koch-Nolte
Plenum Press, New York, 1997

Table 1. Percentages of RT6[+] and CD45RC[+] cells among Thy-1[high], Thy-1[low]
and Thy-1[-] T cells in peripheral blood of 8-week-old AO and BB rats

	% RT6[+]		% CD45RC[+]	
	AO	BB	AO	BB
Thy-1[high]	32 ± 2	4 ± 1	22 ± 3	17 ± 3
Thy-1[low]	72 ± 3	12 ± 1	40 ± 2	40 ± 2
Thy-1[-]	93 ± 2	17 ± 1	86 ± 2	36 ± 2

or whether the acquisition of either or both of these markers is disturbed already in the IPT stage of development, suggesting rather an intrinsic failure to express either or both RT6 and CD45RC.

Our FCM results indicate that the acquisition of RT6 on IPT of BB rats is disturbed, but that CD45RC acquisition on IPT occurs similar to that in control rats. These findings suggest an intrinsic inability for T cells in the BB rat to express RT6.

2. MATERIALS AND METHODS

2.1. Rats

Male BB Rats were raised under clean conventional conditions in our own animal facilities. The original BB/O breeding nucleus was provided by organon pharmaceutics (OSS, the Netherlands). at the time of the study, approximately 90% of the animals developed IDDM between 12 and 20 weeks of age. Male albino oxford (AO) rats were raised under SPF conditions at the central animal laboratories. Immediately after weaning (3 - 4 Weeks) they were brought to our animal facilities, where they were kept under similar conditions as the BB rats

2.2. Monoclonal Antibodies

Monoclonal antibodies (mAb) used in this study were 3G2 (anti-RT6.1 (10)), GY1/12 (anti-RT6.2 (11)), R73 (anti-TCR (12)), MRC OX-22 (anti-CD45RC (13)), and HIS51 (anti-Thy-1.1 (14)). Hybridoma's producing the mAb were kindly provided by Dr. K. Wonigeit (Hannover, FRG), Dr. G. Butcher (Babraham, UK), Dr. T. Hünig (Freiburg, FRG), and the late Dr. A. Williams (Oxford, UK). PE conjugated to HIS51 was purchased from Pharmingen (San Diego, USA). The other mAb were FITC, biotin or APC conjugated by standard procedures. PE- or APC conjugated streptavidin (Pharmingen) was used as a second step reagent when required.

2.3. Flow Cytometry

Preparation and labeling of buffy coat was performed as described elsewhere (9). Cell acquisition was performed on an Epics Elite flow cytometer (Coulter Epics, Hialeah, USA), equipped with a helium-neon and an argon laser and dual laser beam separation. Calibration was performed with QC3[TM] (FITC/PE/PE-Cy5) beads (Flowcytometry Standards Corporation, San Juan, USA). Acquisition and analysis was performed with the Elite software. Per sample, 20,000 to 30,000 events were recorded using a live gate on forward scatter and TCR labeling. Cut off points for unlabeled cells were set according to irrelevant mAb. For the Thy-

1 labeling, the cut off point between labeled and unlabeled cells was determined on blood cells from AO rats thymectomized 6 to 16 weeks earlier. Virtually all rat T cells have lost Thy-1 expression a few weeks after thymectomy (3). Separation between bright and dull labeling was done arbitrarely as described earlier (7).

3. RESULTS AND DISCUSSION

Figure 1A shows that in control AO rats the loss of Thy-1 from T cells is accompanied by the acquisition of RT6. In Table I it can be seen that RT6 expression already starts among Thy-1^{high} T cells (32%), increasing to approximately 72% among Thy-1^{low} T cells and reaching its maximum among mature Thy-1$^-$ T cells (93%). The vast majority of T cells is of the Thy-1$^-$/RT6$^+$ phenotype. In the BB rat, however, there is virtually no acquisition of RT6 expression upon the loss of Thy-1 (Fig. 1A, Table I), and, consequently, only a minority was found to become Thy-1$^-$/RT6$^+$. This finding may indicate that either RT6 acquisition is disturbed on IPT in BB rats, or that most T cells in these rats die around or before the moment where they should start to express RT6. As it was shown earlier that a substantial number of peripheral T cells in the BB rat dies in the Thy-1$^+$ stage of development (7), both options are equally possible.

Upon the loss of Thy-1 expression, T cells of control AO rats also demonstrate acquisition of CD45RC expression (Fig. 1B). However, as compared to the development of RT6 expression, CD45RC expression seems to occur later in development. Percentages of Thy-1^{high} and Thy-1^{low} T cells expressing CD45RC were significantly lower than those ex-

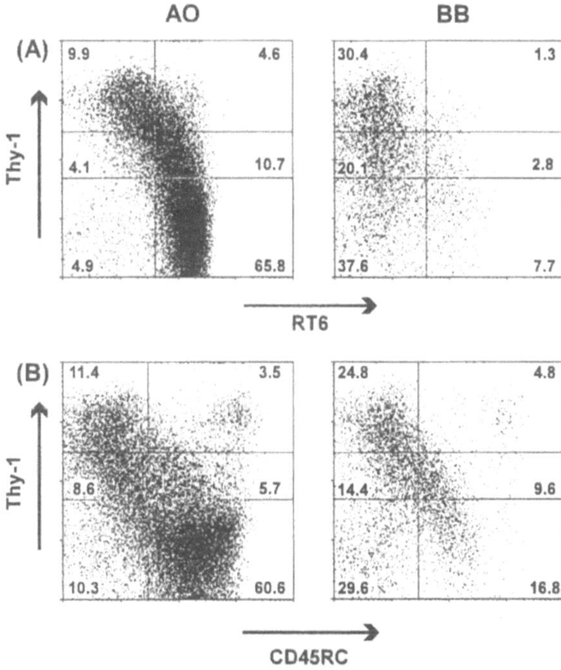

Figure 1. Representative flow cytometry plots of peripheral blood T cells of 8-week-old BB and AO rats. TCR$^+$ cells were gated and checked for (A) Thy-1 and RT6 labeling, and (B) Thy-1 and CD45RC labeling.

pressing RT6 (Table I). This chronological order of acquisition of RT6 and subsequently CD45RC is a confirmation of our earlier findings (1,4) and those of others (15). In the BB rat, the pattern of loss of Thy-1 expression and acquisition of CD45RC expression on IPT (Thy-1^{high} and Thy-1^{low} T cells) seems to resemble the situation in AO rats much more than that seen for RT6 expression. Percentages of cells expressing CD45RC among Thy-1^{high} T cells are not significantly different in AO and BB rats, and those among Thy-1^{low} T cells are identical (Table I). The difference with control rats with respect to CD45RC expression occurs no earlier than in the Thy-1^- stage of development, where in AO rats the majority of T cells is CD45RC$^+$, whereas in BB rats the majority is CD45RC$^-$. These findings suggest that there are no abnormalities with respect to the initiation of CD45RC expression on peripheral T cells in BB rats.

As RT6 expression precedes CD45RC expression in phenotypic T cell development, the lack of RT6 acquisition in the presence of an apparently normal initiation of CD45RC expression on IPT of BB rats seems to indicate that the expression of RT6 is intrinsically disturbed rather than that the cells died before the RT6$^+$ stage of development was reached.

4. REFERENCES

1. Kampinga, J., H. Groen, F.A. Klatter, B. Meedendorp, R. Aspinall, B. Roser, & P. Nieuwenhuis. 1992. Postthymic T cell development in rats: An update. *Biochem. Soc. Trans. 20*:191–7.

2. Thiele, H.-G., F. Koch, & A. Kashan. 1987. Postnatal distribution profiles of Thy-1$^+$ and RT6$^+$ cells in peripheral lymph nodes of DA rats. *Transpl. Proc. 19*:3157–60.

3. Hosseinzadeh, H. & I. Goldschneider. 1993. Recent thymic emigrants in the rat express a unique phenotype and undergo post-thymic maturation in peripheral tissues. *J. Immunol. 150*:1670–9.

4. Kampinga, J., H. Groen, F.A. Klatter, J.M. Pater, A.S. Van Petersen, B. Roser, P. Nieuwenhuis, & R. Aspinall. 1996. *Thymus.* In press

5. Greiner, D.L., E.S. Handler, K. Nakano, J.P. Mordes, & A.A. Rossini. 1986. Absence of the RT-6 T cell subset in diabetes-prone BB/W rats. *J. Immunol. 136*:148–51.

6. Groen, H., J.M.M.M. Van der Berk, P. Nieuwenhuis, & J. Kampinga. 1989. Peripheral T cells in diabetes prone (DP)BB rats are CD45R-negative. *Thymus 14*:145–50.

7. Groen, H., F.A. Klatter, N.H.C. Brons, G. Mesander, P. Nieuwenhuis, & J. Kampinga. 1996. Abnormal thymocyte subset distribution and differential reduction of CD4$^+$ and CD8$^+$ T cell subsets during peripheral maturation in diabetes-prone BB rats. *J. Immunol. 156*:1269–75.

8. Sarkar, P., L. Crisá, U. McKeever, R. Bortell, E. Handler, J.P. Mordes, D. Waite, A. Schoenbaum, F. Haag, F. Koch-Nolte, H.-G. Thiele, D.L. Greiner, & A.A. Rossini. 1995. Loss of RT6 message and most circulating T cells after thymectomy of diabetes-prone BB rats. *Autoimmunity 18*:15–22.

9. Groen, H., F.A. Klatter, N.H.C. Brons, A.S. Wubbena, P. Nieuwenhuis, & J. Kampinga. 1995. High-frequency, but reduced absolute numbers of recent thymic migrants among peripheral blood T lymphocytes in diabetes-prone BB rats. *Cell. Immunol. 163*:113–9.

10. Fangmann, J., R. Schwinzer, H.-J. Hedrich, I. Klöting, & K. Wonigeit. 1991. Diabetes-prone BB rats express the RT6 alloantigen on intestinal intraepithelial lymphocytes. *Eur. J. Immunol. 21*:2011–5.

11. Butcher, G.W. 1987. A list of monoclonal antibodies specific for alloantigens of the rat. *Immunogenetics 14*:163–76.

12. Hünig, T., H.J. Wallny, J. Hartly, A. Lawetsky, & G. Tiefenthaler. 1989. A monoclonal antibody to a constant region of the rat TCR that induces T-cell activation. *J. Exp. Med. 169*:73–86.

13. Spickett, G.P., M.R. Brandon, D.W. Mason, A.F. Williams, & G.R. Woollett. 1983. MRC OX-22, a monoclonal antibody that labels a new subset of T lymphocytes and reacts with the high molecular weight form of the leukocyte-common antigen, *J. Exp. Med. 158*:795–810.

14. Vaessen, L.M.B., P. Joling, F.J. Tielen, & J. Rozing. 1985. New surface antigens on T cells in the early phase of T cell differentiation, detected by monoclonal antibodies. *Adv. Exp. Med. Biol. 186*:251.

15. Haag, F., F. Nolte, A. Lernmark, C. Simrell, and H.-G. Thiele. 1993. Analysis of T-cell surface marker profiles during the postnatal ontogeny of normal and diabetes-prone rats. *Transpl. Proc. 25*:2831–2.

33

EXPRESSION OF RT6 ON ACTIVATED RAT T CELLS

Kurt Wonigeit

Klinik Für Abdominal- Und Transplantationschirurgie
Medizinische Hochschule Hannover
Hannover, Germany

1. ABSTRACT

The gene products of the RT6 system have been demonstrated to be expressed on the majority of resting peripheral T cells but not on fully activated cytotoxic T cells and *in vitro* generated long term T cell lines. This has lead to the conclusion that T cell activation causes a loss of RT6 expression. In the present study this question was examined by analysing the expression of RT6 on T cell populations activated *in vitro* by the T cell mitogen ConA or by allogeneic stimulator cells and subsequently maintained in culture for several weeks. The results showed an initial increase in RT6 expression on most cells during the first two days of culture and a subsequent loss of RT6 expression in many activated cells. Substantial numbers of both CD4$^+$ and CD8$^+$ T cells, however, remained clearly RT6 positive. These cells were present during the entire observation period. The RT6 positive cells did not represent persisting unstimulated cells since they coexpressed CD25 and exhibited typical blast morphology. It is concluded that activation of T cells leads to loss of RT6 expression only in subsets of the CD4 and CD8 T cell populations. The previously reported finding that long term cultures contain only RT6 negative blast cells may indicate preferential survival of RT6 negative blasts rather than obligatory loss of RT6 expression after activation. The possibility is discussed that activated RT6$^-$CD4$^+$ T cells may be inflammatory Th1 cells and activated RT6$^+$CD4$^+$ T lymphoblasts may represent regulatory Th2 cells.

2. INTRODUCTION

The majority of T cells in the peripheral lymphatic organs of the rat express cell surface molecules encoded by the *RT6* gene located on chromosome 1. Although the RT6 system formerly called AgF, ART-2, Pta, and RT-Ly2 has been known already for a long time (1–3), only recently it has been shown that its gene products are ectoenzymes, which may play an important role in the regulation of several lymphocyte functions (4,5). This observation has re-

ADP-Ribosylation in Animal Tissue, edited by Haag and Koch-Nolte
Plenum Press, New York, 1997

newed the interest in the structural and functional analysis of the RT6 system. Our group has re-examined the expression of RT6 and reports in an accompanying paper that it is not restricted to T cells, but that a subset of NK cells, probably representing activated NK cells, also expresses RT6 in low density (6). Furthermore the expression pattern of RT6 on intestinal intraepithelial lymphocytes (IEL) is reviewed, which have previously been shown to express RT6 in much higher density than T cells of the peripheral lymphatic organs (7).

In the present study the effect of *in vitro* activation on the pattern of RT6 expression on T cells from peripheral lymphatic organs has been analysed. Previous studies on the effect of T cell activation have demonstrated that it may lead to a reduction or loss of RT6 expression (8–10). This observation in association with the fact that it has not been possible to generate RT6 positive long term T cell lines or clones *in vitro* has led to the widely accepted view that RT6 expression is lost in all T cells as the result of activation. The experiments presented here demonstrate that this is true only for a subset of activated cells and that another subset continues to express RT6 despite activation.

3. MATERIALS AND METHODS

The inbred rat strain LEW, LEW.1LM2, and LEW.1N were used. All three strains are maintained at the Medizinische Hochschule Hannover and possess the RT6[a] allele. The gene product of RT6[a] displays the serologically defined specificity RT6.1. In the present study this was detected by the monoclonal antibodies 3G2 generated in the author's laboratory and P4/16 generously provided by Dr. G. Butcher (Cambridge/UK). Other antibodies used were directed to $\alpha\beta$TCR (R73), CD4 (W3/25), CD8α (OX8), CD45RC (OX22), CD25 (ART18), and surface Ig (OX12). The staining procedures applied and the performance of the flow cytometry measurements have been described elsewhere (7).

Cell suspensions for the culture experiments were prepared by standard procedures from spleen and lymph nodes. Suspensions enriched for CD4$^+$ T cells were produced by a two step procedure. In the first step, B cells were removed by a nylon wool column; in the second, step the effluent cell population was depleted of CD8$^+$ cells by a panning procedure using the OX8 monoclonal antibody. Cells were cultured in Dulbecco's medium containing glutamine, penicillin-streptomycin (100U/ml each), 2-mercaptoethanol (5×10^{-5}M), and 2% normal rat serum. For ConA stimulation 20×10^6 lymphocytes were cultured in 20ml medium containing 7.5µg ConA per ml in small tissue culture flasks in upright position. For long term culture, rat IL-2 (about 200U/ml medium) was added starting at day 5 after initiation of the cultures. Mixed lymphocyte cultures (MLC) were performed by culturing 20×10^6 responder cells together with 20×10^6 irradiated stimulator cells (2000rd). These cultures were supplemented with exogenous IL-2 starting at day 6 after initiation of the cultures. In order to maintain long term growth the cultures were divided at regular intervals. Usually 5×10^6 lymphoblasts were transferred to new flasks with fresh culture medium containing IL-2.

4. RESULTS

4.1. Sequential Analysis of RT6 Expression on CD4$^+$ T Cells Activated by ConA

Polyclonal activation by the mitogen ConA and subsequent culture with exogenous IL-2 permits the long term culture of rat T cells. Here, this approach has been used to analyse changes in the pattern of RT6 expression associated with the activation process and with subsequent interleukin-2 driven growth. Fig. 1 demonstrates the results obtained with a suspension of lymph node cells highly enriched for CD4⁺ T cells. Unstimulated, these cells exhibited the normal pattern of RT6 expression. A small fraction of cells were clearly RT6 negative, the other cells expressed RT6 in densities varying over a broad range. During the first two days the RT6⁻ population disappeared nearly completely probably indicating the induction of RT6 on previously negative cells. This was followed by the loss of RT6 expression in about one third of the cells, whereas two thirds continued to express RT6. The dichotomy into RT6⁺ and RT6⁻ cells was not caused by the presence of cells which had escaped stimulation for the following reasons: (a) All cells measured showed typical blast morphology; (b) all cells expressed the α chain of the IL-2 receptor (CD25); and (c) all cells lost the expression of CD45RC. A similar experiment performed with unseparated lymph node cells is depicted in fig. 2. The series of flow cytometry measurements started at day 3 after ConA stimulation. At this time the typical reduction of RT6 expression had already occurred which subsequently led to the separation of distinct RT6⁺ and RT6⁻ subsets. In this as well as several other experiments using ConA stimulation, CD4⁺ cells were expanded predominantly. Cells staining for CD8α were mostly CD4⁺CD8⁺ double positive cells expressing the CD8αα homodimer. This phenotype has previously been shown to result from strong activation of some CD4⁺ cells. This and a series of similar experiments demonstrated that a subset of activated CD4⁺ cells continuously expressed RT6 in high density for observation periods up to four weeks.

4.2. RT6 Expression in Alloreactive T Cell Blasts Generated across Full MHC Haplotype Differences in Mixed Lymphocyte Cultures (MLC)

Similar studies were also performed with lymphoblasts generated in mixed lymphocyte cultures. When LEW (RT1^l) or LEW.1LM2 (RT1^{lm2}) spleen or lymph node cells were

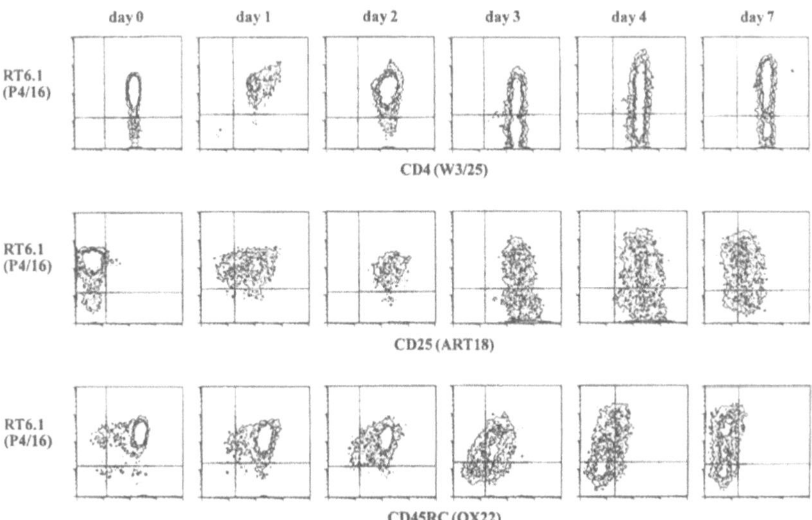

Figure 1. Sequential analysis of changes in RT6 expression induced by ConA stimulation of CD4⁺ lymph node cells of a LEW rat. At various times after stimulation (day 0 to day 7) cells were stained with various mAbs and subsequently analysed with a flow cytometer. Results are given as two colour plots. Staining for RT6.1 was performed with mAb P4/16. This mAb was biotinylated and the bound antibody was visualised with a streptavidin-phycoerythrin conjugate. Binding of mAbs directed to CD4, CD25, and CD45RC was detected by an indirect staining procedure using a FITC labelled goat-anti-mouse Ig. Results given for day 0 were obtained with freshly prepared cells before the addition of ConA. Note the reduction of RT6⁻ cells on day 1 and day 2 and the rapid increase of the RT6⁻ phenotype on day 3.

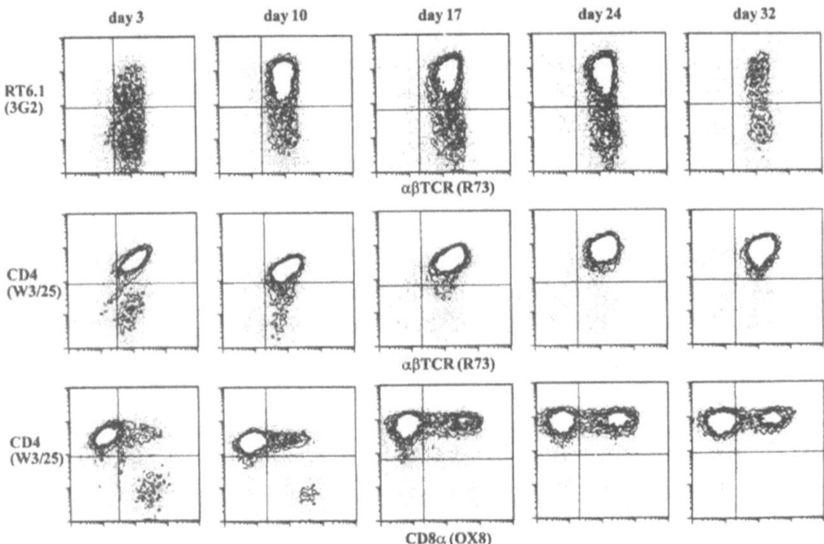

Figure 2. Changes of RT6 expression on LEW lymph node cells activated by ConA and cultured with exogenous IL-2 for 32 days. In all stainings the mAbs given at the y-axis were biotinylated and were detected with a strepta-vidin-phycoerythrin conjugate, whereas the mAbs given at the x-axis were detected by indirect staining with FITC-labelled goat-anti-mouse Ig. A small population of CD8 single positive cells clearly recognised at day 3 and day 10 was lost on day 17 of culture. The remaining CD8$^+$ cells were CD4$^+$CD8$^+$ cells representing highly activated CD4$^+$ cells coexpressing the CD8$\alpha\alpha$ homodimer. In the CD4$^+\alpha\beta$TCR$^+$ population RT6$^+$ and RT6$^-$ lymphoblasts coexisted over the entire observation period. Staining for CD45RC (mAb OX22) showed that substantial numbers of RT6$^+$ blasts reexpressed this marker on days 17, 24, and 32 (data not shown).

cultured with irradiated LEW.1N (RT1^n) stimulator cells usually both CD4$^+$ and CD8$^+$ responder cells became activated since the stimulator cells differed in MHC class I and class II antigens. The fraction of specifically responsive cells in these mixed lymphocyte cultures, however, was much lower than in lymphocyte populations activated by ConA. Nevertheless, populations consisting predominantly of activated cells could be obtained after prolonged periods of culture with exogenous IL-2 since only activated cells are maintained under these conditions. Using this protocol, also among MLC-derived T cell blasts an RT6$^+$ as well as an RT6$^-$ population could be identified (Fig. 3). Both populations contained CD8$^+$ as well as CD4$^+$ cells. Similar results were also obtained with T cells of other rat strains activated *in vitro* by fully MHC-disparate stimulator cells.

5. DISCUSSION

Previous studies on the expression pattern of RT6 on T cells of the peripheral lymphatic organs have revealed the following important features: (a) RT6 expression is associated with a particular maturational process in the peripheral lymphatic organs. This follows from the fact that RT6 is expressed on the vast majority of peripheral T cells but not on thymocytes and on recent thymic emigrants (11,12). (b) Detailed phenotypical studies in animals of different age, in rats with vascularized thymus transplants, and in thymectomized rats have identified two distinct populations of memory T cells differing in the expression of RT6 (13). It is assumed that both populations are derived from anti-

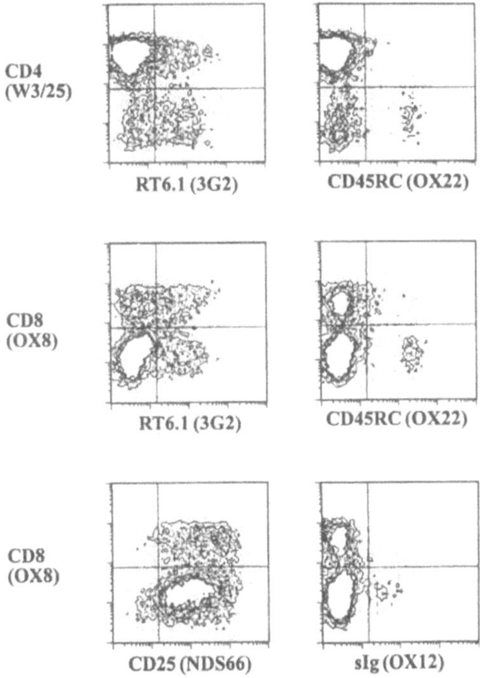

Figure 3. Phenotype LEW.1LM2 spleen cells on day 14 after stimulation with irradiated LEW.1N stimulator cells. Both the CD4$^+$ population and the CD8$^+$ population contain RT6$^+$ and RT6$^-$ cells. CD45RC is only expressed by very few cells mostly consisting of surface Ig positive B cells.

gen-activated T cells having lost RT6 expression. (c) Several studies comparing RT6$^+$ and RT6$^-$ T cells in functional assays have demonstrated clear differences between the two populations (9,10). Most remarkable is the finding that RT6$^+$ T cells are missing in the peripheral lymphatic organs of diabetes-prone BB rats and that the development of disease in this model can be prevented by transfusion of syngeneic RT6$^+$ T cells (14,15). Similar protective effects of CD4$^+$RT6$^+$ T cells have also been documented in other models of autoimmunity (16,17). The three sets of findings summarised here suggest that the expression pattern of RT6 on T cells is different from that of other known T cell markers and that it is related to differention processes of rat T lymphocytes in a complex way.

The experiments reported in this paper addressed the question how mitogen- and antigen-driven T cell activation and subsequent long term culture influence RT6 expression in mature T cells from peripheral lymphatic tissues. The results presented show that the activation protocols used lead to loss of RT6 expression in some T lymphoblasts and maintenance of RT6 expression in others. From this result it can be concluded that loss of RT6 expression is not a general feature of activated T cells but is associated only with certain differentiation pathways of both CD4$^+$ and CD8$^+$ cells. Other differentiation pathways are obviously associated with the continuous expression of RT6. This provides a strong argument against the hypothesis that RT6$^+$ memory cells always develop from RT6$^-$ activated cells (13). The presented data rather suggest that the activation of some naive RT6$^+$ T cells can lead to RT6$^-$ blast cells which then may be able to differentiate further to RT6$^+$ memory cells. The relatively high frequency of lymphoblasts continuously expressing

RT6 in the long term cultures described here may even indicate that this is the major pathway for the development of RT6$^+$ memory cells at least in the CD4 compartment. The fact that, except two RT6 expressing hybridomas, it has not been possible yet to generate RT6$^+$ T cell lines and clones *in vitro* probably is caused by selection processes in long term cultures and may reflect the limitations of the tissue culture protocols currently used. As previously shown by several authors a T cell population prone to develop rapidly an RT6$^-$ phenotype *in vitro* are CD8$^+$ T cells sensitised to allogeneic MHC class I molecules (data not shown). Most likely fully activated CD8$^+$ T cells with this specificity are always RT6 negative (2,18).

Since a small but clearly demonstrable population of RT6$^-$ T cells is present in lymph node cells already before stimulation the possibility has to be considered that these cells are predominantly expanded to generate RT6$^-$ T cell blasts. For the CD4$^+$ subset and ConA stimulation this is unlikely for the following two reasons: (a) The fraction of RT6 negative cells was markedly reduced during the first two days of culture due either to selective loss of these cells or to induction of RT6 expression. The latter possibility is more likely since the cultured cells probably contained a certain percentage of recent thymic emigrants which have been shown to be able to switch to an RT6$^+$ phenotype under *in vitro* conditions (8,10). (b) The occurrence of a substantial number of RT6$^-$CD4$^+$ blasts on day three after ConA stimulation was too rapid to be explained by clonal expansion of the very low numbers of RT6$^-$ cells present in the cultures on day two. Thus, although it cannot be excluded that some RT6$^-$ blasts had been derived from RT6$^-$ precursors, there is no doubt that the majority were the progeny of RT6$^+$ precursor cells. This conclusion is in agreement with the results of other authors (8–10,13,18).

The very rapid and simultaneous switch of a large number of cells from the RT6$^+$ to the RT6$^-$ phenotype on day three after ConA stimulation suggests that the loss of RT6 expression is closely associated with a particular step of the complex activation process and may require that the respective cell has gone into cycle. The most likely mechanisms underlying the switch in phenotype is turning off the transcription of the RT6 gene. Recently it has been shown for cytotoxic T cells of the mouse, however, that a structurally related and also GPI-anchored mono(ADP-ribosyl)transferase is shedded from the cell surface upon crosslinking of T cell receptor molecules (19). The shedding is mediated by a protease or a specific phospholipase either induced or activated by the stimulation process. If a similar mechanism is involved in the modulation of RT6 expression, it would have to be postulated that the activation process induces a similar enzyme activity selectively in RT6$^-$ T cell blasts.

Irrespective of the mechanism of RT6 modulation, the question arises how the RT6 phenotype of the activated cells is determined. At present it is not known whether certain CD4$^+$ clones are predestined to remain RT6 positive and others to become RT6 negative or whether the decision is made only during the activation process itself. Furthermore the functional profiles of *in vitro* generated RT6$^+$ and RT6$^-$ lymphoblasts have not yet been studied. That they may be rather different is suggested by studies demonstrating differential roles of RT6$^+$CD4$^+$ and RT6$^-$CD4$^+$ T cells in several autoimmune models (14–17) and in various cell-mediated immune responses (9,10). From these studies it has been concluded that RT6$^-$CD4$^+$ cells may be inflammatory Th1 cells, whereas RT6$^+$CD4$^+$ T cells may represent regulatory cells of the Th2 type (16,17). If this could be demonstrated also for *in vitro* generated RT6$^-$ and RT6$^+$ T lymphocytes, it would greatly facilitate the further analysis of these cell populations in the rat. Furthermore the possibility would arise that, in addition for the characteristic set of cytokines secreted by Th2 cells, the RT6 molecule itself might be involved in mediating some of the immunoregulatory functions of Th2 cells.

6. ACKNOWLEDGMENTS

The excellent technical assistance of Mrs A. Brinkmann and Mrs A. Dinkel and the secretarial help of Mrs D. Damitz are gratefully acknowledged. Furthermore I wish to thank Dr. R. Schwinzer for helpful discussions. The work was supported by the Deutsche Forschungsgemeinschaft (SFB 244).

7. REFERENCES

1. Lubaroff, D.M., G. Butcher, C. DeWitt, T. Gill III, E. Günther, J. Howard, & K. Wonigeit. 1983. Standardized nomenclature for the rat T-cell alloantigens: Report of the Committee. *Transplant. Proc. 15*: 1683.

2. Wonigeit, K. 1979. Definition of lymphocyte antigens in rats: RT-Ly-1, RT-Ly-2, and a new MHC-linked antigen system. *Transplant. Proc. 11*: 1334–1336.

3. Wonigeit, K. 1979. Characterization of the RT-Ly-1 and RT-Ly-2 alloantigenic systems by congenic rat strains. *Transplant. Proc. 11*: 1631–1635.

4. Takada, T., K. Iida, & J. Moss. 1994. Expression of NAD glycohydrolase by rat mammary adenocarcinoma cells transformed with rat T cell alloantigen RT6.2. *J. Biol. Chem. 269*: 9420–9423.

5. Koch-Nolte, F., F. Haag, R. Kastelein, & F. Bazan. 1996. Uncovered: the family relationship of a T-cell-membrane protein and bacterial toxins. *Immunol. Today 17*: 402–405.

6. Wonigeit, K., A. Dinkel, J. Fangmann, & H. Thude. 1996. Expression of the ectoenzyme RT6 is not restricted to resting peripheral T cells and is differently regulated in normal peripheral T cells, intestinal IEL, and NK cells. In Koch-Nolte, F., and F. Haag (eds). ADP-Ribosylation in animal tissues: Structure, function, and biology of mono(ADP-Ribosyl)transferases and related enzymes. Plenum Publishing Corporation, New York. in press

7. Fangmann, J., R. Schwinzer, & K. Wonigeit. 1991. Unusual phenotype of intestinal intraepithelial lymphocytes in the rat: predominance of T cell receptor a/b$^+$CD2$^-$ cells and high expression of RT6 alloantigen. *Eur. J. Immunol. 21*: 753–760.

8. Hunt, H.D. & D.M. Lubaroff. 1987. Changes in membrane antigen phenotype of T cells during lectin activation. *Transplant. Proc. 19*: 3179–3180.

9. Ernst, D.N. & D.M. Lubaroff. 1984. Membrane antigen phenotype of sensitized T lymphocytes mediating tuberculin-delayed hypersensivity in rats. *Cell. Immunol. 88*: 436–452.

10. Hunt, H.D. & D.M. Lubaroff. 1992. Identification of functional T cell subsets and surface antigen changes during activation as they relate to RT6. *Cell. Immunol. 143*: 194–211.

11. Thiele, H.G., F. Koch, & A. Kashan. 1987. Postnatal distribution profiles of Thy-1$^+$ and RT6$^+$ cells in peripheral lymph nodes of DA rats. *Transplant. Proc. 19*: 3157–3160.

12. Hosseinzadeh, H. & I. Goldschneider. 1993. Recent thymic emigrants in the rat express a unique antigenic phenotype and undergo post-thymic maturation in peripheral lymphoid tissues. *J. Immunol. 150*: 1670–1679.

13. Kampinga, J., H. Groen, F.A. Klatter, J.M. Pater, A.S. van Petersen, B. Roser, P. Nieuwenhuis, & R. Aspinall. 1996. Post-thymic T cell development in the rat. *Thymus* in press

14. Greiner, D.L., E.S. Handler, K. Nakano, J.P. Mordes, & A.A. Rossini. 1986. Absence of the RT6 T cell subset in diabetes-prone BB/W rats. *J. Immunol. 136*: 148–151.

15. Burstein, D., J.P. Mordes, D.L. Greiner, D. Stein, N. Nakamura, E.S. Handler, & A.A. Rossini. 1989. Prevention of diabetes in BB/Wor rat by single transfusion of spleen cells. Parameters that affect degree of protection. *Diabetes 38*: 24–30.

16. Fowell, D. & D. Mason. 1993. Evidence that the T cell repertoire of normal rats contains cells with the potential to cause diabetes. Characterization of the CD4$^+$ T cell subset that inhibits this autoimmune potential. *J. Exp. Med. 177*: 627–636.

17. Beijleveld, L.J.J., C.P.M. Broeren, F.A. Klatter, J. Kampinga, J.G.M.C. Damoiseux, & P.J.V. van Breda Vriesman. 1996. Susceptibility to clinically manifest cyclosporine A (CsA)-induced autoimmune disease is associated with interferon-gamma (IFN-γ)-producing CD45RC$^+$RT6$^-$ T helper cells. *Clin. Exp. Immun. 105*: 486–496.

18. Greiner, D.L., C.W. Reynolds, & D.M. Lubaroff. 1982. Maturation of functional T-lymphocyte subpopulations in the rat. *Thymus 4*: 77–90.
19. Nemoto, E., St. Stohlman, & G. Dennert. 1996. Release of a glycosylphosphatidylinositol-anchored ADP-ribosyltransferase from cytotoxic T cells upon activation. *J. Immunol. 156:* 85–92.

34

EXPRESSION OF THE RT6 MONO(ADP-RIBOSYL)TRANSFERASES IS REGULATED BY TWO PROMOTER REGIONS

Gregor Kuhlenbäumer, Stefan Rothenburg, Martina Matthes,
Christiane Hollmann, Edgar Wingender, Heinz Günter Thiele,
Friedrich Koch-Nolte, and Friedrich Haag

Department of Immunology
University Hospital Hamburg-Eppendorf
Martinistr. 52, D-20246
Hamburg, Germany

1. ABSTRACT

The structure of the *RT6* mono(ADP-ribosyl)transferase gene was studied. Analysis of cDNA clones revealed eight exons and suggested two independent transcriptional start sites. The existence of the downstream initiation site was confirmed by S1-nuclease protection and localized to position +29 of exon 2. The corresponding 5′ flanking regions were found to contain typical promoter structures such as TATA- and CCAAT-boxes. Comparison with sequences deposited in the TRANSFAC database of transcription factor binding sites revealed few putative regulatory elements in the region associated with exon 1 (promoter 1). In contrast, several elements contained in the regulatory regions of other T cell-specific genes, such as *ets*, *lyf-1* and *ikaros* were found in in promoter 2. Analysis of RT6-transcripts showed this region to be the most active promoter in spleen cells of adult rats. Finally, transient transfection assays with reporter gene constructs showed promoter 2 to mediate T-cell specific transcription.

2. INTRODUCTION

RT6 is a 25–30 kd glycosylphosphatidylinositol- (GPI) linked T-cell differentiation antigen. Its expression is restricted to mature peripheral T-cells, intraepithelial lymphocytes and some subsets of natural killer cells (Refs. 1–3, and Wonigeit, this volume). RT6 is a NAD-metabolizing ecto-enzyme belonging to the family of mono-ADP-ribosyltransferases [4–6]. In the rat two highly polymorphic alleles, $RT6^a$ and $RT6^b$, of a single gene locus on chromosome 1 are known [7]. Defects in *RT6* gene structure or expression are found in

ADP-Ribosylation in Animal Tissue, edited by Haag and Koch-Nolte
Plenum Press, New York, 1997

various animal models of autoimmune disease (Refs. 8–10 as well as articles by Thiele, Greiner, Cetkovics and Matthes, this volume).

In the BB-DP rat model for autoimmune insulin dependent diabetes mellitus (IDDM) it has been suggested that a trans-activating factor, necessary for RT6 expression, may be defective (see also Thiele, this volume). As a preliminary step towards testing this hypothesis we have examined the structure of the *RT6* gene [11], and found it to contain two independent promoters.

3. RESULTS

3.1. The Structure of the RT6 Gene

To determine the 5′ end of the gene we employed 5′ rapid amplification of cDNA ends (5′RACE), using adult rat spleen RNA as template [11]. Reverse transcription was done with a gene specific primer derived from exon 7. The exon composition of 18 cloned amplification products is shown in Figure 1A. To determine the gene structure (Fig. 1B), a set of overlapping clones was isolated from a rat genomic library and characterized by restriction mapping and partial sequence analysis. The RT6 gene spans approximately 18 kb. The largest intron has a size of 7,5 kb and exon sizes range from 52 to 694 bp. The entire sequence coding for the mono(ADP-ribosyl)transferase is contained in exon 7. Signal sequences necessary for cell surface expression are encoded by exons 5 (leader peptide) and 8 (GPI attachment signal).

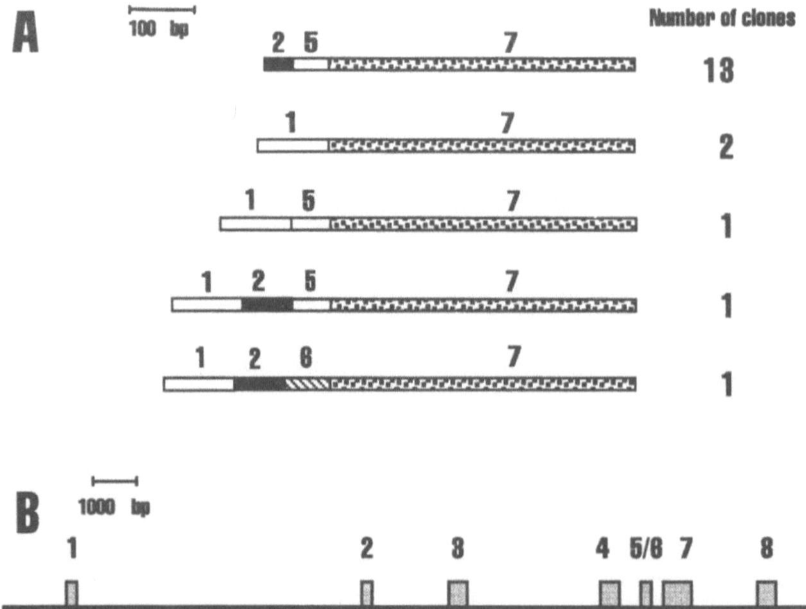

Figure 1. Exon composition of cDNA clones (A) and exon/intron structure (B) of the *RT6* gene. A. *RT6* transcripts generated by 5′RACE-PCR were cloned and analyzed as to their exon composition. B. Position of exons within the *RT6* gene. (With permission from Ref. 11. *Copyright 1996. The American Association of Immunologists.*)

3.2. Determination of Two Transcription Start Sites and Structural Features of the Two Promoter Regions

A new 5'RACE experiment, primed from within exon 1, was performed to determine the position of the transcription start site associated with exon 1 (designated TSS1). A 600 bp fragment of the 5' flanking region was analyzed for putative transcription factor binding sites by comparison with sequences deposited in the TRANSFAC and EMBL/GenBank databases [11, 12]. Sequence anaysis revealed the presence of two putative core promoter elements: a TATA-box around position -30 and an initiator (*Inr*) element at -5 (Fig. 2A). A CCAAT-box (around -420) appears to be 'split off' from the other core promoter elements by the insertion of one an MT repetitive element. Two other potential binding sites for transcription factors were found: one for Ets 1/2 [13], the other for the repressor E4BP4 [14].

The existence of an independent transcription start site (designated TSS2) was localized to position +29 of exon 2, by S1 nuclease protection (data not shown). The genomic sequence upstream of TSS2 contains a TATA-box (-28/-24) and a CCAAT-box (-70/-65) at appropriate positions (Fig. 2B). Several putative binding sites for transcription factors known to be involved in lymphocyte specific expression, e.g. NF-κB, Ikaros [15], Lyf-1 [16], and IRF1/IRF2 [17] were identified. The IRF site lies within a cluster of putative c-Myb sites and is adjacent to a c-Myc/Max site, followed by a high-scoring Ets-1 site.

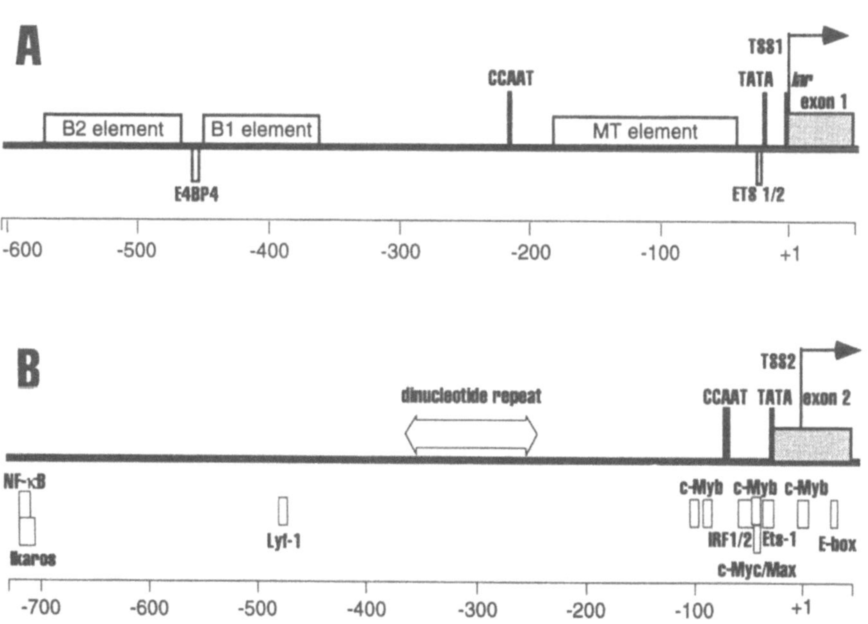

Figure 2. Putative regulatory elements in the 5' flanking regions of TSS1 (A) and TSS2 (B). See text for details. Exons 1 and 2 are denoted by shaded boxes. Core promoter elements are shown above, other predicted regulatory elements are depicted below the line.

3.3. Functional Analysis of the Promoter Region Associated with TSS2 Using Reporter Gene Assays

Quantitative analysis of the exon composition of transcripts amplified by 5´RACE suggested the former to be the most important transcriptional start site in adult rat spleen cells. A fragment from -1200 to +150 with respect to TSS2 was cloned into a promoterless expression vector containing luciferase as a reporter gene. The promoterless vector (pGL3), as well as the same vector containing the SV 40 promoter (pGL3/SV40 prom) were used as negative and positive controls, respectively. The results are shown in Figure 3. Transfection into the RT6.2 expressing T hybridoma cell line EpD3 [18] results in 2,5 fold stronger expression by the *RT6* compared to the SV40 promoter (hatched columns). In the

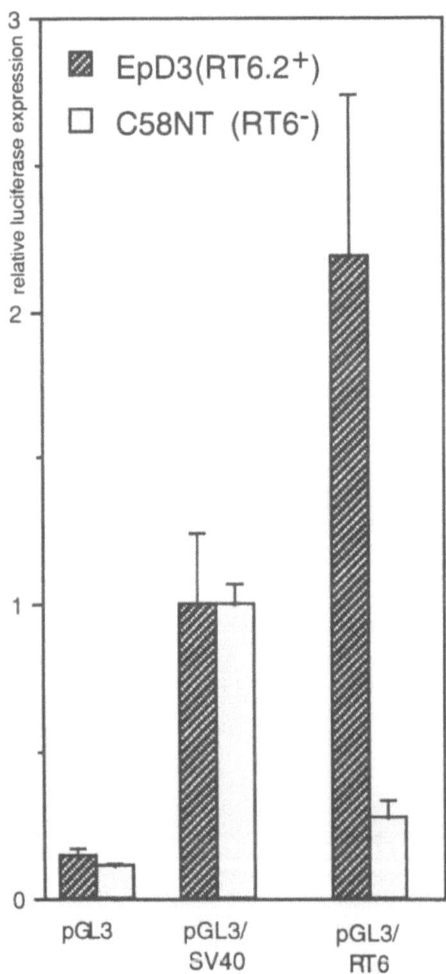

Figure 3. Functional analysis of *RT6* promoter 2. The relative luciferase expression driven by the promoterless vector (pGL3), the SV40 promoter (pGL3/SV40) and the *RT6* promoter 2 (pGL3/RT6) in RT6.2 expressing EpD3 (hatched bars) and RT6-negative C58NT cells (shaded bars) is indicated.

RT6 non-expressing thymoma cell line C58T expression of both constructs is nearly equal (shaded columns).

4. DISCUSSION

The marked heterogeneity of the 5′ region of the *RT6* gene may play a role in the regulation of RT6 expression. The use of alternative transcription start sites is often controlled in a tissue specific or developmentally regulated fashion [19]. The very different structure of the two promoter regions points to differential control of transcription from these sites. Analysis of transcripts by 5′ RACE-PCR show that TSS 2 is the preferential transcription start site in adult rat spleen. Functional analysis of the promoter region 2 shows similar low levels of expression for the RT6 and SV40 promoters after transfection into the RT6 non-expressing cell line C58NT. In contrast, expression driven by the RT6 promoter is much higher than that driven by the SV 40 promoter after transfection into the RT6 expressing T-cell hybridoma EpD3. This points to the presence of T-cell specific enhancer elements in the *RT6* promoter fragment. The results presented in this report open the way for more detailed analysis of *RT6* gene regulation and for the identification of the factor(s) mediating defective *RT6* expression in the BB-DP rat.

5. ACKNOWLEDGMENTS

This work was supported by grant Ha 2369 from the Deutsche Forschungsgemeinschaft DFG to FH and HGT, as well as by a stipend from the Sandoz Foundation for Therapeutic Research to FH.

REFERENCES

1. Thiele, H. G., F. Koch & A. Kashan. 1987. Postnatal distribution profiles of Thy-1+ and RT6+ cells in peripheral lymph nodes of DA rats. *Transplant. Proc. 19*: 3157.
2. Koch, F., F. Haag, A. Kashan & H. G. Thiele. 1990. Primary structure of rat RT6.2, a nonglycosylated phosphatidylinositol-linked surface marker of postthymic T cells. *Proc Natl Acad Sci USA 87*: 964.
3. Fangmann, J., R. Schwinzer, M. Winkler & K. Wonigeit. 1990. Expression of RT6 alloantigens and the T-cell receptor on intestinal intraepithelial lymphocytes of the rat. *Transplant Proc 22*: 2543.
4. Takada, T., K. Iida & J. Moss. 1994. Expression of NAD glycohydrolase activity by rat mammary adenocarcinoma cells transformed with rat T cell alloantigen RT6.2. *J Biol Chem 269*: 9420.
5. Haag, F., V. Andresen, S. Karsten, F. Koch-Nolte & H.-G. Thiele. 1995. Both allelic forms of the rat T cell differentiation marker RT6 display nicotinamide adenine dinucleotide (NAD)-glycohydrolase activity, yet only RT6.2 is capable of automodification upon incubation with NAD. *Eur. J. Immunol. 25*: 2355.
6. Zolkiewska, A., M. S. Nightingale & J. Moss. 1992. Molecular characterization of NAD:arginine ADP-ribosyltransferase from rabbit skeletal muscle. *Proc. Natl. Acad. Sci. USA 89*: 11352.
7. Butcher, G. W., S. Clarke & E. M. Tucker. 1979. Close linkage of peripheral T-lymphocyte antigen A (PtaA) to the hemoglobin variant Hbb on linkage group I of the rat. *Transplant. Proc 11*: 1629.
8. Koch-Nolte, F., J. Klein, C. Hollmann, M. Kühl, F. Haag, R. Gaskins, E. Leiter & H.-G. Thiele. 1995. Defects in the structure and expression of the genes for the T cell marker *Rt6* in NZW and (NZB x NZW) F$_1$ mice. *Int Immunol 7*: 883.
9. Greiner, D. L., H. E.S., K. Nakanko, J. P. Mordes & A. A. Rossini. 1986. Absence of the RT6 T cell subset in diabetes-prone BB/W rats. *J. Immunol. 136*: 148.
10. Prochazka, M., H. R. Gaskins, E. H. Leiter, F. Koch-Nolte, F. Haag & H. G. Thiele. 1991. Chromosomal localization, DNA polymorphism, and expression of Rt-6, the mouse homologue of rat T-lymphocyte differentiation marker RT6. *Immunogenetics 33*: 152.

11. Haag, F., G. Kuhlenbäumer, F. Koch-Nolte, E. Wingender & H.-G. Thiele. 1996. Structure of the gene encoding the rat T cell ecto-ADP-ribosyltransferase RT6. *J. Immunol. 157*: 2022.

12. Wingender, E., P. Dietze, H. Karas & R. Knüppel. 1996. TRANSFAC: A database on transcription factors and their DNA binding sites. *Nucl. Acids Res. 24*: 238.

13. Salmon, P., A. Giovane, B. Wasylyk & D. Klatzmann. 1993. Characterization of the human CD4 gene promoter: transcription from the CD4 gene core promoter is tissue-specific and is activated by Ets proteins. *Proc Natl Acad Sci U S A 90*: 7739.

14. Cowell, I. G., A. Skinner & H. C. Hurst. 1992. Transcriptional repression by a novel member of the bZIP family of transcription factors. *Mol Cell Biol 12*: 3070.

15. Molnar, A. & K. Georgopoulos. 1994. The Ikaros gene encodes a family of functionally diverse zinc finger DNA-binding proteins. *Mol Cell Biol 14*: 8292.

16. Lo, K., N. R. Landau & S. T. Smale. 1991. LyF-1, a transcriptional regulator that interacts with a novel class of promoters for lymphocyte-specific genes. *Mol Cell Biol 11*: 5229.

17. Tanaka, N., T. Kawakami & T. Taniguchi. 1993. Recognition DNA sequences of interferon regulatory factor 1 (IRF-1) and IRF-2, regulators of cell growth and the interferon system. *Mol. Cell. Biol. 13*: 4531.

18. Koch, F., A. Kashan & H. G. Thiele. 1988. Production of a rat T cell hybridoma that stably expresses the T cell differentiation marker RT6.2. *Hybridoma 7*: 341.

19. Takadera, T., S. Leung, A. Gernone, Y. Koga, Y. Takihara, N. G. Miyamoto & T. W. Mak. 1989. Structure of the two promoters of the human lck gene: differential accumulation of two classes of lck transcripts in T cells. *Mol Cell Biol 9*: 2173.

35

"NATURAL" RT6-1 AND RT6-2 "KNOCK-OUT" MICE

Martina Matthes, Christiane Hollmann, Heinrich Bertuleit, Maren Kühl, Heinz-Günter Thiele, Friedrich Haag, and Friedrich Koch-Nolte

Department of Immunology
University Hospital
D-20246 Hamburg, Germany

ABSTRACT

We have screened different mouse strains - including strains with enhanced susceptibility for autoimmune diseases - for deviations of Rt6 gene expression by RT-PCR. Most strains expressed varying amounts of Rt6-1 and Rt6-2. NZW mice, however, do not show any detectable Rt6-2 gene transcripts. BxSB mice show a near complete absence of Rt6-1 gene transcripts. Southern blot and sequence analyses revealed that NZW mice have suffered a deletion of the Rt6.2 gene while the Rt6-1 gene of BxSB mice has been inactivated by a premature stop codon. Thus, these mouse strains represent natural Rt6-2 and Rt6-1 single-gene 'knock-out's, respectively. Since the NZW mouse does not show any gross immunological abnormalities, loss of the Rt6-2 gene by itself is not associated with any obvious immunological phenotype. However, crosses between NZW and certain other mouse strains, e.g. (NZW x NWB)F1 and (NZW x SB)F1 animals, develop a systemic autoimmune disease reminiscent of human lupus erythematosus. Moreover, the BxSB mouse strain is considered to be an independent model for the same disease. It will be of interest to determine whether these spontaneous Rt6 gene defects constitute part of the polygenetic contribution to autoimmune disease in these animals.

BACKGROUND

Defects in the expression of the RT6 T-cell ecto-mono(ADP-ribosyl)transferase have been reported to correlate with enhanced disease susceptibility in the BB rat model for autoimmune diabetes (1–3). In the mouse, RT6-homologous enzymes are encoded by two tandemly linked genes (4, 5). The gene products, Rt6-1 and Rt6-2, show approximately 80% amino acid sequence identity (5). The *Rt6* genes have been mapped to a conserved linkage group on mouse chromosome 7 near the loci for the hemoglobin ß chain

ADP-Ribosylation in Animal Tissue, edited by Haag and Koch-Nolte
Plenum Press, New York, 1997

(*Hbb*), tyrosinase (*Tyr/c*), and the muscle mADPRT (*Art1*) (Fig. 1) (4, 6). Recently, a locus governing SLE disease parameters was mapped near this region of mouse chromosome 7 (7). We here report on two types of spontaneous *Rt6* gene-inactivating mutations that we have discovered in NZW and BxSB mice.

RESULTS

Southern Blot analyses indicate that RT6 is encoded by a single copy gene in the rat and by two gene copies in the mouse. We cloned and sequenced both *Rt6* genes of the mouse. The nucleotide sequences of the *Rt6-1* and *Rt6-2* coding regions show 87% sequence identity, the deduced amino acid sequences 79% identity to one another and 71% identity to rat RT6 (5, 8). Northern blot analyses reveal that transcription of the mouse *Rt6* genes is restricted to T cell rich tissues similar to the restricted expression pattern in the rat. RT-PCR analysis indicate that most mouse strains tested express varying levels of both genes, *Rt6-1 and Rt6-2*. Notable exceptions are the NZW and BxSB mice which show marked reductions of *Rt6-2* or *Rt6*-1-specific transcripts, respectively.

We investigated whether these *Rt6*-expression patterns are due to allelic mutations by Southern blot and sequence analyses. Southern hybridisations show that while most inbred and wild mice carry copies of both *Rt6* genes, NZW mice do not show any *Rt6-2*-

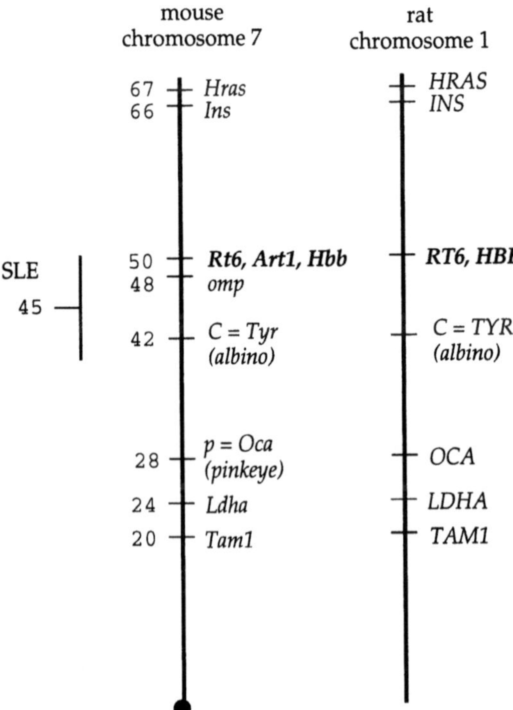

Figure 1. Schematic diagram of the chromosomal localization of mouse Rt6 and neighboring loci. The genes for mouse Rt6-1 and Rt6-2 have been mapped in close vicinity to the loci for hemoglobin ß and the skeletal muscle mADPRT (*Art1*) (4, 6, 10). A locus governing disease parameters in lupus-prone mice (SLE) has recently been mapped in the chromosomal vicinity of the Rt6 locus (7).

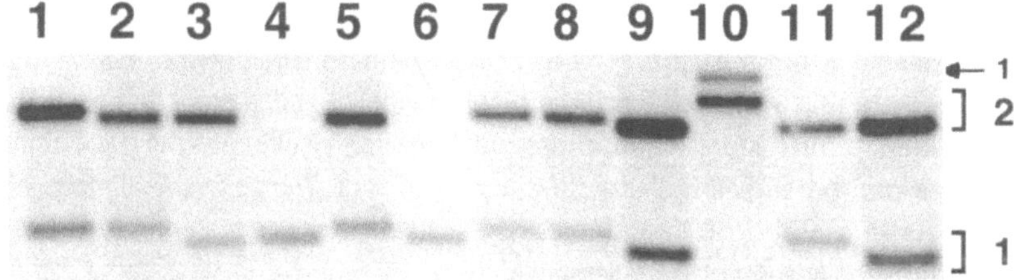

Figure 2. Southern blot analysis of mouse Rt6 genes. Genomic DNAs were digested with DraI and subjected to Southern blot analysis with radiolabeled *Rt6-2*-specific cDNA probe. DNAs were from the following mouse strains: lane 1: C3H, 2: BALB/c, 3: C57BL/10, 4: NZW, 5: MRL, 6: SB, 7: DBA/1, 8: DBA/2, 9: *Mus spretus*, 10: *Mus caroli*, 11: *Mus musculus* , 12: *Mus domesticus* . Arrow and brackets on the right show the assignment of the bands to *Rt6-1* or *Rt6-2* on the basis of differential hybridization at high stringency. Reprinted from *Molecular Immunology Vol. 33*, C. Hollmann, F. Haag, M. Schlott, A. Damaske, H. Bertuleit, M. Matthes, M. Kühl, H. G. Thiele & F. Koch-Nolte, Molecular characterization of mouse T-cell ecto-ADP-ribosyltransferase Rt6: cloning of a second functional gene and identification of the Rt6 gene products, pp. 807–817, (1996), with kind permission from Elsevier Science Ltd, The Boulevard, Langford Lane, Kidlington OX5 1GB, UK.

specific bands (Fig. 2) (9). Thus, the absence of *Rt6-2* -specific transripts in NZW mice evidently is due to a deletion of the *Rt6-2* gene in this strain.

Sequence analyses revealed that the *Rt6-1* gene of BXSB mice contains an exchange of arginine 161 versus a premature stop codon (Fig. 3). The low *Rt6-1* transcript levels observed in BXSB mice is in accord with observations that human diseases which arise as a consequence of premature stop codons are usually accompanied by strongly reduced levels of the corresponding mRNAs.

DISCUSSION

We have discovered two types of Rt6 gene-inactivating mutations in laboratory mice: 1) the *Rt6-2* gene is deleted in NZWmice; 2) a premature stop codon occurs in the *Rt6-1* coding region of BXSB mice. Our results show that animals considered as models for autoimmune systemic lupus erythematodes such as the F1 descendants from crosses of

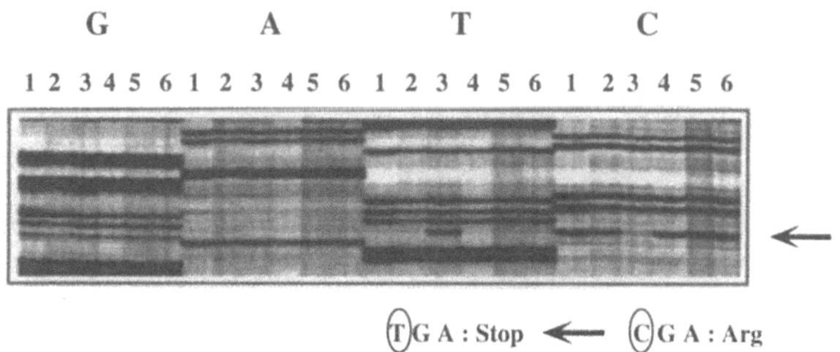

Figure 3. Nonsense mutation in the coding region of the Rt6-1 gene of BXSB mice. The coding region of the Rt6-1 gene was amplified from genomic DNA and sequenced directly (i.e. without any cloning steps). DNAs were from the following mouse strains: lane 1: DBA/2, 2: BALB/c, 3: BXSB, 4: 129/Sv, 5: NZW, 6: *Mus musculus*. The G-T transversion is marked by an arrow.

NZW and SB, or NZW and NZB, as well as the BXSB strain carry a defective Rt6 gene. The coincidence of *Rt6* gene defects with polygenetically influenced autoimmune disorders could indicate that *Rt6* gene variants may be one of the factors which influence the development and progression of autoimmune disease. However, the fact that parental NZW mice are normal shows that functional loss of one *Rt6* gene *alone* does not lead to obvious defects in viability or immunological capacity. It is conceivable that the physiological functions of Rt6-1 can be compensated for by Rt6-2 and vice versa. To further investigate the role of *Rt6* a mouse model which is deficient for both *Rt6* genes would be of interest. Appropriate experiments to establish such a model are underway in our laboratory.

ACKNOWLEDGMENT

This work was supported by grants Ko1144/1-2 and No310/1-1 from the Deutsche Forschungsgemeinschaft to F.K-N.

REFERENCES

1. Greiner, D. L., E. S. Handler, K. Nakano, J. P. Mordes & A. A. Rossini. 1986. Absence of the RT-6 T cell subset in diabetes-prone BB/W rats. *J. Immunol. 136*: 148–151.
2. Greiner, D. L., J. P. Mordes, E. S. Handler, M. Angelillo, N. Nakamura & A. A. Rossini. 1987. Depletion of RT6.1+ T lymphocytes induces diabetes in resistant biobreeding/Worcester (BB/W) rats. *J. Exp. Med. 166*: 461–475.
3. Fowell, D. & D. Mason. 1993. Evidence that the T cell repertoire of normal rats contains cells with the potential to cause diabetes. Characterization of the CD4+ T cell subset that inhibits this autoimmune potential. *J. Exp. Med. 177*: 627–36.
4. Prochazka, M., H. R. Gaskins, E. H. Leiter, F. Koch-Nolte, F. Haag & H. G. Thiele. 1991. Chromosomal localization, DNA polymorphism, and expression of Rt-6, the mouse homologue of rat T-lymphocyte differentiation marker RT6. *Immunogenetics 33*: 152–156.
5. Hollmann, C., F. Haag, M. Schlott, A. Damaske, H. Bertuleit, M. Matthes, M. Kühl, H. G. Thiele & F. Koch-Nolte. 1996. Molecular characterization of mouse T-cell ecto-ADP-ribosyltransferase Rt6: cloning of a second functional gene and identification of the Rt6 gene products. *Mol. Immunol. 33*: 807–817.
6. Koch-Nolte, F., M. Kühl, F. Haag, M. Cetkovich-Cvrlje, E. H. Leiter & H. G. Thiele. 1996. Assignment of the human and mouse genes for muscle ecto mono ADP-ribosyltransferase to a conserved linkage group on human Chromosme 11p15 and mouse Chromosome 7. *Genomics 36*: 215–216.
7. Drake, C. G., S. J. Rozzo, H. F. Hirschfeld, N. P. Smarnworawong, E. Palmer & B. L. Kotzin. 1995. Analysis of the New Zealand Black contribution to lupus-like renal disease. Multiple genes that operate in a threshold manner. *J. Immunol. 154*: 2441–2447.
8. Koch, F., F. Haag & H. G. Thiele. 1990. Nucleotide and deduced amino acid sequence for the mouse homologue of the rat T-cell differentiation marker RT6. *Nucleic Acids Res. 18*: 3636.
9. Koch-Nolte, F., J. Klein, C. Hollmann, M. Kühl, F. Haag, H. R. Gaskins, E. H. Leiter & H. G. Thiele. 1995. Defects in the structure and expression of the genes for the T cell marker Rt6 in NZW and (NZB x NZW)F1 mice. *Int. Immunol. 7*: 883–890.
10. Butcher, G. W., S. Clarke & E. M. Tucker. 1979. Close linkage of peripheral T-lymphocyte antigen A (PtaA) to the hemoglobin variant Hbb on linkage group I of the rat. *Transplant. Proc 11*: 1629–1630.

36

AN ADP-RIBOSYLTRANSFERASE FROM BOVINE ERYTHROCYTES APPARENTLY SPECIFIC FOR CYSTEINE RESIDUES

Simon van Heyningen and Barbara A. Saxty[*]

Department of Biochemistry
The University of Edinburgh
Hugh Robson Building, George Square
Edinburgh EH8 9XD, Great Britain

SUMMARY

An NAD^+:cysteine glycohydrolase purified from bovine erythrocytes had a specific activity of 1900 (nmol nicotinamide released).$min^{-1}.mg^{-1}$, a K_m for cysteine of 4.0 mM, and an M_r of 45 000. The enzyme also catalysed the dose-dependent ADP-ribosylation of several bovine erythrocyte proteins, including a doublet of high M_r and proteins of M_r 60 000, 55 000, and 29 000. ADP-ribosylation of the M_r 55 000 protein was blocked by pre-treatment of the erythrocyte membranes with N-ethylmaleimide, and ADP-ribose was released by treatment with mercuric ions, but not with hydroxylamine. The enzyme therefore appears to be a cysteine-specific ADP-ribosyltransferase.

BACKGROUND

Much research on physiological mono ADP-ribosylation in eukaryotic cells has been driven by the desire to find analogues of the bacterial toxins, whose ADP-ribosyltransferase activities have, in general, been better characterised [1,2]. Endogenous enzymes that have the same activity as many of these toxins have been prepared and investigated. The fact that pertussis toxin (secreted by *Bordetella pertussis*, and thought to be involved with the pathology of whooping cough) modifies cysteine-352 of the $G_i\alpha$ protein of the adenylate cyclase complex [3] makes it therefore reasonable to suppose that there might be endogenous cysteine-specific ADP-ribosyltransferases in eukaryotic cells. However es-

---

[*] Present address: Kennedy Institute of Rheumatology, 6 Bute Gardens, Hammersmith, London W6 7DW

ADP-Ribosylation in Animal Tissue, edited by Haag and Koch-Nolte
Plenum Press, New York, 1997

sentially only one report of such an activity had been published at the time we started our research in this area [4].

The experiments described in this paper (which have been published in detail in the *Biochemical Journal* [5]) are about the isolation of such an enzyme from bovine erythrocytes. We chose this source because of its ready availability and comparative simplicity. Since assays by direct measurement of ADP-ribosylation of cysteine residues on proteins are difficult and slow, the purification was initially worked out using the much simpler assay for cysteine-dependent NAD^+ glycohydrolase activity [6]. Subsequent experiments showed that enzyme purified this way also had ADP-ribosyltransferase activity.

RESULTS

Table 1 shows the result of the purification procedure, which used three steps [5]: precipitation with ammonium sulphate (40% of saturation), chromatography on a cysteine-sepharose column, and chromatography on a matrex Orange column (for proteins that bind NAD^+).

The purified enzyme was inactive in the absence of cysteine, and showed a single band (silver stained) on polyacrylamide gel electrophoresis in the presence of SDS corresponding to an M_r of 45 000. It was unstable on storage except in the presence of 300 g.l^{-1} glycerol and protease inhibitors and in the absence of thiol. Even at -16°, activity was rapidly lost in the presence of cysteine.

Possible alternative ADP-ribose acceptors such as glycine, arginine methyl ester, or *S*-carboxymethylcysteine were ineffective, but the enzyme showed saturation kinetics with cysteine as a substrate: the apparent K_m values were 8 μM for NAD^+ and 4.0 mM for cysteine.

The enzyme also catalysed the incorporation of radioactive label from [*adenylate*-^{32}P]NAD^+ into proteins of washed inside-out erythrocyte membranes, presumably due to ADP-ribosylation. Interpreting these results is complicated by the fact that enzyme-catalysed ADP-ribosylation is always seen against a significant background of endogenous activity. Thus, even in the absence of added enzyme, proteins of M_r 55 000 and 60 000 and a double at high M_r were labelled. In the presence of enzyme the intensity of labelling of all these bands increased in a dose-dependent manner, and a further faint band at M_r 29 000 appeared. However, a salt wash of the membranes abolished the labelling of all bands other than that at M_r 55 000, so subsequent experiments concentrated only on this band. The enzyme-labelled proteins were not recognised on Western blots by anti-peptide antibodies to $G_i\alpha$ or $G_s\alpha$. No labelling of any membrane protein was observed in the presence of free [^{32}P]ADP-ribose rather than NAD^+, (showing that the enzyme-catalysed effect was

Table 1. Purification of a cysteine-dependent NAD^+ glycohydrolase[*]

	Total protein (mg)	Total activity (nmol.min^{-1})	Specific activity (nmol.min^{-1}.mg^{-1})	Fold purification	Yield, %
Cell lysate	16250	690	0.04	1	100
40% NH$_4$(SO$_4$)$_2$	120	400	3	75	58
cysteine-Sepharose	0.6	152	250	6250	22
Matrex Orange	0.06	116	1900	47500	16

[*]Cysteine-dependent activity is the difference between the glycohydrolase activity (assayed essentially by the method of Moss and Vaughan [6]), in the presence and absence of 100 mM cysteine.

not due to release of ADP-ribose from NAD^+ by the glycohydrolase activity and its subsequent non-enzymic reaction with the protein).

Since the enzyme had been purified by assaying for cysteine-dependent activity, it seemed reasonable to suppose that ADP-ribosylation was on a cysteine residue. Pretreatment of the membranes with 1 mM N-ethylmaleimide (which alkylates available thiols) abolished the effect of enzyme on the labelling of the M_r 55 000 protein, while leaving the background labelling unchanged (Table 2). Furthermore, treatment of the ADP-ribosylated material with mercuric ion reduced the intensity of labelling to background (i.e. abolished the enzyme-catalysed effect), whereas treatment with neutral hydroxylamine had no effect. This behaviour is known to be characteristic of ADP-ribosylated cysteine residues rather than arginine residues [7,8], and was also observed after ADP-ribosylation of the same membranes with pertussis toxin. These experiments suggest that the enzyme-catalysed labelling is of cysteine residue, whereas the background labelling is either of different residues or of different proteins at the same M_r, or both.

Finally, protein of M_r 55 000 that had been labelled in the presence or absence of enzyme was eluted from a polyacrylamide gel, and treated with 10 mM mercuric chloride. The released products were subjected to HPLC on an S5-SAX column: there were none after labelling without enzyme, but, after labelling in the presence of enzyme, a radiolabelled product whose elution time was coincident with that of free ADP-ribose was released. This is strong evidence for ADP-ribosylation of a cysteine residue.

DISCUSSION

NAD^+ glycohydrolase activity is ADP-ribosyltransferase activity with water as a second substrate. It is therefore not surprising that most ADP-ribosyltransferases also have glycohydrolase activity, and that such activity has often been used in their purification. The enzyme prepared in this work catalyses NAD^+ hydrolysis only in the presence of thiols, and therefore appears to be a cysteine-specific ADP-ribosyltransferase. Indeed its K_m for cysteine of 4.4 mM is much smaller than the reported K_m for cysteine of pertussis toxin [9]. This paper present strong evidence that cysteine residues in proteins are also substrates, as they are for the toxin. The enzyme has a relative molecular mass of about 45 000, and so must be different from the enzyme with similar activity reported in human erythrocytes, which had M_r 28 000 [4].

ADP-ribose-cysteine links are common in whole cells; indeed as common as ADP-ribose-arginines [1], but many of them e.g. in β-actin [10] and in glyceraldehyde-3-phosphate

Table 2. ADP-ribosyl linkage tests. This table shows the amount of labelling of the M_r 55 000 protein after various treatments and in the presence or absence of the enzyme

	Label remaining after treatment, %	
	Without enzyme	With enzyme
N-ethylmaleimide	90	15
neutral hydroxylamine	100	80
mercuric acetate	90	60
0.5M NaCl, pH 7.4	100	100

dehydrogenase [11] have been shown to occur non-enzymically. It is very difficult totally to rule out a non-enzymic reaction in the experiments presented here, but the absence of any labelling following incubation with [^{32}P]ADP-ribose argues strongly against it.

The identification of this enzyme strengthens the feeling that the ADP-ribosyltransferase activities of the bacterial toxins are also found in even simple eukaryotic cells like erythrocytes. It will not be possible to understand their function in more detail until more is known of the enzyme and of the protein substrates.

REFERENCES

1. Jacobson, M.K. & E.L. Jacobson. 1989. *ADP-ribose Transfer Reactions: Mechanisms and Biological Significance*. Springer Verglag, New York.
2. Takada, T., K. Iida, & J. Moss. 1995. Conservation of a common motif in enzymes catalysing ADP-ribose transfer. *J. Biol. Chem. 270*, 541–544.
3. West, R.E., J. Moss, M. Vaughan, T. Lui, & T.-Y. Lui. 1985. Pertussis toxin catalysed ADP-ribosylation of transducin: cysteine 347 is the ADP-ribose acceptor. *J. Biol. Chem. 260*, 14428–14430.
4. Tanuma, S., K. Kawashima, & H. Endo. 1987. An NAD:cysteine ADP-ribosyltransferase is present in human erythrocytes. *J. Biochem. 101*, 821–824.
5. Saxty, B.A. & S. van Heyningen. 1995. The purification of a cysteine-dependent NAD$^+$ glycohydrolase from bovine erythrocytes and evidence that it exhibits a novel ADP-ribosyltransferase activity. *Biochem. J. 310*, 931–937.
6. Moss, J. & M. Vaughan. 1984. Toxin ADP-ribosyltransferases that act on adenylate cyclase systems. *Meth. Enzymol. 106*, 411–418.
7. Richter, C., K.H. Winterhalter, S. Baumhüter, H. Lotscher, & S. Moser. 1983. ADP-ribosylation in inner membrane of rat liver mitochondria. *Proc. Natl. Acad. Sci. USA. 80*, 3188–3192.
8. Cervantes-Laurean, D., D.E. Minter, M.K. Jacobson, & E.L. Jacobson. 1993. ADP-ribose modification of protein lysine residues. *Biochemistry. 32*, 1528–1534.
9. Lobban, M.D. & S. van Heyningen. 1988. Thiol reagents are substrates for the ADP-ribosyltransferase activity of pertussis toxin. *FEBS Lett. 233*, 229–232.
10. Just, I., P. Wollenberg, J. Moss, & K. Aktories. 1994. Cysteine-specific ADP-ribosylation of actin. *Eur. J. Biochem. 221*, 1293–1295.
11. McDonald, L.J., L.A. Wainschel, N.J. Oppenheimer, & J. Moss. 1992. Amino acid-specific ADP-ribosylation: structural characterisation and chemical differentiation of ADP-ribose-cysteine adducts formed non-enzymically and in pertussis toxin-catalysed reactions. *Biochemistry. 31*, 11881–11887.

ENDOGENOUS ADP-RIBOSYLATION OF PHOSPHOPROTEIN B-50/GAP-43 AND OTHER NEURONAL SUBSTRATES

H. Zwiers,[1] M. D. Hollenberg,[2] K. N. McLean,[1] and K. D. Philibert[1]

Endocrine Research Group
Departments of Physiology and Biophysics
[1]Medical Biochemistry
[2]Pharmacology and Therapeutics
The University of Calgary
Faculty of Medicine
Health Sciences Centre
3330 Hospital Drive N.W.
Calgary, Alberta T2N 4N1
Canada

1. INTRODUCTION

Numerous brain-localized proteins are known to be subject to a regulatory ADP-ribosylation reaction. Such proteins whose functions have been studied extensively include tubulin, a cytoskeletal protein involved in the growth of axons and dendrites [1], actin, another cytoskeletal protein involved in neuron motility and receptor organization [2], and the family of G-proteins performing regulatory functions in cell signal transduction mechanisms [3].

Our laboratory is interested in the function of a neuronal tissue specific phosphoprotein, named B-50 [4,5]. We originally discovered B-50 as a protein kinase C substrate, for which phosphorylation in synaptic membranes was inhibited by ACTH-derived neuropeptides [6]. Furthermore, the structure-activity relationship for the potency of ACTH-peptides to affect animal behaviour correlated strongly with the ability to inhibit B-50 phosphorylation. We postulated that B-50 could play a role in the molecular mechanisms underlying the ability of neuropeptides to affect brain activities [6]. B-50 has since been identified independently in a number of distinct contexts, and is also known in the literature as GAP-43 [7]; neuromodulin [8]; or F1 [9] protein. There is a great deal of experimental evidence suggesting that B-50 is exclusively located in nervous tissue, and, at the subcellular level, in neuron axons [10], in the presynaptic membrane of nerve terminals [11], and in nerve growth cones [12]. This localization suggests a role for the protein in neurite outgrowth; a

ADP-Ribosylation in Animal Tissue, edited by Haag and Koch-Nolte
Plenum Press, New York, 1997

recent study of transgenic mice lacking B-50/GAP-43 strongly suggested that neuronal pathfinding is indeed abnormal in such mice [13]. At the neurochemical level, a great deal of information is available indicating that B-50 is a modulator of signal transduction pathways. It is (a) a modulator of phosphatidyl inositol phosphate-kinase [reviewed in 14]; (b) capable of binding Calmodulin at very low [Ca^{2+}] [15]; and (c) a regulator of the G-protein subunit, G_o [16]. There were two reasons why we suspected that B-50 might be subject to an ADP-ribosylation reaction. First, as mentioned above, there was the suggestion that B-50 is a regulator of G-proteins, and it is well known that ADP-ribosylation may play an important role in G-protein function [3]. Second, preliminary NMR studies of pure B-50 suggested the presence of aromatic residues that could not be ascribed to the primary structure of the protein, but rather could be present as a hitherto novel covalent modification of B-50 [17,18]. In the present paper, we describe evidence for ADP-ribosylation of B-50 and other brain Calmodulin binding proteins. Also, we present evidence suggesting a role for ADP-ribosylation in the regulation of the intracellular localization of B-50.

2. MATERIALS AND METHODS

In this section we will outline in brief the materials and methods used that have been described in greater detail elsewhere [19, 20, 21].

2.1. Purified Proteins

B-50 was purified from bovine brain using alkaline extraction at pH 11.5, selective heat precipitation, ammonium sulphate precipitation, reverse phase HPLC, Calmodulin (CaM) Sepharose chromatography, and finally, repeat reverse phase HPLC on a Waters μ Bondapack C18 column. The final B-50 preparation is homogeneous on SDS-PAGE with a yield of approx. 5 mg protein per kg. bovine brain starting material [19]. BICKS was purified from bovine brain as described for B-50, and in fact, copurified until the first HPLC step, where a slightly broader pool was taken than for B-50 isolation only. BICKS also binds to CaM-Sepharose at low calcium concentration (<1 μM) and elutes at higher Ca^{2+} (1 mM). We found that inclusion of 0.5 M NaCl in the Calcium-elution buffer greatly improved the yield of BICKS (see also Coggins et al, 1993 for details). H-20 was identified as a novel Calmodulin binding protein from bovine brain [22]. It was initially found on HPLC tracings as a minor peak (fig. 1 Ref. 23) during the purification of B-50 and BICKS. Its apparent relative molecular weight is 20kDa by SDS-PAGE. Amino acid sequencing of tryptic fragments has revealed it to be a novel brain Calmodulin binding protein[22]. MARCKS was purified as detailed elsewhere [24].

2.2. ADP-Ribosylation Reactions

ADP-ribosylation of B-50 in synaptosomal plasma membranes (SPM) was performed as described [19]; the ADP-ribosylation of BICKS [23]; and of MARCKS [24] have been previously described. H-20 was ADP-ribosylated by adding 2 μg of this protein to 60μg synaptosomal membranes and allowing the ADP-ribosylation reaction to proceed as described for B-50 [19].

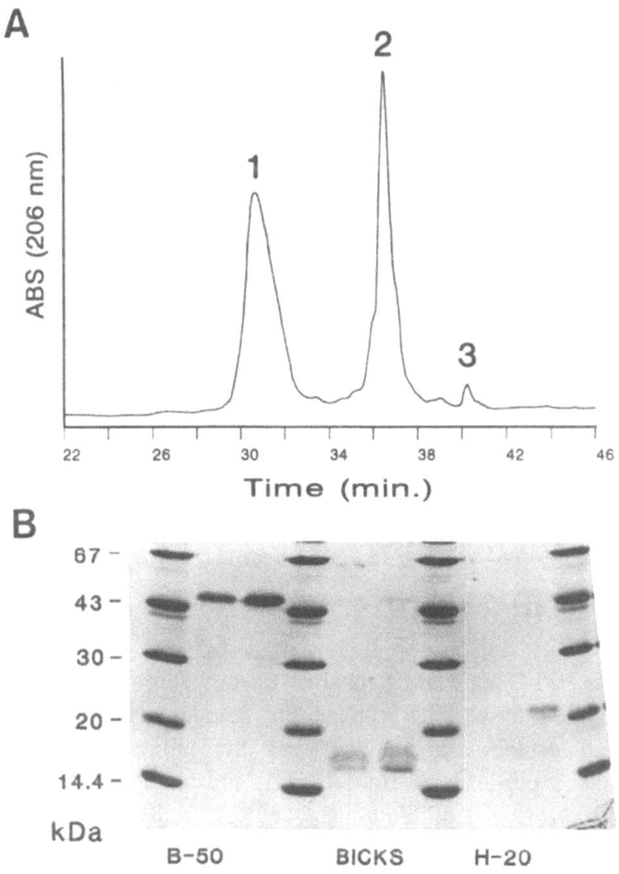

Figure 1. Properties of Some Calmodulin Binding Proteins from Brain. A. Reverse phase HPLC chromotography of bovine brain calmodulin-binding proteins. Proteins recovered from the Calmodulin-Sepharose affinity column were subjected to HPLC analysis. B. Analysis by SDS-PAGE of Peaks 1, 2 and 3 recovered after HPLC chromatography (Fig A). Proteins resolved by SDS-PAGE were detected by Fast Green staining. The positions of the molecular weight (KDa) markers are shown on the left.

2.3. Effects of Phosphorylation and ADP-Ribosylation on B-50 Localization

Carefully washed synaptic membranes (100 µg of protein) were incubated for the phosphorylation reaction in the presence of 10 mM Tris-HCl (pH 7.5), 10 mM $MgCl_2$, 0.1 mM $CaCl_2$, and 10 µCi [γ-^{32}P]ATP at a final concentration of 7.5 µM [25]. Incubation was for 45s at 37°C. The reaction was stopped by placing the sample on ice. The reaction mixture was quickly divided into two equal-sized samples; one sample was fractionated into a soluble and a particulate fraction by centrifugation in a microfuge for 2 min, while the other sample sat on ice. In the case of the ADP-ribosylation reaction synaptic membranes (100µg protein) were incubated in the presence of 50 mM Tris-HCl (pH 7.5), 1 mM APAD, 0.1 mM Gpp(NH)p, and 10 µCi [adenylate-^{32}P]NAD$^+$ at a final concentration of 10 µM [19]. Following incubation at 37° C for 60 minutes, incubates were put on ice. Samples were quickly aliquoted (50 µg SPM each), after which one sample was fractionated by centrifugation in a microfuge for 2 min while the other sat on ice. Upon careful separation of the fractionated aliquots into the soluble (supernatant) and particulate (pellet) fractions, the samples were subsequently frozen and lyophilized. The time course of B-50 translocation by ADP-ribosylation was determined essentially as described above, except that the incubations at 37°C were allowed to proceed for 0, 5, 10, and 30 min, and all pellets were washed prior to lyophilization. Incorporation of ^{32}P-ADP-ribose was quantified by densitometry of the autoradiogram and excision and liquid scintillation counting of B-50 gel pieces, while protein translocation was quantified by densitometry of the stained (Fast Green) 2D gels. Proteins resolved by 2D gel electrophoresis were visualized as described [26]. In brief, lyophilized samples were resuspended in 60 µL buffer (6.6M urea, 0.3% Triton X-100, 0.3% ampholines (pH 4–6 and 3.5–10), and 5% sucrose) and run on a 5% polyacrylamide isoelectric focusing (IEF) gel, with hemoglobin (5 µg/lane) added to the sample as a colored tracer for the first dimension. IEF gel lanes were carefully excised and run on an 11% SDS-polyacrylamide gel for the second dimension. Proteins were stained with Fast Green and the gels were subjected to autoradiography (-80°C using an intensifying screen) for 12 to 72 hours.

3. RESULTS AND DISCUSSION

Reverse phase HPLC and CaM-Sepharose affinity chromatography were used to purify novel brain Calmodulin binding proteins. Fig. 1A shows the HPLC tracing of the bovine brain derived protein fraction that has the unique property of binding to CaM-Sepharose in the presence of 1mM EGTA, and eluting when $2mMCa^{2+}$ + 0.5 M NaCl is added. The absorbance at 206 nm showed essentially 3 peaks, which, by analysis on SDS-PAGE with Fast Green, (fig. 1B), were found to consist of B-50 (Peak 1); BICKS protein (Peak 2), and H-20 protein (Peak 3). These three purified brain Calmodulin binding proteins, as well as the MARCKS protein, were used as substrates for an ADP-ribosylation reaction, using crude rat brain homogenate, crude bovine brain homogenate, and a rat brain synaptosomal membrane fraction as a source of ADP-ribosylating activity.

As shown in figure 2, two dimensional analysis is a very useful technique for the identification and characterization of ADP-ribosylation reaction products. When the synaptosomal plasma membranes alone were subjected to an ADP-ribosylation reaction, numerous labelled spots were visible. Of the several major endogeneously labelled spots detected upon ADP-ribosylation of synaptic plasma membranes, it was possible to identify four constituents as: *a*, B-50, *b*, MARCKS, *c*, actin, and *d*, tubulin. Careful comparison of the silver-stained gel and the autoradiogram revealed that approximately 10% of the protein spots had accumulated radioactivity, suggesting that a considerable portion of all the brain membrane proteins were subject to an ADP-ribosylation modification. In Table 1, we have summarized our results for the brain Calmodulin binding proteins B-50, BICKS, H-20 and MARCKS.

As can be seen, all four Calmodulin binding proteins were labelled in an ADP-ribosylation reaction. It can be pointed out that these four proteins are also protein kinase C substrates.

Relatively little is currently known about the molecular consequences of protein ADP-ribosylation. We have recently shown B-50 to be ADP-ribosylated on cysteine residues [20]. This finding is significant, in that B-50 membrane association occurs primarily through the palmitoylation of B-50's only 2 cysteines [27,28]. The convergence of multiple

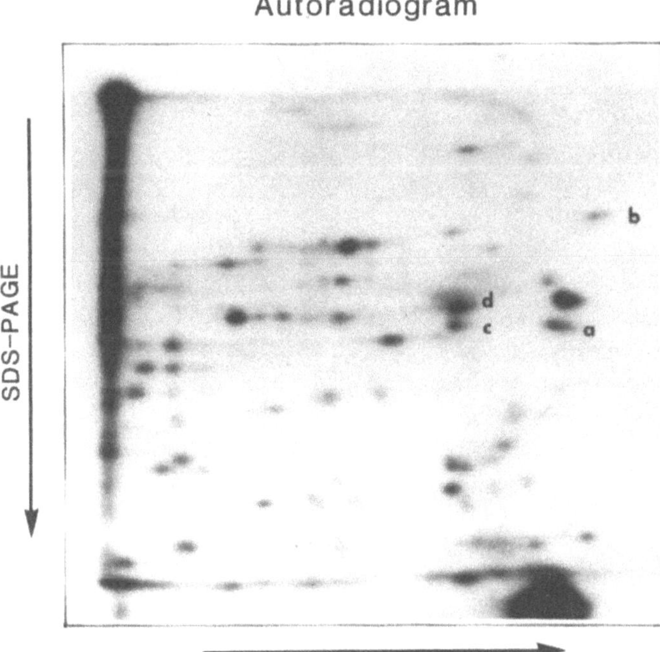

Figure 2. Autoradiogram of 2D separated rat brain SPM proteins subjected to an ADP-ribosylation reaction. The letters denote proteins for which a positive identification has proved possible: a = B-50, b = MARCKS, c = Actin, d = tubulin.

modifications on the same cysteine residues suggested that ADP-ribosylation might be involved in membrane attachment of B-50. We used a simple fractionation assay wherein synaptosomal plasma membranes (SPM), which were carefully washed to remove unbound B-50, were incubated either in the presence of ^{32}P-ATP, or in the presence of [adenylate^{32}-P] NAD+. Samples were centrifuged to collect the soluble and particulate fractions. Endogenous phosphorylation by protein kinase C (PKC) at serine 41 [25], as ex-

Table 1. Properties of some calmodulin binding proteins from brain

	B-50(GAP-43)	BICKS (Neurogranin;RC3)	H-20	MARCKS
Calmodulin				
with Ca^{2+}	-	-	+	+
without Ca^{2+}	+	+	+	-
ADP-ribosylation	+	+	+	+
Kinase C Phosphorylation	+	+	+	+
Phosphorylation and Calmodulin binding	+	+	-	+
Phosphorylation and membrane attachment	-	-	-	+
Unique Neuronal location	+	+	N.D.	+
Acid stability	+	+	Partial	+
Subcellular localization	Axonal	Dendritic	Myelin	Widespread

pected had no effect on B-50 localization, as most of the B-50 protein and all of the phosphate-labelled B-50 were recovered in the pellet fraction (Fig. 3A and 3B). However, when synaptosomal plasma membranes were incubated in the presence of the ADP-ribose donor ^{32}P-NAD$^+$, a major portion (~50%) of the labeled B-50 recovered was found in the supernatant (Fig. 3C and 3D). The percentage of ^{32}P-labelled ADP-ribosylated B-50 recovered in the supernatant (soluble fraction) rose from a level of approximately 50% to a level of 92 ± 2.8% (mean ± S.E.M., N=8) of the total radiolabelled B-50 present in the entire preparation (i.e. pellet plus all supernatant fractions, including membrane washes). The process appeared to be specific for B-50, (identified with pure protein standards, [19]), since another protein subject to ADP-ribosylation, with a slightly more acidic isoelectric point and a slightly higher molecular weight (previously identified in [19]) remained firmly associated with the membrane fraction. The small percentage of radioactively labelled B-50 remaining in the particulate fraction, even after extensive washing might be accounted for by the presence of B-50 ADP-ribosylated at a distinct C-terminal domain not involved in membrane attachment [20]. Therefore, we concluded that ADP-ribosylation specifically releases B-50 from the synaptic membrane.

To investigate further the relationship between ADP-ribosylation and membrane release, time course studies were performed to monitor the release of B-50 protein from the membrane during the course of the ADP-ribosylation reaction (Figure 4). The initial levels of B-50 protein released into the supernatant in all trials was variable, representing about

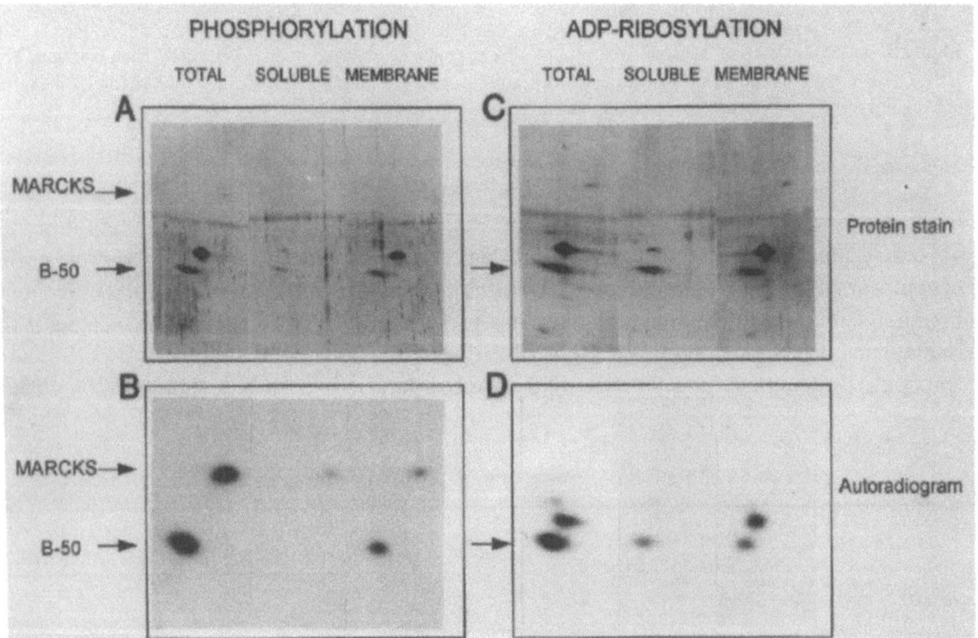

Figure 3. ADP-ribosylation abolishes membrane association of B-50. Synaptosomal plasma membranes were incubated in the presence of [γ-^{32}P]ATP or [adenylate-^{32}P]NAD$^+$. Protein staining of two-dimensional electrophoresis of endogenously phosphorylated proteins (A) reveals that the majority (> 90%) of B-50 protein remains associated with the membrane fraction. Autoradiography (B) indicates that all of the ^{32}P-labelled B-50 is associated with the membrane. Following endogenous ADP-ribosylation, a major proportion of B-50 protein (C) is recovered in the soluble fraction without changing the distribution of another major unknown labelled protein substrate, with slightly higher MW and more acidic pI than B-50. Recovery of labelled B-50 (D) mimics the protein distribution.

2% of the total amount of B-50 in the membranes. After 5 min, slightly over 6% of the total B-50 protein had been released from the membrane (3-fold increase, fig. 4A) while the increase of ADP-ribosylated B-50 in the supernatant had increased by about 7-fold over basal (Fig. 4B). It appeared that at all time points, the quantity of B-50 released from the membrane was proportional to its level of ADP-ribosylation. The high variation at the zero time point was probably due to experimental difficulties inherent in the procedure that required the washing of the membranes, resuspension in buffer, and a separation of the supernatant from the pellet in a minimal time span. If the driving force for the release of B-50 from the membranes is depalmitoylation of B-50, then according to our model

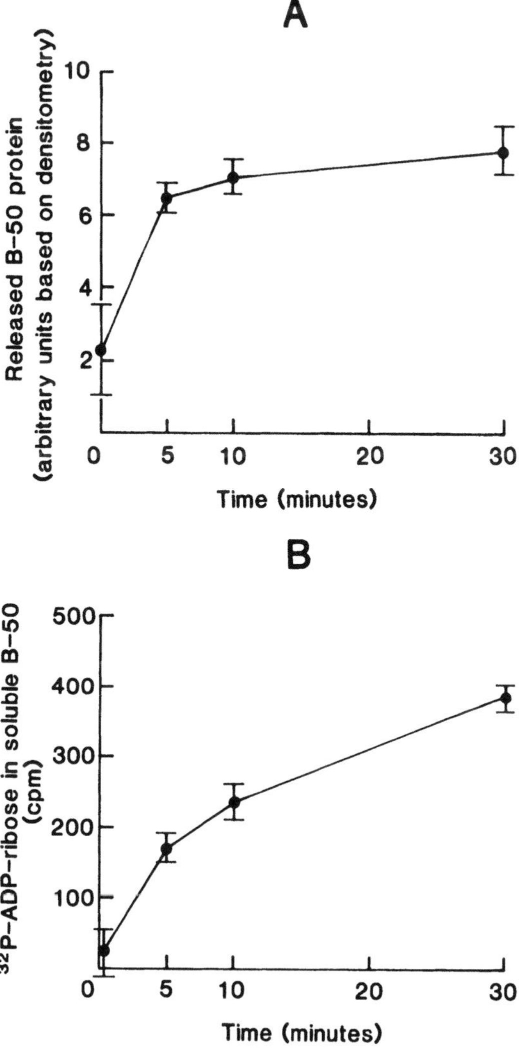

Figure 4. Time course of B-50 ADP-ribosylation and membrane release. Thoroughly washed synaptic membranes were incubated in the presence of ^{32}P-NAD$^+$ for 0, 5, 10 and 30 minutes prior to fractionation by centrifugation, lyophilization, and 2D gel electrophoresis as described in Materials and Methods. Protein recovery of release B-50 (4A), quantified with video camera / digital image analysis computer densitometry, was initially (Time = 0 min) ≈ 2% of the total B-50. The level of B-50 protein release from the membrane increased with time. In comparison, the initial incorporation of ADP-ribose (4B) was roughly 25 CPM which increased to 425 cpm after 30 minutes. The data represent the mean ± S.E.M. of four experiments performed in quadruplicate.

(Fig. 5, step 1) this process would probably be constitutively active, so as to yield soluble unconjugated B-50. Our preliminary results can be best explained by an ongoing process of depalmitoylation of B-50 causing release (consistent with reports in the literature [29,30] for other proteins), followed by ADP-ribosylation of reactive, unconjugated Cys3/Cys4 residues. A physiological function for ADP-ribosylation may be that it would prevent the palmitoylation reaction which would subsequently lead to the incorporation of B-50 with the membrane. In this scheme, ADP-ribosylation would play a regulatory role in the dynamic processes involved in attachment/detachment of the important regulatory Calmodulin binding protein B-50 to the synaptic membrane. We suggest that the intracellular localization of the other three Calmodulin binding proteins BICKS, H-20 and MARCKS also may be affected by ADP-ribosylation.

In summary and in conclusion, in this paper, we have shown that a group of brain Calmodulin binding proteins are subject to an ADP-ribosylation reaction. The generally accepted covalent modification as a result of this reaction is mono-ADP-ribosylation, which might have a regulatory function for all of these signal transduction proteins. For one of the four proteins studied here, B-50, we presented evidence for a role of ADP-ribosylation in the regulation of its intracellular localization.

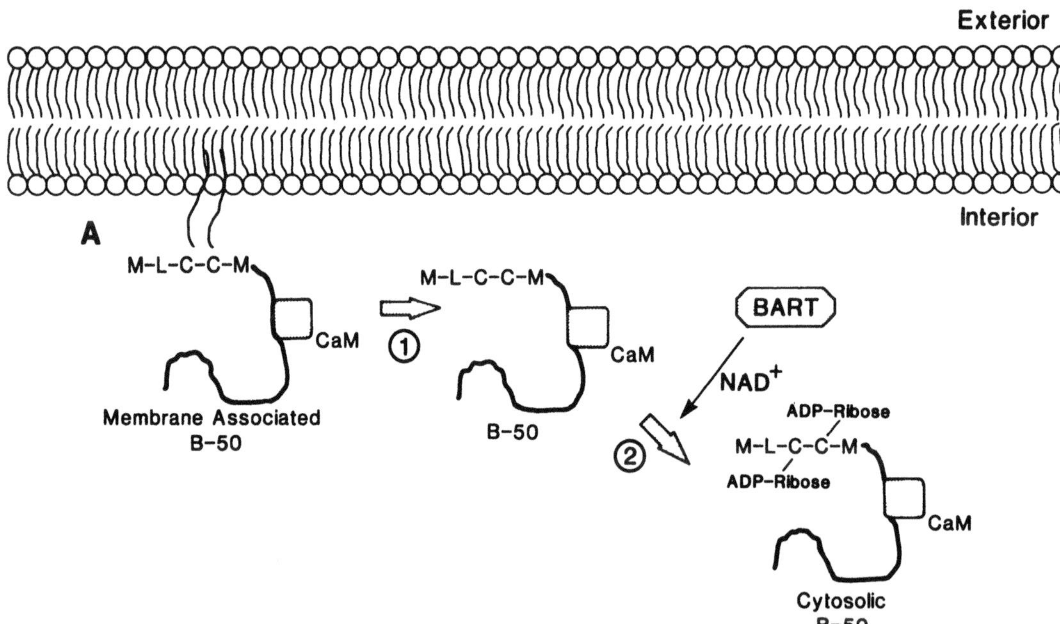

Figure 5. Schematic representation of proposed model. At rest (A), B-50 is firmly anchored to the cytosolic face of the synaptic membrane by palmitoylation of one or both N-terminal cysteine residues, shown here with both sites palmitoylated. Due to enzymatic depalmitoylation [29, 30] B-50 looses its membrane anchors (reaction 1). B-50 ADP-ribosyltransferase (BART) transfers an ADP-ribose from a nicotinamide adenine dinucleotide (NAD+) donor to one or both underivitized B-50 cysteines (reaction 2). BART might be related to a recently purified and characterized cysteine modifying ADP-ribosyl transferase from bovine erythrocytes [31]. The detachment of palmitic acid is accompanied by the translocation of B-50 from the membrane to the cytosol, whereby B-50 becomes susceptible to ADP-ribosylation. Our working hypothesis is that ADP-ribosylation of B-50 at Cys_3, Cys_4 interferes with the process of palmitoylation of B-50 by which the protein could reattach to the membrane. B-50 is portrayed associated with Calmodulin (CaM) to illustrate our findings that ADP-ribosylation does not appear to alter the kinetics of CaM binding with B-50 [20].

4. ACKNOWLEDGMENT

This work was supported by funds from the Medical Research Council of Canada. H. Zwiers is a senior scholar of the Alberta Heritage Foundation for Medical Research. We are indebted to Daryl Beers for her secretarial assistance.

REFERENCES

1. Scaife, R.M., L. Wilson & D.L. Purich. 1991. Microtubule protein ADP-ribosylation *in vitro* leads to assembly inhibition and rapid depolymerization. *Biochem 31:* 310–316.

2. Aktories, K., & A. Wegner. 1992. Mechanisms of the cytopathic action of actin-ADP-ribosylating toxins. *Mol. Microbiol. 6:* 2905–2908.

3. Williamson, K.C. and J. Moss. 1990. Mono-ADP-ribosyltransferases and ADP-ribosylarginine Hydrolases: A mono-ADP-ribosylation Cycle in Animal Cells. In: Moss, J., & M. Vaughan (eds): *ADP-ribosylating Toxins and G Proteins: Insights into Signal Transductions,* American Society for Microbiology, Washington, D.C., pp. 493–510.

4. Zwiers, H., V.M. Wiegant, P. Schotman & W.H. Gispen. 1978. ACTH-induced inhibition of endogenous rat brain protein phosphorylation in vitro: structure-activity. *Neurochem.* Res. *3:* 455–463.

5. Zwiers, H., P. Schotman & W.H. Gispen. 1980. Purification and some characteristics of an ACTH-sensitive sensitive protein kinase and its substrate protein. *J. Neurochem. 34:* 1689–1699.

6. Zwiers, H., & P.J. Coggins. 1991. B-50 structure, processing and interaction with ACTH. *Progress in Brain Research. 89:* 3–16.

7. Skene, J.H.P., & M. Willard. 1981. Changes in Axonally transported proteins during Axon Regeneration in Toad Retinal Ganglion Cells. *J. Cell Biology. 89:* 86–95.

8. Andreasen, T.J., C.W. Luetje, W. Heideman & D.R. Storm. 1983. Purification of a novel calmodulin binding protein from bovine cerebral cortex membranes. *Biochemistry. 22:* 4615–4618.

9. Chan, S.Y., K. Murakami & A. Routtenberg. 1986. Phosphoprotein F1: Purification and characterisation of a brain kinase C substrate related to plasticity. *J. Neurosci. 6:* 3618–3627.

10. Goslin, K., D.J. Schreyer, J.H.P. Skene & G.A. Banker. 1988. Development of neuronal polarity: GAP-43 distinguishes axonal from dendritic growth cones. *Nature. 336:* 672–674.

11. Gispen, W.H., J.J.M. Leunissen, A.B. Oestreicher, A.J. Verkleij & H. Zwiers. 1985. Presynaptic localization of B-50 phosphoprotein: the ACTH-sensitive protein kinase substrate involved in rat brain phosphoinositide metabolism. *Brain Res.. 328:* 381–385.

12. Strittmatter, S.M., M. Igarashi & M.C. Fishman. 1994. GAP-43 amino terminal peptides modulate growth cone morphology and neurite outgrowth. *J. Neurosci. 14:* 5503–5513.

13. Strittmatter, S.M., C. Fankhauser, P.C. Huang, H. Mashimo & M.C. Fishman. 1995. Neuronal path-finding is abnormal in mice lacking the neuronal growth cone protein GAP-43. *Cell. 80:* 445–452.

14. Coggins, P.J., & H. Zwiers. 1991. B-50 (GAP-43): Biochemistry and Functional Neurochemistry of a Neuron-Specific Phosphoprotein. *Journal of Neurochemistry. 56:* 1095–1106.

15. Alexander, K.A., B.M. Cimler, K.E. Meier, & D.R. Storm. 1987. Regulation of calmodulin binding to P-57. *J. Biol. Chem. 262:* 6108–6113.

16. Strittmatter, S.M., D. Valenzuela, T.E. Kennedy, E.J. Neer & M.C. Fishman. 1990. G_0 is a major growth cone protein subject to regulation by GAP-43. *Nature. 344:* 836–841.

17. Coggins, P.J., D.D. McIntyre, H.J. Vogel, & H. Zwiers. 1989. Neuronal protein B-50: a proton nuclear-magnetic-resonance study. *Biochemical Society Transactions 629th Meeting, London. 17:* 785–787.

18. Zhang, M. & H.J. Vogel. Nuclear magnetic resonance studies of the structure of B50/neuromodulin and its interaction with calmodulin. *Biochem. Cell Biol. 72:* 109–116.

19. Coggins, P.J., K. McLean, A. Nagy & H. Zwiers. 1993. ADP-ribosylation of the neuronal phosphoprotein B-50/GAP-43. *J. Neurochem. 60:* 368–371.

20. Philbert, K. & H. Zwiers. 1995. Evidence for multi-site ADP-ribosylation of neuronal phosphoprotein B-50/GAP-43. *Mol. Cell. Biochem. 149:* 183–190.

21. Zwiers, H., & K.D. Philbert. 1996. ADP-ribosylation. In: *Neuromethods: Posttranslational Modifications: Techniques and Protocols.* Boulton, A.A., G.B. Baker & H. C. Hemmings, Eds. Humana Press Inc; Totowa, N.J. in press.

22. Zwiers, H., K. McLean, K. Philbert & K.A. Sharkey. 1996. A Novel Calmodulin Binding Protein H-20 Related to B-50/GAP43 and BICKS/RC2 is Primarily Located in Nervous Tissue, and is Highly Enriched in Myelin. *J. Neurochem. 66*: 27th ASN Meeting, Philadelphia, PA. S5A.

23. Coggins, P.J., K. McLean & H. Zwiers. 1993. Neurogranin, a B-50/GAP-43-immunoreactive C-kinase substrate (BICKS), is ADP-ribosylated. *FEBS. 335*: 109–113.

24. Chao, D., D.L. Severson, H. Zwiers & M.D. Hollenberg. 1995. Radiolabelling of bovine myristoylated alanine-rich protein kinase C substrate (MARCKS) in an ADP-ribosylation reaction. *Biochem. Cell Biol. 72*: 391–396.

25. Coggins, P.J. & H. Zwiers. 1989. Evidence for a single protein kinase C-mediated phosphorylation site in rat brain protein B-50. *J. Neurochem. 53*: 1895–1901.

26. Palkiewicz, P., H. Zwiers & F.L. Lorscheider. 1994. ADP-ribosylation of brain neuronal proteins is altered by *in vitro* and *in vivo* exposure to inorganic mercury. *J. Neurochem. 62*: 2049–2052.

27. Skene, J.H.P. & I. Virag. 1989. Posttranslational membrane attachment and dynamic fatty acylation of a neuronal growth cone protein, GAP-43. *J. Cell Biol. 108*: 613–624.

28. Zuber, M.X., S.M. Strittmatter & M.C. Fishman. 1989. A membrane-targeting signal in the amino terminus of the neuronal protein GAP-43. *Nature. 341*: 345–348.

29. Milligan, G., M. Parenti & A.I. Magee. 1995. The dynamic role of palmitoylation in signal transduction. *TIBS. 20*: 181–186.

30. Linder, M.E., C. Kleuss, & S.M. Mumby. 1995. Palmitoylation of G-Protein α Subunits. *Methods in Enzymology. 250*: 314–329.

31. Saxty, B.A. & S. van Heyningen. 1995. The Purification of a Cysteine-Dependent NAD+ Glycohydrolase Activity from Bovine Erythrocytes and Evidence that it Exhibits a Novel ADP-ribosyltransferase Activity. *Biochem. 310*: 931–937.

ENDOGENOUS MONO-ADP-RIBOSYLATION IN RETINA AND PERIPHERAL NERVOUS SYSTEM

Effects of Diabetes

Alfredo Gorio, Maria Lucia Donadoni, Cristina Finco, and
Anna Maria Di Giulio

Laboratory for Research on Pharmacology of Neurodegenerative Disorders
Department Medical Pharmacology
Via Vanvitelli 32
20129 Milano, Italy

ABSTRACT

The extranuclear endogenous mono-ADP-ribosylation of proteins was monitored in cellular preparations of retina, superior cervical ganglion, dorsal root ganglia and peripheral nerve. At least 6 protein fractions are ADP-ribosylated in the crude extract fraction from retina control preparations, while in diabetic rats the number of retina labeled proteins and the extent of labeling are highly reduced. In the superior cervical ganglion labeling was present in 10 proteins, in diabetics it was greatly decreased. Treatment of diabetic rats with silybin, a flavonoid mono-ADP-ribosyltransferase inhibitor, did not affect hyperglycemia, but prevented the alteration of extent of protein ADP-ribosylation. These data suggest that proteins of retina and peripheral ganglia are excessively ADP-ribosylated *in vivo*. The effects of silybin treatment on excessive mono-ADP-ribosylation of proteins was associated with the prevention of reduction of substance P-like immunoreactivity levels, that is typical of diabetic neuropathy. In the membrane fraction of sciatic nerve Schwann cells, at least 9 proteins were ADP-ribosylated, diabetes caused a marked increase of labeling. A comparable increase involving the same proteins is triggered by chronic nerve injury and by corticosteroid treatment. Silybin treatment of diabetic rats prevented such an increase. We propose that the inhibition of excessive protein mono-ADP-ribosylation by silybin prevented the onset of diabetic neuropathy. While the effects on Schwann cells is likely indirect and secondary to the improvement of diabetic axonopathy.

ADP-Ribosylation in Animal Tissue, edited by Haag and Koch-Nolte
Plenum Press, New York, 1997

BACKGROUND

Among the functional and morphological alterations of peripheral nerves in diabetes mellitus, there are reductions in nerve conduction velocity, axonal transport, and dwindling with selective changes in transport of cytoskeletal axonal components (1, 2, 3). Among the neuropeptides most affected by diabetic neuropathy is substance P (4), the major neuropeptide of the sensory system that is particularly affected by diabetes (1). In the sciatic nerve of diabetic rats, substance P axonal transport is impaired markedly (5). Results obtained in our laboratory suggest that protein mono-ADP-ribosylation might be one of the cellular mechanisms abnormally triggered by diabetes and capable of causing diabetic neuropathy. The α-subunits of G proteins are selectively ADP-ribosylated by cholera and pertussis toxins (6, 7, 8). In many biological systems, there are endogenous mono-ADP-ribosyltransferases that regulate transfer of ADP-ribose from NAD to receptive proteins *in vivo* and even in absence of cholera and pertussis toxin (9, 10, 11). Several membrane and cytoplasmic proteins are subjected to these modifications (9), but also G proteins are a potential target (12). These data suggest that mono-ADP-ribosylation might be an endogenous regulatory mechanism of protein functions and in particular of receptor signal transduction. We have previously shown that Gi/Go systems are impaired in retina and striatum of diabetic animals, the functional reduction was correlated with lower pertussis toxin-mediated labeling and no changes in mRNA (13, 14, 15). These data suggested that endogenous mono-ADP-ribosylation might be involved as a mechanism underlying diabetes induced dysfunction. Because *in vivo* treatment with insulin normalizes blood glucose levels and prevents the onset of diabetic complications (1, 14, 16), we performed a series of studies treating diabetic animals with either insulin or silybin, an inhibitor of protein mono-ADP-ribosylation. Diabetes was induced by causing degeneration of β-pancreatic cells with alloxan (17), thereafter animals developed peripheral neuropathy that was not related to the toxic effects of alloxan upon the nervous system (18).

EXPERIMENTAL PROCEDURES

Diabetes was induced in 250 g Sprague-Dawley rats by means of a single subcutaneous injection of alloxan (100 mg/kg). The onset of diabetes was established by determining plasma glucose levels (GOP PAP test, Boehringer, Mannheim, Germany) 1 week after alloxan injection. Diabetic rats with glycemia below 450 mg/dl of plasma did not enter the study. Some animals were treated daily with insulin and others with silybin given in drinking water at the dose of 50 mg/kg. For further details on the experimental procedures see references 19, 20 and 21.

RESULTS

Endogenous ADP-ribosylation was measured in absence of exogenous toxins in retina and superior cervical ganglion crude extracts of control and insulin or silybin treated animals. When retina preparations were incubated with [^{39}P]- NAD, 6 proteins were labeled in controls. Their molecular weights were 85 K, 70 K, 55 K, 41 K, 39 K, 36 K (Fig.1, lane A). In diabetic rat retinas, labeling was reduced to fewer protein bands; in the 85 K there was an incorporation higher than in controls, while labeling was barely de-

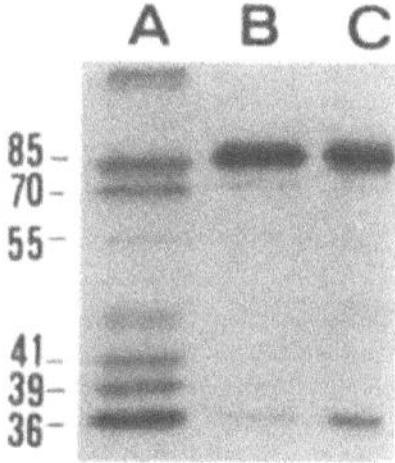

Figure 1. ADP-ribosylation of retina crude extract proteins from controls (A), untreated (B) and insulin-treated (C) diabetic rats. Following incubation with [^{32}P]- NAD, the labeled proteins were identified by electrophoresis and subsequent autoradiography. Note the reduced number of ADP-ribosylated proteins in the diabetic group (B) and the selective improvement by insulin on the 36 K band (C). Reproduced from reference 19 with permission of the editor.

tectable for bands of 70 K and 36 K and absent in others (Fig. 1, lane B). Treatment with insulin of diabetic rats reduced glucose blood levels to control values, but the ADP-ribosylation was just normalized. Only labeling of the 36 K protein band was greatly improved; the other proteins were ADP-ribosylated similarly to the retinas from untreated diabetics (Fig. 1, lane C). When retina crude extract and 1 µM silybin were mixed with the ADP-ribosylation medium, the *in vitro* reaction was inhibited (Fig. 2). The figure shows that the addition of silybin (C+) to the incubation medium did not alter the crude extract protein composition (C). The autoradiogram showed that several bands (C") were labeled by incubation with [^{32}P]- NAD and that the addition of silybin (C"+) inhibited ADP-ribosylation of most proteins, with the exception of the > 100 K band (Fig. 2). The incubation of control superior cervical ganglion crude extracts (Fig. 3, side R, lane A) with (^{32}P) NAD promoted a measurable extent of labeling in 10 proteins of the following molecular weight > 100 K, > 100 K, 100 K, 74 K, 56 K, 48 K, 44 K, 41 K, 33 K, 15.5 K.

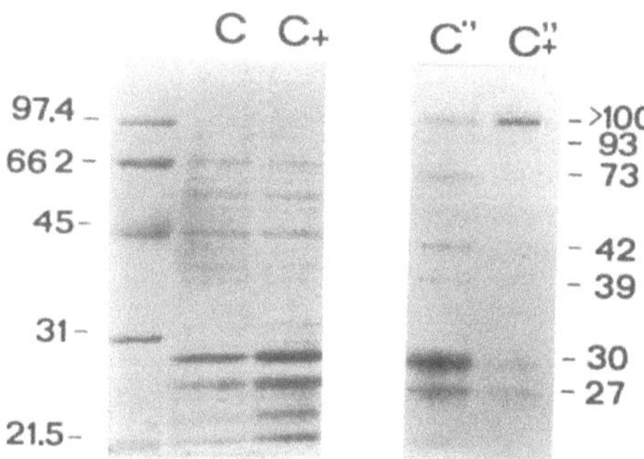

Figure 2. Coomassie blue staining (C and C+) and ADP-ribosylation autoradiograms (C" and C"+) of retina crude extract proteins. Proteins were identified by gel electrophoresis and then autoradiographed to identify the proteins labeled by ADP-ribosylation. At least six proteins were mono-ADP-ribosylated, whereas addition of silybin (1 µM) to the assay medium inhibited labeling of most proteins, only the >100 K band still incorporates ADP-ribose (C"+). Reproduced from reference 20 with permission of the editor.

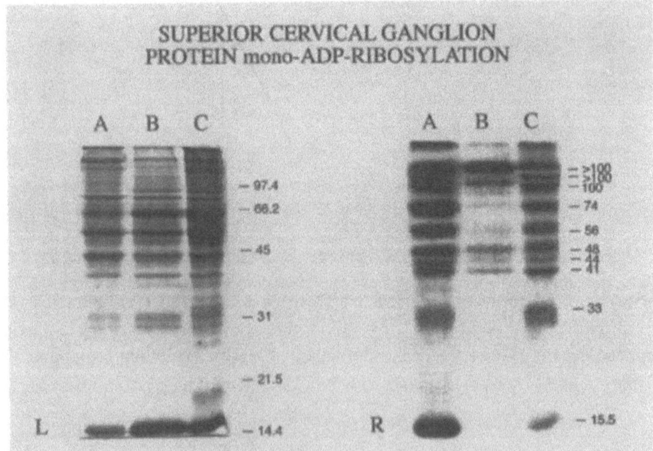

Figure 3. Superior cervical ganglia crude extract proteins were identified by gel electrophoresis and Coomassie blue staining (L) and autoradiography after mono-ADP-ribosylation assay (R). Ganglionic preparations were collected from control (A), diabetic (B) and silybin-treated diabetic (C) rats. In diabetic preparations, the extent of protein mono-ADP-ribosylation was reduced markedly for 74 K, 56 K, 48 K, 44 K and 15.5 K proteins. Treatment with silybin markedly improved the labeling of these proteins (C). Reproduced from refence 21 with permission of the editor.

In the ganglion of untreated diabetic rats (Fig. 3, side R, lane B) the extent of labeling was reduced in 5 proteins, 74 K, 56 K, 48 K, 44 K, 15.5 K. The inhibition of ADP-ribosylation by *in vivo* treatment of diabetics with silybin (Fig. 3, side R, lane C) allowed a good restoration of the extent of protein labeling in the superior cervical ganglion. The improvement was present in any single protein and ranged from 35% to 100%. Sciatic nerve membrane fraction Coomassie blue staining is shown on the left of Fig. 4, while

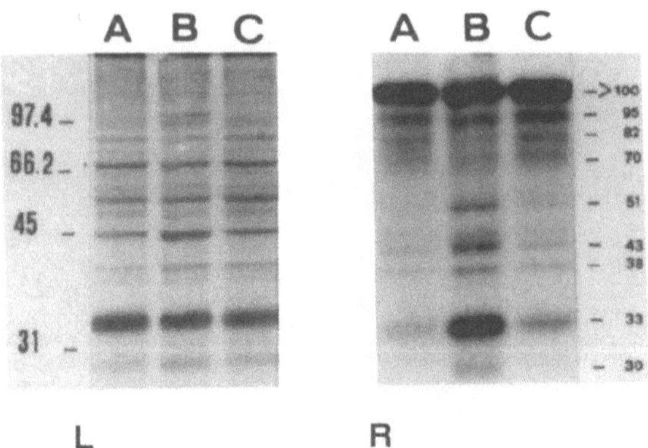

Figure 4. Sciatic nerve membrane rich fraction proteins were identified by gel electrophoresis and Coomassie blue staining (L) and autoradiography after mono-ADP-ribosylation assay (R). Sciatic nerve preparations were collected from control (A), diabetic (B) and silybin-treated diabetic (C) rats. At least 9 proteins were labeled in controls but, in diabetics, the extent of protein mono-ADP- was increased in several proteins; treatment with silybin restored the normal extent of protein labeling. Reproduced from reference 21 with permission of the editor.

the ADP-ribosylation assay autoradiogram is on the right. The pattern of protein labeling shows that 6 bands incorporate ADP-ribose in controls (Fig. 4, right side, lane A), with molecular weights > 100 K, > 100 K, 94 K, 51 K, 43 K and 39 K. In the sciatic nerve of diabetics two bands were increased, the lower > 100 K and 43 K and two others decreased, 94 K and 51 K (lane B), treatment of diabetics with silybin restored normal labeling (lane C). Figure 5 shows the ADP-ribosylation of the 7 day resected sciatic nerve distal stump that is now constituted mainly by Schwann cells and no axons are present. The number of labeled proteins increased markedly after injury (C_L) with a pattern similar to diabetes; corticosteroid treatment had a similar effect (C+) and the lesion further increased ADP-ribosylation (C_L+). Thereby diabetes, injury and corticosteroids had the same effects upon Schwann cell protein ADP-ribosylation. The levels of substance P decreased significantly in 14 week-diabetic rats; the control content was 2.5 ± 0.11 ng/mg protein, whereas in untreated diabetics it was reduced to 0.9 ± 0.05. Treatment with silybin restored normal substance P values, 2.2 ± 0.13.

DISCUSSION

These results show that alteration of protein mono-ADP-ribosylation occur in experimental diabetes and suggest that such protein post-translational modification may be involved in the mechanisms leading to the onset of diabetic neuropathy. Treatment of diabetic rats with silybin, an inhibitor of protein mono-ADP-ribosylation, prevented the biochemical changes and the marked reduction of substance P content typical of diabetes. Insulin treatment, however, was very selective and normalized ADP-ribosylation of a 36–38 K protein that is likely a G inhibitory protein (19). Indeed insulin normalizes

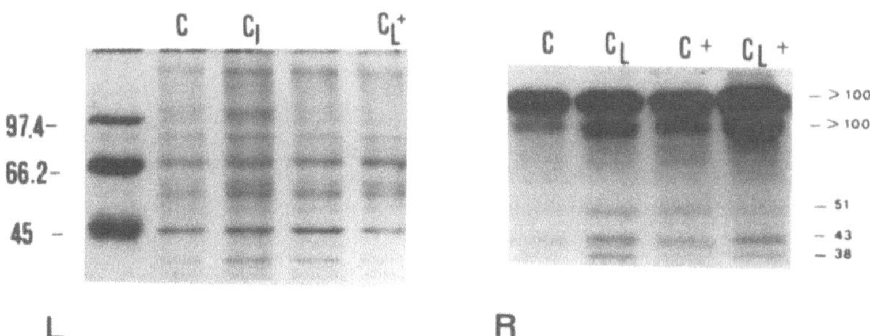

Figure 5. Membrane-rich fractions from unlesioned sciatic nerve (C) and from the non-regenerating distal stump (C_L) and with corticosterone tretment (C+; C_L+). The proteins were identified by gel electrophoresis and Coomassie blue staining (L) and autoradiography after mono-ADP-ribosylation assay (R). At least 7 days after axotomy, the distal stump lacked the axonal part that was fully degenerated and contained only Schwann cells. The distal stump of the lesioned nerve displayed the same increase of protein ADP-ribosylation as in diabetics, and corticosterone treatment had the same effect as well. The lack of some protein bands (see fig. 4) might have been due to the much younger age of these animals, the chronic diabetics being 14 weeks older. Reproduced from reference 21 with editor permission.

adenylate cyclase altered by diabetes (19). The excessive ADP-ribosylation of the other proteins was not modified by insulin .

Both retina and peripheral ganglia of diabetic animals showed high levels of endogenous mono-ADP-ribosylation under our testing conditions, thereby it appears that, in diabetes, ADP-ribosyltransferase activity is higher than normal. A selected population of proteins was excessively ADP-ribosylated, thereby causing neuronal dysfunction. In the central nervous system these changes were absent (manuscript in preparation), suggesting that neurons not protected by the blood brain barrier are more susceptible to the consequences of hyperglycemia. In the sciatic nerve, however, our testing conditions showed that diabetes markedly increases incorporation of ADP-ribose in at least 4 proteins. The nerve lesion test revealed that responsible for such an increase were the denervated Schwann cells, since injury and diabetes affected the same proteins in a similar manner. On the other hand it is well known that diabetic neuropathy is an axonopathy with secondary effects on Schwann cells. Silybin treatment prevented the protein changes induced by diabetes, the incorporation of ADP-ribose was brought back to normal either in the case of reduced incorporation as for peripheral ganglia and in the case of augmented labeling as for sciatic nerve. The effect on Schwann cells of the diabetic rats was likely secondary to the silybin-promoted improvement of the sciatic nerve function as indicated by the normalization of nerve substance P content and transport (20).

REFERENCES

1. Brown M.J. & A.K. Asbury. 1984. Diabetic neuropathy. *Ann. Neurol. 15*:2–12
2. Norido F., R. Canella, R. Zanoni & A. Gorio. 1984. Development of diabetic neuropathy in C57BL/KS (db/db) mouse. *Exptl. Neurol. 83*:221–232
3. Vitadello M., G. Filliatreu, J.L. Dupont, R. Hassig, A. Gorio & L. Di Giamberardino. 1985. Altered axonal transport of cytoskeletal proteins in the mutant diabetic mouse. *J. Neurochem. 45*:860–868
4. Di Giulio A.M., E. Lesma & A. Gorio. 1995. Diabetic neuropathy in the rat. I. ALCAR augments levels and axoplasmic transport of substance P. *J. Neuroci. Res. 40*:414:415
5. Robinson J.P., G.B. Willars., D.R. Tomlison & P. Keen. 1987. Axonal transport and tissue contents of substance P in rats with long-term streptozotocin-diabetes. Effects of the aldose reductase inhibitor "Statil". *Brain Res. 426*:339–348
6. Stryer L. & H.R. Bourne. 1986. G proteins: A family of signal transducers. *Annu. Rev. Cell Biol. 2*:264–272
7. Graziano M.P. & A. G. Gilman. 1987. Guanine nucleotide-binding regulatory proteins: mediators of transmembrane signaling. *Trends Pharmacol. Sci. 8*:478–481
8. Birnbauer L. 1990. G proteins in signal trasduction. *Ann. Rev. Pharmacol. Toxicol. 30*:675–705
9. Duman S.R., R.Z. Terwiller, and E.J. Nestler. 1991. Endogenous ADP-ribosylation in brain: Initial characterization of substrate proteins. *J. Neurochem. 57*:2124–2132
10. Matsuyama S. & S. Tsuyama. 1991. Mono-ADP-ribosylation in brain: Purification and characterization of ADP-ribosyltransferases affecting actin from rat brain. *J. Neurochem. 57*:597–601
11. Williams M.B., X. Li, X. Gu & R.S. Jope. 1992. Modulation of endogenus ADP-ribosylation in rat brain. *Brain Res. 592*:49–56
12. Jacquemin C., H. Thibout, B. Lambert & C. Correze. 1986. Endogenous ADP-ribosylation of Gs sub-unit and autonomous regulation of adenylate-cyclase. *Nature, 323*:182–184
13. Abbracchio, M.P., F. Cattabeni, A.M. Di Giulio, C. Finco, A.M. Paoletti, B. Tenconi, & A. Gorio. 1991. Early alterations of Gi/Go protein-dependent trasductional processes in the retina of diabetic animals. *J. Neurosci. Res. 29*:196–200
14. Abbracchio M.P., M. Di Luca, A.M. Di Giulio, F. Cattabeni, B. Tenconi & A. Gorio. 1989. Denervation and hyperinnervation in the nervous system of diabetic animals. III. Functional alterations of G proteins in diabetic encephalopathy. *J. Neurosi. Res. 24*:517–523
15. Finco C., M.P. Abbracchio, M.L. Malosio, F. Cattabeni, A.M. Di Giulio, B. Paternieri, P. Mantegazza & A. Gorio. 1992. Diabetes-induced alterations of central nervous system G proteins. ADP-ribosylation, immunoreactivity, gene expression studies in rat striatum. *Mol. Chem. Neuropathol. 17*:259–272

16. Gorio A., A.M. Di Giulio, M.L. Donadoni, B. Tenconi, E. Germani, A. Bertelli & P. Bertelli. 1992. Early neurochemical changes in the autonomic neuropathy of the gut in experimental diabetes. *Int. J. Pharmacol. Res. XII*:217–224

17. Dunn J.S., H.L. Sheehan & N.G. McLechtie. 1943. Necrosis of the islet of pancreas produced experimentally. *Lancet 1*:484–487

18. Eliasson S.G.. 1964. Nerve conduction changes in experimental diabetes. *J. Clin. Invest. 43*: 2353–2358

19. Gorio A., M.L. Donadoni & A.M. Di Giulio. 1995. Nitric oxide-sensitive protein ADP-ribosylation is altered in rat diabetic neuropathy. *J. Neurosc. Res. 40* : 420–426

20. Donadoni M.L., R. Gavezzotti, Borella F. A.M. Di Giulio & A. Gorio. 1995. Experimental diabetic neuropathy. Inhibition of protein mono-ADP-ribosylation preventa reduction of substance P axonal transport. *J. Pharmacol. Exptl. Therap. 274*:570–576

21. Gorio A., M.L. Donadoni, Finco C., Borella F. & Di Giulio A.M.. 1996. Alterations of protein mono-ADP-ribosylation and diabetic neuropathy. A novel pharmacological approach. *Europ. J. Pharmacol.* In press

THE α7 INTEGRIN AS A TARGET PROTEIN FOR CELL SURFACE MONO-ADP-RIBOSYLATION IN MUSCLE CELLS

Anna Zolkiewska and Joel Moss

Pulmonary-Critical Care Medicine Branch
National Heart, Lung, and Blood Institute
National Institutes of Health
Bethesda, Maryland 20892–1434

1. ABSTRACT

A membrane-associated arginine-specific mono-ADP-ribosyltransferase was purified 215,000-fold from rabbit skeletal muscle and its gene was isolated from a skeletal muscle cDNA library. The enzyme was a glycosylphosphatidyl-inositol-linked protein, present on the surface of differentiated skeletal muscle myoblasts (myotubes). Following incubation of cultured, intact myotubes with [adenylate-^{32}P]NAD and analysis by SDS-PAGE, a major radiolabeled protein of 97/140 kDa (reduced/ nonreduced conditions) was observed. It was identified as integrin α7 based on its size, binding to a laminin affinity column, immunoprecipitation with a monoclonal antibody, and partial amino acid sequencing.

Since ADP-ribosylarginine hydrolase, the enzyme responsible for cleavage of the ADP-ribosylarginine bond and a component with the transferase of a putative ADP-ribosylation cycle, is cytosolic, whereas the transferase is attached via a GPI-anchor to the cell surface, the processing of ADP-ribosylated integrin α7 was investigated. ^{32}P label was rapidly removed from [^{32}P]ADP-ribosylated integrin α7, a process inhibited by free ADP-ribose or p-nitrophenylthymidine-5′-monophosphate, alternative substrates for 5′-nucleotide phosphodiesterase. The processed integrin α7 was not susceptible to subsequent ADP-ribosylation, although the amount of surface integrin α7 remained constant. During the processing, no loss of label was observed from integrin α7 radiolabeled with [^{14}C]NAD, containing ^{14}C in the nicotinamide-proximal ribose, consistent with a degradation of the ADP-ribose moiety by a cell surface 5′-nucleotide phosphodiesterase. Thus, cell surface ADP-ribosylation, in contrast to intracellular ADP-ribosylation, is not readily

ADP-Ribosylation in Animal Tissue, edited by Haag and Koch-Nolte
Plenum Press, New York, 1997

reversed by the presently known ADP-ribosylarginine hydrolase and seems to operate outside the postulated ADP-ribosylation cycle.

2. BACKGROUND

Skeletal muscle arginine-specific ADP-ribosyltransferase was the first mammalian transferase to be sequenced and molecularly characterized[1]. Surprisingly, it turned out to be a cell-surface protein, anchored in the membrane via a glycosylphosphatidylinositol tail, present as a covalent modification at the C terminus of the protein[2,3]. Similar structure and extracellular localization were later found for other trans- ferases[3–5].

Integrins are a family of heterodimeric integral membrane proteins that mediate interactions of cells with extracellular matrix and cell-cell communications[6]. Integrin-mediated signaling is involved in regulation of cell growth, shape, migration, differentiation, and death[7,8]. Activation of integrins leads to phosphorylation of a number of cellular proteins, elevation of cytosolic calcium levels, cytoplasmic alkalinization, changes in phospholipid metabolism, and, ultimately, changes in gene expression. Moreover, signaling pathways activated by integrins synergize with other receptor pathways to enhance or restrain signals triggered by each receptor[7,8].

Integrin α7 was the major cell surface substrate for the ADP-ribosyltransferase in cultured intact skeletal muscle cells[2]. Integrin α7, as a dimer with integrin β1, binds to laminin, an extracellular matrix (ECM) protein abundant in muscle basement membrane[9,10], but the signaling pathways triggered by the activation of integrin α7β1 are not known. In adult skeletal muscle, integrin α7 is concentrated in the neuromuscular and myotendinous junctions. It is alternatively spliced in both its cytoplasmic and extracellular domains[11–13]. The localization of integrin α7 and its structural diversity, generated in part by alternative splicing, reflect a rich potential for participating in the transduction of signals from ECM, with ADP-ribosylation being one of the mechanisms regulating these events.

Coexpression of integrin α7 and ADP-ribosyltransferase in skeletal and cardiac muscle, similar regulation of their expression during the process of myogenic differentiation, and selective ADP-ribosylation of integrin α7 in intact cells suggested that this modification may play a significant role in the regulation of muscle cell function by extracellular matrix. Therefore, we characterized the ADP-ribosylation of integrin α7 and cellular processing of the modified protein, and addressed the question of the role of this post-translational modification in integrin function.

3. RESULTS

3.1. Purification, Cloning, and Expression of Arginine-Specific Mono-ADP-Ribosyltransferase

Using sequential chromatographic procedures, a membrane-associated, arginine-specific ADP-ribosyltransferase was purified ~215,000-fold from rabbit skeletal muscle[1]. Following the purification, ADP-ribosyltransferase apparently represented ~90% of the total protein. Its specific activity was 68 μmol·min^{-1}·mg^{-1} when assayed with 2 mM NAD and 20 mM agmatine as substrates[1].

On the basis of the amino acid sequences of HPLC-purified tryptic peptides, degenerate oligonucleotide primers were synthesized and used in a polymerase chain reaction (PCR) to generate a cDNA probe for screening of a rabbit skeletal muscle cDNA library. A composite cDNA sequence, obtained from library screening and rapid amplification of the 5′-end of the cDNA (5′-RACE), contained a 981-base-pair open reading frame, encoding a 36,134-Da protein[1]. The deduced amino acid sequence included sequences of all seven tryptic peptides of the transferase. Expression of the cloned cDNA in mammalian cells[3] produced an active arginine-specific ADP-ribosyltransferase. The deduced amino acid sequence of the transferase contained two potential sites for N-linked glycosylation: Asn[65] and Asn[253]. A transferase-specific oligoprobe, when hybridized with RNA isolated from different rabbit tissues, recognized a ~4-kb mRNA expressed primarily in skeletal and cardiac muscle, but not in smooth muscle, brain, lung, kidney, spleen, or liver[1].

The hydrophilicity plot of the transferase showed very hydrophobic N and C termini, with a hydrophilic center. This pattern is characteristic of glycosylphosphatidylinositol (GPI)-linked proteins, where hydrophobic tails represent signal sequences that are removed during the maturation of GPI-anchored proteins and are not required for enzymatic activity[14,15]. Indeed, a truncated form of the rabbit transferase (amino acids 24–303), expressed in *E.coli,* displayed ADP-ribosyltransferase activity, whereas a full-length clone was not active[1]. Mammalian cells transfected with a truncated transferase cDNA, without a C-terminal signal sequence required for the attachment of a glycosylphosphatidylinositol, expressed a soluble, secreted form of transferase[3]. Similar structure and extracellular localization were later found for other trans- ferases[3–5].

3.2. ADP-Ribosylation in Skeletal Muscle Cells

To identify endogenous substrates of the muscle ADP-ribosyltransferase, cultured mouse skeletal muscle myoblasts (C2C12 or G8 cell lines) were used as a model system, which enabled study of the modification in intact cells. Using agmatine (an arginine analog) as the acceptor of ADP-ribose, it was found that the appearance of transferase activity correlated with cell differentiation and formation of multinucleated myotubes[2]. More than 95% of the transferase activity was associated with the surface of myotubes. Treatment of intact cells with phosphatidylinositol-specific phospholipase C (PI-PLC) from *B. thuringiensis* released ~85% of the transferase activity to the medium, consistent with the presence of a GPI-anchor in the structure of the C2C12 enzyme[2].

After incubation of intact myotubes with [*adenylate*-^{32}P]NAD and separation of cellular proteins by SDS-PAGE under reducing conditions, a major 97-kDa radiolabeled protein was observed[2]. The same protein was labeled when [*adenylate*-^{32}P]NAD was replaced with [*adenine*-U-^{14}C]NAD, but no labeling was observed with [*carbonyl*-^{14}C]NAD or [*adenylate*-^{32}P]ADP-ribose, consistent with an enzymatic addition of ADP-ribose. More than 90% of the radioactivity associated with the 97-kDa protein was released by snake venom phosphodiesterase and identified as 5′-AMP, the expected product of a mono-ADP-ribosylated protein. The chemical stability of the radiolabel associated with the 97-kDa protein was consistent with an arginine-ADP-ribose linkage[2].

Incubation of cell lysate or cell membranes with [*adenylate*-^{32}P]NAD resulted in much less pronounced labeling of the 97-kDa protein and strong labeling of other cellular proteins. Treatment of intact cells with PI-PLC prior to the incubation of intact myotubes with [*adenylate*-^{32}P]NAD prevented subsequent labeling[2], consistent with GPI-anchoring of the enzyme involved in modification of the 97-kDa protein. The appearance of the radiolabeled protein during myogenic differentiation correlated with the expression of ADP-ribosyltransferase activity. Based on these observations, it was concluded that the 97-kDa protein was modified at the surface of intact myotubes by the GPI-anchored, arginine-specific mono-ADP-ribosyltransferase.

3.3. Identification of the Mono-ADP-Ribosylated Protein in Myotubes as Integrin α7

Electrophoretic mobility of the protein radiolabeled in intact myotubes in SDS gel was 97–102 kDa under reducing conditions and 140 kDa under nonreducing conditions. These values were similar to those reported for the mobilities of integrin α7, isolated from skeletal muscle cells on the basis of its binding to laminin, an extracellular matrix protein. Indeed, the ADP-ribosylated protein specifically bound to a laminin affinity column in the presence of divalent cations and was eluted with a buffer containing EDTA (Fig. 1). By N-terminal and partial internal sequencing, the [32P]ADP-ribosylated protein was confirmed to be mouse integrin α7[2]. Accordingly, an antibody against a C-terminal peptide of integrin α7 immunoprecipitated the radiolabeled protein from cell lysate.

It has been reported that integrin α7 forms a heterodimer with integrin β1. Indeed, integrin β1 co-purified with integrin α7 on the laminin column and [32P]ADP-ribosylated integrin α7 co-immunoprecipitated with integrin β1, when using anti-integrin β1 antibody. However, in spite of an apparent close association of the two integrins and the transferase in intact cells, no ADP-ribosylation of integrin β1 was observed.

3.4. Characterization of the ADP-Ribosylation of Integrin α7

Depending on the NAD concentration, two forms of integrin α7, with different number of ADP-ribosylated sites, were resolved by SDS-PAGE[16]. Based on the electrophoretic mobility of the radiolabeled fragments of integrin α7 under reducing and non-reducing conditions, it was concluded that at micromolar NAD, a single site (or sites) located in the 39-kDa segment between amino acids 575 and 886 was the preferred target

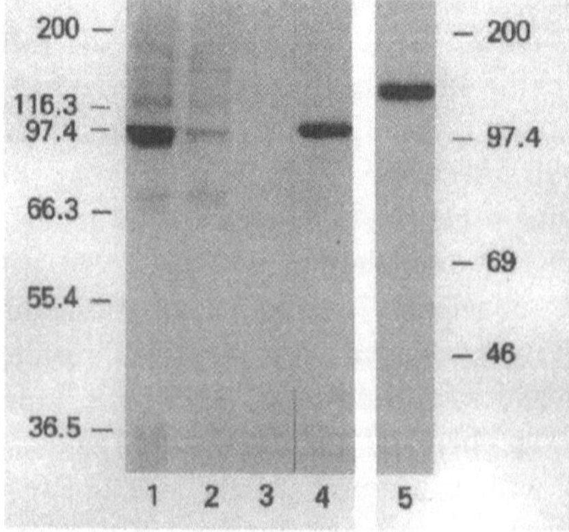

Figure 1. Identification of the 97-kDa protein ADP-ribosylated in intact myotubes as integrin α7. Following labeling of intact C2C12 myotubes with [adenylate-32P]NAD, proteins were extracted with octyl glucoside and purified on a laminin affinity column. Autoradiogram of SDS polyacrylamide gels under reducing (lanes 1–4) and nonreducing conditions (lane 5) of the same fractions of cell extract (lane 1), flow-through (lane 2), last wash (lane 3), and eluate (lanes 4 and 5) is shown. Data are from ref. 2.

for the ADP-ribosyltransferase. At higher NAD concentrations, the 63-kDa N-terminal region of integrin α7, which presumably contained the ligand binding site, was also modified (Fig. 2).

The two ADP-ribosylated forms of integrin α7 were readily extracted from cells with Triton X-100, consistent with their loose interaction with the cytoskeleton. The relative amounts of radioactivity associated with the two species before and after purification on laminin affinity column were similar, confirming that ADP-ribosylation of the 63-kDa N-terminal segment of integrin α7 did not prevent its binding to immobilized laminin. Both ADP-ribosylated forms of α7 protein were immunoprecipitated with an antibody against integrin β1, demonstrating that regardless of different extents of ADP-ribosylation, their association with the β1 subunit was not changed[16].

3.5. Processing of the ADP-Ribosylated Integrin α7 in Myotubes

The extent of intracellular protein ADP-ribosylation has been proposed to be controlled by cytosolic ADP-ribosyltransferases and ADP-ribosylarginine hydrolases. In contrast, cell surface ADP-ribosylation in skeletal muscle myotubes was processed by a 5′-nucleotide phosphodiesterase and seemed to operate outside the postulated ADP-ribosylation cycle[17, 18].

Differentiated C2C12 cells expressed a high level of extracellular phosphodiesterase activity, which was estimated as 1.27 ± 0.11 nmol·min^{-1}·mg protein^{-1}. Extracellular NAD and ADP-ribose, free or linked to integrin α7, were subjected to rapid degradation. Therefore, the presence of ADP-ribose or *p*-nitrophenylthymidine-5′-monophosphate (NP-TMP), alternate substrates for a 5′-nucleotide phosphodiesterase, in addition to [*adenylate*-^{32}P]NAD, was required to observe significant [^{32}P]ADP-ribosylation of integrin α7. ^{32}P label associated with integrin α7 was almost completely lost during one hour after termination of the labeling in the absence of ADP-ribose or NP-TMP, but was well preserved in their presence. When cells were labeled with [^{14}C]NAD, however, with the radiolabel present in the nicotinamide-proximal ribose, it remained fully associated with integrin α7 (Fig. 3). Consistently, the sites in integrin α7, that were previously ADP-ribo-

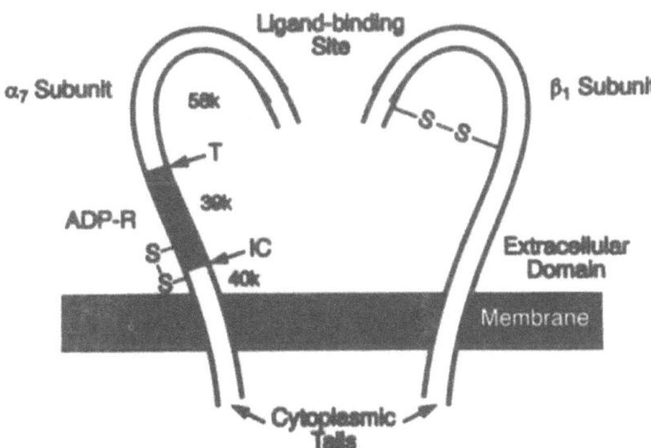

Figure 2. The structure of integrin α7β1 dimer. The site of the cleavage by intracellular proteases (IC) and the preferential site for cleavage by trypsin (TR) in the integrin α7 subunit are shown. The middle 39-kDa domain of integrin α7, containing ADP-ribosylation sites modified at micromolar NAD concentrations, is highlighted.

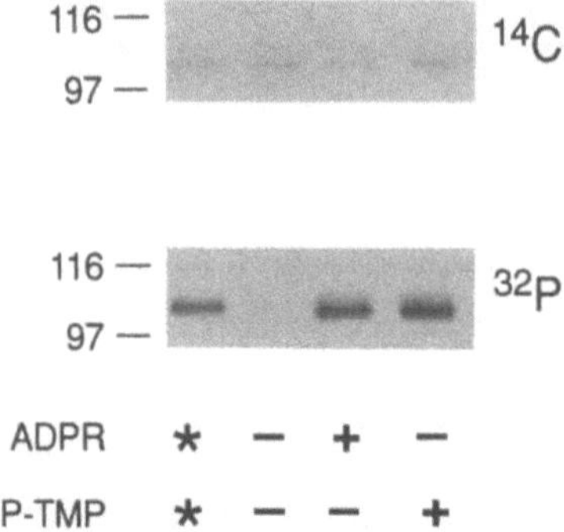

Figure 3. Processing of ADP-ribosylated integrin α7 by a putative cell surface 5′-nucleotide phosphodiesterase. C2C12 myotubes were incubated for 30 min in DPBS with 1 mM ADP-ribose and 16.9 μM [*ribose*-U-^{14}C]NAD or [*adenylate*-^{32}P]NAD. Cellular proteins were analyzed by SDS-PAGE and autoradiography, either at the termination of the labeling (.) or 1 h later, following incubation in DPBS only, or with either 1 mM ADP-ribose or NP-TMP, as indicated. Data are from ref. 16.

sylated and then modified by phosphodiesterase action, were effectively blocked, for at least one hour, to subsequent ADP-ribosylation. During that time, the amount of integrin α7 at cell surface remained constant, as assessed by cell surface biotinylation and immunoprecipitation with integrin-specific antibodies.

4. DISCUSSION

Selective ADP-ribosylation of integrin α7 in intact skeletal muscle myotubes, along with coexpression of integrin α7 and the transferase in skeletal and cardiac muscle, suggests that this modification may play a significant role in the regulation of integrin α7 signaling. It has been demonstrated that human kidney cells and human melanoma cells, which were devoid of endogenous integrin α7 and could not migrate on laminin substrate, gained motility on laminin, when transfected with the α7 gene[19]. Adhesion of mouse C2C12 myoblasts was inhibited 70 % by rat monoclonal antibodies against integrin α7[20]. It was argued, therefore, that integrin α7, together with integrin β1, could be a functional receptor for laminin.

Integrin α7 mRNA is alternatively spliced in the regions corresponding both to its cytoplasmic[11–13] and extracellular domains[11], and the expression of each splice variant is differently regulated during the process of myogenic differentiation. Since the switch in the expression of integrin α7 isoforms correlates with the onset of transferase expression, it is possible that ADP-ribosylation provides a mechanism to regulate some unique functions performed by myotube-specific integrin α7 isoforms. These may include the formation, maintainance or signaling at neuromuscular and myotendinous junctions, to which integrin α7 has been preferentially localized in adult muscle tissue.

The turnover of this extracellular ADP-ribosylated protein seems to be different from that proposed for intracellular ADP-ribosylated proteins. A cell surface 5′-nucleotide phosphodiesterase, and not an ADP-ribosylarginine hydrolase, was involved in the processing of ADP-ribosylated integrin α7. As a result of stepwise degradation of ADP-ribose, integrin α7 was heterogeneously modified, with each form of modification having potentially a different effect on the structure of the protein. It remains to be established what is the exact role of each adduct on the ligand binding and signaling properties of integrin α7.

5. REFERENCES

1. Zolkiewska, A., M. S. Nightingale, and J. and Moss. 1992. Molecular characterization of NAD:arginine ADP-ribosyltransferase from rabbit skeletal muscle. *Proc. Natl. Acad. Sci. USA. 89*: 11352–11356.
2. Zolkiewska, A., and J. Moss. 1993. Integrin α7 as substrate for a glycosylphospha- tidylinositol-anchored ADP-ribosyltransferase on the surface of skeletal muscle cells. *J. Biol. Chem. 268*: 25273–25276.
3. Okazaki, I. J., A. Zolkiewska, M. S. Nightingale, and J. Moss. 1994. Immunological and structural conservation of mammalian skeletal muscle glycosylphosphatidyl- inositol-linked ADP-ribosyltransferases. *Biochemistry. 33*: 12828–12836.
4. Tsuchiya, M., N. Hara, K. Yamada, H. Osago, and M. Shimoyama. 1994. Cloning and expression of cDNA for arginine-specific ADP-ribosyltransferase from chicken bone marrow cells. *J. Biol. Chem. 269*: 27451–27457.
5. T. Davis, and S. Shall. 1995. Sequence of a chicken erythroblast mono(ADP- ribosyl)transferase-encoding gene and its upstream region. *Gene. 164*: 371–372.
6. R. O. Hynes. 1992. Integrins: versatility, modulation, and signaling in cell adhesion. *Cell. 69*: 11–25.
7. E. A. Clark, and J. S. Brugge. 1995. Integrins and signal transduction pathways: the road taken. *Science. 268*: 233–239.
8. Rosales, C., V. O'Brien, L. Kornberg, and R. Juliano. 1995. Signal transduction by cell adhesion receptors. *Biochim. Biophys. Acta. 1242*: 77–98.
9. von der Mark, H., J. Durr, A. Sonnenberg, K. von der Mark, R. Deutzmann, and S. L. Goodman. 1991. Skeletal myoblasts utilize a novel β1-series integrin and not α6β1 for binding to the E8 and T8 fragments of laminin. *J. Biol. Chem. 266*: 23593- 23601.
10. Song, W. K., W. Wang, R. F. Foster, D. A. Bielser, and S. J. Kaufman. 1992. H36-α7 is a novel integrin alpha chain that is developmentally regulated during skeletal myogenesis. *J. Cell Biol. 117*: 643–657.
11. Ziober, B. L., M. P. Vu, N. Waleh, J. Crawford, C.-S. Lin, and R. H. Kramer. 1993. Alternative extracellular and cytoplasmic domains of the integrin α7 subunit are differentially expressed during development. *J. Biol. Chem. 268*: 26773–26783.
12. Collo, G., L. Starr, and V. Quaranta. 1993. A new isoform of the laminin receptor integrin α7β1 is developmentally regulated in skeletal muscle. *J. Biol. Chem. 268*: 19019–19024.
13. Song, W. K., W. Wang, H. Sato, D. A. Bielser, and S. J. Kaufman. 1993. Expression of α7 integrin cytoplasmic domains during skeletal muscle development: alternate forms, conformational change, and homologies with serine/threonine kinases and tyrosine phosphatases. *J. Cell Sci. 106*: 1139–1152.
14. Udenfriend, S., and K. Kodukula. 1995. How glycosylphosphatidylinositol-anchored membrane proteins are made. *Annu. Rev. Biochem. 64*: 563–591.
15. Tartakoff, A. M., and N. Singh. 1992. How to make a glycoinositol phospholipid anchor. *Trends in Biochem. Sci. 17*: 470–473.
16. Zolkiewska, A., and J. Moss. 1995. Processing of ADP-ribosylated integrin α7 in skeletal muscle myotubes. J. Biol. Chem. *270*: 9227–9233.
17. Williamson, K. C., and J. Moss. 1990. Mono-ADP-ribosyltransferases and ADP- ribosylarginine Hydrolases: a Mono-ADP-ribosylation cycle in animal cells. In *ADP- Ribosylating Toxins and G proteins: Insights into Signal Transduction*: 493–510.
18. Moss, J., S. J. Stanley, M. S. Nightingale, J. J. Murtagh, L. Monaco, K. Mishima, H.-C. Chen, K. C. Williamson, and S.-C. Tsai. 1992. Molecular and immunological characterization of ADP-ribosylarginine hydrolases. J. Biol. Chem. 267: 10481–10488.
19. Echtermeyer, F., S. Schöber, E. Pöschl, H. von der Mark, and K. von der Mark. 1996. Specific induction of cell motility on laminin by α7 integrin. *J. Biol. Chem. 271*: 2071–2075.
20. Yao, C.-C., and R. H. Kramer. 1996. The role of α7 integrin in myoblast motility and myogenesis. *J. Dent. Res. 75* (IADR Abstracts): 266.

40

REGULATORY ROLE OF ARGININE-SPECIFIC MONO(ADP-RIBOSYL)TRANSFERASE IN MUSCLE CELLS

Donald J. Graves,[1] Ted W. Huiatt,[2] Hao Zhou,[1] Hui-Yi Huang,[1] Suzanne W. Sernett,[2] Richard M. Robson,[2] and Kathryn K. McMahon[3]

[1]Department of Biochemistry and Biophysics
Iowa State University
Ames, Iowa 50011
[2]Muscle Biology Group
Departments of Animal Science and of Biochemistry and Biophysics
Iowa State University
Ames, Iowa 50011
[3]Department of Pharmacology
Texas Tech University
Health Sciences Center
School of Medicine
Lubbock, Texas 79430

ABSTRACT

Earlier we demonstrated that meta-iodobenzylguanidine (MIBG), a specific inhibitor of arginine mono-ADP-ribosylation blocks proliferation and differentiation of chick skeletal myogenic cells in culture (*Exp. Cell Res.*, 1992, *201*: 33–42). Membrane fractions from 4-day, myotube cultures of embryonic chick muscle cells were incubated with ^{32}P-NAD$^+$. Several proteins were labeled, but labeling of two bands of about 53 and 36 kDa appeared to be due to arginyl ADP-ribosylation. Immunoprecipitation with D3 monoclonal antibody to the intermediate filament protein desmin, SDS-PAGE and autoradiography demonstrated that the 53 kDa band contained desmin, and that this desmin is ADP-ribosylated by the endogenous arginine-specific mono(ADP-ribosyl)transferase (*Exp. Cell Res.*,1996, in press). Desmin is the muscle-specific intermediate filament protein, and it appears to be one of the first muscle-specific proteins expressed during terminal myogenic differentiation. We have examined whether desmin can be ADP-ribosylated in muscle cells by use of polyclonal antibodies for ADP-ribosylated arginyl residues. We have found that soluble desmin is present in 5–6 day myogenic cell cultures and that this desmin contains ADP-ribose, demonstrating that desmin is ADP-ribosylated in skeletal muscle cells. We also

ADP-Ribosylation in Animal Tissue, edited by Haag and Koch-Nolte
Plenum Press, New York, 1997

found that purified avian desmin contains antigenic material that reacts with these antibodies. In both cases, NaCl had no effect on the reactivity, but NH$_2$OH did, which is consistent with an arginine-ADPR linkage. In summary, these results suggest that ADP-ribosylation is an important regulatory mechanism in differentiating muscle cells, and that the intermediate filament protein desmin is an important substrate for modification in muscle cells.

BACKGROUND

The involvement of ADP-ribosylation reactions in myogenic cell differentiation was first suggested by the work of Shall and associates (1). They found that 3-aminobenzamide, an inhibitor of ADP-ribosylation reactions, reversibly blocked the differentiation-specific increase in phosphocreatine kinase activity and fusion of myoblasts in primary cultures of chick myogenic cells. It was suggested that the inhibitor caused these changes by interfering with poly-ADP-ribosylation reactions, but the concentration of inhibitor used could affect both poly- and mono-ADP-ribosylation reactions (2).

Our laboratory reported that enzymes are present in skeletal muscle that catalyze arginine-specific mono-ADP-ribosylation and de-ADP-ribosylation reactions (3, 4). To test the involvement of arginine-specific mono-ADP-ribosylation in skeletal muscle differentiation, we measured arginine-specific ADP-ribosyltransferase and hydrolase activities in primary cultures of embryonic chick myogenic cells (5, 6). Further, we examined the effects on terminal differentiation in these cultures of the compound *meta*-iodobenzylguanidine (MIBG) (6), which is a potent inhibitor of arginine-specific mono-ADP-ribosylation (7, 8). MIBG reversibly blocked both proliferation and differentiation and the increase in arginine-specific ADP-ribosyltransferase associated with terminal myogenic differentiation. These results suggest that mono-ADP-ribosylation of arginyl residues in proteins is important in regulating muscle development. However, what biological signals trigger ADP-ribosylation reactions in skeletal muscle, what proteins are modified, and how does modification change function are all important unanswered questions.

One approach that we have taken to answer some of these questions has been to identify protein substrates in skeletal muscle cells in culture. We focused initially on membrane fractions, as previous studies in our lab (3, 6) and others have indicated that the majority of the muscle ADP-ribosyltransferase is associated with membranes. In addition, we initially used myotube cultures, because transferase activity increases dramatically after myoblast fusion (6). By examination of membrane fractions from myotubes in 96-h cultures of chick myogenic cells, we found that the endogenous arginine-specific ADP-ribosyltransferase modified several proteins, one of which is desmin (9). Desmin is the muscle-specific subunit protein of intermediate (10-nm diameter) filaments. In mature muscle cells, desmin functions to connect myofibrils to each other and to other cellular organelles (see 10, 11 for review). It is one of the first muscle-specific structural proteins synthesized during terminal differentiation of skeletal myoblasts (12), and desmin intermediate filaments undergo rearrangement as part of the late stages of myofibrillogenesis in myotubes after fusion (13, 14). However, the precise role of desmin in myogenesis is not well understood. Earlier, we found that incorporation of one mole of ADP-ribose per mole of 53 kDa desmin subunit by an arginine-specific mono(ADP-ribosyl)transferase was sufficient to block desmin's ability to assemble into 10-nm filaments *in vitro* (15). Incubation of modified desmin with the arginine-specific ADP-ribosyl hydrolase reverses the effect of the modification on filament formation, in keeping with a regulatory mechanism (16).

The question remains, does modification of desmin occur in the intact muscle cell? In this paper, we report that desmin purified from smooth muscle cells, and crude desmin isolated from chick embryonic primary muscle cell cultures react with antibodies to ADP-ribosylated arginine residues. The antibody reactivity is significantly reduced by treatment of the protein with NH_2OH, a reagent known to cleave ADP-ribosylated arginine (17), but not by NaCl. These results suggest that desmin is an important *in vivo* substrate, but the present work does not indicate what triggers modification of desmin nor what the effects of modification are on the structure and function of desmin during muscle cell development.

EXPERIMENTAL PROCEDURES

Myogenic cell cultures were derived from 14-d chick embryos and grown as described (18). For isolation of soluble desmin, 5-d myotube cultures were rinsed three times with PBS containing 2 mM $MgCl_2$ and 1 mM EGTA and then extracted with 100 mM KCl, 2 mM $MgCl_2$, 1 mM EGTA, 10 mM Pipes, pH 6.8, 0.5 % Triton X-100, 1 mM phenylmethylsulfonyl fluoride and 10 mM leupeptin (6 ml per 150 mm plate) for 20 min at 4 °C essentially as described in (19) for extraction of soluble vimentin from cell cultures. The extract was clarified by centrifugation at 160,000 x *g* for 30 min at 4 °C. The supernatant was immunoprecipitated with D3 desmin antibodies (20) by addition of 1/10 volume of 2x concentrated D3 hybridoma (Developmental Studies Hybridoma Bank) cell culture supernatant followed by overnight incubation at 4 °C. Protein A-agarose was added and the mixture was incubated for 3 h at 4 °C followed by centrifugation at 15,000 x *g* for 20 min. The pellet was washed twice with ice-cold extraction buffer, resuspended in 0.2 % SDS, 0.75 mM Tris-HCl, pH 8.8, heated at 100 °C for 10 min, and then centrifuged in a microfuge. Aliquots of the supernatant were treated with NH_2OH or NaCl in the presence of 0.05 M Hepes, pH 7.5, for 12 h at 37 °C. Protein was precipitated with TCA, resuspended in sample buffer, and analyzed by SDS-PAGE (21).

Desmin was purified from turkey gizzard according to (22). Desmin was dialyzed extensively against 10 mM Tris-HCl, pH 8.5, and clarified by centrifugation before use. Aliquots were treated with NH_2OH or NaCl as described above.

Immunoblotting was done by standard procedures (23) using electrophoretic transfer, 1% nonfat dry milk for blocking and antibody dilutions, and an Amersham ECL chemiluminescence kit for detection. The rabbit polyclonal R-28 antibody, which reacts with ADP-ribose groups on proteins, was prepared by using ADP-ribosylated polyarginine as an antigen (24).

RESULTS

Purified desmin has the ability to assemble into 10-nm diameter intermediate filaments in physiological-like solutions (22). As shown in Fig. 1, ADP-ribosylation reversibly blocks the ability of purified desmin to assemble into filaments *in vitro*. After incorporation of two moles of ADP-ribose per mole of 53 kDa desmin, the desmin no longer assembles under typical filament forming conditions (compare Fig 1A and B), whereas removal of the ADP-ribose from the modified desmin by hydrolase treatment restores the ability to form filaments (Fig. 1C).

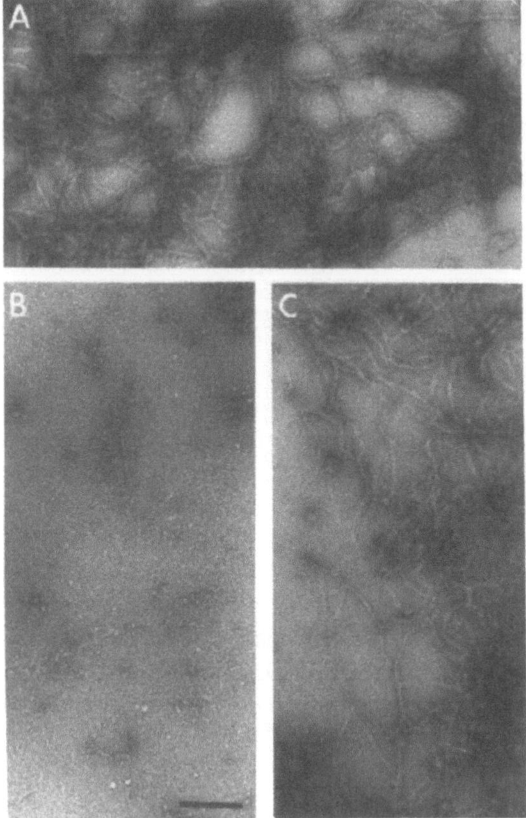

Figure 1. Electron micrographs of purified gizzard desmin showing the effect of ADP-ribosylation on desmin assembly. (A) Untreated desmin control; (B) ADP-ribosylated desmin, containing 2 moles of ^{32}P-ADP-ribose per mole of 53 kDa desmin subunit; (C) ADP-ribosylated desmin after removal of essentially all of the ^{32}P-ADP-ribose by treatment with purified ribosylarginine hydrolase for 6 h at 30°C. For ADP-ribosylation, desmin (0.25 mg/ml) in 10 mM Tris-HCl, pH 8.5, was incubated with recombinant muscle arginine-specific ADP-ribosyltransferase (20μg/ml) and [^{32}P]-NAD$^+$ for 4 h at 30°C, and the amount of incorporation was measured by a filter paper assay essentially as described (16). For assembly, desmin samples were dialyzed overnight against 170 mM NaCl, 0.2 mM MgCl$_2$, 5 mM DTT, 10 mM Tris-HCl, pH 7.5, at 4°C. Samples were negatively stained with 2% aqueous uranyl acetate and examined in a JEOL 100CX-II electron microscope operated at 80 kV. Bar = 200 nm.

Fig. 2 shows the amino terminal sequence of the head domain of desmin, which contains the sites of ADP-ribosylation. Amino acid sequencing of peptides derived from a sample of desmin containing 1.2 moles of ADP-ribose per mole of 53 kDa desmin showed that the protein contained two modification sites (16). The primary site is arginine 48 (0.84 moles of ADP-ribose) and the secondary site is arginine 62 (0.36 moles of ADP-ribose). These results suggest that the primary effect of modification on the physical properties is due to modification of arginine 48.

The purified desmin preparation used for the above ADP-ribosylation experiments was analyzed to determine if it contained any antigenic material that would react with antibodies specific for ADP-ribosylated arginine residues. Unexpectedly, we found that R-28 antibodies against ADP-ribosylated arginine reacted with the purified desmin that had been subjected to SDS-PAGE and blotted onto nitrocellulose membrane (Fig. 3). When

<pre>
1   20
S-Q-S-Y-S-S-S-Q-R-V-S-S-Y-R-R-T-F-G-G-G-

21   40
T-S-P-V-F-P-R-A-S-F-G-S-R-G-S-G-S-S-V-T-

41   *   60
S-R-V-Y-Q-V-S-**R**-T-S-A-V-P-T-L-S-T-F-R-T-

61 *   69
T-**R**-V-T-P-L-R-T-Y--Rod and Tail Domain
</pre>

Figure 2. Location of the arginines in the N-terminal head domain of desmin that are ADP-ribosylated by skeletal muscle arginine-specific mono(ADP-ribosyl)transferase. The sequence of chicken gizzard desmin is from Geisler and Weber (35). Amino acid residues are represented by the single-letter code. The two ADP-ribosylation sites are shown by an asterisk (*) above the arginine residues, which are also indicated in bold letters.

the purified desmin was first treated with neutral NH_2OH or with NaCl, only the NH_2OH treatment removed the antigenic material. These results suggest that desmin isolated from turkey gizzard smooth muscle contains some ADP-ribosylated arginine residues. However, it is not clear why this sample readily formed filaments *in vitro*, because our previous results showed that ADP-ribosylation blocks the ability of desmin to form filaments. Several explanations are possible. One, the site or sites modified in the purified gizzard desmin are different than those described above (Fig. 2), and modification at this other site(s) does not affect filament formation. Two, the same sites are modified, but sub-stoichiometric labeling is present in the desmin sample. In this case, reaction at the above

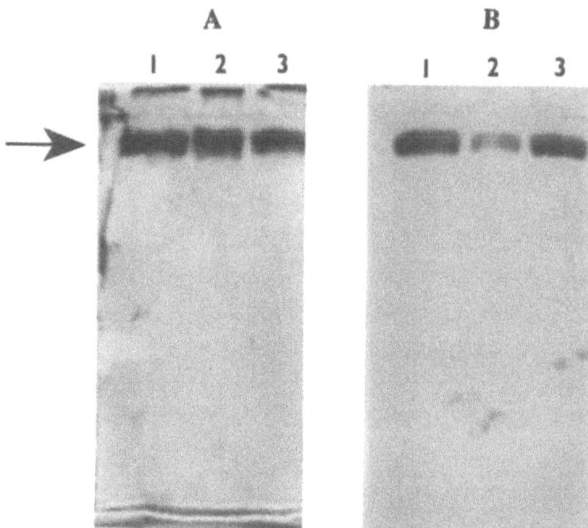

Figure 3. Immunoblot analysis of desmin from adult smooth muscle with antibodies to ADP-ribosylated arginine. Desmin purified from turkey gizzard was resolved by 10% SDS-PAGE under reducing conditions, transferred to a nitrocellulose membrane, and immunoblotted with R-28 polyclonal antibodies specific for ADP-ribosylated arginine. Panel **A** is the Commassie blue stained gel; panel **B** is the chemiluminescence image of the Western blot of a duplicate gel, labeled with R-28 antibody. Lane 1, desmin control; lane 2, desmin after treatment with neutral 0.5 M hydroxylamine at 37 °C for 6 h; and lane 3, desmin after treatment with 0.5 M NaCl at 37°C for 6 h. The migration position of desmin is indicated by the arrow.

sites could occur *in vitro* with the arginine-specific ADP-ribosyltransferase and block filament formation. Three, the bound antigenic material in the blotted desmin is not ADP-ribose, but is derived from it, and this derivative does not block filament formation. Dr. McMahon has shown that cleavage of bound ADP-ribosylarginine derivatives by snake venom phosphodiesterase generates a ribosyl phosphate arginyl derivative that also reacts with the antibodies (24). Work is in progress in our laboratory to distinguish among these possibilities.

We have previously demonstrated that a 53 kDa band was labeled in membrane fractions isolated from chick myotube cultures that had been incubated with ^{32}P-NAD$^+$, either in the presence or absence of added exogenous ADP-ribosyltransferase (9). Furthermore, this band was labeled with desmin-specific antibodies, and ^{32}P-labeled protein could be immunoprecipitated from these labeled membrane preparations with the desmin antibodies. While these results demonstrate that desmin can be ADP-ribosylated in such isolated membrane preparations, they do not show that desmin can be ADP-ribosylated in intact muscle cells. Thus, we isolated a soluble, desmin-containing fraction from 5-d myotube cultures and attempted to determine if this desmin was ADP-ribosylated. Five to six days in culture is approximately the time at which the intermediate filaments in myotubes are rearranged from a pattern parallel to the nascent myofibrils to a transverse pattern in association with myofibrillar Z-lines (13). Because, from our *in vitro* studies with desmin (15), we would expect that ADP-ribosylated desmin would not be able to form filaments, we prepared a soluble, cytoplasmic extract of the cell cultures, immunoprecipitated this extract with the D3 monoclonal desmin antibody, and analyzed the resulting immunoprecipitate by Western blotting with the R-28 antibody to ADP-ribosyl arginine (Fig. 4). The R-28 antibody reacted with a band at about 53 kDa in the immunoprecipitate. The reactivity was removed by treatment with neutral NH$_2$OH, but not with NaCl, in agreement with

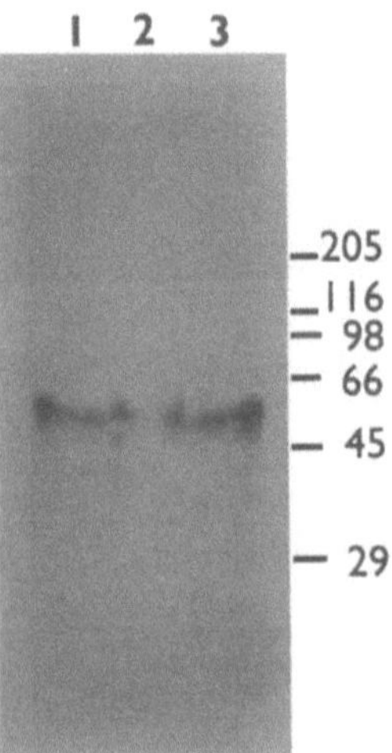

Figure 4. Immunoblot analysis of soluble desmin isolated from chick skeletal muscle myotube cultures with R-28 antibodies, showing the presence of ADP-ribosylated desmin in the cultures. High-salt extracts of 5-d cultures of embryonic chick myogenic cells were immunoprecipitated with D3 monoclonal antibody to chicken desmin. The immunoprecipitate was separated by SDS-PAGE, electroblotted on nitrocellulose, and reacted with R-28 antibodies. Reaction was detected by chemiluminescence. Lane 1, untreated immunoprecipitate; lane 2, immunoprecipitate treated with with neutral 0.5 M hydroxylamine at 37°C for 12 h.; and lane 3, immunoprecipitate after treatment with 0.5 M NaCl at 37°C for 12 h. Positions corresponding to the migration on the gel of standards of the indicated size are shown on the right.

the presence of an ADP-ribosyl-arginine linkage. This result suggests that desmin can be ADP-ribosylated in *intact* skeletal muscle cells.

DISCUSSION

Substantial evidence exists to demonstrate that arginine-specific mono-ADP-ribosylation is an important process in skeletal muscle. The inhibitor of protein arginine-specific mono-ADP-ribosylation, MIBG, blocks proliferation and differentiation of myoblasts, and terminal myoblast differentiation is accompanied by a marked increase in mono(ADP-ribosyl)transferase activity (6). Evidence has been provided that suggests that protein substrates exist both inside and outside the muscle cell, and that ADP-ribosyltransferases responsible for the modification reactions are located in membranous, cytosolic, and nuclear fractions (6). An ADP-ribosylarginine hydrolase is also found in the cystosol, suggesting that reversible ADP-ribosylation reactions could serve an important regulatory function (4). Other processing may also occur and affect function. For example, it has been demonstrated that integrin α7 is a major cell surface substrate for the glycosylphosphatidylinositol-anchored arginine-specific mono(ADP-ribosyl)transferase (25), and that the bound ADP-ribose is not acted on by the hydrolase but by a cell surface phosphodiesterase (26).

We have found that desmin is both an *in vitro* and an *in vivo* subsrtate for the arginine-specific mono(ADP-ribosyl)transferase. The incorporation of one to two moles of ADP ribose/mole of desmin in its head region at arginine 48 and 62 is sufficient to block the formation of intermediate filaments (9, 15, 16). The ADP-ribosylarginine hydrolase removes ADP-ribose and restores desmin's ability to form filaments.

Could ADP-ribosylation regulate organization of cytoskeletal structures in intact muscle cells? There is good evidence for effects of ADP-ribosylation on another cytoskeletal filament system, namely actin filaments, in some cells. *Botulinum* C2 and *Perfringens* iota toxin, for instance, are known to disrupt the cytoskeletal framework of host cells by modifying monomeric actin (27, 28). Analysis of the structure of actin suggests modification by the toxins inhibits polymerization by blocking the nucleotide binding site (29–32). Modification of actin is not restricted to these toxins, as it has been shown that chicken heterophil arginine-specific ADP-ribosyltransferase modifies actin *in vitro* and *in situ* (33, 34). In this latter case different arginyl residues are modified, but these reactions also affect actin filament formation. Arginine 206, which is located in subdomain 4 of actin, is modified and this could block actin-actin interactions necessary for polymerization (31, 32).

Our work presented herein with desmin suggests that ADP-ribosylation of desmin can occur in intact muscle cells. Soluble desmin extracted from 5-day myotubes in primary chick skeletal muscle cell cultures reacted positively with antibodies for ADP-ribosylated arginine. The reactivity with the antibody is reduced significantly by treatment with hydroxylamine but not by sodium chloride, consistent with ADP-ribosylation of one or more an arginine residues. Similar results were found with purified desmin prepared from adult turkey gizzard smooth muscle. However, the precise nature of the antigenically reactive material in the purified desmin has not been identified. It could be ADP-ribose or a metabolite derived from ADP-ribose. Chemical characterization is in progress to define the structure of the antigenically positive material and to define where it is located within the sequence of the turkey gizzard desmin. Additional studies are also in progress to determine the precise physiological function of ADP-ribosylation of desmin in muscle tissue.

ACKNOWLEDGMENTS

This research was supported in part by grants from the Muscular Dystrophy Association and the American Heart Association, Iowa Affiliate, and by USDA NRICGP Award 92–37206–8051. This is Journal Paper NO. J-16885 of the Iowa Agriculture and Home Economics Experiment Station, Ames, IA, Projects 2677, 2127, 3169, and 3349. We thank Jo Philips for expert technical assistance.

REFERENCES

1. Farzaneh, F., R. Zalin, D., Brill, and Shall. 1982.DNA strand breaks and ADP-ribosyltransferase activation during cell differentiation. *Nature 300*: 362–366.

2. Duncan, M. R., P. R. Rankin, R. L. King, M. K. Jacobson, & R. T. Dell'Orco. 1988. Stimulation of mono (ADP-ribosylation) by reduced extracellular calcium levels in human fibroblasts. *J. Cell Physiol. 134*: 161–165.

3. Soman, G., J. R. Mickelson, C. F. Louis, & D. J. Graves. 1984. NAD: guanidino group specific mono ADP-ribosyltransferase activity in skeletal muscle. *Biochem. Biophys. Res. Commun. 120*: 973–980.

4. Chang, Y., G. Soman, & D. J. Graves. 1986. Identification of an enzymatic activity that hydrolyzes protein-bound ADP-ribose in skeletal muscle. *Biochem. Biophys. Res. Commun. 139*: 932–939.

5. Kim, E.-S., & D. J. Graves. 1990. Development of a high-performance liquid chromatography assay method and characterization of adenosine diphosphate-ribosylarginine hydrolase in skeletal muscle. *Anal. Biochem. 187*: 251–257.

6. Kharadia, S. V., T. W. Huiatt, H.-Y. Huang, J. E. Peterson, & D. J. Graves. 1992. Effect of an arginine-specific ADP-ribosyltransferase on differentiation of embryonic chick skeletal muscle cells in culture. *Exp. Cell Res. 201*: 33–42.

7. Loesberg, C., H. Van Rooji, & L. A. Smets. 1990. Meta-iodobenzylguanidine (MIBG), a novel high-affinity substrate for *Cholera toxin* that interferes with cellular mono(ADP-ribosylation). *Biochim. Biophys. Acta. 1037*: 92–99.

8. Smets, L. A., C. Loesberg, M. Janssen, & H. Van Rooij. 1990. Intracellular inhibition of mono(ADP-ribosylation) by meta-iodobenzylguanidine:specificity, intracellular concentration and effects on glucocorticoid-mediated cell lysis. *Biochim. Biophys. Acta 1054*: 49–55.

9. Huang, H.-Y., H. Zhou, T. W. Huiatt, & D. J. Graves. 1996. Target proteins for arginine-specific mono(ADP-ribosyl)transferase in membrane fractions from chick skeletal muscle cells. *Exp. Cell Res.* in press.

10. Robson, R. M. 1989. Intermediate filaments. *Curr. Opin. Cell Biol. 1*: 36–43.

11. Fuchs, E., & K. Weber. 1994. Intermediate filaments: structure, dynamics, function, and disease. *Annu. Rev. Biochem. 63*: 345–382.

12. Lin, Z., M.-H. Lu, T. Schultheiss, J. Choi, S. Holtzer, C. DiLullo, D. A. Fischman, & H. Holtzer. 1994. Sequential appearance of muscle-specific proteins in myoblasts as a function of time after cell division: Evidence for a conserved myoblast differentiation program in skeletal muscle. *Cell Motil. Cytoskel. 29*: 1–19.

13. Bennett, G. S., S. A. Fellini, Y. Toyama, & H. Holtzer. 1979. Redistribution of intermediate filament subunits during skeletal myogenesis and maturation in vitro. *J. Cell Biol. 82*: 577–5

14. Sjoberg, G., W.-Q. Jiang, N. R. Ringertz, U. Lendahl, & T. Sejersen. 1994. Colcalization of nestin and vimentin/desmin in skeletal muscle cells demonstrated by three-dimensional fluorescence digital imaging microscopy. *Exp. Cell Res. 214*: 447–458.

15. Huang, H.-Y., D. J. Graves, R. M. Robson, & T. W. Huiatt. 1993. ADP-ribosylation of the intermediate filament protein-desmin and inhibition of desmin assembly in vitro by muscle ADP-ribosyltransferase. *Biochem. Biophys. Res. Commun. 197*: 570 577.

16. Zhou, H., T. W. Huiatt, R. M. Robson, S. W. Sernett, & D. J. Graves. 1996. Characterization of ADP-ribosylation sites on desmin and restoration of desmin intermediate filament assembly by de-ADP-ribosylation. *Arch. Biochem. Biophys.* submitted for publication.

17. Payne, D. M., E. L. Jacobson, J. Moss, & M. K. Jacobson. 1985. Modification of proteins by mono(ADP-ribosylation) in vivo. *Biochemistry 24*: 7450–7459.

18. Fuller , J. C., Jr., S. L. Nissen, & T. W. Huiatt. 1993. Use of ^{18}O-labeled leucine and phenylalanine to measure protein turnover in muscle cell cultures and possible futile cycling during aminoacylation. *Biochem. J. 294*: 427–433.

19. Isaacs, W. B., R. K. Cook, J. C. Van Atta, C. M. Redmond, & A. B. Fulton. 1989. Assembly of vimentin in cultured cells varies with cell type. *J. Biol. Chem. 264*: 17953–17960.

20. Danto, S. I., & D. A. Fischman. 1984. Immunocytochemical analysis of intermediate filaments in embryonic heart cells with monoclonal antibodies to desmin. *J. Cell Biol. 98*: 2170–2191.

21. Laemmli, U. K. 1970. Cleavage of structural proteins during the assembly of the head of bacteriophage T4.

22. Huiatt, T. W., R. M. Robson, N. Arakawa, & M. H. Stromer. 1980. Desmin from avian smooth muscle. Purification and partial characterization. *J. Biol. Chem. 255*: 6981–6989.

23. Harlow, E., & Lane, D. 1988. *Antibodies, a Laboratory Manual.* Cold Spring Harbor Laboratory. Cold Spring Harbor, New York.

24. McMahon, K. K., C. J. Schwab, K. J. Piron, Colville, & A. T. Fullerton. 1996. Detection of endogenous mono-ADP-ribosylated proteins via anti-ADP-ribosylarginine immunoreactivity. Manuscript in preparation.

25. Zolkiewska, A., & J. Moss. 1993. Integrin α7 as substrate for a glycosylphosphatidylinositol-anchored ADP-ribosyltransferase on the surface of skeletal muscle cells. *J. Biol. Chem. 268*: 25273–25276.

26. Zolkiewska, A., & J. Moss. 1995. Processing of ADP-ribosylated integrin α7 in skeletal muscle myotubes. *J. Biol. Chem. 270*: 9227–9233.

27. Reimer, K. H., P. Presek, B. Boschek, & K. Aktories. 1987. Botulinum C2 toxin ADP-ribosylates actin and disorganizes the microfilament network in intact cells. *Eur. J. Cell Biol. 43*: 134–140.

28. Wiegers, W., I. Just, H. Muller, A. Hellwig, P. Traub, & K. Aktories. 1991. Alteration of cytoskeleton of mammalian cells cultured in vitro by Clostridium botulinum C2 toxin and C3 ADP-ribosyltransferase. *Eur. J. Cell Biol. 54*: 237–245.

29. Vandekerkhove, J., B. Schering, M. Barmann, & K. Aktories. 1987. Clostridium perfringens iota toxin ADP-ribosylates skeletal muscle actin in arginine 177. *FEBS Lett. 225*: 48–52.

30. Vandekerkhove, J., B. Schering, M. Barmann, & K. Aktories. 1988. Botulinum C2 toxin ADP-ribosylates cytoplasmic β/γ-actin in arginine 177. *J. Biol. Chem. 263*: 696–700.

31. Kabsch, W., H. G. Mannherz, D. Suck, E. F. Pai, & K. C. Holmes. 1990. Atomic structure of the actin:DNase I complex. *Nature 347*: 37–44.

32. Holmes, K. C., D. Popp, W. Gebhard and W. Kabsch. 1990. Atomic model of the actin filament. *Nature 347*: 44–49.

33. Terashima, M., K. Mishima, K. Yamada, M. Tsuchiya, T. Wakutani, & M. Shimoyama. 1992. ADP-ribosylation of actins by arginine-specific ADP-ribosyltransferase purified from chicken heterophils. *Eur. J. Biochem. 204*: 305–311.

34. Terashima, M., C. Yamamori, & M. Shimoyama. 1995. ADP-ribosylation of Arg 28 and Arg 206 on the actin molecule by chicken ADP-ribosyltransferase. *Eur. J. Biochem. 231*: 242–249.

35. Geisler, N., & K. Weber. 1982. The amino acid sequence of chicken muscle desmin provides a common structural model for intermediate filament proteins. *EMBO J. 1*:1649–1656.

41

ACTIVATION OF TOXIN ADP-RIBOSYLTRANSFERASES BY THE FAMILY OF ADP-RIBOSYLATION FACTORS

Martha Vaughan and Joel Moss

Pulmonary-Critical Care Medicine Branch
National Heart, Lung, and Blood Institute
National Institutes of Health, Building 10, Room 5N307
Bethesda, Maryland

ABSTRACT

ADP-ribosylation factors or ARFs are 20-kDa guanine nucleotide-binding proteins, initially identified as stimulators of cholera toxin-catalyzed ADP-ribosylation of Gsα. We now know that ARFs play a critical role in many vesicular trafficking events and ARF activation of a membrane-associated phospholipase D (PLD) has been recognized. ARF is active and associates with membranes when GTP is bound. The active state is terminated by hydrolysis of bound GTP, producing inactive ARF•GDP. The nucleotide effect on ARF association with membranes is related to alteration in orientation of the N-terminal myristoyl moiety that is important for ARF function. Cycling of ARF between active and inactive states involves guanine nucleotide-exchange proteins (GEPs) that accelerate replacement of bound GDP with GTP and GTPase-activating proteins (GAPS) that are responsible for ARF inactivation.

Six mammalian ARFs have been identified by cDNA cloning. Class I ARFs 1 and 3 have been studied most extensively. Their activation (GTP binding) is catalyzed by a GEP now purified from spleen cytosol. In crude preparations, GEP was inhibited by brefeldin A (BFA), which in cells causes apparent disintegration of Golgi. Demonstration that the ~60 kDa purified GEP was not inhibited by BFA means that contrary to earlier belief, there must be another protein to mediate BFA inhibition. GEP activity was greatly enhanced by phosphatidyl serine. The purified GEP, equally active with ARFs 1 and 3, was inactive with ARFs 5 and 6 (Classes II and III); myristoylated ARFs were better substrates than were their non-myristoylated counterparts.

ARF GAP purified from bovine spleen cytosol in our laboratory had much broader substrate specificity than the GEP. It used both ARFs 5 and 6 at least as well as ARFs 1 and 3; myristoylation was without effect. It also accelerated GTP hydrolysis by certain ARF mutants and an ARF-like protein (ARL1) that does not have ARF activity. The puri-

ADP-Ribosylation in Animal Tissue, edited by Haag and Koch-Nolte
Plenum Press, New York, 1997

fied GAP also differed from the GEP in its rather specific requirement for phosphatidyli-nositol bisphosphate. This was also observed with a seemingly different ARF GAP that was purified and subsequently cloned in Cassel's laboratory. Activation and inactivation of ARFs present many potential sites for physiological regulation and, therefore, for pathological disruption of ARF function.

INTRODUCTION

ARF was first described and purified as a 20-kDa membrane-associated protein that was required for the cholera toxin-catalyzed ADP-ribosylation of $G_{s\alpha}$, the stimulatory GTP-binding protein of adenylyl cyclase (1). ARF proteins were later purified from sol-uble cell fractions and are now known to be present in all eukaryotic cells, where they have a vital role in many pathways of vesicular transport.

ARF ACTIVATION OF CHOLERA TOXIN

It was established by 1988 that ARF•GTP interacts directly with the catalytic frag-ment of cholera toxin to enhance its activity in all reactions, i.e., NAD glycohydrolase, ADP-ribosylation of arginine in proteins other than $G_{s\alpha}$ as well as auto-ADP-ribosylation of the toxin itself, and ADP-ribosylation of free arginine or simple guanidino compounds, e.g., agmatine (2). ARF with GTP bound appears to serve as an allosteric activator of the toxin, lowering the apparent K_m for both NAD, the ADP-ribose donor, and agmatine, which was used as ADP-ribose acceptor in the kinetic experiments (3).

ARF also activates the *E. coli* heat-labile toxins (LT) that resemble cholera toxin closely in structure and activity. A mutant *E. coli* toxin that lacks ADP-ribosyltransferase activity was used to determine whether the enzymatically active form of the toxin is re-quired for interaction with ARF (4). Both cholera toxin and LT are active only after the catalytic A subunit is proteolytically cleaved at a specific site so that subsequent reduction of a disulfide bond linking the two fragments releases an active A1 protein. It was shown that the mutant LT could inhibit ARF activation of cholera toxin only after proteolysis and reduction to produce the molecule that, if it were lacking the single amino acid replace-ment, would have been catalytically active. These studies provided evidence that the al-teration in toxin structure required to generate a functional ADP-ribosyltransferase is also necessary to generate the ARF interaction site (4). In addition, they demonstrated that the mutation abolishing catalytic activity does not interfere with that site.

ARF STRUCTURE

Six mammalian ARFs have been identified by cDNA cloning (5). They fall into three classes based on size, amino acid sequence, gene structure, and phylogenetic analysis. All contain several short amino acid sequences that are involved in GTP binding and hydrolysis. These are identical in all mammalian ARFs and have only a few conservative replacements in other ARFs (5). Several authors have noted that in the myristoylation of an N-terminal gly-cine, as well as in these "signature" sequences, ARFs are more similar to the α-subunits of heterotrimeric G proteins than they are to the many ~20-kDa Ras-like proteins. As ARF has been found in the primitive parasite *Giardia lamblia*, which reportedly lacks Gα subunits, it has been suggested that ARF may be an ancestor of both of these families of GTPases (6).

Class I ARFs are seemingly the most abundant of the mammalian ARFs and most resemble the ARFs initially described in other species. More recently, however, evidence for the presence of Class II and III ARFs in *Drosophila melanogaster* (7) and a class III ARF in *Saccharomyces cerevisiae* has appeared (8). Elucidation of the function of Class I ARFs, especially ARFs 1 and 3 in vesicular transport from endoplasmic reticulum to Golgi and between Golgi cisternae, is proceeding rapidly. ARF6 (Class III) has been suggested for an analogous role in the endocytic pathway (9). At present, no clues to the function(s) of ARFs 4 and 5 (Class II) have emerged, although it seems most likely that they play similar parts in vesicular trafficking at other specific sites in the cell. ARF was first implicated in vesicular transport in the Golgi as a result of the discovery that deletion of the two *S. cerevisiae* ARF genes known at that time was lethal (10). Some spontaneous revertants contained a functioning *ARF1* gene that resulted from recombination with the *ARF 2* gene. The ability to complement the lethal double ARF mutant in yeast has since become a second criterion (activation of cholera toxin ADP-ribosyltransferase is the first) for designation as an ARF.

ACTIVATION OF ARF

ARF proteins move between cytosolic and membrane compartments depending on whether GTP or GDP is bound. With GDP bound, ARF is inactive and is found in the cytosol. After the exchange of GDP for GTP, ARF•GTP becomes membrane-bound, or in a model system, becomes associated with phospholipid. In this cycle of activation and inactivation, ARF resembles numerous other GTP-binding proteins, or GTPases, that function in so many cellular regulatory and signalling pathways (11).

The GTPase activity is critical in these cycles because, in addition to serving the role of "timer" in turning off the activity of ARF (or other GTP-binding protein), it provides a directional or vectorial component to the pathway in which it operates (12). As the ARF proteins lack detectable GTPase activity it was necessary to postulate the existence of a GTPase-activating protein or GAP that would stimulate the hydrolysis of GTP bound to ARF. Several GAPs specific for other GTP-binding proteins, e.g., Ras and Rho, have been identified and characterized. In addition, it has been shown that hydrolysis of bound GTP by some GTP-binding proteins is accelerated when the protein interacts with its effector (i.e., the effector serves also as a GAP), thus linking the lifetime of the activated GTPase protein to the performance of its function (12). ARF GAP has been isolated and cloned only relatively recently (13, 14).

Reactivation of ARF•GDP requires the release of GDP, followed by binding of GTP, which is present in cells at a higher concentration than GDP. The spontaneous release of GDP from ARF is very low at normal cellular concentrations of Mg^{2+} and guanine nucleotide-exchange proteins or GEPs that accelerate GDP release have been found in both membrane and cytosolic fractions of cells. ARF GEPs have been purified from rat brain and spleen cytosol. Although the former behaved initially as a >600-kDa protein with activity that was inhibited by brefeldin A, the purified GEP appeared to be a ~60-kDa monomeric protein and was insensitive to brefeldin A (15). The GEP in rat spleen in cytosol was not associated with a large protein complex and was not inhibited by brefeldin A (16). It was extensively purified as a protein of ~55 kDa, with activity dependent on Mg^{2+} and dramatically stimulated by phosphatidylinositol or phosphatidylserine. This GEP activated ARFs 1 and 3 (Class I), as did the GEP purified from rat brain. It did not activate ARF proteins lacking N-terminal myristoylation, nor ARFs from Class II or III (16).

ARF FUNCTION IN CELLS

For several years there were no clues to a physiological function for ARF, despite a growing recognition of the ubiquitous distribution of these proteins and the remarkable degree of conservation of structure, from which a critical role in the cell might be inferred. There is still no convincing evidence that ARF has a role in endogenous ADP-ribosylation in eukaryotic cells, nor have ARF proteins been implicated in the pathological action of cholera or related toxins. Although their interaction has not been ruled out, subcellular localization of toxins on the one hand, and of ARFs on the other, may make that unlikely. It was hoped, nevertheless, that with understanding of the mechanism by which toxin activation occurred, insight might be gained into the mechanism(s) by which ARF acts in cells. This has not, thus far, been the case.

In fact, relatively recently it has been established that the capacity to activate cholera toxin and at least one, presumably physiological, ARF function, i.e., activation of phospholipase D (17–19), are properties of different parts of the molecule (20). For these studies, advantage was taken of the existence of a family of ARF-like proteins or ARLs first identified as a result of attempts to clone an ARF cDNA from *Drosophila melanogaster* (21). ARL cDNAs have now been cloned from several species and share with ARFs 30–60% identity of deduced amino acid sequences. Recombinant ARLs expressed in *E. coli* do not, however, activate cholera toxin or rescue the yeast *arf1⁻ arf2⁻* lethal double mutant. They do bind and hydrolyze GTP. To identify the sequence(s) critical for cholera toxin activation, chimeric proteins were prepared in which the amino-terminal 73 amino acids of human (h) ARF1 and hARL1 were exchanged. Deduced sequences (each 181 amino acids) of hARF1 and hARL1 are 57% identical (68% identical in the first 73 positions). Recombinant (r)ARF1, rF73/L (N-terminal 73 residues of hARF1 and C-terminal 108 residues of hARL1), and rL73/F (first 73 residues of hARL1 plus last 108 residues of hARF1) were synthesized in *E. coli* as non-myristoylated proteins and purified (20).

Activation of cholera toxin and of a partially purified ARF-dependent phospholipase D from rat brain by the recombinant proteins was compared, in assay mixtures of identical composition with the exception of the addition of [adenine-U-^{14}C]NAD to toxin assays or dipalmitoylphosphatidyl-[^{3}H]choline as phospholipase D substrate, in some experiments. The same conclusions were reached using this and standard assays for each of the enzymes. Comparison of different assay conditions is important because ARF activities (including GTP binding) are dramatically influenced by specific phospholipids and detergents. Cholera toxin was activated, in a concentration-dependent fashion by rARF1 and rL73F, but not by rF73L, which also failed to activate the toxin in the presence of other phospholipids or with higher concentrations of GTPγS. In contrast, rF73L activated phospholipase D to essentially the same maximal extent as did rARF1. It was, however, clearly less potent. This may mean that not all of the sequence required for optimal phospholipase D activation is present in the first 73 amino acids of ARF1. Alternatively, the chimera may associate with phospholipid less well than does rARF1 (20).

Although it seems clear that an ARF domain important for phospholipase D activation resides in the N-terminal 73 amino acids, on-going studies are aimed at more precise localization. This part of the molecule contains sequence that corresponds to the Ras so-called effector loop and is 84% identical in hARF1 and hARL1 (amino acids 37 to 55).

This relatively high degree of identity may account for the observation that rL73F caused limited, but significant, GTP-dependent activation phospholipase D (20). Overall, the direct demonstration that different domains of the ARF molecule are involved in the activation of cholera toxin and phospholipase D, invites speculation that ARF may have two different actions within the cell. Studies by a number of investigators that cannot be discussed here have described a critical role for ARF in the formation of membrane transport vesicles. Rothman and his coworkers have been major contributors, both experimentally and conceptually, to a current understanding of intracellular vesicular transport, which is summarized in a recent review (22).

Initiation of vesicle formation from endoplasmic reticulum or Golgi requires first binding of ARF•GTP (probably a Class I ARF), followed by binding of a large molecular complex of seven proteins termed "coatomer" (23). With these two cytosolic components bound, a membrane bud begins to form. A vesicle is finally completed and released by fusion of the membrane at the base of the bud, which requires only the addition of fatty acyl coenzyme A (22). It is not known whether this action of ARF in vesicle budding involves its function as an activator of phospholipase D, or reflects a second activity that remains to be defined mechanistically. There is evidence from cross-linking experiments of an interaction of ARF and the β-COP protein in coatomer (22).

Although activation of phospholipase D could occur at the site of vesicle formation, it has seemed somewhat easier, with the limited information available, to picture it occurring as a part of the interaction that leads to vesicle fusion with the target membrane. The action of phospholipase D, which catalyzes the production of phosphatidic acid (and water-soluble choline) from phosphatidylcholine, could lead to production of second messengers or activators of specific enzymes. Activation of phosphatidylinositol 4'-phosphate 5'-kinase by phosphatidic acid could increase amounts of phosphatidylinositol 4', 5'-bisphosphate which activates phospholipase D and also activates ARF GAP (13), resulting in formation and release of ARF•GDP, followed by coatomer, leaving an un-coated vesicle presumably ready for fusion. That last step might, in addition, be promoted by so-called fusogenic lipids such as phosphatidic acid and diglyceride that can be formed from it by the action of phosphatidic acid phosphatase. These components, both proteins and phospholipids, could be part of a positive feed-back loop at the target membrane, as suggested by Liscovitch et al. (24).

CONCLUSIONS

Information relevant to the role of ARF in vesicular transport and as an activator of phospholipase D has been accumulating relatively rapidly in recent years. Nevertheless, much remains to be learned about the actions of ARF proteins and their relations to ARF structure. Perhaps one of the most interesting questions is why they are activators of toxin ADP-ribosyltransferases.

REFERENCES

1. Kahn, R. A., & A. G. Gilman. 1984. Purification of a protein cofactor required for ADP-ribosylation of the stimulatory regulatory component of adenylate cyclase by cholera toxin. *J. Biol. Chem. 259*: 6228–6234.
2. Tsai, S.-C., M. Noda, R. Adamik, P. P. Chang, H.-C. Chen, J. Moss, & M. Vaughan. 1988. Stimulation of choleragen enzymatic activities by GTP and two soluble proteins purified from bovine brain. *J. Biol. Chem. 263*: 1768–1772.

3. Noda, M., S.-C. Tsai, R. Adamik, J. Moss, & M. Vaughan. 1990. Mechanism of cholera toxin activation by a guanine nucleotide-dependent 19-kDa protein. *Biochim. Biophys. Acta. 1034*: 195–199.

4. Moss, J., S. J. Stanley, M. Vaughan, & T. Tsuji. 1993. Interaction of ADP- ribosylation factor with *Escherichia coli* enterotoxin that contains an inactivating lysine 112 substitution. *J. Biol. Chem. 268*: 6383–6387.

5. Moss, J., & M. Vaughan. 1993. Mini-review: ADP-ribosylation factors, 20,000 M_r guanine nucleotide-binding protein activators of cholera toxin and components of intracellular vesicular transport systems. In: *Cellular Signalling*. Pergamon Press Ltd., Great Britain.

6. Murtagh, J. J., Jr., M. R. Mowatt, C.-M. Lee, F.-J. S. Lee, K. Mishima, T. E. Nash, J. Moss, & M. Vaughan. 1992. Guanine nucleotide-binding proteins in the intestinal parasite *Giardia lamblia. J. Biol. Chem. 267*: 9654–9662.

7. Lee, F.-J. S., L. A. Stevens, L. A. Hall, J. J. Murtagh, Jr., Y. L. Kao, J. Moss, & M. Vaughan. 1994. Characterization of class II and class III ADP-ribosylation factor genes and proteins in *Drosophila melanogaster. J. Biol. Chem. 269*: 21555–21560.

8. Lee, F.-J., L. A. Stevens, Y. L. Kao, J. Moss, & M. Vaughan. 1994. Characterization of a glucose-repressible ADP-ribosylation factor 3 (ARF3) from *Saccharomyces cerevisiae. J. Biol. Chem. 269*: 20931–20937.

9. D'Souza-Schorey, C., G. Li, M. I. Colombo, & P. D. Stahl. 1995. A regulatory role for ARF6 in receptor-mediated endocytosis. *Science 267*: 1175–1178.

10. Botstein, D., N. Segev, T. Stearns, M. A. Hoyt, J. Holden, & R. A. Kahn. 1988. Diverse biological functions of small GTP-binding proteins in yeast. *Cold Spring Harbor Symposia on Quantitative Biology, Volume LIII*: 629–636.

11. Moss, J., & M. Vaughan. 1995. Mini-review: Structure and function of ARF proteins: activators of cholera toxin and critical components of intracellular vesicular transport processes. *J. Biol. Chem. 270*: 12327–12330.

12. Bourne, H. R., D. A. Sanders, & F. McCormick. 1991. The GTPase superfamily: conserved structure and molecular mechanism. *Nature 349*: 117–127.

13. Makler, V., E. Cukierman, M. Rotman, A. Admon, & D. Cassel. 1995. ADP-ribosylation factor-directed GTPase-activating protein. *J. Biol. Chem. 270*: 5232–5237.

14. Cukierman, E., I. Humer, M. Rotman, D. Cassel. 1995. The ARF1 GTPase-activating protein: zinc finger motif and Golgi complex localization. *Science 270*: 1999–2002.

15. Tsai, S.-C., R. Adamik, J. Moss, & M. Vaughan. 1994. Identification of a brefeldin A-insensitive guanine nucleotide-exchange protein for ADP-ribosylation factor in bovine brain. *Proc. Natl. Acad. Sci. USA 91*: 3063–3066.

16. Tsai, S.-C., R. Adamik, J. Moss, & M. Vaughan. 1996. Purification and characterization of a guanine nucleotide-exchange protein for ADP-ribosylation factor from spleen cytosol. *Proc. Natl. Acad. Sci. USA 93*: 305–309.

17. Brown, H. A., S. Gutowski, C. R. Moomaw, C. Slaughter, & P. Sternweis. 1993. ADP-ribosylation factor, a small GTP-dependent regulatory protein, stimulates phospholipase D activity. *Cell 75*: 1137–1144.

18. Cockcroft, S., G. M. H. Thomas, A. Fensome, B. Geny, E. Cunningham, I. Gout, I. Hiles, N. F. Totty, O. Truong, & J. J. Hsuan. 1994. Phospholipase D: a downstream effector of ARF in granulocytes. *Science 263*: 523–526.

19. Massenburg, D., J.-S. Han, M. Liyanage, W. A. Patton, S. G. Rhee, J. Moss, & M. Vaughan. 1994. Activation of rat brain phospholipase D by ADP-ribosylation factors 1, 5, and 6: separation of ADP-ribosylation factor-dependent and oleate-dependent enzymes. *Proc. Natl. Acad. Sci. USA 91*: 11718–11722.

20. Zhang, G.-F., W. A. Patton, F.-J. S. Lee, M. Liyanage, J.-S. Han, S. G. Rhee, J. Moss, & M. Vaughan. 1995. Different ARF domains are required for the activation of cholera toxin and phospholipase D. *J. Biol. Chem. 270*: 21–24.

21. Tamkun, J. W., R. A. Kahn, M. Kissinger, B. J. Brizuela, C. Rulka, M. P. Scott, & J. A. Kennison. 1991. The arflike gene encodes an essential GTP-binding protein in *Drosophila. Proc. Natl. Acad. Sci. USA 88*: 3120–3124.

22. Rothman, J. E., & F. T. Wieland. 1996. Protein sorting by transport vesicles. *Science 272*: 227–234.

23. Waters, M. G., T. Serafini, & J. E. Rothman. 1991. 'Coatomer': a cytosolic protein complex containing subunits of non-clathrin-coated Golgi transport vesicles. *Nature 349*: 248–251.

24. Liscovitch, M., V. Chalifa, P. Pertile, C.-S. Chen, & L. C. Cantley. 1994. Novel function of phosphatidylinositol 4,5-bisphosphate as a cofactor for brain membrane phospholipase D. *J. Biol. Chem. 269*: 21403–21406.

42

POSSIBLE ROLE OF BARS-50, A SUBSTRATE OF BREFELDIN A-DEPENDENT MONO-ADP-RIBOSYLATION, IN INTRACELLULAR TRANSPORT

Maria Giuseppina Silletta, Maria Di Girolamo, Giusy Fiucci,
Roberto Weigert, Alexander Mironov, Maria Antonietta De Matteis,
Alberto Luini, and Daniela Corda

Department of Cell Biology and Oncology
Istituto di Ricerche Farmacologiche "Mario Negri"
Consorzio Mario Negri Sud, Via Nazionale
66030 Santa Maria Imbaro (Chieti), Italy

1. ABSTRACT

Brefeldin A (BFA), a fungal metabolite that inhibits membrane transport, potently stimulates an endogenous ADP-ribosylation reaction that selectively modifies two cytosolic proteins of 38 and 50 kDa on an amino acid residue different from those used by all known mADPRTs. The 38-kDa substrate was identified as the glycolytic enzyme glyceraldehyde-3-phosphate dehydrogenase (GAPDH), whereas the 50-kDa substrate (BARS-50) was characterized as a novel guanine nucleotide binding protein. Thus, BARS-50 is able to bind GTP and its ADP-ribosylation is inhibited by the $\beta\gamma$ subunit of GTP-binding (G) proteins. Moreover, BARS-50 was demonstrated to be a group of closely related proteins that appear to be different from all the known G proteins. A partially purified BARS-50 was obtained from rat brain cytosol, which was then used for microsequencing and in functional studies. A similar procedure led to the purification of native (non-ADP-ribosylated) BARS-50.

The possible role of the BFA-dependent ADP-ribosylation and of BARS-50 in the maintenance of Golgi structure and function was addressed by examining which of the effects of BFA may be modified by inhibiting this reaction. We find that the BFA-dependent transformation of the Golgi stacks into a tubular reticular network is prevented when the BFA-dependent ADP-ribosylation activity was blocked by specific inhibitors thus indicating that BFA-dependent ADP-ribosylation of cytosolic proteins participate in the dynamic regulation of intracellular transport.

ADP-Ribosylation in Animal Tissue, edited by Haag and Koch-Nolte
Plenum Press, New York, 1997

2. INTRODUCTION

The fungal toxin brefeldin A (BFA) is a well known inhibitor of vesicular transport[1]. Its effects include the inhibition of constitutive secretion, the disappearance of non-clathrin-coated vesicles, the induction of tubules formation by which the Golgi complex redistributes in the endoplasmatic reticulum[1,2]. These effects are in part mediated by the ability of BFA to inhibit an ARF-specific GTP/GDP membrane-bound exchange factor and thus to dissociate the small GTP-binding protein ARF and coatomer from the Golgi membrane[3]. It is unlikely however, that this mechanism accounts for all the complex effects of BFA, thus other putative mechanisms of action of the toxin have been investigated[4–6].

We have recently reported that BFA is able to activate an endogenous mADPRT that might be an important element of the toxin mechanism of action[4,5]. The BFA-dependent mADPRT (BART) has two specific substrates of 38 and 50 kDa respectively, and, since it acts on a amino acid residue different from those (arginine, cysteine, asparagine, diphthamide) used by all the known mADPRTs, it is likely to be a member of a novel class of enzymes[4,7]. Both effects of BFA on ARF binding and on BART are mediated by a BFA-binding site that has identical ligand selectivity[4,5]. Moreover, BFA analogs that do not alter the organization of the Golgi complex are also unable to stimulate BART[5]. These observations strongly suggest that BART and its specific substrates are involved in the mechanism of action of BFA. Thus, it is relevant to identify the substrates and better understand the role of the ADP-ribosylation reaction in the control of Golgi dynamics.

We have previously identified the 38 kDa substrate as one of the isoforms of the glycolytic enzyme glyceraldehyde-3-phosphate dehydrogenase (GAPDH)[4,5]. This is a multifunctional protein that has been implicated in several diverse cellular activities, including the binding to small GTP binding proteins and the induction of membrane fusion[4,8,9]. These activities would be consistent with a role of GAPDH in vesicular transport, but no evidence to this point is yet available.

As discussed in the following, the 50-kDa substrate (BARS-50) has recently been characterized by us as a novel GTP-binding protein possibly involved in the action of BFA and thus relevant in the control of the organization of the Golgi complex[6]. This finding is in line with the evidence that heterotrimeric G proteins might take part in the modulation of coat dynamics and vesicular transport [10,11].

3. RESULTS AND DISCUSSION

3.1. Characteristics of the Mono-ADP-Ribosylation of BARS-50

The BFA-dependent ADP-ribosylation of two specific substrates, GAPDH and BARS-50, in postnuclear of FRTL5 thyroid cells is shown in Fig. 1A.

This reaction is dose- and time-dependent (Fig. 2)[6]. In line with the data previously reported for GAPDH in RBL cell postnuclears, the lowest effective concentration of BFA was 2 µg/ml and the BFA EC50 was 15 µg/ml[4–6] .The majority if not all of BARS-50 can be ADP-ribosylated under saturating conditions[6]. This contrasts strikingly with the case of the other substrate GAPDH, of which only 1–3% can be modified by the BFA-dependent mono-ADP-ribosylation under identical experimental conditions. Unless this fraction of GAPDH represents an as yet unidentified isoform of the protein, these data suggest that GAPDH is a less efficient substrate of the BFA-dependent ADP-ribosylation reaction, as

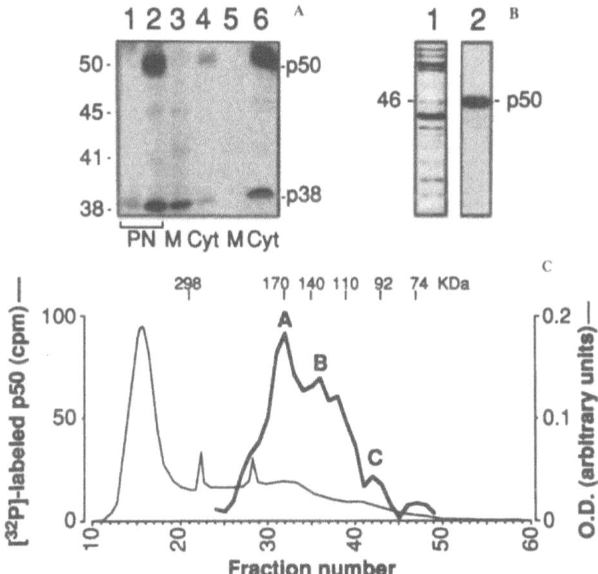

Figure 1. (A) Stimulation of the mono-ADP-ribosylation of GAPDH (p38) and BARS-50 (p50) induced by BFA in FRTL5 cells. Post-nuclear preparations (PNS) (lanes 1 and 2), membranes (lane 3), and cytosol (lane 4) were ADP-ribosylated with [32P]-NAD in the absence (lane 1) or presence (lane 2–4) of BFA. Membranes (lane 5) and cytosol (lane 6) obtained by centrifugation of PNS after the ADP-ribosylation reaction, show that both proteins are associated with the cytosol[6]. (B) Silver staining detection of partially purified BARS-50. ADP-ribosylated cytosol from rat brain was precipitated with ammonium sulfate, eluted sequentially by hydrofobic and hydrossiapatite columns and separated by SDS-PAGE[6]. BARS-50 protein was identified by both silver-staining (lane 1) and by autoradiography (lane 2). (C) Elution pattern of ADP-ribosylation BARS-50 by size-exclusion chromatography. Cytosol from rat brain was ADP-ribosylated and was applied at a flow rate of 0.4 ml/min to a Superose 12 HR 10/30 (Pharmacia) equilibrated with 25 mM Tris (pH 8), 150 mM KCl, 5% glicerol, 1 mM DTT. Fractions of 0.2 ml were collected. The protein eluted as two main peaks of 180–170 kDa and of 130–120 kDa. Similar results were obtained when BFA-induced ADP-ribosylation followed the separation by gel filtration of the native protein. The experimental procedures are described in detail in Ref. 6.

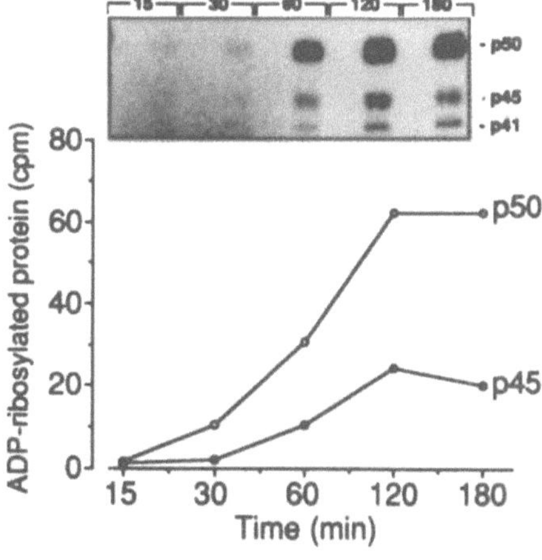

Figure 2. Time-course of the BFA-induced ADP-ribosylation of BARS-50 and p45. Membrane (10 µg/sample) and cytosol (50 µg) were incubated with [32P]-NAD in the presence of BFA (30 µg/ml) for the indicated times at 37°C. Samples were analyzed by SDS-PAGE. The lower panel shows the analysis of the bands corresponding to p45 and BARS-50 proteins quantified by an InstantImager (Packard); the upper panel shows the autoradiography of the same samples[6]. The experimental procedures are described in detail in Ref. 6.

compared to BARS-50 or, alternatively, that the ADP-ribosylated GAPDH has properties different from the non-ADP-ribosylated form. In CHO cells a dominant mutation in GAPDH has recently been found to confer morphological changes in late endocytic compartments that are in some ways reminiscent of the effects of BFA[6].

BARS-50 is a relatively unstable protein, as shown by the appearance of two degradation products (p41 and p45) in some of the postnuclears prepared in the absence of protease inhibitors (Fig. 2)[6]. Another peculiarity of BARS-50 is the apparent Mr which changes according to the presence of urea in the SDS-PAGE gel; thus, in the absence or presence of urea the Mr was 46 and 50 kDa, respectively[6]. (Fig. 1A and 1B, with and without urea, respectively). Other proteins in the system (such as GAPDH, the second BFA-specific substrate, or heterotrimeric G proteins) were not affected by the presence of urea, whereas the degradation product p45, showed a urea-dependent Mr shift very similar to that of BARS-50 (apparent Mr of 42 kDa in gels without urea)[6].

BARS-50 exists in the cytosol (~0.005% of the cytosolic proteins) as part of complexes of 170 and 130 kDa. A minor fraction of the protein (<5%) is also found in membranes in the presence of low concentration of salts, suggesting that BARS-50 might cycle between the two compartments[6]. The molecular size of the ADP-ribosylated BARS-50 determined by gel filtration revealed that the protein reproducibly eluted in two main peaks of ~170 kDa and of ~130 kDa, and a minor peak of ~100 kDa (Fig. 1C). The nature of the elements in the complex is still under investigation.

BARS-50 appears to be a group, rather than a single species, of BFA-specific substrates. Thus, upon 2D-IEF-PAGE analysis, the ADP-ribosylated BARS-50 was resolved into a cluster comprising at least 11 distinguishable spots with pI values ranging from 6.55 to 6.1 (Fig. 3). Also the degradation products p45 and p41 generated a similar cluster of at least 9 spots with pI values ranging from 6.65 to 6.2 (Fig. 3). From these data it can be

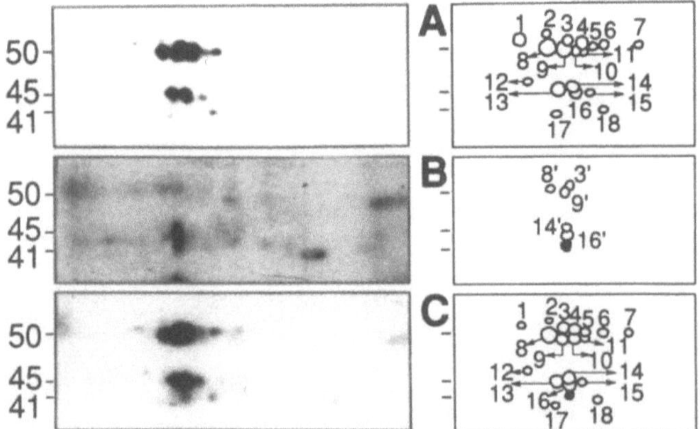

Figure 3. BARS-50 either ADP-ribosylated (A) or AAGTP-labeled (B) analyzed by 2D gel electrophoresis. Membranes (10 µg) and cytosol (50 µg) from FRTL5 cells were incubated with [^{32}P]-NAD plus BFA (A), or with AAGTP (B), or, samples separately labeled as in (A) and in (B) were combined and analyzed (C). Proteins were analyzed by IEF-SDS-PAGE and by autoradiography. Panels on the right are a schematic representation of labeled spots, which were resolved by autoradiography at different exposure times to obtain the best resolution of spots labeled with different efficiencies. Namely, spots with heavier labeling were resolved after a short exposure time, the autoradiography was then repeated for a longer time, to resolve weaker labeling (such as 1 and 12). Each experiments was repeated four times, with similar results. The Mr (kDa) are indicated. Taken from Ref. 6. The experimental procedures are described in detail in Ref. 6.

proposed that BARS-50 is either a class of proteins comprising many isoforms or a single protein undergoing different and multiple postranslational modifications.

3.2. Characteristics of the Mono-ADP-Ribosylation Reaction

The BFA-induced labelling of GAPDH and BARS-50 is due to the enzymatic transfer of ADP-ribose from NAD to the acceptor amino acid[4–6]. Thus, the digestion with snake venom phosphodiesterase of ADP-ribosylated BARS-50 (transferred to nitrocellulose) produced 5'-AMP, a result consistent with the mono-ADP-ribosylation of the protein[6]. The non-enzymatic mono-ADP-ribosylation of BARS-50 due to NAD-glycohydrolase-dependent formation of ADP-ribose from NAD and consequent non-enzymatic formation of ADP-ribose-protein adducts, could be excluded since BARS-50 was not ADP-ribosylated in the presence of ADP-ribose under conditions supporting the NAD-dependent reaction[4,6]. As mentioned in the introduction, the specific aminoacid involved in this reaction has not been identified; it is not an arginine, cysteine, or asparagine since the stability of the BARS-50/ADP-ribose bond to either NH_2OH, $HgCl_2$, acid or base is different from that reported for the above adducts[6]. These characteristics are shared with the ADP-ribosylated GAPDH, further indicating that the two proteins are indeed substrates of the same enzyme[4–6].

3.3. BARS-50 Is a Novel GTP Binding Protein

Some of the biochemical features of BARS-50 were suggestive of the possibility that it could be a GTP binding protein: its Mr is similar to that of the α subunits of G proteins; its ability to be ADP-ribosylated, similarly to several small and heterotrimeric G proteins; the role of G proteins in BFA-sensitive steps of vesicular transport[10,11]. Several approaches were undertaken to assess this possibility[6]. First, the BFA-dependent ADP-ribosylation of BARS-50, but not that of GAPDH, was shown to be inhibited by guanine nucleotides, indicating that the substrate, rather than the enzyme was the guanine nucleotide-sensitive component of the reaction[6]. BARS-50 was then shown to be able to directly bind GTP in blot overlay and photolabelling studies (Fig. 4)[6]. Thus, cytosolic proteins

Figure 4. Direct binding of GTP to BARS-50. (A) Binding of [α[32]P]-GTP to FRTL5 cytosolic proteins separated by SDS/PAGE in the presence (A) or absence (B) of urea and transferred to nitrocellullose. The blot-overlay studies were performed as described[6] [α[32]P]-GTP-labeled proteins (lane 3 in A and lane 4 in B), ADP-ribosylated BARS-50 (lane 2 in A and lane 3 in B), membrane proteins from FRTL5 cells ADP-ribosylated by pertussis toxin (lanes 1), and [α[32]P]-GTP-labeled FRTL5 membrane proteins (lane 2 in B). (C) Photoaffinity labeling of FRTL5 cytosol (200 mg protein/lane) with increasing concentration of [^{32}P]-AAGTP (10 nM lane 2; 50 nM lane 3; 100 nM lanes 4, 6–8; 500 nM lane 5). Samples also included 100 μM GTP[γS] (lane 6), 1 mM ATP (lane 7), and 100 μM guanosine 5'-[β,γ-imido] triphosphate (lane 8). PNS ADP-ribosylated by BFA is in lanes 1 (same sample at two exposure times). Proteins were then analyzed by SDS/PAGE and autoradiography. The experimental procedures are described in detail in Ref. 6.

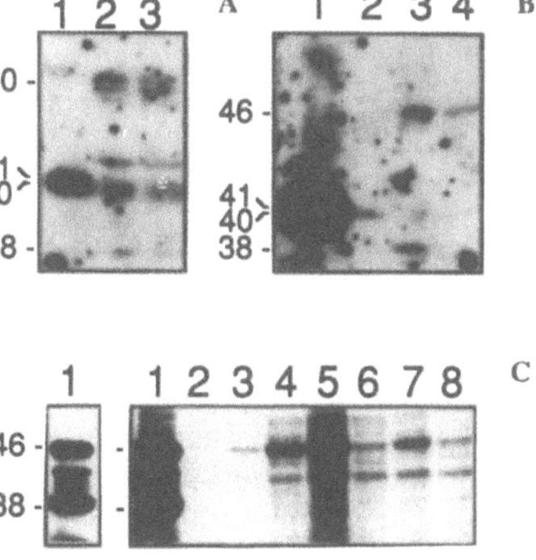

separated by SDS-PAGE (in the presence and absence of urea) and blotted onto nitrocellulose were overlayed with [32P]αGTP and a protein migrating at 50 kDa and 46 kDa in the two conditions was shown to bind GTP and thus was identified as BARS-50 (Fig. 4A and 4B)[6].

The direct binding of GTP to native BARS-50 was demonstrated in a series of photolabelling experiments employing [32P]-p3–1,4-azidoanilido-p1–5'-GTP (AAGTP), which showed the comigration of a AAGTP-photolabeled protein with BARS-50, in both the presence and absence of urea (Fig. 4C)[6]. The interaction of AAGTP with BARS-50 is specific since it is inhibited by GTPγS and GppNHp, but not by ATP (Fig. 4C). The photolabelling of BARS-50 could also be resolved in 2D gels: (Fig. 3B) in this case there was a precise comigration of the AAGTP-labeled spots with the same ADP-ribosylated proteins; for instance, the three AAGTP-labeled spots at 50 kDa (3', 8', 9') comigrated precisely with the 3, 8 and 9 ADP-ribosylated proteins (Fig. 3C). Not all the ADP-ribosylated spots resolved by 2D gels were however photolabelled. A calculation of the number of AAGTP-labeled and ADP-ribosylated protein molecules in the same spot shows that ADP-ribosylation is at least 20 fold more efficient than photolabeling for BARS-50. This is expected since the affinity of AAGTP photoincorporation is about 2.5% into tubulin[12], or Gα[13] and might in part explain why only some of the ADP-ribosylated spots appear to correspond to AAGTP-labeled proteins. The direct and specific binding of GTP to BARS-50 was also confirmed by affinity chromatography: the protein was applied to an agarose-GTP column, then eluted with either 1 mM GTP or 1 mM ATP. About 10% of native BARS-50 was eluted in the presence of GTP, indicating its specific interaction with the protein[6].

These series of experiments clearly show that BARS-50 directly binds GTP. This protein is not however a classical heterotrimeric G protein, since it does not comigrate with ADP-ribosylated αs or αi, nor is it a substrate of pertussis or cholera toxin-induced ADP-ribosylation, nor is it recognized by several antibodies raised against the α subunits of heterotrimeric G proteins[6].

The toxin-dependent ADP-ribosylation of the α subunits of GTP-binding proteins is in part modulated by their interaction with the βγ subunit, which might increase (αs) or prevent (αi) the transfer of ADP-ribose[14]. Bovine brain βγ subunit completely inhibited the ADP-ribosylation of BARS-50 (Fig. 5). It could be demonstrated that the βγ subunit does not interact with the membrane-associated ADPRT, nor has it any effect on the ADP-ribosylation of GAPDH, indicating that it interacts with BARS-50 (either directly or through some associated proteins)[6]. There is an increasing number of proteins that are able to interact with βγ, including β adrenergic receptor kinase, phospholipase Cβ and adenylyl cyclase[15]. Thus BARS-50 could share some structural similarities with these proteins or with the α subunits of heterotrimeric G proteins or alternatively, it could be characterized by pleckstrin homology (PH) domain, a specific sequence involved in protein-protein interaction and able to bind βγ [16,17].

3.4. Partial Purification of BARS-50

In order to clarify the nature and function of BARS-50, the protein was partially purified from ADP-ribosylated rat brain cytosol[18]. The strategy followed included an ammonium sulfate precipitation of the [32P]-ADP-ribosylated BARS-50 followed by hydrophobic (15-fold enrichment, 70–80% yield) and hydroxylapatite chromatography (5-fold enrichment, 70–80% yield). This resulted in a 225 fold enriched preparation of BARS-50, which was detectable by silver staining (Fig. 1B). The purified BARS-50 was then either employed in functional studies or for microsequencing. In the latter case, the

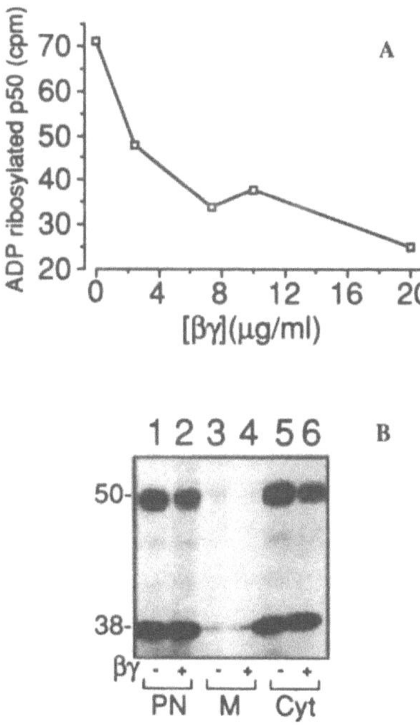

Figure 5. Inhibitory effect of βγ subunit on BFA-induced ADP-ribosylation of BARS-50. (A) Dose-dependence of the brain Gβγ-induced inhibition of BFA-dependent ADP-ribosylation. (B) ADP-ribosylation of PNS (1 and 2), of membranes (3 and 4) and cytosol (5 and 6) obtained by centrifugation of FRTL5 PNS ADP-ribosylated in the absence (1, 3, 5) or in the presence (2, 4 and 6) of brain Gβγ (10 μg/ml). Samples were analyzed by SDS-PAGE and autoradiography. The Mr (kDa) are indicated. The experimental procedures are described in detail in Ref. 6. Taken from Ref. 6.

protein was further purified by gel electrophoresis and eluted from a 2D gel. The sequences of 2 peptides of 10 and 11 aminoacid was obtained and, since there was no homology with known proteins, it can be concluded that indeed BARS-50 is a novel protein[18].

In parallel, the purification of native (non-ADP-ribosylated) BARS-50 was obtained by an analogous procedure.

3.5. ADP-Ribosylation and Intracellular Transport

The BFA-activated ADP-ribosylation might represent an additional mechanism of action of the toxin, parallel to the inhibition of the GTP-GDP exchange on ARF[1]. To examine whether indeed an ADP-ribosylation reaction is involved in the maintenance of Golgi structure we have used two approaches, both aimed at blocking endogenous ADP-ribosylation, to then examine if the structural effects of BFA are prevented[19–21].

One approach consisted in the identification of a series of specific inhibitors of the BFA-dependent ADP-ribosylation[5,21,22]. These included nicotinamide, dicumarol, and ilimaquinone. In particular, the marine toxin ilimaquinone potently inhibited the ability of BFA to rapidly transform Golgi stacks into a tubular-reticular network[20,21]. Similarly, also nicotinamide and dicumarol inhibited the BFA effects on Golgi structure in RBL cells[20,21]. The potencies of these inhibitors in antagonizing the BFA effects on Golgi assembly, correlated extremely well with their potencies in inhibiting the BFA-dependent ADP-ribosylation of GAPDH and BARS-50, strengthening the causal relation between the two phenomena[21]. Moreover, long treatments with the inhibitors caused *per se* the progressive breakdown first of the tubular interconnections between Golgi stacks, then of the stacks themselves, resulting in near-complete vesiculation of the complex, suggesting that ADP-ribosylation is indeed involved in the maintainance of the Golgi structure[21].

The second approach used consisted in depleting cells of NAD^+, the ADP-ribose donor in ADP-ribosylation reactions. This was achieved by permeabilizing RBL cells with streptolysin O, a procedure that results in the loss of all low MW cytosol components[20,21]. Under these conditions, the BFA-dependent alteration in the organelle's structure was inhibited. The readdition of NAD^+ was sufficient to trigger Golgi's disassembly[20,21]. These results indicate that NAD^+ and ADP-ribosylation participate in the mechanisms controlling the structure/function of the Golgi complex. A detailed accounts of these approaches is presented elsewhere in this book[23,24].

4. CONCLUSION

In conclusion, we have described a novel mechanism of action of BFA, the mono-ADP-ribosylation of two specific cytosolic substrates, BARS-50 and GAPDH, that might be relevant in inducing the toxin-dependent Golgi disassembly[4–6]. In particular, we have shown that BARS-50 consist of a family of cytosolic proteins that can bind GTP and, most likely, the $\beta\gamma$ subunit of G proteins. The interaction with $G\beta\gamma$ might be of functional significance in the context of membrane transport, since trimeric G proteins have been implicated in the formation of coated buds and vesicles[10,11,25]. Moreover, $\beta\gamma$ subunits have been shown to interfere with the activation and binding of ARF to Golgi membranes[11], and, most important, to interact directly with ARF^{26}. Thus, the effect of $\beta\gamma$ on BARS-50 might result in the regulation of ARF binding. The state of ADP-ribosylation of BARS-50 could be part of this regulatory mechanism. There is not yet experimental proof for the hypotheses presented above, of a direct effect of BARS-50 on ARF function. However, we have obtained evidence that involve the ADP-ribosylation reaction in the control of intracellular transport, by demonstrating that inhibitors of the ADP-ribosylation prevent the effects of BFA and that NAD^+ is suffucient to restore the BFA-dependent Golgi disassembly in permeabilized cells[19–21]. We propose that NAD^+, GAPDH and BARS-50 are all elements possibly involved in the control of the Golgi apparatus structure and function.

5. ACKNOWLEDGMENTS

This study was supported in part by the Italian Association for Cancer Research (AIRC) and the Italian National Research Council (Progetto Finalizzato ACRO ctr. n. 95.00558.PF39 and Convenzione CNR-Consorzio Mario Negri Sud).

6. REFERENCES

1. Klausner, R. D., J. G. Donaldson, & J. Lippincott-Schwartz. 1992. Brefeldin A: insights into the control of membrane traffic and organelle structure. *J. Cell Biol. 116*: 1071–1080.
2. Pelham, H. R. B. 1991. Multiple targets for brefeldin A. *Cell 67*: 449–451.

3. Donaldson, J. G., D. Cassel, R. A. Kahn, & R. D. Klausner. 1992. ADP-ribosylation factor, a small GTP-binding protein, is required for binding of the coatomer protein β-COP to Golgi membranes. *Proc. Natl. Acad. Sci. USA.* *89*: 6408–6412.

4. De Matteis, M. A., M. Di Girolamo, A. Colanzi, M. Pallas, G. Di Tullio, L. J. McDonald, J. Moss, G. Santini, S. Bannykh, D. Corda, & A. Luini. 1994. Stimulation of endogenous ADP-ribosylation by brefeldin A. *Proc. Natl. Acad. Sci. USA 91*: 1114–1118.

5. Colanzi, A., M. Di Girolamo, G. Santini, G. Sciulli, S. Santarone, M. Pallas, G. Di Tullio, S. Bannykh, D. Corda, M. A. De Matteis, & A. Luini. 1994. Brefeldin A, an inhibitor of vesicular traffic, stimulates the ADP-ribosylation of two cytosolic proteins. In *GTPase-Controlled Molecular Machines*, eds. D. Corda, H. Hamm, & A. Luini. Ares-Serono Symposia Publications, Rome, pp. 197–217.

6. Di Girolamo, M., M. G. Silletta, M. A. De Matteis, A. Braca, A. Colanzi, D. Pawlak, M. M. Rasenick, A. Luini, & D. Corda. 1995. Evidence that the 50-kDa substrate of brefeldin A-dependent ADP-ribosylation binds GTP and is modulated by the G-protein βγ subunit complex. *Proc. Natl. Acad. Sci. USA.* *92*: 7065–7069.

7. De Murcia, G., M. Jacobson, & S. Shall. 1995. Regulation by ADP-ribosylation. *Trends Cell Biol. 5*: 78–81.

8. Kots, A. Y., A. V. Skurat, E. A. Sergienko, T. V. Bulargina, & E. S. Severin. 1992. Nitroprusside stimulates the cysteine-specific mono (ADP-ribosylation) of glyceraldehyde-3-phosphate dehydrogenase from human erythrocytes. *FEBS Lett. 300*: 9–12.

9. Lopez Vinals, A. E., R. N. Farias, & R. D. Morero. 1987. Characterization of the fusogenic properties of glyceraldehyde-3-phosphate dehydrogenase: fusion of phospholipid vesicles. *Biochem. Biophys. Res. Commun. 143*: 403–409.

10. Donaldson, J. G., J. Lippincott-Schwartz, & R. D. Klausner. 1991. Guanine nucleotides modulate the effects of brefeldin A in semipermeable cells: regulation of the association of a 110-kD peripheral membrane protein with the Golgi apparatus. *J. Cell Biol. 112*: 579–588.

11. Donaldson, J. G., R. A. Kahn, J. Lippincott-Schwartz, & R. D. Klausner. 1991. Binding of ARF and β-COP to Golgi membranes: possible regulation by a trimeric G protein. *Science 254*: 1197–1199.

12. Rasenick, M. M., & N. Wang. 1988. Exchange of guanine nucleotides between tubulin and GTP-binding proteins that regulate adenylate cyclase: cytoskeletal modification of neuronal signal transduction. *J. Neurochem. 51*: 300–311.

13. Gordon, J. H., & M. M. Rasenick. 1988. In situ binding of a photo-affinity GTP analog to synaptic membrane G-proteins. Distribution of bound GTP analog reflects the status of adenylate cyclase. *FEBS Lett 235*: 201–206.

14. Gilman, A. G. 1987. G proteins: transducers of receptor-generated signals. *Annu. Rev. Biochem. 56*: 615–649.

15. Clapham, D. E., & E. J. Neer. 1993. New roles for G-protein βγ-dimers in transmembrane signalling. *Nature 365*: 403–406.

16. Gibson, T. J., M. Hyvönen, A. Musacchio, M. Saraste, & E. Birney. 1994. PH domain: the first anniversary. *Trends Biochem. Sci. 19*: 349–353.

17. Touhara, K., J. Inglese, J. A. Pitcher, G. Shaw, & R. J. Lefkowitz. 1994. Binding of G protein βγ-subunits to pleckstrin homology domains. *J. Biol. Chem. 269*: 10217–10220.

18. Silletta, M. G., M. Di Girolamo, S. Mironov, G. Fiucci, A. Colanzi, M. A. De Matteis, A. Luini, & D. Corda. 1995. Purification and functional studies on BARS-50, the substrate of brefeldin A-dependent ADP-ribosylation. *Mol. Biol. Cell 6* (Suppl.): 352a.

19. Santini, G., M. G. Sciulli, A. Colanzi, A. Mironov, S. Santarone, G. Innamorati, A. Fusella, M. Di Girolamo, D. Corda, M. A. De Matteis, & A. Luini. 1994. Cellular effects of inhibitors of BFA-stimulated ADP-ribosyltransferase. *Mol. Biol. Cell 5*: 242a.

20. Mironov, A., R. Weigert, A. Colanzi, S. Flati, G. Di Tullio, A. Fusella, M. Di Girolamo, D. Corda, M. A. De Matteis, & A. Luini. 1995. Role of the brefeldin A-sensitive ADP-ribosyltransferase in the structure and function of the Golgi complex. *Mol. Biol. Cell 6* (Suppl.): 329a.

21. Mironov, A., A. Colanzi, S. Flati, I. Santone, A. Fusella, R. Polishchuk, A. Mironov Jr, G. Di Tullio, R. Weigert, V. Malhorta, D. Corda, M. A. De Matteis, & A. Luini. Role of NAD^+ and ADP-ribosylation in the maintenance of the Golgi structure. (Submitted).

22. Weigert, R., A. Colanzi, G. Sciulli, A. Mironov, G. Santini, M. Di Girolamo, J. G. Donaldson, D. Corda, M. A. De Matteis, & A. Luini. Pharmacologic inhibitors of the brefeldin A-induced ADP-ribosylation. (Submitted).

23. Weigert, R., A. Colanzi, C. Limina, C. Cericola, G. Di Tullio, A. Mironov, G. Santini, G. Sciulli, D. Corda, M. A. De Matteis, & A. Luini. 1996. Characterization of the endogenous mono ADP-ribosylation stimutated by brefeldin A. This book.

24. Colanzi, A., A. Mironov, R. Weigert, C. Limina, S. Flati, C. Cericola, G. Di Tullio, M. Di Girolamo, D. Corda, M. A. De Matteis, & A. Luini. 1996. Brefeldin A-induced ADP-ribosylation in the structure and function of the Golgi complex. This book.

25. Colombo, M. I., L. S. Mayorga, P. J. Casey, P. D. Stahl. 1992. Evidence of a role for heterotrimeric GTP-binding proteins in endosome fusion. *Science 255*: 1695–1697.

26. Colombo, M. I., J. Inglese, C. D'Souza-Schorey, W. Beron, & P. D. Stahl. 1995. Heterotrimeric G proteins interact with the small GTPase ARF. Possibilities for the regulation of vesicular traffic. *J. Biol. Chem. 270*: 24564–24571.

BREFELDIN A-INDUCED ADP-RIBOSYLATION IN THE STRUCTURE AND FUNCTION OF THE GOLGI COMPLEX

Antonino Colanzi, Alexander Mironov, Roberto Weigert, Cecilia Limina, Silvio Flati, Claudia Cericola, Giuseppe Di Tullio, Maria Di Girolamo, Daniela Corda, Maria Antonietta De Matteis, and Alberto Luini

Department of Cell Biology and Oncology
Istituto di Ricerche Farmacologiche "Mario Negri"
Consorzio Mario Negri Sud, Via Nazionale
66030 S.Maria Imbaro
(Chieti) Italy

1. ABSTRACT

Brefeldin A (BFA) is a fungal metabolite that exerts generally inhibitory actions on membrane transport and induces the disappearance of the Golgi complex. Previously we have shown that BFA stimulates the ADP-ribosylation of two cytosolic proteins of 38 and 50 KD. The BFA-binding components mediating the BFA-sensitive ADP-ribosylation (BAR) and the effect of BFA on ARF binding to Golgi membranes have similar specificities and affinities for BFA and its analogues, suggesting that BAR may have a role in the cellular effects of BFA. To investigate this we used the approach to impair BAR activity by the use of BAR inhibitors. A series of BAR inhibitors was developed and their effects were studied in RBL cells treated with BFA. In addition to the common ADP-ribosylation inhibitors (nicotinamide and aminobenzamide), compounds belonging to the cumarin (novobiocin, cumermycin, dicumarol) class were active BAR inhibitors. All BAR inhibitors were able to prevent the BFA-induced redistribution of a Golgi marker (Helix pomatia lectin) into the endoplasmic reticulum, as assessed in immunofluorescence experiments. At the ultrastructural level, BAR inhibitors prevented the tubular-vesicular transformation of the Golgi complex caused by BFA. The potencies of these compounds in preventing the BFA effects on the Golgi complex were similar to those at which they inhibited BAR. Altogether these data support the hypothesis that BAR mediates at least some of the effects of BFA on the Golgi structure and function.

ADP-Ribosylation in Animal Tissue, edited by Haag and Koch-Nolte
Plenum Press, New York, 1997

2. INTRODUCTION

BFA is an antiviral agent of fungal origin with striking effects on membrane transport, including the inhibition of constitutive secretion, the disappearance of non-clathrin coated secretory vesicles and the induction of long membranous tubules through which the Golgi compartment redistributes into the endoplasmic reticulum[1,2]. These dramatic morphological changes are preceded by the release of a set of specific proteins from the Golgi complex including the non clathrin coat proteins ADP-ribosylation factor (ARF, a small ras-like GTPase) and β–COP, a component of the cytosolic protein complex "coatomer"[3,4,5]. Recent biochemical studies have shown that BFA inhibits the GDP-GTP exchange on ARF catalyzed by a Golgi protein, suggesting that ARF or an ARF-involving machinery is the primary site of action of BFA and a key regulator in membrane transport pathways[6,7]. It is still unclear, however, whether the complex cellular actions of BFA can be entirely explained by its effect on ARF. ARF, in addition to being involved in membrane transport, has long been known to act as a cofactor in the ADP-ribosylation of Gsα (the α subunit of the GTP-binding protein coupled in a stimulatory fashion to adenylyl cyclase) by cholera toxin, and to be able to interact directly with both cholera toxin and Gsα[8,9]. ADP-ribosylation is a postranslational modification of proteins that is involved in the action of bacterial toxins and in the regulation of cellular processes, and is due to the enzymatic transfer of ADP-ribose from NAD^+ to specific amino acid residues[10].

We have shown that BFA markedly stimulates the ADP-ribosylation of two cytosolic proteins of 38 and 50 kDa (p38 and p50)[11,12]. The p50 substrate (BARS-50) binds GTP and is regulated by βγ subunits of trimeric G proteins; it has therefore been proposed to be a novel G protein involved in membrane transport[13]. BFA activates ADP-ribosylation both in intact and Triton-solubilized Golgi membranes through a site with a ligand selectivity identical to that involved in the BFA-dependent inhibition of ARF binding [12,13], suggesting that ADP-ribosylation plays a role in the cellular actions of BFA. We examined whether ADP-ribosylation is involved in the maintenance of Golgi structure by studying the cellular effects of BAR inhibitors[14.] We found that treatment of cells with such inhibitors strongly inhibit the ability of BFA to induce the redistribution of Golgi complex into the endoplasmic reticulum. These results implicate that the BFA-dependent ADP-ribosylation may take part in controlling the structure of the Golgi complex[14].

3. RESULTS AND DISCUSSION

To study the role of ADP-ribosylation in the maintenance of Golgi morphology we used the approach to investigate whether pharmacologic inhibitors of ADP-ribosylation inhibit the Golgi disassembly induced by BFA. We have recently found that, in addition to the well known ADP-ribosylation inhibitors nicotinamide and related compounds, several compounds belonging to the coumarin group act as selective inhibitors of the BFA-dependent ADP-ribosylation *in vitro* [15,16]. They do not, for instance, interfere with the activity of the ADP-ribosylating enzymes cholera and pertussis toxin or with the action of cytosolic NAD glycohydrolases[16]. Dicumarol (known as an inhibitor at the NAD^+-binding site of the oxido-reductase that catalyzes the reduction of vitamin K)[17] is a relatively potent and non-toxic representative of this class of compounds. Fig. 1 shows that a concentration of dicumarol (200 μM) with near-maximal effects in ADP-ribosylation assays *in vitro* [15,16] markedly inhibited the ability of BFA to induce the redistribution of the Golgi complex as revealed with Helix pomatia lectin. The effects were significantly stronger

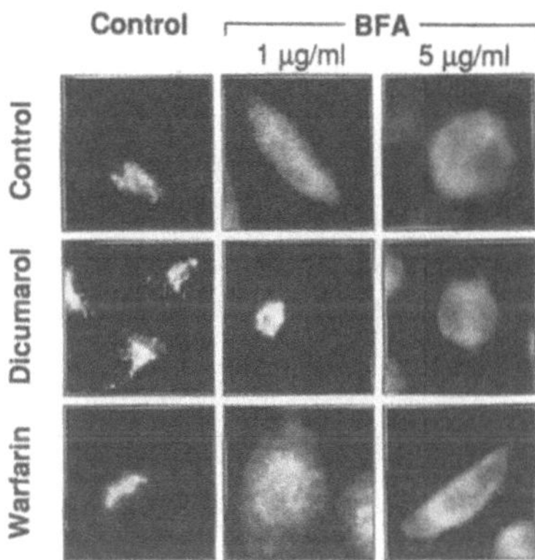

Figure 1. ADP-ribosylation inhibitors prevent the BFA-induced redistribution of Golgi markers into the ER. RBL cells were grown in MEM supplemented with 16% foetal calf serum and plated at a density of 10^5 cells/well in chamber slides (Lab-Tek, Nunc). The cells were pretreated with control buffer, or 200 µM dicumarol, or 400 µM warfarin for 5 min before treatment with 1 µg/ml or 5 µg/ml BFA for 10 min. The cells were fixed in 4% paraformaldehyde and permeabilized with 0.05% saponin, 0.2% BSA in PBS for 30 min at room temperature and stained with FITC-conjugated Helix pomatia lectin (100 µg/ml in PBS containing 0.2% BSA) for 45 min. Slides were mounted in Moviol 4–88 (Calbiochem) and examined using a Zeiss Axiophot photomicroscope equipped with a Plan-Neofluar 40X objective.

when dicumarol was administered a few minutes before BFA (possibly because the drug needs time to reach its intracellular targets). Dicumarol did not exhibit evident cellular toxicity in intact cells (cells remained viable for up to several hours by the Trypan blue test and morphological criteria), and its inhibitory effects were readily reversible (one hour after washout of dicumarol BFA fully recovered its cellular effects)[14]. Moreover, BFA, at higher concentrations, or when applied for a longer time, completely overcame the inhibition by dicumarol (Fig. 1); this indicates that the dicumarol-induced block is not due to toxicity and that the machinery responsible for BFA-induced Golgi redistribution maintains the potential to function in the presence of the inhibitor[14].

Two other members of the cumarin-like family of inhibitors (cumermycin and novobiocin) had effects similar to those of dicumarol at 100 and 300 µM, respectively (at these concentrations they also near-maximally inhibited ADP-ribosylation *in vitro*)[15,16], but exhibited significant mitochondrial toxicity and could only be used for shorter periods of time. In contrast, several other coumarin analogues including cumarin itself and warfarin (Fig.1) that were devoid of ADP-ribosylation inhibitory activity[15,16], were also unable to antagonize the effects of BFA on Golgi morphology. This demonstrates a remarkable structural specificity in the action of the cumarinic inhibitors.

The class of ADP-ribosylation inhibitors that includes the widely used but rather impotent and nonspecific nicotinamide and derivatives also antagonized the morphological effects of BFA *in vivo*, albeit at much higher concentrations (not shown)[12,14]. Altogether, therefore, there was an excellent correlation between the potencies of the above compounds as inhibitors of *in vitro* ADP-ribosylation and as antagonists of the structural effects of BFA *in vivo* .

The effects of ADP-ribosylation inhibitors were also examined at the ultrastructural level. The exposure to 1 or to 5 µg/ml BFA for 15 min caused the disorganization and tubular-vesicular transformation of the Golgi complex. Dicumarol strongly inhibited these alterations and, in fact, afforded a remarkable preservation of the stack structure[14]. Their effects were dose-dependent and dependent on the dose of BFA, since higher concentrations of the toxin overcame the inhibition by dicumarol, as was also shown in immunofluorescence experiments.

The fact that ADP-ribosylation inhibitors prevent the effects of BFA on the architecture of the Golgi complex is a valid indication of the role of ADP-ribosylation in regulating the Golgi structure only if the effects of these inhibitors can be demonstrated to be specific. Two lines of evidence show that this is the case: 1) the inhibitors are not obviously toxic and their effects could be overcome by increasing the concentration of BFA. This shows that the drugs do not paralyze nonspecifically the machinery responsible for Golgi disassembly, which remains able to respond to higher concentrations of the toxin. 2) There was an excellent correlation between the potency of the inhibitors in blocking ADP-ribosylation *in vitro* and that in preventing Golgi disassembly by BFA *in vivo*[14,15,16], suggesting that these two effects are mediated by the same binding site.

Collectively these results strongly support a key role for ADP-ribosylation in the machinery controlling the Golgi architecture.

3. ACKNOWLEDGMENTS

This study was supported in part by the Italian Association for Cancer Research (AIRC) and the Italian National Research Council (Progetto Finalizzato ACRO ctr. n. 95.00558.PF39, and Convenzione CNR-Consorzio Mario Negri Sud).

4. REFERENCES

1. Klausner, R. D., J. G. Donaldson, & J. Lippincott-Schwartz. 1992. Brefeldin A: insights into the control of membrane traffic and organelle structure. *J. Cell Biol. 116*: 1071–1080.
2. Price, S. R., M. Nightingale, S.-C. Tsai, K. C. Williamson, R. Adamik, H.-C. Chen, J. Moss, & M. Vaughan. 1988. *Proc. Natl. Acad. Sci. USA 85*: 5488–5491.
3. Donaldson, J. G., J. Lippincott-Schwartz, G. S. Bloom, T. E. Kreis, & R. D. Klausner. 1990. Dissociation of a 110-kD peripheral membrane protein from the Golgi apparatus is an early event in brefeldin A action. *J. Cell Biol. 111*: 2295–2306.
4. Donaldson, J. G., J. Lippincott-Schwartz, & R. D. Klausner. 1991. Guanine nucleotides modulate the effects of brefeldin A in semipermeable cells: regulation of the association of a 110-kD peripheral membrane protein with the Golgi apparatus. *J. Cell Biol.* 112: 579–588.
5. Donaldson, J. G., R. A. Kahn, J. Lippincott-Schwartz, & R. D. Klausner. 1991. Binding of ARF and β-COP to Golgi membranes: possible regulation by a trimeric G protein. *Science 254*: 1197–1199.
6. Donaldson, J. G., D. Finazzi, & R. D. Klausner. 1992. Brefeldin A inhibits Golgi membrane-catalysed exchange of guanine nucleotide onto ARF protein. *Nature 360*: 350–352.
7. Helms, J. B., & J. E. Rothman. 1992. Inhibition by brefeldin A of a Golgi membrane enzyme that catalyses exchange of guanine nucleotide bound to ARF. *Nature 360*: 352–354.
8. Kahn, R. A., & A. G. Gilman. 1984. Purification of a protein cofactor required for ADP-ribosylation of the stimulatory regulatory component of adenylate cyclase by cholera toxin. *J. Biol. Chem. 259*: 6228–6234.
9. Tsai, S.-C., M. Noda, R. Adamik, J. Moss, & M. Vaughan. 1987. Enhancement of choleragen ADP-ribosyltransferase activities by guanyl nucleotides and a 19-kDa membrane protein. *Proc. Natl. Acad. Sci. USA. 84*: 5139–5142.
10. Ueda, K., & O. Hayaishi. 1985. ADP-ribosylation. *Annu. Rev. Biochem. 54*: 73–100.
11. De Matteis, M. A., M. Di Girolamo, A. Colanzi, M. Pallas, G. Di Tullio, L. J. McDonald, J. Moss, G. Santini, S. Bannykh, D. Corda, & A. Luini. 1994. Stimulation of endogenous ADP-ribosylation by brefeldin A. *Proc. Natl. Acad. Sci. USA 91*: 1114–1118.
12. Colanzi, A., M. Di Girolamo, G. Santini, G. Sciulli, S. Santarone, M. Pallas, G. Di Tullio, S. Bannykh, D. Corda, M. A. De Matteis, & A. Luini. 1994. Brefeldin A, an inhibitor of vesicular traffic, stimulates the

ADP-ribosylation of two cytosolic proteins. In *GTPase-Controlled Molecular Machines*, eds. D. Corda, H. Hamm, & A. Luini. Ares-Serono Symposia Publications, Rome, pp. 197–217.

13. Di Girolamo, M., M. G. Silletta, M. A. De Matteis, A. Braca, A. Colanzi, D. Pawlak, M. M. Rasenick, A. Luini, & D. Corda. 1995. Evidence that the 50-kDa substrate of brefeldin A-dependent ADP-ribosylation binds GTP and is modulated by the G-protein βγ subunit complex. *Proc. Natl. Acad. Sci. USA.* *92*: 7065–7069.

14. Mironov, A., A. Colanzi, S. Flati, I. Santone, A. Fusella, R. Polishchuk, A. Mironov Jr, G. Di Tullio, R. Weigert, D. Corda, M. A. De Matteis, & A. Luini. Role of NAD^+ and ADP-ribosylation in the maintenance of the Golgi structure. (Submitted).

15. Weigert, R., A. Colanzi, C. Limina, C. Cericola, G. Di Tullio, A. Mironov, G. Santini, G. Sciulli, D. Corda, M. A. De Matteis, & A. Luini. 1996. Characterization of the endogenous mono ADP-ribosylation stimulated by brefeldin A. This book.

16. Weigert, R., A. Colanzi, G. Sciulli, A. Mironov, G. Santini, M. Di Girolamo, J. G. Donaldson, D. Corda, M. A. De Matteis, & A. Luini. Pharmacologic inhibitors of the brefeldin A-induced ADP-ribosylation. (Submitted).

17. Ma, Q., K. Cui, F. Xiao, A. Y. H. Lu, & C. S. Yang. 1992. Identification of a glycine-rich sequence as an NAD(P)H-binding site and tyrosine 128 as a dicumarol-binding site in rat liver NAD(P)H: quinone oxidoreductase by site-directed mutagenesis. *J. Biol. Chem.* *267*: 22298–22304.

CHARACTERIZATION OF THE ENDOGENOUS MONO-ADP-RIBOSYLATION STIMULATED BY BREFELDIN A

Roberto Weigert, Antonino Colanzi, Cecilia Limina, Claudia Cericola,
Giuseppe Di Tullio, Alexander Mironov, Giovanna Santini, Gina Sciulli,
Daniela Corda, Maria Antonietta De Matteis, and Alberto Luini

Istituto di Ricerche Farmacologiche "Mario Negri"
Consorzio Mario Negri Sud
Department of Cell Biology and Oncology-66030
S. Maria Imbaro (Chieti) Italy

1. ABSTRACT

We have recently described a novel enzymatic mono-ADP-ribosyl transfer reaction
induced by brefeldin A, a well characterized inhibitor of vesicular traffic, which selectively modifies two cytosolic proteins of 38 kDa (p38) and 50 kDa (BARS-50). p38 was
identified as glyceraldehyde-3-phosphate dehydrogenase (GAPDH), a glycolytic enzyme
and a multifunctional protein involved in several cellular processes; BARS-50 might be a
novel G protein, since it is able to bind GTP and the $\beta\gamma$ subunit of G proteins. We have
characterized this enzymatic activity and screened in vitro the effects of different drugs
belonging to the coumarine (dicumarol, coumermicin A1 and novobiocin) and quinone
(ilimaquinones, benzoquinones and naphtoquinones) class. These drugs blocked the BFA-
dependent mono-ADP-ribosylation, showed remarkable effects on Golgi morphology in
control cells, and antagonized the tubular reticular redistribution of the Golgi complex in
brefeldin A treated cells (see papers of Corda and Colanzi in this issue) suggesting a possible role for ADP-ribosylation in both the cellular effects of brefeldin A and the maintenance of the structure/function of the Golgi complex.

2. INTRODUCTION

Brefeldin A (BFA), a fungal macrocyclic lactone originally discovered for its ability
to inhibit constitutive protein secretion [3-7] causes an impressively rapid and extensive disruption of Golgi morphology, consisting of the transformation of Golgi stacks into a tubular-reticular network as early as 30 seconds after its application [8]. This is followed by the

ADP-Ribosylation in Animal Tissue, edited by Haag and Koch-Nolte
Plenum Press, New York, 1997

formation of long microtubule-dependent tubules emanating from the Golgi region and by the redistribution of most of the Golgi resident proteins into the endoplasmic reticulum (ER) [8,9,10,11]. It has been hypothesized that the structural effects of BFA are due to the release of coat proteins, in particular the coatomer, from Golgi membranes [12,13]. However, although the coatomer might indeed play a role in Golgi structure [14,15,16], the evidence that its detachment from the organelle is the sole cause of BFA-induced Golgi disruption is far from direct or clear. It seems more likely that multiple factors are involved in the dynamic control of the Golgi shape.

We have recently reported that BFA potently stimulates an endogenous ADP-ribosylation reaction that selectively modifies two cytosolic proteins of 38 and 50 kDa [1]. BFA activates ADP-ribosylation both in intact and Triton-solubilized Golgi membranes through a site with a ligand selectivity identical to that involved in the BFA-dependent inhibition of ADP-ribosylation factor (ARF) binding [2], suggesting that ADP-ribosylation plays role in the cellular actions of BFA. In this study, we characterize the BFA-dependent ADP-ribosylation reaction and examine a series of compounds with regard to their ability to inhibit ADP-ribosylation in vitro [20]. We find that agents that inhibit the BFA-dependent mono-ADP-ribosylation also prevent the effect of BFA on the Golgi structure [21,22], indicating that, together with coat protein, mono-ADP-ribosylation plays a role in the mainteinance of the Golgi apparatus.

3. EXPERIMENTAL PROCEDURE

Total membranes and cytosol (or postnuclear supernatant (PNS) alone) were incubated at 37°C for 1h with BFA in a reaction buffer containing 50 mM phosphate buffer pH 7.4, 30 μM NAD^+, 0.01 μCi/μl [^{32}P] NAD^+, 2.5 mM MgCl, 5 mM DTT and 10 mM thymidine (final volume of 50 μl). The reaction was stopped by the addition of sample buffer, the samples were boiled for 5 minutes and processed for SDS-page electrophoresis. Proteins resolved on polyiacrylamide gels were transferred on nitrocellulose by electroblotting and radiolabelled proteins were detected by electronic autoradiography (Instant Imager, Canberra Packard). Alternatively purified GAPDH was incubated with total membranes as described above, the samples were pelletted with a microfuge (12000 rpm for 10 min.) and the supernatant was treated with 100 μl of ice-cold 3 mM NAD, 3 mM AMP, 1.8 mM sodium pyrophosphate. Samples were then loaded on a 96 well Bio Rad Dot Blot pre-loaded with ice-cold 20% TCA (200 μl per well). GAPDH was allowed to adhere to nitrocellulose through a gentle filtration first, then was washed twice with ice-cold TCA 20% (100 μl per well) and 4 times with low salt TBS (20 mM Tris-HCl pH 7.5, 500 mM NaCl). Nitrocellulose was then boiled in distilled water for 10 min and stained with ponceau red. Radiolabelled GAPDH was detected and quantified as described above.

4. RESULTS AND DISCUSSION

BFA was previously reported to induce a dose- and time-dependent enzymatic mono ADP-ribosylation of two cytosolic proteins of 38 kDa (GAPDH) and 50 kDa (BARS-50) respectively [1]. The enzyme catalyzing this reaction is an integral membrane protein as shown by high salt extraction and detergent treatment. In order to test the enzyme specificity for BFA, a large number of BFA analogs, including the BFA stereoisomer, were tested in a standard ADP-ribosylation assay. None of these analogs were able to induce the mono-ADP-ribosylation of both GAPDH and BARS-50. Since the BFA analogs, including the stereisomer, display the same lack of activity on Golgi structure, the BFA-binding components involved in the effect on the Golgi complex and those involved in the mono-ADP-ribosylation are likely the same. A kinetic characterization of the activity showed two populations with different affinities for BFA (EC_{50}: 17 µg/ml for the first population) and with the same affinity for both NAD^+ (EC_{50}: 75 µM) and one of the substrates (GAPDH, EC_{50}: 15 µM) [20]. The enzymatic activity was found to be ubiquitarily distributed in rat tissues (with the kidney as the only exception) and in subcellular fractions enriched respectively in Golgi membranes, endoplasmic reticulum and mitochondria [20]. Strikingly, the enzymatic activity was present in all the cell lines tested with the exception of MDCK, PTK1, CHO D.1.4 and CHO G.2.1. [20], previously described as resistant to the effect of BFA on the Golgi complex [17,18,19] (fig.1).This finding is obviously in line with the BFA dependent mono-ADP-ribosylation being a possible intracellular mediator of the BFA action in living cells.

In addition to the known ADP-ribosylation inhibitors (nicotinamide and aminobenzamides EC_{50}: 10 mM), we also screened and characterized other blockers. Two classes of drugs displayed inhibitory activity, one containing the coumarine (dicumarol EC_{50}: 200 µM, coumermicin A1 EC_{50}:50 µM and novobiocin EC_{50}:350 µM), and the other the quinone (ilimaquinone EC_{50}:50 µM, benzoquinone EC_{50}:30 µM and naphtoquinone EC_{50}:30 µM) moiety [20,21,22]. These inhibitors acted quite selectively on the BFA dependent mono ADP-ribosylation, as they were inactive in inhibiting both cholera and pertussis toxin and a brain NAD-glycohydrolase. The mechanism of action of the coumarinic agents, and specifically of dicumarol, was characterized. First, since dicumarol is a well known inhibitor of diaphorases and competes for the NADH or NADPH binding site (EC50 3–10 µM), we checked whether other coumarinic inhibitors reported to compete for the same sites (warfarin, hydroxycoumarine and others) were effective in inhibiting the BFA dependent mono-ADP-ribosylation. Strikingly, these drugs were totally inactive. The analysis of

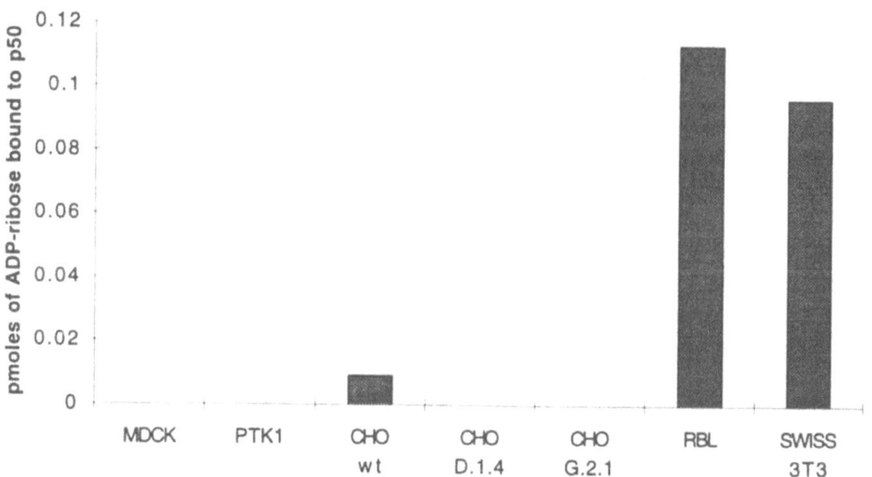

Figure 1. 50 µg of PNS prepared from different the cell lines were incubated with 90 µg/ml of BFA and 100 µM NAD+ for 4h at 37°C. Labeling of p50 was analyzed by SDS-page and Instant Imager analysis. MDCK, PTK1, CHO D.1.4, CHO G.2.1 are BFA insensitive cell lines.

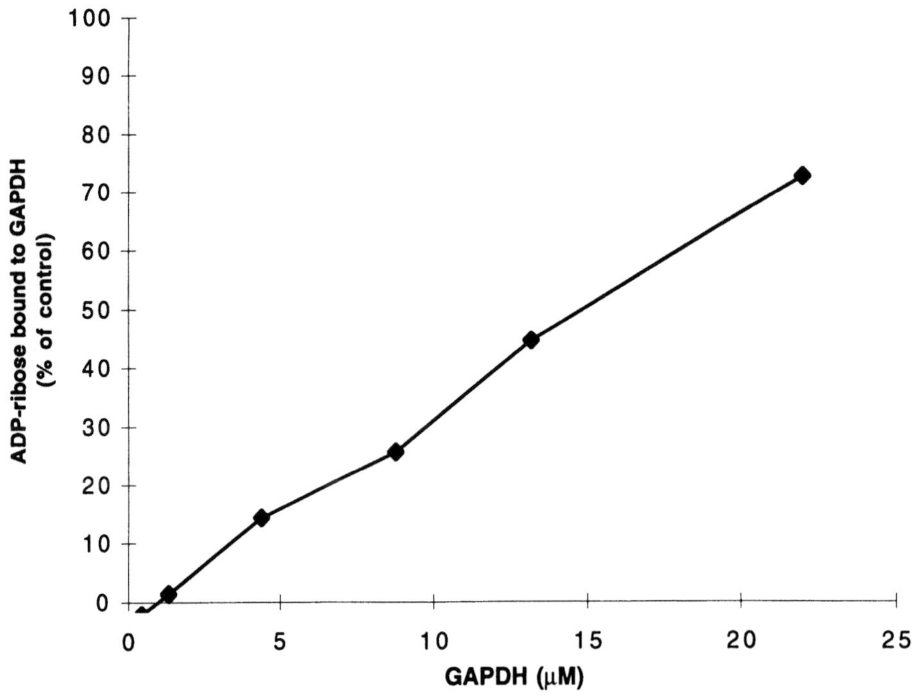

Figure 2. Different concentrations of GAPDH were incubated for 4h in a standard ADP-ribosylation assay in the presence or in the absence of 200 µM dicumarol. The ADP-ribosylation of GAPDH is expressed as % of control (absence of dicumarol).

their molecular structures revealed that the active coumarines have at least two α-pyrone rings and two oxydril groups as a common structural motif. We also carried out an analysis of the mechanism of action of dicumarol. Kinetic evidence indicated that dicumarol does not compete with NAD^+ and BFA or with the enzyme; dicumarol appears instead to bind on the ADP-ribosylation site on GAPDH (fig. 2).

Preliminary data performed with the other coumarines and ilimaquinone suggest a similar mechanism of action. Strikingly, these inhibitors were also found to prevent in intact cells the BFA induced redistribution of the Golgi complex into the endoplasmic reticulum with the same potencies found to be effective in inhibiting the BFA-dependent mono-ADP-ribosylation. This strong correlation suggests a possible role for ADP ribosylation in both the cellular effects of BFA and the maintenance of the structure/function of the Golgi complex.

5. ACKNOWLEDGMENTS

This study was supported in part by the Italian Association for Cancer Research (AIRC) and the Italian National Research Council (Progetto Finalizzato ACRO ctr. N° 95.00558.PF39, and Convenzione CNR-Consorzio Mario Negri Sud).

6. REFERENCES

1. De Matteis, M. A., M. Di Girolamo, A. Colanzi, M. Pallas, G. Di Tullio, L. J. McDonald, J. Moss, G. Santini, S. Bannykh, D. Corda, & A. Luini. 1994. Stimulation of endogenous ADP-ribosylation by brefeldin A. *Proc. Natl. Acad. Sci. USA. 91*: 1114–1118.

2. Di Girolamo M., M. G. Silletta, M. A. De Matteis, A. Braca, A. Colanzi, D. Pawlak, M. M. Rasenick, A. Luini, & D. Corda. 1995. Evidence that the 50-kDa substrate of brefeldin A-dependent ADP-ribosylation binds GTP and is modulated by the G-protein bg subunit complex. *Proc. Natl. Acad. Sci. USA. 92*: 7065–7069

3. Takatsuki, A., & G. Tamura. 1985. Brefeldin A, a specific inhibitors of intracellular translocation of vesicular stomatitis virus G protein: intracellular accumulation of high-mannose type G protein and inhibition of its cell surface expression. *Agric. Biol. Chem. 49*: 899–902.

4. Misumi, Y., Y. Misumi, K. Miki, A. Takatsuki, G. Tamura, and Y. Ikehara. 1986. Novel blockade by Brefeldin A of intracellular transport of secretory proteins in cultured rat hepatocytes. *J. Biol. Chem. 261*: 11398–11403.

5. Magner, J. A., and E. Papagiannes. 1988. Blockade by Brefeldin A of intracellular transport of secretory proteins in mouse pituitary cells: effects on the biosynthesis of thyrotropin and free a-subunits. *Endocrinology 122*: 912–920.

6. Fujiwara, T., K. Oda, S. Yokota, A. Takatsuki, & Y. Ikehara. 1988. Brefeldin A causes disassembly of the Golgi complex and accumulation of secretory proteins in the endoplasmic reticulum. *J. Biol. Chem. 263*: 18545–18552.

7. Doms, R. W., G. Russ, & J. W. Yewdell. 1989. Brefeldin A redistributes resident and itinerant Golgi proteins to the endoplasmic reticulum. *J. Cell Biol. 109*: 61–72.

8. Pavelka, M., & A. Ellinger. 1993. Early and late transformations occurring at organelles of the Golgi area under the influence of brefeldin A: an ultrastructural and lectin cytochemical study. *J. Histochem. Cytochem. 41*: 1031–1042.

9. Lippincott-Schwartz, J., L. C. Yuan, J. S. Bonifacino, & R. D. Klausner. 1989. Rapid redistribution of Golgi proteins into the ER in cells treated with Brefeldin A: evidence for membrane cycling from Golgi to ER. *Cell 56*: 801–813.

10. Lippincott-Schwartz, J., J. G. Donaldson, A. Schweizer, E. G. Berger, H. P. Hauri, L. C. Yuan, & R. D. Klausner. 1990. Microtubule-dependent retrograde transport of proteins into the ER in the presence of Brefeldin A suggests an ER recycling pathway. *Cell 60*: 821–836.

11. Alcalde, J., P. Bonay, A. Roa, S. Vilaro, & I. V. Sandoval. 1992. Assembly and disassembly of the Golgi complex: two processes arranged in a cis-trans direction. *J. Cell Biol. 116*: 69–83.

12. Donaldson, J. G., J. Lippincott-Schwartz, & R. D. Klausner. 1991. Guanine nucleotides modulate the effects of brefeldin A in semipermeable cells: regulation of the association of a 110-kD peripheral membrane protein with the Golgi apparatus. *J. Cell Biol. 112*: 579–588.

13. Klausner, R. D., J. G. Donaldson, & J. Lippincott-Schwartz. 1992. Brefeldin A: insights into the control of membrane traffic and organelle structure. *J. Cell Biol. 116*: 1071–1080.

14. Guo, Q., E. Vasile, & M. Krieger. 1994. Disruptions in Golgi structure and membrane traffic in a conditional lethal mammalian cell mutant are corrected by ε-COP. *J. Cell Biol. 125*: 1213–1224.

15. Misteli, T., & G. Warren. 1994. COP-coated vesicles are involved in the mitotic fragmentation of Golgi stacks in a cell-free system. *J. Cell Biol. 125*: 269–282.

16. Misteli, T., & G. Warren. 1995. Mitotic disassembly of the Golgi apparatus in vivo. *J. Cell Sci. 108*: 2715–2727.

17. Hunziker, W., J. A. Whitney, & I. Mellman. 1991. Selective inhibition of transcytosis by brefeldin A in MDCK cells. *Cell 67*: 617–627.

18. Yan, J.-P., M. E. Colon, L. A. Beebe, & P. Melançon. 1994. Isolation and characterization of mutant CHO cell lines with compartment-specific resistance to brefeldin A. *J. Cell Biol. 126*: 65–75.

19. Ktistakis N. T., M. G. Roth, G. S. Bloom. 1991. Ptk1 cells contain a nondiffusible, dominant factor that makes the Golgi apparatus resistant to brefeldin A. *J. Cell Biol. 113*: 1009–1023.

20. Weigert, R., A. Colanzi, G. Sciulli, A. Mironov, G. Santini, M. Di Girolamo, J. G. Donaldson, D. Corda, M. A. De Matteis, & A. Luini. Pharmacologic inhibitors of the brefeldin A-induced ADP-ribosylation. (Submitted).

21. Mironov, A., A. Colanzi, S. Flati, I. Santone, A. Fusella, R. Polishchuk, A. Mironov Jr, G. Di Tullio, R. Weigert, V. Malhorta, D. Corda, M. A. De Matteis, & A. Luini. Role of NAD+ and ADP-ribosylation in the maintenance of the Golgi structure. (Submitted).

22. Colanzi, A., A. Mironov, R. Weigert, C. Limina, S. Flati, C. Cericola, G. Di Tullio, M. Di Girolamo, D. Corda, M. A. De Matteis, & A. Luini. 1996. Brefeldin A-induced ADP-ribosylation in the structure and function of the Golgi complex. This book.

45

MODULATORY ROLE OF GTP-BINDING PROTEINS IN THE ENDOGENOUS ADP-RIBOSYLATION OF CYTOSOLIC PROTEINS

Maria Di Girolamo, Rosita Lupi, Maria Giuseppina Silletta,
Sabrina Turacchio, Cristiano Iurisci, Alberto Luini, and Daniela Corda

Department of Cell Biology and Oncology
Istituto di Ricerche Farmacologiche "Mario Negri"
Consorzio Mario Negri Sud
Via Nazionale
66030 Santa Maria Imbaro (Chieti), Italy

1. ABSTRACT

The endogenous ADP-ribosylation of cytosolic proteins and the pattern of NAD degradation were analyzed in subcellular fractions of rat liver in order to investigate the modulation of these reactions by GTP-binding (G) proteins.

We could show that intracellular membranes from rat liver have a guanine nucleotide- and divalent cation-dependent pyrophosphatase activity able to rapidly degrade NAD to AMP. This enzymatic activity was investigated by two different approaches: the degradation of $[^{32}P]$-NAD in the presence of intracellular membranes and the mono-ADP-ribosylation of cytosolic proteins. Divalent cations, preferentially Zn^{2+} and Mn^{2+}, were required for the pyrophosphatase activity, since in the presence of the Zn^{2+} chelator TPEN (N,N,N´,N´-tetrakis(2-pyridyl-methyl)ethylenediamine) or EDTA, the NAD degradation was inhibited by about 50%. Accordingly, in the presence of TPEN the endogenous ADP-ribosylation of cytosolic proteins was enhanced, whereas Zn^{2+} caused a significant inhibition of this reaction.

GDPβS was able to strongly activate the mono-ADP-ribosylation of cytosolic proteins. This effect was abolished by GTPγS, suggesting that a G protein, or rather one of the subunits of a heterotrimeric G protein, is involved in the modulation of the pyrophosphatase and consequently, of endogenous ADP-ribosylation. We propose that a regulatory pathway involving a heterotrimeric G protein modulates enzymes affecting the NAD turnover and availability of NAD for endogenous mADPRTs.

ADP-Ribosylation in Animal Tissue, edited by Haag and Koch-Nolte
Plenum Press, New York, 1997

2. INTRODUCTION

In eucaryotic cells NAD is consumed by multiple classes of enzymatic activities. Poly(ADP-ribosyl)polymerases catalyze the transfer of ADPR residues to specific acceptor proteins generating polymers of ADPR that are involved in the DNA repair process[1]. mADPRTs catalyze the transfer of a single ADPR residue from NAD to one or more specific amino acid residues of acceptor proteins, such as the GTP binding proteins when the reaction is catalyzed by specific bacterial toxins[2], or integrin[3], desmin[4], B50/GAP43[5], BiP/GRP78[6] when the reaction is catalyzed by endogenous mADPRTs.

Another class of enzymes able to hydrolyze NAD are the NADases that cleave the glycosylic linkage between nicotinamide and ADPR[7]. The majority of eukaryotic NADases are membrane-associated enzymes, localized at the outer surface of the plasma membrane, often anchored via a GPI linkage. The mammalian NADases which are not anchored via GPI linkage, display ADPR cyclase and cADPR hydrolase activities in addition to NADase activity[8].

We have evidence that, in addition to mADPRT and NADase, also a NAD pyrophosphatase able to produce AMP from NAD is involved in the turnover of NAD and thus in the NAD availability for endogenous ADP-ribosylation reaction[9].

3. RESULTS AND DISCUSSION

Enriched preparations of Golgi membranes obtained from rat liver were employed in studies aimed at investigating the modulation of enzymatic activities involved in endogenous ADP-ribosylation reactions.

Golgi membranes incubated for 30 min with [^{32}P]-NAD led to a progressive decrease in NAD and to the production of AMP and of an additional, as yet unidentified, metabolite (named "X"). Under these experimental conditions ADPR represented only 2 to 5% of total cpm (Fig. 1A)[9].

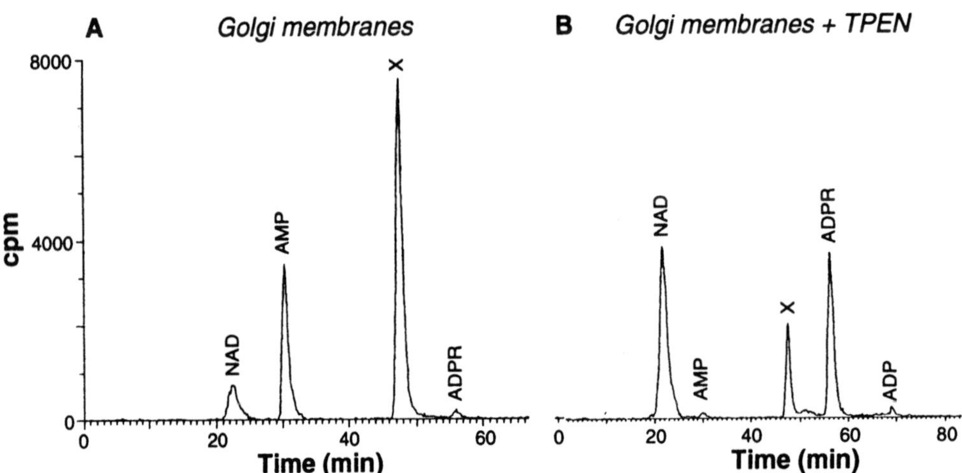

Figure 1. HPLC elution pattern of NAD metabolites obtained upon incubation with Golgi membranes (A) and TPEN (B). The elutions times are 22 min for NAD, 31 min for AMP, 47 min for "X" and 57 min for ADPR. The experimental procedure are described in detail in ref. 9.

These results originated the hypothesis that two different hydrolytic enzymes able to consume NAD are associated with intracellular membranes[9]. The first enzyme cleaves the pyrophosphoric bond and yields nicotinamide mononucleotide and AMP, the second one cleaves the glycoside link and produces nicotinamide and ADPR[9].

The first enzyme could be identified as a nucleotide pyrophosphatase (E.C. 3.6.1.9), a glycoprotein that is likely to exist as a dimer of identical subunits, each with a M.W. of 115/135 kDa depending on the tissue examined[10]. This enzymatic activity is sensitive to reducing reagents and requires divalent cations[11]. This enzyme is located on the external surface of plasma membranes, and in rough endoplasmic reticulum[12]. Our data suggest that this NAD pyrophosphatase might be also associated with the Golgi apparatus and with endocytic vesicles. Although the biological role of this enzyme is still unknown, it has been suggested that it might control the glycosylation process in mammalian cells[12].

The second enzyme could be identified as a NADase (E.C. 3.2.2.5) that has been reported to be ubiquitous since it was found in organisms ranging from bacteria to mammals. All mammalian NADases described so far are membrane-associated, with the exception of one described in seminal fluid. NADase activities were also reported in microsomal and mitochondrial membranes[13].

Although NADases have been known for more than 50 years, the physiological function of these enzymes in cellular metabolism remains unknown. Some NADases are bifunctional enzymes that catalyze both the synthesis and hydrolysis of cADPR. Under defined experimental conditions, we could observe a very limited formation of cADPR, which suggested that the NADase activity present in Golgi membranes is able to catalyze both the synthesis and degradation of cADPR. The levels of endogenous cADPR were previously determined in rat heart, brain and liver and found to be 208, 550 and 674 nM, respectively[14], in agreement with our observation that a NADase which has also cyclase activity is present in rat liver[9].

NAD-PDE and NADase might play a role in the endogenous ADP-ribosylation reaction, since they both participate in the NAD turnover; in particular the NAD-PDE is very active in competing with the endogenous mADPRTs for NAD. In line with this hypothesis, high levels of endogenous enzymic and non-enzymic ADP-ribosylation of cytosolic proteins were observed only when the NAD-phosphodiesterase was inhibited by TPEN (100 μM) or by EDTA (100 μM)[9] (as determined by HPLC analysis). The NAD degradation was inhibited by about 50%, in the presence of TPEN or EDTA, whereas the products of NAD, AMP and "X", were present in low percentage and ADPR was at least 30% of the total cpm (Fig. 1B)[9].

Interestingly, not only divalent cations, but also guanine nucleotides, were able to modulate the NAD phosphodiesterase activity, thus suggesting that this enzyme can be regulated by a GTP binding protein.

In the presence of 200 μM GDPβS, the NAD phosphodiesterase was inhibited[9]. This result paralleled the data obtained with TPEN and EDTA. The HPLC analysis revealed that the production of AMP and of the "X" metabolite was inhibited by 50%, and that ADPR was produced and employed for non-enzymic ADP-ribosylation of cytosolic proteins. Free ADPR is characterized by a reactive aldehyde that might be involved in covalent non-enzymic modifications of proteins[15]. Protein glycation by hexoses has been implicated in the pathophysiology of many diseases and in the aging process.

We have reported that brefeldin A (BFA) is able to activate an endogenous mono-ADPRT[16,17]. GDPβS led to an increased labeling of the two cytosolic proteins, GAPDH and BARS-50, that are enzymatically mono-ADP-ribosylated by the BFA-dependent mADPRT[16–18]. We hypothesize that the increased availability of NAD, due to the inhibition of NAD phosphodiesterase by GDPβS, might be employed, not only by NADases producing free ADPR, but also by the BFA-dependent mADPRT and eventually by other endogenous mADPRT. Thus, the modulation of NAD levels due to the NAD phosphodiesterase activity can be considered a relevant step in the endogenous enzymic and non-enzymic mono-ADP-ribosylation.

4. ACKNOWLEDGMENTS

This study was supported in part by the Italian Association for Cancer Research (AIRC) and the Italian National Research Council (Progetto Finalizzato ACRO ctr. n. 95.00558.PF39, and Convenzione CNR-Consorzio Mario Negri Sud).

5. REFERENCES

1. Ueda, K., & O. Hayaishi. ADP-ribosylation. 1985. *Annu. Rev. Biochem. 54*: 73–100.
2. Moss, J., & M. Vaughan. 1988. ADP-ribosylation of guanyl nucleotide-binding proteins by bacterial toxins. *Adv. Enzymol. 61*: 303–379.
3. Zolkiewska, A., & J. Moss. 1993. Integrin α7 as substrate for a glycosyl-phosphatidylinositol-anchored ADP-ribosyltransferase on the surface of skeletal muscle cells. *J. Biol Chem. 268*: 25273–25276.
4. Huang, H.-Y., D. J . Graves, R. M. Robson, & T. W. Huiatt. 1993. ADP-ribosylation of the intermediate filament protein desmin and inhibition of desmin assembly in vitro by muscle ADP-ribosyltransferase. *Biochem. Biophys. Res. Commun. 197*: 570–577.
5. Coggins, P. J., K. McLean, A. Nagy, & H. Zwiers. 1993. ADP-ribosylation of the neuronal phosphoprotein B-50/GAP-43. *J. Neurochem.. 60*: 368–371.
6. Ledford, B. E.. & G. H. Leno. 1994. ADP-ribosylation of the molecular chaperone GRP78/BiP. *Mol. Cell. Biol. 138:* 141–148.
7. Price, S. R., P. H. Pekala. Pyridine nucleotide-linked glycohydrolases. In: *Pyridine Nucleotide Coenzymes: chemical, biochemical, and medical aspects.* Dolphin, D., O. Avramovic & R. Poulson (eds.). Wiley-Interscience, New York, pp. 513–548.
8. Kim, H., E. L. Jacobson, & M. K. Jacobson. 1993. Synthesis and degradation of cyclic ADP-ribose by NAD glycohydrolases. *Science 261*: 1330–1333.
9. Di Girolamo, M., R., Lupi, M., G., Silletta, S., Turacchio, C. Iurisci, A.,Luini & D., Corda. Modulation of NAD availability by Golgi membrane-associated enzymes. (Submitted)
10. Rebbe, N. F., B. D. Tong, E. M. Finley, & S. Hickman. 1991. Identification of nucleotide pyrophosphatase/alkaline phosphodiesterase I activity associated with the mouse plasma cell differentiation antigen PC-1. *Proc. Natl. Acad. Sci. USA. 88*: 5192–5196.
11. Yano, T., I. Funakoshi, & I. Yamashina. 1985. Purification and properties of nucleotide, pyrophosphatase from human placenta. *J. Biochem. 98*: 1097–1107.
12. Bischoff, E., T. A.,Tran-Thi, & K. F. A.Decker. 1975. Nucleotide pyrophosphatase of rat liver: a comparative study on the enzymes solubilized and purified from plasma membrane and endoplasmic reticulum. *Eur. J. Biochem. 51*: 353–361.
13. Frei, B., & C. Richter. 1988. Mono(ADP-ribosylation) in rat liver mitochondria. *Biochemistry 27*: 529–535.
14. Walseth, T. F., R. Aarhus, R. J. Zeleznikar, Jr., & H. C. Lee. 1991. Determination of endogenous levels of cyclic ADP-ribose in rat tissues. *Biochim Biophys Acta 1094*: 113–120.
15. Jacobson, E. L., D. Cervantes-Laurean, & M. K. Jacobson. 1994. Glycation of proteins by ADP-ribose. *Mol Cell Biochem 138*: 207–212.
16. De Matteis, M. A., M. Di Girolamo, A. Colanzi, M. Pallas, G. Di Tullio, L. J. McDonald, J. Moss, G. Santini, S. Bannykh, D. Corda, & A. Luini. 1994. Stimulation of endogenous ADP-ribosylation by brefeldin A. *Proc. Natl. Acad. Sci. USA 91*: 1114–1118.
17. Colanzi, A., M. Di Girolamo, G. Santini, G. Sciulli, S. Santarone, M. Pallas, G. Di Tullio, S. Bannykh, D. Corda, M. A. De Matteis, & A. Luini. 1994. Brefeldin A, an inhibitor of vesicular traffic, stimulates the

ADP-ribosylation of two cytosolic proteins. In *GTPase-Controlled Molecular Machines*, eds. D. Corda, H. Hamm, & A. Luini. Ares-Serono Symposia Publications, Rome, pp. 197–217.

18. Di Girolamo, M., M. G. Silletta, M. A. De Matteis, A. Braca, A. Colanzi, D. Pawlak, M. M. Rasenick, A. Luini, & D. Corda. 1995. Evidence that the 50-kDa substrate of brefeldin A-dependent ADP-ribosylation binds GTP and is modulated by the G-protein βγ subunit complex. *Proc. Natl. Acad. Sci. USA 92*: 7065–7069.

46

ADP-RIBOSYLATING AND GLUCOSYLATING TOXINS AS TOOLS TO STUDY SECRETION IN RBL CELLS

Ulrike Prepens, Ingo Just, Fred Hofmann, and Klaus Aktories

Institut für Pharmakologie und Toxikologie der Albert-Ludwigs-Universität
   Freiburg
Hermann-Herder-Str. 5
79104 Freiburg i. Br., Germany

ABSTRACT

The influence of different ADP-ribosylating and glucosylating cytotoxins on stimulated protein tyrosine phosphorylation and secretion in rat basophilic leukemia (RBL) cells was studied.

Treatment of RBL cells with *Clostridium botulinum* C2 toxin, which specifically ADP-ribosylated monomeric G-actin and caused complete depolymerization of the actin cytoskeleton in intact cells, inhibited $Fc_\varepsilon RI$ receptor-mediated tyrosine phosphorylation of various proteins in a time- and concentration-dependent manner with maximal effects at 100 ng/ml C2I and 200 ng/ml C2II. C2 toxin (10 ng/ml C2I and 20 ng/ml C2II) increased antigen- or calcium ionophore (A23187)-stimulated [^{3}H]serotonin release maximally by about 3fold.

Clostridium botulinum C3, which ADP-ribosylated Rho in intact RBL cells, had no effect on protein tyrosine phosphorylation and stimulated secretion. In contrast, the cytotoxic *Clostridium difficile* toxin B (ToxB), which glucosylated the Rho-subtype family members RhoA and Cdc42, blocked or reduced antigen- or calcium ionophore-mediated [^{3}H]serotonin release, respectively, and decreased tyrosine phosphorylation of a 110 kDa protein.

The data indicate that different actin pools control tyrosine phosphorylation and secretion in RBL cells and suggest that Rho subfamily proteins regulate secretion independently of the actin cytoskeleton.

BACKGROUND

Mast cells and rat basophilic leukemia cells (RBL 2H3 hm1) are activated by antigen-induced cross-linking of the high affinity IgE-receptor, $Fc_\varepsilon RI$ [1]. Receptor aggregation

ADP-Ribosylation in Animal Tissue, edited by Haag and Koch-Nolte
Plenum Press, New York, 1997

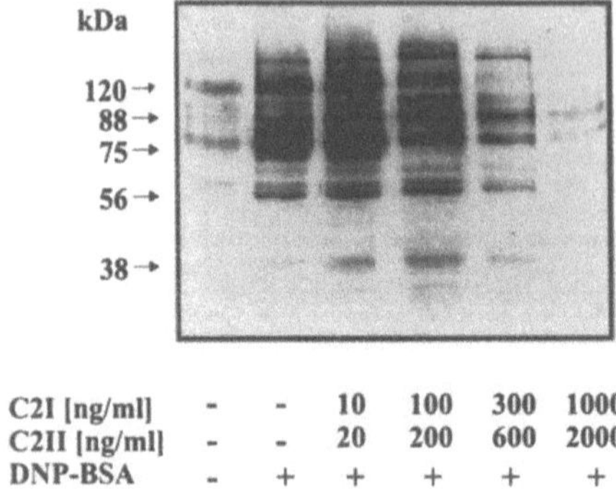

<table>
<tr><td>C2I [ng/ml]</td><td>-</td><td>-</td><td>10</td><td>100</td><td>300</td><td>1000</td></tr>
<tr><td>C2II [ng/ml]</td><td>-</td><td>-</td><td>20</td><td>200</td><td>600</td><td>2000</td></tr>
<tr><td>DNP-BSA</td><td>-</td><td>+</td><td>+</td><td>+</td><td>+</td><td>+</td></tr>
</table>

Figure 1. Effect of *C. botulinum* C2 toxin on protein tyrosine phosphorylation in RBL cells. RBL cells were incubated overnight with anti-DNP-IgE [0.2 µg/ml]. After incubation without (-) or with C2 toxin (+, at indicated concentration) for 2h, cells were washed and stimulated for 3 min at 37°C without (-) or with (+) DNP-BSA [500 ng/ml]. Thereafter lysates were separated by polyacrylamide gel electrophoresis. After immunoblotting tyrosine phosphorylated proteins were detected with anti-phosphotyrosine antibodies and the ECL Western blotting detection system (Amersham). Chemoluminogram is shown.

initiates a cascade of biochemical events which result in secretion of inflammatory mediators. Studies of the signal transduction pathways in RBL cells show that antigen binding stimulates tyrosine phosphorylation of multiple proteins (Lyn, Syk, β,γ-subunits of $Fc_\varepsilon RI$, PLCγ1, p125^{FAK}) [2], followed by an increase in intracellular Ca^{2+} and activation of protein kinase C [3], phospholipase (PL)D and/or A_2 [4,5]. Furthermore, stimulation of the IgE receptor is accompanied by changes in RBL cell morphology, enhanced cell spreading/flattening and an increase in F-actin [1,2]. The downstream events linking receptor activation to specific membrane and cytoskeletal responses are not known with certainty.

Actin, the main component of the microfilamental cytoskeleton, is involved in the organization of cell architecture and various cellular motile functions such as locomotion, endo- and exocytosis [6]. The ADP-ribosyltransferase C2 from *Clostridium botulinum* [7], which easily enters intact cells, selectively modifies monomeric actin at arginine-177. C2 toxin consists of two separate components, the binding component C2II (100 kDa) and the enzyme component C2I (45 kDa). C2II promotes the translocation of C2I in eucaryotic cells. The covalent modification of G-actin by C2I finally leads to a complete depolymerization of the actin cytoskeleton.

The low molecular mass GTP-binding proteins of the Rho subfamily (RhoA/B/C, Rac 1/2, Cdc42Hs) are involved in the regulation of the actin filament network and in the formation of stress fibers (Rho), membrane ruffles (Rac) and microspikes or lamellipodia (Cdc42) [8]. Furthermore Rho subfamily proteins probably act as molecular switches in various signal transduction pathways, i.e. the regulation of PLD [9], phosphatidylinositol-4-phosphate-5-kinase [10] or phosphatidylinositol-3-kinase [11,12]. Rho proteins (RhoA/B/C) are the specific substrates for C3, another ADP-ribosyltransferase from *Clostridium botulinum* [13]. The ADP-ribosylation at asparagine-41 renders the GTP-binding protein biologically inactive and leads to the destruction of the actin cytoskeleton. Comparable effects are induced by the monoglucosyltransferases from *Clostridium difficile*, the enterotoxin A and the cytotoxin B, which cause antibiotic-associated diarrhea and pseudomembraneous colitis [14]. Both block the biological activity of the Rho subfamily proteins by glucosylation at threonine-37/35 and prevent subsequent ADP-ribosylation by C3-like transferases [15,16].

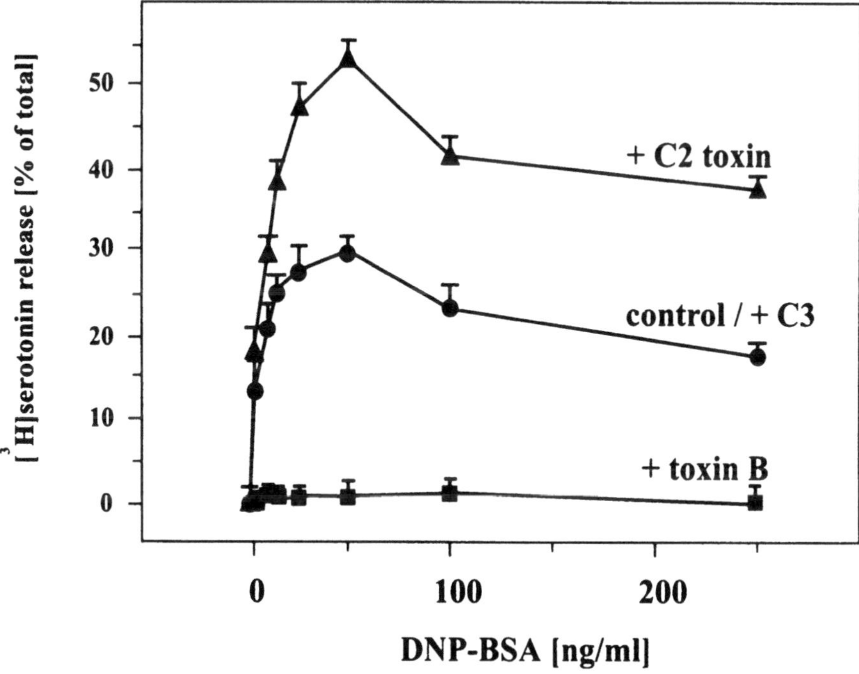

Figure 2. Effects of *C. botulinum* C2 toxin, C3 exoenzyme or *C. difficile* toxin B on DNP-BSA-stimulated [³H]serotonin release. RBL cells were incubated overnight with [³H]serotonin [2μCi/ml] and anti-DNP-IgE [0.2 μg/ml] following treatment with buffer (control), toxin B [3 ng/ml, 2h], C2 toxin [100 ng/ml C2I plus 200 ng/ml C2II] or C3 exoenzyme [40 μg/ml, 24 h]. Medium was removed and cells were washed three times. Serotonin release was induced by addition of DNP-BSA at the indicated concentration. After 20 min the supernatant was removed and the amount of [³H]serotonin was determined by scintillation counting. Data are given as percent of total amount of serotonin after correction for spontaneous release and are means ± S.E. of three different experiments.

RESULTS

The antigen-induced protein tyrosine phosphorylation of total cell lysates from C2 toxin-treated RBL cells was significantly reduced (i.e. at about 75 kDa and 120 kDa) in comparison to untreated cells.

In RBL cells loaded with [³H]serotonin and anti-DNP IgE, incubation with C2 toxin previous to antigen (DNP-BSA) or calcium ionophore (A23187) stimulation resulted in an enhanced secretory response without changes in basal mediator release. Significant toxin effects were observed after 2 hours at concentrations of 100 ng/ml C2I plus 200 ng/ml C2II (tyrosine phosphorylation) or 10 ng/ml C2I plus 20 ng/ml C2II (secretion). The single toxin components, C2I or C2II, had no inhibitory or stimulatory influence, whereas the complete toxin acted in a time- and concentration-dependent manner.

Clostridium difficile toxin B glucosylated at least two proteins with Mr of about 22,000 in RBL cells. By applying two-dimensional polyacrylamide gel electrophoresis and immunoblotting, we identified RhoA and Cdc42 as substrates of toxin B (not shown).

In toxin B-treated cells DNP-BSA- or A23187-induced [³H]serotonin release was blocked or reduced, respectively. Maximal effects were observed at 0.2 ng/ml toxin B for 2 hours. Furthermore, toxin B strongly decreased the antigen-stimulated tyrosine phosphorylation of a 110 kDa protein (not shown).

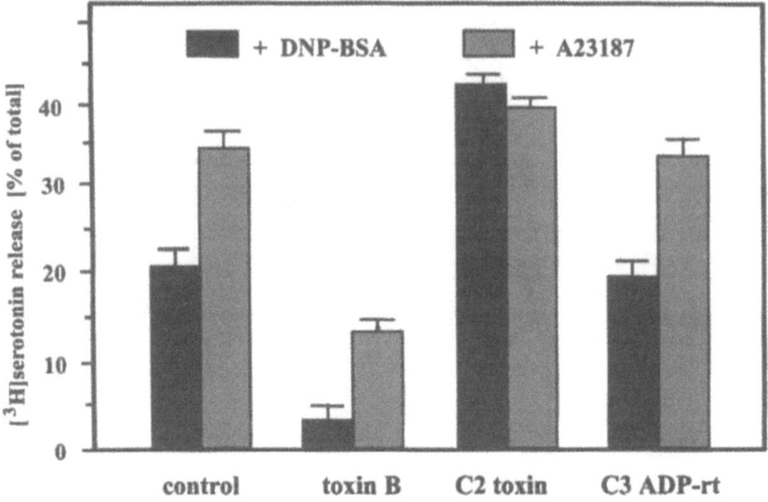

Figure 3. Effects of different clostridial cytotoxins on DNP-BSA- or A23187-stimulated [³H]serotonin release. RBL cells were primed with [³H]serotonin and anti-DNP-IgE overnight following treatment with buffer (control), toxin B [3 ng/ml, 2h], C2 toxin [100 ng/ml C2I plus 200 ng/ml C2II, 2h] or C3 exoenzyme [40µg/ml, 24h]. Thereafter medium was removed and degranulation was initiated by incubation with DNP-BSA [30 ng/ml] or A23187[300 nM]. After 20 min the amount of [³H]serotonin released was determined by scintillation counting. Data are given as percent of total amount of serotonin after correction for spontaneous release and are means of three different experiments.

Treatment of RBL cells with the ADP-ribosyltransferase C3 did neither affect stimulated protein tyrosine phosphorylation nor [³H]serotonin release. C3-catalyzed ADP-ribosylation of Rho in intact cells was confirmed by subsequent ADP-ribosylation of lysates from C3 toxin-treated cells, which showed a reduced incorporation of [³²P]ADP-ribose in comparison to control cells (not shown).

DISCUSSION

Here we report that the ADP-ribosyltransferase C2 from *C. botulinum* inhibits stimulated protein tyrosine phosphorylation and enhances antigen- and calcium ionophore-induced serotonin release from RBL 2H3 hm1 cells. The actin cytoskeleton seems to be differently involved in both events. It is suggested that a distinct actin pool plays a crucial role in the regulation of granule movement to the plasma membrane, vesicle docking and membrane fusion. Probably an intact actin filament network acts as a barrier thereby preventing degranulation. Tyrosine phosphorylation may depend on a different actin pool less sensitive towards C2-induced depolymerization, because higher concentrations of C2 were necessary to block tyrosine phosphorylation.

The inhibition of IgE-stimulated degranulation by *C. difficile* toxin B is most likely caused by the glucosylation of the Rho subfamily proteins RhoA and Cdc42 and appears to be independent of the actin cytoskeleton. Furthermore, this family of GTP-binding proteins is probably involved in early signal transduction processes of the Fc$_\varepsilon$RI pathway. Recent studies with mast cells indicate the involvement of Rho and Rac in exocytic processes [17]. However, our study with C3, which ADP-ribosylated at least 90 % of Rho in intact RBL cells, rather suggests an essential role for Cdc42 in RBL cell secretion. Whether the toxin B-induced decrease in protein tyrosine phosphorylation is due to a direct or indirect effect is subject of current investigations.

REFERENCES

1. Beaven, M. A., & H. Metzger. 1993. Signal transduction by Fc receptors: the Fc$_\varepsilon$RI case. *Immunol. Today.* *14:* 222–226.
2. Pfeiffer, J. R., & J. M. Oliver. 1994. Tyrosine kinase-dependent assembly of actin plaques linking Fc$_\varepsilon$RI cross-linking to increased cell substrate adhesion in RBL-2H3 tumor mast cells. *J. Immunol. 152:* 270–279.
3. Fasolato, C., J. Hoth, & R. Penner. 1993. A GTP-dependent step in the activation mechanism of capacitative calcium influx. *J. Biol. Chem. 268:* 20737–20740.
4. Lin, P., G.A. Wiggan, & A.M. Gilfillan. 1991. Activation of phospholipase D in a rat mast (RBL 2H3) cell line. *J. Immunol. 146:* 1609–1616.
5. Nakatani, Y., M. Murakami, I. Kudo, & K. Inoue. 1994. Dual regulation of cytosolic phospholipase A$_2$ in mast cells after cross-linking of Fcε-receptor. *J. Immunol. 153:* 796–803.
6. Bershadsky, A.D., & J.M.Vasiliev. 1988. *Cytoskeleton.* Plenum Press, New York.
7. Aktories, K., M. Bärmann, I. Ohishi, S. Tsuyama, K.H. Jakobs, & E. Habermann. 1986. Botulinum C2 toxin ADP-ribosylates actin. *Nature. 322:* 390–392.
8. Nobes, C.D., & A. Hall.1995. Rho, Rac and Cdc42 GTPases regulate the assembly of multimolecular focal complexes associated with actin stress fibers, lamellipodia and filopodia. *Cell. 81:* 53–62
9. Malcolm, K.C., A.H. Ross, R.-G. Qiu, M. Symons, & J.H. Exton. 1994. Activation of rat liver phospholipase D by the small GTP-binding protein RhoA. *J.Biol.Chem. 269:* 25951–25954.
10. Chong, L.D., A. Traynor-Kaplan, G.M. Bokoch, & M.A. Schwartz. 1994. The small GTP-binding protein Rho regulates a phosphatidylinositol 4-phosphate 5-kinase in mammalian cells. *Cell 79:* 507–513.
11. Zhang, J., W.G. King, S. Dillon, A. Hall, A., L. Feig, & S.E. Rittenhouse. 1993. Activation of platelet phosphatidylinositide 3-kinase requires the small GTP-binding protein Rho. *J.Biol.Chem. 268:* 22251–22254.
12. Hawkins, P.T., A. Eguinoa, R.-G. Qiu, D. Stokoe, F.T. Cooke, R. Walters, S. Wennström, L. Claesson-Welsh, T. Evans, M. Symons, & L. Stephens. 1995. PDGF stimulates an increase in GTP-Rac via activation of phosphoinositide 3-kinase. *Curr.Biol. 5:* 393–403.
13. Sekine, A., M. Fujiwara, & S. Narumiya. 1989. Asparagine residue in the rho gene product is the modification site for botulinum ADP-ribosyltransferase. *J.Biol.Chem. 264:* 8602–8605.
14. Bongaerts, G.P.A., & D.M. Lyerly. 1994. Role of toxins A and B in the pathogenesis of *Clostridium difficile* disease. *Microb.Pathog. 17:* 1–12.
15. Just, I., J. Selzer, M. Wilm, C. von Eichel-Streiber, M. Mann, & K. Aktories. 1995. Glucosylation of Rho proteins by *Clostridium difficile* toxin B. *Nature 375:* 500–503.
16. Just, I., M.Wilm, J. Selzer, G. Rex, C. von Eichel-Streiber, M. Mann, & K. Aktories. 1995. The enterotoxin from *Clostridium difficile* (ToxA) monoglucosylates the Rho proteins. *J.Biol.Chem. 270:* 13932–13936.
17. Price, L.S., J.C. Norman, A.J. Ridley, & A. Koffer. 1995. The small GTPases Rac and Rho as regulators of secretion in mast cells. *Curr Biol 5:* 68–73.

47

CELL SURFACE DYNAMICS OF GPI-ANCHORED PROTEINS

Frederick R. Maxfield[1] and Satyajit Mayor[2]

[1]Department of Biochemistry
Cornell University Medical College
1300 York Avenue
New York, NY 10021
[2]National Centre for Biological Sciences
TIFR Centre
Post Box 1234, IISC Campus
Bangalore 560012
India

ABSTRACT

Several cell surface eukaryotic proteins have a glycosylphosphoinositol lipid (GPI) modification at the carboxy-terminal end that serves as their sole means of membrane anchoring. In this report we review recent observations regarding the surface dynamics of GPI-anchored proteins. We discuss the association of GPI-anchored proteins with caveolae at the cell surface and their role in signal transduction as determined by the ability of GPI-anchored proteins to form detergent-insoluble complexes enriched in several cytoplasmic proteins including non-receptor type tyrosine kinases and caveolin/VIP-21, a component of the striated coat of caveolae. We have shown by immunofluorescence and electron microscopy that GPI-anchored proteins are not constitutively concentrated in caveolae but may be enriched in these structures only after cross-linking. While caveolae occupy only a small fraction of the cell surface (< 4%) almost all of the GPI-anchored protein at the cell surface becomes incorporated into detergent-insoluble low-density complexes, suggesting that these proteins are intrinsically detergent-insoluble in the milieu of the plasma membrane, and their co-purification with caveolin is not reflective of their native distribution. The finding that GPI-anchored proteins are not normally clustered over caveolae raised questions about the involvement of caveolae in the internalization of GPI-anchored proteins. In recent studies we have found that GPI-anchored proteins are internalized into *bona fide* endosomes wherein they appear to be sorted from bulk membrane components. The implications of these observations on the biology of GPI-anchored proteins are discussed.

ADP-Ribosylation in Animal Tissue, edited by Haag and Koch-Nolte
Plenum Press, New York, 1997

INTRODUCTION

A diverse set of cell surface eukaryotic proteins including several receptors, enzymes, and adhesion molecules have a glycolipid modification at the carboxy-terminal end (Fig 1A). This is a post-translational modification that serves as a membrane anchor and involves the replacement of the carboxy-terminal peptide sequence of the protein by a glycosyl-inositol phospholipid (GPI) moiety [1–5]. The structure and biosynthesis of the GPI moiety are now well understood, and it appears that all GPI-anchors have a common core glycan that bridges an ethanolamine residue in amide linkage with the protein and an inositol phospholipid [2, 4, 6, 7]. The main differences between GPI-anchors are due to numerous types of side chain modifications decorating the core glycan structure and the presence of various lipid backbones including glycerolipid and ceramide-based structures attached to a variety of mainly saturated acyl and alkyl chains [7] (Fig. 1B).

In spite of the extensive biochemical information on the GPI-moiety of GPI-anchored proteins, the functions of this ubiquitous protein modification are less understood, although it has been implicated in a variety of cell biological processes [8]. The GPI-anchor has been proposed to act as an apical targeting signal for proteins in some epithelial cell

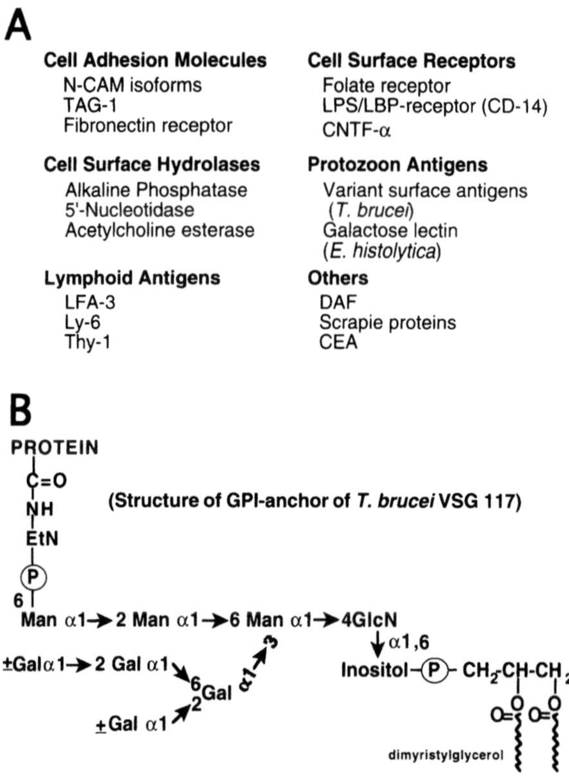

Figure 1. Functional diversity of GPI-anchored proteins and complete structure of a GPI-anchor. A). Representative list of GPI-anchored proteins. This list is meant to give an idea of the different types of proteins which are GPI-anchored. More complete listings of GPI-anchored proteins are available [2, 4, 7]. B). Complete structure of a GPI-anchor present on a variant surface glycoprotein from *Trypanosoma brucei* . Structure drawn according to reference 57.

types via its association with putative glycolipid rafts in the trans-Golgi network [9, 10]. GPI-anchoring has also been shown to be important for the intracellular signaling capacity of several proteins especially in lymphocytes. In most cases the cross-linking of the protein is a prerequisite for their signaling function [11, 12].

CAVEOLAR ASSOCIATION AND DETERGENT INSOLUBILITY

In immunolocalization studies, GPI-anchored proteins have been found to be clustered at the cell surface, and a significant fraction of the clusters are localized to 50–60 nm caveolin/VIP-21-coated membrane invaginations called caveolae [13, 14]. GPI-anchored proteins, and a subset of membrane lipids, (sphingomyelin, acidic and neutral glycolipids, and cholesterol), and caveolin/VIP-21 have been shown to be largely insoluble in non-ionic detergents (mainly cold Triton X-100) [15–18]. Together these properties have been used by many investigators to purify caveolae or 'caveolin/VIP-21-rich membranes' (reviewed in refs. 15, 19). Many cytoplasmically-oriented signaling molecules, including heterotrimeric GTP-ases, small GTP-ases, and non-receptor type protein-tyrosine kinases (PTKs) have been localized to caveolae, primarily in aggregates derived with Triton X-100 (reviewed in ref. 11). Based on this co-purification GPI-anchored proteins have been proposed to mediate intracellular signaling in caveolae due to their association with PTKs in these structures [20, 11, 19].

However, we have recently shown using fluorescently-labelled monoclonal antibodies to different GPI-anchored proteins, that these proteins are not constitutively concentrated in caveolae; they are enriched in these structures only after cross-linking with polyclonal secondary antibodies [21]. Similar results have been obtained by the use of fluorescein-modified folic acid as a monovalent ligand for the GPI-anchored folate receptor showing that the diffuse distribution of these proteins is not a consequence of antibody binding [22]. Analyses of the cell-surface distribution of GPI-anchored folate receptor and alkaline phosphatase by electron microscopy have also confirmed that these proteins are not constitutively enriched in caveolae [21, 23]. Thus, multimerization of GPI-anchored proteins regulates their sequestration in caveolae, but in the absence of agents that promote clustering they are diffusely distributed over the plasma membrane [21]. (see Fig. 2)

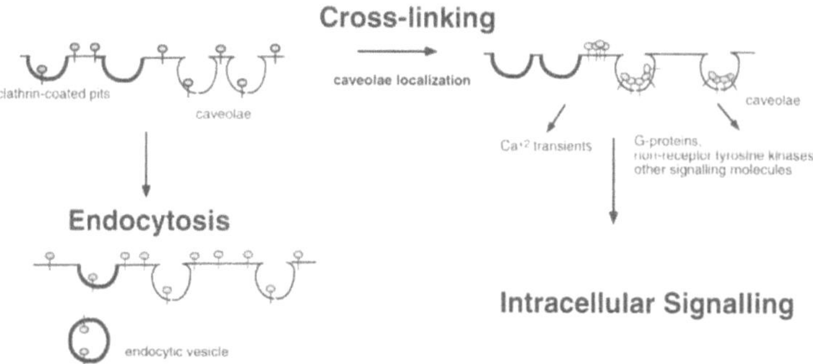

Figure 2. Dynamics of GPI-anchored proteins at the cell surface. A schematic of the cell-surface dynamics of GPI-anchored proteins wherein in their native state _GPI-anchored proteins are free to diffuse in the plane of the bilayer. Consequently, they are neither enriched nor depleted in coated or non-coated pits and are taken into the cell by mechanisms of bulk membrane endocytosis. Upon crosslinking by antibodies or by physiological agents (see text) these proteins will be clustered and preferentially localize to caveolae.

The finding that GPI-anchored proteins are not normally clustered over caveolae raised questions about how these proteins become enriched in the caveolae preparations obtained from detergent-extracted cells. We have recently analyzed the effect of Triton X-100 on the surface distribution of GPI-anchored proteins [22]. The data show that GPI-anchored proteins at the cell surface are almost completely insoluble in Triton X-100 and become quantitatively incorporated in low density complexes although only 4% of the surface is occupied by caveolae in fibroblast cells. Under these conditions transmembrane proteins such as the transferrin receptor and most other polypeptide-based membrane-anchored proteins are readily solubilized [24, 18, 25].

These observations are contrary to the widespread interpretation that the low density Triton X-100 insoluble membranes represent a purified caveolae preparation (or a caveolin/VIP-21-rich preparation: 50–100 fold enriched) thereby accounting for the 100–200 fold enrichment of the GPI-anchored proteins contained in them [20, 26, 19, 25]. The almost complete insolubility of the GPI-anchored proteins along with their lack of enrichment in the relatively small punctate areas occupied by caveolae after detergent-treatment showed that these proteins are detergent-insoluble independent of their association with caveolae. The lack of association with caveolin/VIP-21 is consistent with recent studies in lymphocytes where GPI-anchored Thy-1 and the ganglioside GM1 were detergent-insoluble and almost quantitatively formed 'low-density complexes' even though the cells lack detectable caveolin/VIP-21 or caveolae [27]. As suggested by [27] this could be due to the formation of specific detergent-insoluble glycolipid domains. Alternatively, these proteins may be intrinsically detergent-insoluble in the milieu of the plasma membrane (see below).

Multiple GPI-anchored proteins and glycolipids are simultaneously enriched in the detergent resistant complexes [26–28, 19, 25]. This has been interpreted to suggest that multiple GPI-anchored proteins are associated with each other (or a common receptor) in these complexes. In direct contradiction of these observations we have shown that multiple GPI-anchored proteins on the same cell did not co-cluster nor get recruited to caveolae unless they were independently cross-linked indicating that the GPI-anchored proteins are not associated with each other at the cell surface [21]. Similar observations have been made in lymphocytes where Thy-1 and GM-1 did not co-cluster after cross-linking with multivalent agents [27].

INSOLUBILITY, MICRODOMAINS, AND SIGNALING VIA GPI-ANCHORED PROTEINS

It has been shown [29] that the insolubility of GPI-anchored proteins and membrane lipids in Triton X-100 can be reconstituted in the absence of any special structures or protein(s). The primary requirement for detergent insolubility is the presence of a significant mole fraction of high-melting temperature lipids such as saturated acyl chain-containing phospholipids as well as an optimal concentration of cholesterol or neutral glycolipids. Furthermore, detergent insolubility of GPI-anchored proteins requires the presence of the appropriate lipid milieu in the same bilayer as these proteins (D. Brown, personal communication; [29]). Thus, the insolubility of GPI-anchored proteins in Triton X-100 depends mainly on the acyl or alkyl chain composition of the membrane lipids, cholesterol or neutral glycolipid content, and probably the degree of saturation of the acyl or alkyl moiety of the GPI-anchor. Saturated alkyl/acyl chains appear to be the predominant components of the lipid portion of GPI-anchors [7].

Many cytoplasmically oriented proteins such as PTKs and heterotrimeric G proteins have been found in detergent insoluble complexes [30, 26, 31, 28, 19, 32]. As pointed out by Lisanti and co-workers these proteins share a common lipid modification motif, Met-Gly-Cys, which is the site for N-myristylation (on the N-terminal Gly after cleavage of Met) and palmitylation (on the Cys) at the N-terminus and may direct association with the detergent-insoluble complexes [33]. In confirmation of this hypothesis, the association of two PTKs, p56^{lck} and p59^{fyn}, with the detergent-insoluble complexes was found to be due to the N-terminal myristylation at the Gly residue and palmitylation at the Cys 3 residue [34]. Furthermore, another PTK, p60^{src}, is N-myristylated but is not palmitylated (since it has the sequence Met Gly Ser at the N-terminus) and was found not to be associated with these detergent-insoluble complexes [34]. However, a mutant version of p60^{src} that contains the Met Gly Cys motif at its N-terminus is N-myristylated at the Gly residue and palmitylated at Cys, and was found to be associated with the detergent-resistant complex. These data are consistent with the idea that association with the detergent-insoluble membranes requires closely juxtaposed saturated fatty acyl chains and the presence in the same bilayer of the various saturated acyl/alkyl chain-containing proteins and lipids [29, 35].

These cytoplasmic lipid-modified proteins share the property of detergent-insolubility with GPI-anchored proteins and association with the low-density complexes. However, we have shown that in the case of the GPI-anchored proteins, this does not reflect an initial concentration in any specialized structures. Rather, it is a measure of their distribution into detergent-resistant structures after detergent treatment. It is possible that a similar redistribution occurs with these cytoplasmic proteins. These observations indicate that there is little evidence for the existence of complexes of GPI-anchored proteins and PTKs at the cell surface prior to detergent-extraction. Further studies will be required to distinguish whether cross-linked GPI-anchored proteins could cause the formation of membrane domains which assemble signaling molecules or if GPI-anchored proteins and cytoplasmic signaling molecules exist in complexes in micro-domains which are unresolvable at the resolution of current localization techniques.

It is possible that complexes similar to those formed after detergent-extraction may be formed in cells as a regulated event and therefore will require physiological triggers. Antibody-mediated cross-linking of GPI-anchored proteins, although not entirely physiological, is a means for activating signaling functions of GPI-anchored proteins (reviewed in [11, 8, 12]). The aggregated state of these lipid-linked proteins could generate a relatively stable membrane domain where signaling proteins such as N-terminal myristylated and Cys-3-palmitylated PTKs could be recruited and consequently activated. While no physiological agents have been shown to directly cross-link GPI-anchored proteins it is conceivable that GPI-anchored cell adhesion proteins may be cross-linked at cell-cell junctions due to the end to end association of homotypic dimers as in the case of Tag-1 [36, 37]. This could then trigger signaling events across cell-cell boundaries. It would be interesting to determine if such complexes may be formed in the membrane of caveolae since, after cross-linking, GPI-anchored proteins are relatively enriched in these structures in cells that have caveolae [21, 23].

In specific cases, there appear to be direct lectin-like interactions between GPI-anchored proteins (e. g., GPI-anchored isoform of the FcγRIII receptor (CD16), the lipopolysaccharide binding protein (CD14), or urokinase-type plasminogen activator receptor (uPAR)) and transmembrane proteins which are capable of intracellular signaling (e. g., complement receptor type 3 (CR3)) [38, 39]. The binding of soluble ligands to their respective receptors leads to large scale clusters of the GPI-linked receptor and the transmembrane proteins, followed by an activation of intracellular signaling. However, in these cases it is not clear whether any of the intracellular signaling is mediated via direct clustering of the GPI-anchored proteins.

ENDOCYTIC TRAFFIC OF GPI-ANCHORED PROTEINS

Numerous GPI-anchored proteins are internalized and recycled back to the cell surface [40–45]. It has been proposed that GPI-anchored proteins are internalized via the pinching off of caveolae in a process called potocytosis [46, 47]. However, these studies were based on the use of crosslinking antibodies or multivalent ligands that would cluster the GPI-anchored proteins into caveolae. The recent findings that these proteins are diffusely distributed at the cell surface, being neither enriched nor excluded from coated and non-coated pits at the cell surface, reopen the question about the mechanisms and pathways involved in the trafficking of GPI-anchored proteins in the endocytic process.

Regardless of the initial step in internalization, the endocytic trafficking of GPI-anchored proteins has some features that are different from other membrane components. For example, the recycling of folate receptors in MA104 cells [41] has been reported to be significantly slower than the typical recycling rates of lipids or recycling receptors [48, 49]. This could be consistent either with a specialized endocytic pathway [46] or with altered kinetics of passage through the typical endocytic recycling itinerary.

Recent data from our laboratory suggest that similar to other cell surface molecules, after internalization GPI-anchored proteins are found in early sorting endosomes wherein recycling components (*e.g.*, the Tf-receptor or the low density lipoprotein receptor) are sorted from lysosomally-directed components (*e.g.*, acid-released ligands such as low density lipoprotein or α_2-macroglobulin). From our studies we cannot exclude the possibility that GPI-anchored proteins are delivered to sorting endosomes via parallel endocytic pathways [50], different from the pathway of receptor-mediated endocytosis. However, our data showing extensive co-localization of GPI-anchored proteins with transferrin and internalized FITC-dextran as well as the electron microscopic studies of [51, 52] show that internal GPI-anchored proteins are mainly in endosomes. Recent data from our laboratory also show that GPI-anchored proteins recycle from endosomes back to the cell surface at about 1/3 the rate of transferrin receptors or bulk membrane components. This retention/sorting is dependent on the level of cholesterol in membranes. When cholesterol levels are reduced by 40%, GPI-anchored proteins exit from endosomes at the same rate as transferrin receptors or bulk membrane (Mayor & Maxfield, unpublished).

IMPLICATIONS FOR GPI-ANCHORED PROTEIN FUNCTION

The ability of GPI-anchored proteins to be sorted from bulk membrane in endosomes has significant implications for GPI-anchor function. The retention in endosomes will expose these proteins to the acidic milieu of both sorting and recycling endosomes for at least three fold longer times than recycling receptors with conventional transmembrane tails. Furthermore, this retention may take place in specialized membrane domains. The implications of these observations for individual proteins whose biological roles are dependent on GPI-anchoring are discussed below.

GPI-anchored cellular scrapie protein is processed to the infectious proteinase K-resistant prion form more efficiently than a form of the scrapie protein with a proteinaceous transmembrane anchor. This processing is dependent on acidic pH in endosomes and is in-

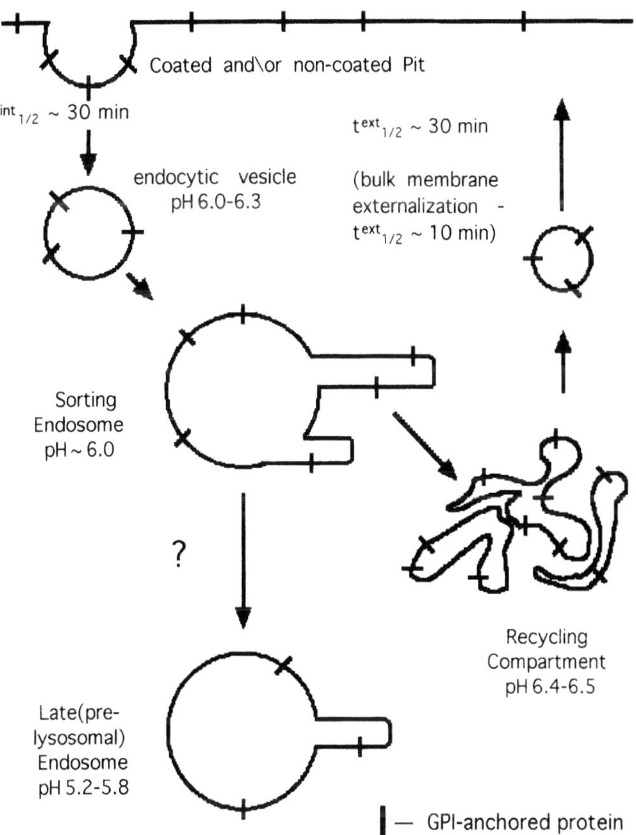

Figure 3. Endocytic route of GPI-anchored proteins. This schematic illustrates various fates of internalized GPI-anchored proteins. These proteins are first internalized into sorting endosomes and then recycled back to the plasma membrane via a peri-centriolar recycling compartment. The sorting of recycled molecules from lysosomally-destined molecules takes place in tubulo-vesicular endosomes, referred to as sorting endosomes, and recycling molecules then accumulate in the recycling compartment *en route* to the cell surface [49]. The late endosome represents a compartment that contains endocytosed molecules destined for the lysosome but not recycling receptors, such as LDL receptors or transferrin receptors [58]. Although GPI-anchored proteins follow the same pathway as recycling receptors and bulk membrane they are significantly slowed in their exit from the recycling compartment (half times $(t_{1/2}) \sim 30$ min vs. 10 min) and possibly sorting endosomes.

hibited by the depletion of cholesterol in membranes [53]. These observations are consistent with our results on GPI-anchor function. We propose that the GPI-anchor causes extensive retention of the cellular scrapie protein in acidic endosomes. If the infectious (prion) form of the scrapie protein is also present in these endosomes, this cholesterol-dependent retention mechanism could facilitate the efficient conversion of native scrapie protein into prions possibly in distinct membrane domains or aggregates.

Using [^{3}H]folic acid as a ligand for the folate receptor, Anderson, Kamen and colleagues [41] have shown that at steady state half the recycling pool of folate receptors are located at the cell surface while the other half are sequestered in an acid-resistant intracellular compartment(s). Using a fluorescent folic acid analogue, PLF, we have shown that the internal (acid-resistant) pool of receptors are in endosomes which can be accessed by recycling ligands such as transferrin. The kinetics of approach to steady state of receptor occupancy for the fluorescent ligand, PLF, is identical to that of the radioactive ligand, arguing strongly that the intracellular distribution of the fluorescent analogue accurately reflects the intracellular location of the folate receptor.

Considering the kinetics of internalization and export, morphological distribution and steady state kinetic parameters, we conclude that folate receptors are internalized as bulk membrane constituents into acidic sorting endosomes and recycling endosomes. In these endosomes they are retained longer than bulk membrane and other recycling receptors, consequently being exposed to the acid milieu of endosomes for relatively (three fold) longer periods. Recently Lacey and coworkers have shown that the GPI-anchored form is more efficient at folate uptake than the proteinaceous transmembrane anchored receptor with a functional endocytosis signal [54]. We suggest that it is the retention/sorting of the GPI-anchored folate receptor in acidic endosomes that makes the GPI-anchored folate receptors more efficient than the protein-anchored type. Furthermore, our data also provide an explanation for the drastic reduction in folate uptake efficiency in cells depleted of cholesterol [55, 56]; we find that GPI-anchored proteins including folate receptors are not sorted from bulk membrane constituents in cholesterol depleted cells.

CONCLUSIONS

In this report we have critically examined the association of GPI-anchored proteins, caveolae and cytoplasmic signaling molecules, and discuss a novel function of GPI-anchoring of proteins, namely, retention relative to bulk membrane constituents in endosomes. This has important implications for the biology of many of the GPI-anchored proteins described to date. We suggest that the ability of GPI-anchored proteins to sort from bulk membrane proteins and lipids in a manner dependent on the cholesterol content of membranes provides greater access to specialized environments in endosomes, and at least in the case of a few well studied examples this sorting may be functionally significant (e.g., folate transport via the GPI-anchored folate receptor and scrapie formation).

REFERENCES

1. Cross, G. A. M. 1990. Glycolipid anchoring of plasma membrane proteins. *Ann. Rev. Cell Biol.*, 6:1–34.
2. Englund, P. T. 1993. The structure and biosynthesis of glycosyl phosphatidylinositol protein anchors. [Review]. *Annu Rev Biochem*, 62:121–38.
3. Ferguson, M. A. J. and A. F. Williams 1988. Cell-surface anchoring of proteins via glycosyl phosphatidylinositol structures. *Ann. Rev. Biochem.*, 57:285–320.
4. Field, M. C. and A. K. Menon 1992. Glycolipid-anchoring of cell surface proteins. *In* Lipid modification of proteins. Schlesinger, M. J., Schlesinger, M. J.s. CRC Press, Boca Raton Ann Arbour London Tokyo. 83–134.
5. Low, M. G. 1989. The glycosyl-phosphatidylinositol anchor of membrane proteins. *Biochim. Biophys. Acta, 988:427–454.*
6. Mayor, S. and A. K. Menon 1990. Structural analysis of the glycosylinositol phospholipid anchors of membrane proteins. *Methods: A Companion to Methods in Enzymology*, 1:297–305.
7. McConville, M. J. and M. A. Ferguson 1993. The structure, biosynthesis and function of glycosylated phosphatidylinositols in the parasitic protozoa and higher eukaryotes. *Biochem. J.*, 294:30–204.
8. Ferguson, M. A. J. 1994. What can GPI do for you. *Parasitology Today*, 10:48 52.
9. Lisanti, M. P. and E. Rodriguez-Boulan 1990. Glycophospholipid membrane anchoring provides clues to the mechanism of protein sorting in polarized epithelial cells. *Trends Biochem. Sci.*, 15:113–118.
10. Simons, K. and A. Wandinger-Ness 1990. Polarized sorting in epithelia. [Review]. *Cell , 62:207–10.*
11. Brown, D. 1993. The tyrosine kinase connection: how GPI-anchored proteins activate T cells. [Review]. *Current Opinion in Immunology, 5:349–54.*

12. Robinson, P. J. 1991. Phosphatidylinositol membrane anchors and T-cell activation. *Immunol. Today*, 12:35–41.

13. Dupree, P., R. G. Parton, G. Raposo, T. V. Kurzchalia and K. Simons 1993. Caveolae and sorting in the trans-Golgi network of epithelial cells. *Embo J*, 12:1597–605.

14. Rothberg, K. G., J. E. Heuser, W. C. Donzell, Y.-S. Ying, J. R. Glenney and R. G. W. Anderson 1992. Caveolin, a protein component of caveolae membrane coats. *Cell*, 68:673–682.

15. Brown, D. 1992. Interactions between GPI-anchored proteins and membrane lipids. *Trends in Cell Biology*, 2:338–343.

16. Brown, D. A. and J. K. Rose 1992. Sorting of GPI-anchored proteins to glycolipid-enriched membrane subdomains during transport to the apical cell surface. *Cell*, 68:533–44.

17. Glenney, J. R. and L. Zokas 1989. Novel tyrosine kinase substrates from rous sarcoma virus transformed cells are present in the membrane skeleton. *J. Cell Biol.*, 108:2401–2408.

18. Hooper, N. M. and A. J. Turner 1988. Ectoenzymes of the kidney microvillar membrane: differential solubilization by detergents can predict a glycosyl-phosphatidylinositol membrane anchor. *Biochem. J.*, 250:865–869.

19. Lisanti, M. P., Z. L. Tang and M. Sargiacomo 1993. Caveolin forms a hetero-oligomeric protein complex that interacts with an apical GPI-linked protein: implications for the biogenesis of caveolae. *Journal of Cell Biology*, 123:595–604.

20. Anderson, R. G. 1993. Caveolae: where incoming and outgoing messengers meet. [Review]. *Proceedings of the National Academy of Sciences of the United States of America*, 90:10909–13.

21. Mayor, S., K. G. Rothberg and F. R. Maxfield 1994. Sequestration of GPI-anchored proteins in caveolae triggered by cross-linking. *Science*, 264:1948–51.

22. Mayor, S. and F. R. Maxfield 1995. Insolubility and redistribution of GPI-anchored proteins at the cell surface after detergent treatment. *Molecular Biology of the Cell*, 6:929–44.

23. Parton, R. G., B. Joggerst and K. Simons 1994. Regulated internalization of caveolae. *J. Cell Biol.*, 127:1199–1215.

24. Hooper, N. M. and A. Bashir 1991. Glycosyl-phosphatidylinositol-anchored membrane proteins can be distinguished from transmembrane polypeptide-anchored proteins by differential solubilization and temperature-induced phase separation in Triton X—114. *Biochem J.*, 280:745–751.

25. Sargiacomo, M., M. Sudol, Z. Tang and M. P. Lisanti 1993. Signal transducing molecules and glycosyl-phosphatidylinositol-linked proteins form a caveolin-rich insoluble complex in MDCK cells. *Journal of Cell Biology*, 122:789–807.

26. Chang, W. J., Y. S. Ying, K. G. Rothberg, N. M. Hooper, A. J. Turner, H. A. Gambliel, G. J. De, S. M. Mumby, A. G. Gilman and R. G. Anderson 1994. Purification and characterization of smooth muscle cell caveolae. *Journal of Cell Biology*, 126:127–38.

27. Fra, A. M., E. Williamson, K. Simons and R. G. Parton 1994. Detergent-insoluble glycolipid microdomains in lymphocytes in the absence of caveolae. *J. Biol. Chem*, 269:30745–30748.

28. Lisanti, M. P., P. E. Scherer, J. Vidugiriene, Z. Tang, V. A. Hermanowski, Y. H. Tu, R. F. Cook and M. Sargiacomo 1994. Characterization of caveolin-rich membrane domains isolated from an endothelial-rich source: implications for human disease. *Journal of Cell Biology*, 126:111–26.

29. Schroeder, R., E. London and D. A. Brown 1994. Interactions between saturated acyl chains confer detergent resistance on lipids and glycosylphosphatidylinositol (GPI)-anchored proteins: GPI-anchored proteins in liposomes and cells show similar behavior. *Proc. Natl. Acad. Sci. (USA)*, 91:12130–12134.

30. Bohuslav, J., T. Cinek and V. Horejsi 1993. Large, detergent-resistant complexes containing murine antigens Thy-1 and Ly-6 and protein tyrosine kinase p56lck. *European Journal of Immunology*, 23:825–31.

31. Cinek, T. and V. Horejsi 1992. The nature of large noncovalent complexes containing glycosyl-phosphatidylinositol-anchored membrane glycoproteins and protein tyrosine kinases. *Journal of Immunology*, 149:2262–70.

32. Stefanova, I., V. Horejsi, I. J. Ansotegui, W. Knapp and H. Stockinger 1991. GPI-anchored cell-surface molecules complexed to protein tyrosine kinases. *Science*, 254:1016–1019.

33. Lisanti, M. P., P. E. Scherer, Z. L. Tang and M. Sargiacomo 1994. Caveolae, caveolin and caveolin-rich membrane domains: a signalling hypothesis. *Trends in Cell Biology*, 4:231–235.

34. Shenoy, S. A., D. J. Dietzen, J. Kwong, D. C. Link and D. M. Lublin 1994. Cysteine3 of Src family protein tyrosine kinase determines palmitoylation and localization in caveolae. *Journal of Cell Biology*, 126:353–63.

35. Yu, J., D. A. Fishman and T. L. Steck 1973. Selective solubilization of proteins and phospholipids from red blood cell membranes by nonionic detergents. *J. Supramol. Struct.*, 1:233–248.

36. Felsenfeld, D. P., M. A. Hynes, K. M. Skoler, A. J. Furley and T. M. Jessell 1994. TAG-1 can mediate homophilic binding, but neurite outgrowth on TAG-1 requires an L1-like molecule and beta 1 integrins. *Neuron*, 12:675–90.

37. Furley, A. J., S. B. Morton, D. Manalo, D. Karagogeos, J. Dodd and T. M. Jessell 1990. The axonal glycoprotein TAG-1 is an immunoglobulin superfamily member with neurite outgrowth-promoting activity. *Cell*, *61:157–70.*

38. Kindzelskii, A. L., Z. O. Laska, R. 3. Todd and H. R. Petty 1996. Urokinase-type plasminogen activator receptor reversibly dissociates from complement receptor type 3 (alpha M beta 2' CD11b/CD18) during neutrophil polarization. *Journal of Immunology*, 156:297–309.

39. Zarewych, D. M., A. L. Kindzelskii, R. 3. Todd and H. R. Petty 1996. LPS induces CD14 association with complement receptor type 3, which is reversed by neutrophil adhesion. *Journal of Immunology*, 156:430–3.

40. Borchelt, D. R., A. Taraboulos and S. B. Prusiner 1992. Evidence for synthesis of scrapie prion proteins in the endocytic pathway. *Journal of Biological Chemistry*, 267:16188–99.

41. Kamen, B. A., M. T. Wang, A. J. Streckfuss, X. Peryea and R. G. Anderson 1988. Delivery of folates to the cytoplasm of MA104 cells is mediated by a surface membrane receptor that recycles. *J Biol Chem*, 263:13602–9.

42. Keller, G.-A., M. W. Siegel and I. W. Caras 1991. Endocytosis of glycophospholipid-anchored and transmembrane forms of CD4 by different endocytic pathways. *EMBO J.*, 11:863–874.

43. Lisanti, M. P., I. W. Caras, T. Gilbert, D. Hanzel and B. E. Rodriguez 1990. Vectorial apical delivery and slow endocytosis of a glycolipid-anchored fusion protein in transfected MDCK cells_. *Proc. Natl. Acad. Sci. USA*, 87:7419–7423.

44. Rothberg, K. G., Y.-S. Ying, J. F. Kolhouse, B. A. Kamen and R. G. W. Anderson 1990. The glycophospholipid-linked folate receptor internalizes folate without entering the clathrin-coated pit endocytic pathway. *J. Cell Biol.*, 110:637–649.

45. Taraboulos, A., D. Serban and S. B. Prusiner 1990. Scrapie prion proteins accumulate in the cytoplasm of persistently infected cultured cells. *J. Cell Biol.*, 110:2117–2132.

46. Anderson, R. G. W. 1993. Potocytosis of small molecules and ions by caveolae. *Trends Cell Biol.*, 3:69–72.

47. Turek, J. J., C. P. Leamon and P. S. Low 1993. Endocytosis of folate-protein conjugates: ultrastructural localization in KB cells. *Journal of Cell Science*, 106:423–430.

48. Koval, M. and R. E. Pagano 1989. Lipid recycling between the plasma membrane and intracellular compartments: transport and metabolism of fluorescent sphingomyelin analogues in cultured fibroblasts. *J. Cell Biol.*, 108:2169–2181.

49. Mayor, S., J. P. Presley and F. R. Maxfield 1993. Sorting of membrane components from endosomes and subsequent recycling to the cell surface occurs by a bulk flow process. *J. Cell Biol.*, 121:1257–1269.

50. Lamaze, C. and S. L. Schmid 1995. The emergence of clathrin-independent pinocytic pathways. *Current Opinion in Cell Biology*, 7:573–580.

51. Birn, H., J. Selhub and E. I. Christensen 1993. Cell fractionation and electron microscope studies of kidney folate-binding protein. *American Journal of Physiology*, 264:C302-C310.

52. Rijnboutt, S., G. Jansen, G. Posthuma, J. B. Hynes, J. H. Schornagel and G. J. Strous 1996. Endocytosis of GPI-linked membrane folate receptor-α. *J. Cell Biol.*, 132:35–47.

53. Taraboulos, A., M. Scott, A. Semenov, D. Avraham, L. Laszlo and S. B. Prusiner 1995. Cholesterol depletion and modification of COOH-terminal targeting sequence of the prion protein inhibit formation of the scrapie isoform. *Journal of Cell Biology*, 129:121–32.

54. Ritter, T. E., O. Fajardo, H. Matsue, R. G. Anderson and S. W. Lacey 1995. Folate receptor targeted to clathrin-coated pits cannot regulate vitamin uptake. *Proc Natl Acad Sci (U S A)*, 92:3824–3828.

55. Chang, W.-J., K. G. Rothberg, B. A. Kamen and R. G. W. Anderson 1992. Lowering cholesterol content of MA104 cells inhibits receptor-mediated transport of folate. *J. Cell Biol.*, 118:63–69.

56. Rothberg, K. G., Y.-S. Ying, B. A. Kamen and R. G. W. Anderson 1990. Cholesterol controls the clustering of the glycophospholipid-linked membrane receptor for 5-methyltetrahydrofolate. *J. Cell Biol.*, 111:2931–2938.

57. Ferguson, M. A. J., S. W. Homans, R. A. Dwek and T. W. Rademacher 1988. Glycosyl-phosphatidylinositol moiety that anchors *Trypanosoma brucei* variant surface glycoprotein to the membrane. *Science*, 239:753–759.

58. Dunn, K. W. and F. R. Maxfield 1992. Delivery of ligands from sorting endosomes to late endosomes occurs by maturation of sorting endosomes. *J. Cell. Biol.*, 117:301–310.

48

SIGNAL TRANSDUCTION VIA GPI-ANCHORED MEMBRANE PROTEINS

Peter J. Robinson

Transplantation Biology Group
MRC Clinical Sciences Centre
Royal Postgraduate Medical School
Hammersmith Hospital, Du Cane Road
London W12 0NN
England

INTRODUCTION

Eukaryotic cells carry upon their cell surfaces a number of proteins which are anchored in the membrane *via* glycosylphosphatidylinositol (GPI) membrane anchors[1]. These proteins are attached to the cell membrane by hydrophobic attraction of the phospholipid tails of the GPI-anchor with lipids in the outer face of the plasma membrane bilayer. Consequently, they neither span the cell membrane nor do they interact directly with intracellular components. During intracellular transport, GPI anchors recruit specialized lipids and glycolipids[2,3,4]. Acquisition of these lipids renders GPI-anchored proteins insoluble in a number of non-ionic detergents, and enables their enrichment in cell lysates by density-gradient centrifugation. These associated lipids, which include sphingomyelin and cholesterol, may alter the cell-surface distribution of GPI-linked molecules relative to their non-GPI-linked counterparts, and this is currently the subject of intensive study[5]. These GPI-rich detergent-insoluble complexes also contain a number of other proteins which are not GPI-linked, and it is thought that these other molecules may play a key role in the functional properties of GPI anchors. It was shown using photobleaching recovery (FRAP) or single particle tracking (SPT) techniques[6] that GPI-linked molecules are not freely mobile in the cell membrane, an expected finding for a molecule which is associated only with the outer lipid leaflet of the plasma membrane. This finding implies that the mobility of Thy-1 is influenced by its association with immobile components of the membrane, and thus, indirectly, with the cytoskeleton. These observations implicate transmembrane molecules in the interactions of GPI-linked molecules with cytoplasmic and cytoskeletal components.

One feature shared by several structurally and functionally distinct GPI-linked molecules is their ability to interact with host cell signal transduction pathways. A number of T

ADP-Ribosylation in Animal Tissue, edited by Haag and Koch-Nolte
Plenum Press, New York, 1997

lymphocyte GPI-linked molecules, such as Thy-1 and Ly-6, were identified as mitogenic targets before they were known to be GPI-anchored. Most of the early work in this area has been reviewed in detail elsewhere [7]. Since then many further examples of GPI-mediated signalling phenomena have been described, but exactly which cellular pathways are involved remains unclear. A consistent finding, however, is that the primary signalling event is initiated by clustering of GPI-linked molecules on the plasma membrane. In this article I shall attempt to discuss some current ideas on how such signals may occur and what might be their physiological significance.

PARAMETERS OF GPI-MEDIATED SIGNALLING

It has been known for some years that antibodies against GPI-linked cell surface molecules are mitogenic for lymphoid cells and more recently responses have been described in several other cell types[7]. To achieve a mitogenic response, which is usually accompanied by cytokine production, in lymphoid cells requires the binding of a specific, preferably soluble, antibody directed against the GPI-anchored molecule, followed by a second cross-linking anti-immunoglobulin antibody and an additional stimulus such as the phorbol ester e.g. phorbol myristate acetate (PMA). The requirement for phorbol ester appears, however, to be a downstream event in the signalling cascade because changes in intracellular free calcium occur with antibody cross-linking alone, conditions under which cell proliferation does not usually take place. Clustering of GPI-linked molecules on the cell surface appears, therefore, to be sufficient for the initial stimulatory event[8]. Immobilisation of the first antibody to a solid phase drastically reduces its effectiveness as a stimulator, probably because this inhibits its ability to induce formation of clusters on the cell membrane.

Precisely what the next step in the signalling cascade is remains unknown. Phosphorylation of cellular substrates and changes in intracytoplasmic free calcium concentration have both been used as an early readout for GPI-mediated signalling. Most mitogenic stimuli in lymphoid cells, routinely show an easily measurable calcium transient occurring seconds after stimulation This technique has been difficult to apply to GPI-induced signals. In whole mouse T-cell populations, we were unable to detect significant GPI-mediated calcium signals by flow cytometry or bulk fluorimetric analysis although successful use of these methods to measure calcium changes has been reported by others[9].

We have recently used fluorescence of the calcium-sensitive dye Fluo-3 in individual human vascular endothelial cells to show that anti Thy-1 antibodies induce calcium ion changes at different times after addition of a second, cross-linking, antibody[10]. Freshly isolated human umbilical vein endothelial cells express high levels of Thy-1 when induced with tumour promoters. When the inducer is removed, Thy-1 expression remains high for 1–2 days. When cells are treated with monoclonal anti-Thy-1 antibodies followed by anti-mouse immunoglobulin and examined by quantitative laser microscopy, transient increases in cytoplasmic free calcium concentrations of up to fourfold were detected. Individual cells responded asynchronously, with response times in the range of 1–5 min. after addition of the second antibody. In some cases individual cells were found to respond more than once to a single stimulus. This result is significant, because the heterogeneous nature of the response differs markedly from the effect of thrombin on the same cells. With thrombin, all cells show a strong sustained calcium signal which begins 10–20 seconds after addition of the stimulus.

INVOLVEMENT OF NON-GPI-LINKED MEMBRANE PROTEINS IN THE SIGNALLING PROCESS

The mechanism by which GPI-linked molecules transduce activation signals is currently unknown. However, a number of recent observations give some clues as to the pathways involved. First, a number of additional molecules have been found to co-isolate with GPI-anchored molecules in immunoprecipitation experiments. Perhaps the most consistent finding with respect to associated molecules has been the co-isolation with GPI-anchors of members of the src-homology kinase family such as lck, fyn, and lyn[11]. These kinases, all of which associate with the inner membrane leaflet *via* an N-terminal myristoyl group, can be isolated using antibodies directed against various GPI-linked proteins and have been shown to be functionally active in *in vitro* kinase assays, where they have a tendency to autophosphorylate. A second important observation is the apparent insolubility of GPI-linked molecules in several non-ionic detergents commonly used to solubilize biological membranes. Lipid analysis of the insoluble fraction shows enrichment of sphingomyelin and cholesterol, probably occurring as a consequence of removal of phospholipids by the detergent[2,4]. The fact that cell lysates prepared using detergents such as Triton-X100 and NP-40 show co-association of GPI-anchors and src-family kinase activity, has led to the hypothesis that a physical association may exist between these two types of molecules in the cell membrane, even though they occupy opposite leaflets of the lipid bilayer. GPI-linked molecules and kinase activity are both enriched in the detergent-insoluble fraction. However, when detergent lysates are fractionated by density gradient centrifugation, the lighter fractions containing both GPI-anchors and kinases show the majority of the kinase not to be GPI-associated. It is therefore possible that immunoisolation does not reflect an association in the intact membrane. More recently, however, it has been shown that membrane vesicles shed from cell membranes can also be fractionated by density and also show co-enrichment of GPI-anchors and kinase activity (S.Ilangumaran, personal comunication). If this is a consistent finding it would strengthen the argument that a physical association between GPI-anchors and kinases also exists *in vivo*.

Does a physical association between kinase molecules and a GPI-anchored receptor necessarily indicate that the kinase plays a role in signal transduction? There are other factors which need to be considered. First, receptor molecules such as CD4 and CD8, which are know to associate tightly with lck through their cytoplasmic domains, do not transduce activation signals as a result of antibody-induced clustering. This indicates that clustering of kinase molecules on the cytoplasmic face of the plasma membrane is not sufficient to signal to the cell. Second, the activity of src-homology kinases is tightly regulated *in vivo* by the presence of a regulatory tyrosine residue, (tyrosine-505 in lck), phosphorylation of which maintains the kinase in an inactive configuration. Most of the kinase is held in this inactive form, and the action of a phosphotyrosine phosphatase is necessary to remove this regulatory phosphate and activate the enzyme. In lymphoid cells, a strong candidate is the abundant cell surface glycoprotein CD45. CD45 is crucial in the regulation of lymphocyte activation by immunological stimuli[12], and its deficiency prevents activation of lymphocytes *via* the antigen receptor and *via* Thy-1[13]. It is possible, therefore, that cross-linking of GPI-anchored molecules may cause recruitment of CD45 into GPI-rich areas of the cell membrane where it may activate kinases locally. There have been reports of physical interactions between CD45 and Thy-1[14], and we also have some evidence for co-association between these two molecules obtained by cross-linking followed by immunoisolation. Under normal circumstances, CD45 does not show detergent insolubility of the type mentioned previously, nor does it appear to migrate into GPI-rich areas of the plasma membrane. However, there is one report of a postranslational modification on the extracellular portion of the CD45 molecule involving the addition of a sphingolipid[15], which may facilitate its enrichment in sphingolipid-rich areas of the membrane generated by clustering of GPI-anchors. Clearly, this is one area of GPI-anchor research which deserves further study.

A number of other proteins have been reported to co-isolate with GPI-anchors. Little is known about their identity or their relevance to signalling events. Apart from the src-homology kinases already described another molecule, caveolin, has attracted great interest[16]. Caveolin is major structural component of caveolae, which are plasma membrane invaginations thought to be involved in both endocytosis and signalling. As with the kinase connection, little evidence as yet demonstrates involvement of caveolin in GPI-mediated signalling, and it has been suggested that its apparent association with GPI-anchors in some cell types may be an antibody-induced rather than a natural liason[17]. There are a few examples of GPI-linked molecules which appear to function primarily as signal-transducing receptors. Most notable in this category are the GPI-linked ciliary neurotrophic factor receptor[18], which uses a 130 kilodalton glycoprotein for signal transduction, and the glial-derived neurotrophic factor receptor[19] which uses an undefined transducer molecule. The macrophage lipopolysaccharide receptor CD14 probably also employs a similar mechanism but, as with the glial receptor, no transducer has yet been identified. In these three examples signal transduction clearly does not take place through the GPI-anchor, prompting the question why are they GPI-linked. One suggestion is that the anchor may play a facilitative role by increasing the mobility of the receptor and thereby enhancing its ability to capture ligand more efficiently.

PHYSIOLOGICAL IMPLICATIONS

What are the physiological consequences, if any, of GPI-mediated signal transduction? I would like to suggest two possibilities. First, an efficient clustering of receptors has been shown to be essential for detectable signals to occur. *In vitro*, antibodies have been used to induce this. However, circulating antibody is unlikely to be present *in vivo* where the receptor is a self component, since these receptors are unlikely to induce an antibody immune response. This limits antibody mediated effects to GPI-linked molecules of either parasitic or viral origin. It has been shown that GPI-linked parasite coat proteins can transfer to mammalian cells, and that transferred GPI-anchored molecules can mediate intracellular calcium fluxes in an antibody-dependent fashion[20]. It would therefore be of great interest to see whether host cells from individuals with acute parasitic infections can be signalled by antibodies against parasite coat proteins. Similarly, GPI-linked molecules encoded by certain flaviviruses induce a strong antibody response in infected individuals (M.Jacobs, personal communication). These antibodies may be able to activate infected cells cause cell activation and cytokine release. Second, clustering of GPI-linked molecules may occur as a result of interactions with ligands on another cell or present in the circulation. Such a mode of stimulation may be physiologically similar to that induced by antibodies and may be an important way in which cell-cell contact mediates growth control. In both antibody-mediated and natural ligand scenarios, GPI-induced signals are unlikely to operate alone. Interaction of a GPI-anchored cell surface molecule with its natural ligand , for example, is more likely to provide a costimulatory signal which amplifies a physiological response by lowering the threshold for cell activation, rather than acting as a primary signal. In T lymphocytes, it is clear that GPI-mediated signals do not constitute a primary route for cell activation, the TCR/CD3 antigen receptor complex being the primary signalling receptor. However, under physiological conditions antigen is often in short supply, so that additional co-receptor pathways become increasingly important in determining whether triggering of the T cell takes place.

An alternative way for GPI anchors to modulate cell physiology may be by direct uptake, or transfer of large amounts of GPI-anchored parasite coat proteins or glycolipids of parasite origin. These molecules, produced during acute phase infection, may directly alter the metabolic state of the cell. In acute malaria infections, GPI-anchored coat proteins appear to transfer into cells of the host immune system and profoundly alter the activation state of host macrophages[21]. The ability to alter the response of host cells is likely to have important consequences for the parasite-host relationship, as well as interactions with the immune system. This potential signalling pathway may function independently of, or together with, cross-linking of receptors by antibody. Other recently described examples of transfer of GPI-linked molecules between cells both *in vitro* and *in vivo*[22] indicate the generality of the GPI transfer process and highlights its potential importance for signal transduction.

REFERENCES

1. Ferguson, M.A. and Williams, A.F.(1988) Cell-surface anchoring of proteins via glycosyl-phosphatidylinositol structures. *Annu.Rev.Biochem.* 57:285–320

2. Hanada, K., Nishijima, M., Akamatsu, Y., and Pagano, R.E.(1995) Both sphingolipids and cholesterol participate in the detergent insolubility of alkaline phosphatase, a glycosylphosphatidylinositol-anchored protein, in mammalian membranes. *J.Biol.Chem.* 270:6254–6260

3. Brown, D.A. and Rose, J.K.(1992) Sorting of GPI-anchored proteins to glycolipid-enriched membrane subdomains during transport to the apical cell surface. *Cell* 68:533–544

4. Brown, D.A.(1992) Interactions between GPI-anchored membrane proteins and membrane lipids. *Trends in Cell Biology* 2:338–343

5. Mayor, S. and Maxfield, F.R.(1995) Insolubility and redistribution of GPI-anchored proteins at the cell surface after detergent treatment. *Mol.Cell Biol.* 6:929–944

6. Edidin,M., Kuo, S.C. and Sheetz, M.P. (1991) Lateral movement of membrane glycoproteins restricted by dynamic cytoplasmic barriers. *Science* 254:1379–1382

7. Robinson, P.J. Phosphatidylinositol membrane anchors and T-cell activation.(1991) *Immunol.Today* 12:35–41

8. Kroczek, R.A., Gunter, K.C., Germain, R.N., and Shevach, E.M.(1986) Thy-1 functions as a signal transduction molecule in T lymphocytes and transfected B cells. *Nature* 322:181–184

9. Barboni, E., Gormley, A.M., Pliego Rivero, F.B., Vidal, M., and Morris, R.J. (1991)Activation of T lymphocytes by cross-linking of glycophospholipid-anchored Thy-1 mobilizes separate pools of intracellular second messengers to those induced by the antigen-receptor/CD3 complex. *Immunology* 72:457–463

10. Mason, J.C., Yarwood, H., Tarnok, A., Sugars, K., Harrison, A.A., Robinson, P.J., and Haskard, D.O.(1996) Human Thy-1 is cytokine-inducible on vascular endothelial cells and is a signalling molecule regulated by protein kinase C. *J.Immunol.* In press

11. Shenoy Scaria, A.M., Kwong, J., Fujita, T., Olszowy, M.W., Shaw, A.S., and Lublin, D.M.(1992) Signal transduction through decay-accelerating factor. Interaction of glycosyl-phosphatidylinositol anchor and protein tyrosine kinases p56lck and p59fyn. *J.Immunol.* 149:3535–3541.

12. Pingel, J.T. and Thomas, M.L.(1989) Evidence that the leukocyte-common antigen is required for antigen-induced T lymphocyte proliferation. *Cell* 58:1055–1065

13. Pingel, J.T., Cahir McFarland, E.D., and Thomas, M.L.(1994) Activation of CD45-deficient T cell clones by lectin mitogens but not anti-Thy-1. *Int.Immunol.* 6:169–178

14. Volarevic, S., Burns, C.M., Sussman, J.J., and Ashwell, J.D.(1990) Intimate association of Thy-1 and the T-cell antigen receptor with the CD45 tyrosine phosphatase. *Proc.Natl.Acad.Sci.U.S.A.*87:7085–7089

15. Takeda, A. (1993) Sphingolipid-like molecule linked to CD45, a protein tyrosine phosphatase Adv.Lipid Res.26:293–317

16. Parton, R.G. and Simons, K.(1995) Digging into caveolae. *Science* 269:1398–1399

17. Mayor, S., Rothberg, K.G., and Maxfield, F.R.(1994) Sequestration of GPI-anchored proteins in caveolae triggered by cross-linking. *Science* 264:1948–1951

18. Davis, S., Aldrich, T.H., Stahl, N., Pan, L., Taga, T., Kishimoto, T., Ip, N.Y., and Yancopoulos, G.D.(1993) LIFR and gp130 as heterodimerizing signal transducers of the tripartite CNTF receptor. *Science* 260:1805–1810

19. Treanor, J.J.S., Goodman, L., deSauvage, F., Stone, D.M. *et.al.*(1996) Characterization of a multicomponent receptor for GDNF. *Nature* 382: 80–83

20. van den Berg, C.W., Cinek, T., Hallett, M.B., Horejsi, V., and Morgan, B.P.(1995) Exogenous glycosyl-phosphatidylinositol-anchored CD59 associates with kinases in membrane clusters on U937 cells and becomes Ca2+ signalling competent. *J.Cell Biol.* 131:669–677

21. Schofield, L. and Hackett, F.(1993) Signal transduction in host cells by a glycosylphosphatidylinositol toxin of malaria parasites. *J.Exp.Med.* 177:145–153

22. Ilangumaran, S., Robinson, P.J. and Hoessli, D.C.(1996) Transfer of exogenous glycosylphosphatidylinositol(GPI)-linked molecules to plasma membranes *Trends in Cell Biology* 6: 163–167

49

ADP-RIBOSE IN GLYCATION AND GLYCOXIDATION REACTIONS[*]

Elaine L. Jacobson, Daniel Cervantes-Laurean, and Myron K. Jacobson

Department of Clinical Sciences and Division of Medicinal Chemistry and
   Pharmaceutics
College of Pharmacy, A323A ASTeCC
University of Kentucky
Lexington, Kentucky 40506-0286

1. ABSTRACT

Glycation is initiated by reaction of a reducing sugar with a protein amino group to generate a Schiff base adduct. Following an Amadori rearrangement to form a ketoamine adduct, a complex chemistry involving oxidation often leads to protein glycoxidation products referred to as advanced glycosylation end products (AGE)[†]. The AGE include protein carboxymethyllysine (CML) residues and a heterogeneous group of complex modifications characterized by high fluorescence and protein-protein cross links. The sugar sources for the glycoxidation of intracellular proteins are not well defined but pentoses have been implicated because they are efficient precursors for the formation of the fluorescent AGE, pentosidine. ADP-ribose, generated from NAD by ADP-ribose transfer reactions, is a likely intracellular source of a reducing pentose moiety. Incubation of ADP-ribose with histones results in the formation of ketoamine glycation conjugates and also leads to the rapid formation of protein CML residues, histone H1 dimers, and highly fluorescent products with properties similar to the AGE. ADP-ribose is much more efficient than other possible pentose donors for glycation and glycoxidation of protein amino groups. Recently developed methods that differentiate nonenzymic modifications of proteins by ADP-ribose from enzymic modifications now allow investigations to establish whether some protein modifications by monomers of ADP-ribose *in vivo* represent glycation and glycoxidation.

---

[*] This investigation was supported in part by National Institutes of Health Grant CA43894
[†] Abbreviations:AGE, advanced glycation end products; CML, N^ε-(carboxymethyl) lysine; G-ADP, 3"-ADP glyceric acid; ADP-ribose, adenosine diphosphoribose.

ADP-Ribosylation in Animal Tissue, edited by Haag and Koch-Nolte
Plenum Press, New York, 1997

2. INTRODUCTION

Protein amino groups are sites of reaction with reducing sugars that result in protein modification by ketoamine adducts[1–5]. Since ketoamine adducts are not very stable, many are precursors to more complex protein modifications[1, 6]. Glucose is assumed to be the major source of glycation and glycoxidation of extracellular proteins *in vivo* based on its abundance and association with diabetic complications[7–9], yet intracellular sugar sources for the glycoxidation of proteins are not well studied. Since pentoses are efficient precursors of AGE[6, 7] and ADP-ribose contains a highly reactive reducing pentose moiety shown to participate in glycation[10, 11], studies have been conducted to examine the potential of ADP-ribose as a participant in intracellular glycoxidation of proteins and for the development of methods to identify products specific to ADP-ribose glycation and glycoxidation of proteins. ADP-ribose is generated from NAD by multiple metabolic pathways[10–12]. The removal of ADP-ribose from proteins modified by the action of protein-mono-ADP-ribosyltransferases[13] and the turnover of cyclic ADP-ribose[14, 15] may be major sources. Additionally, oxidative stresses and other conditions that cause DNA strand breaks stimulate the synthesis of nuclear polymers of ADP-ribose, which are rapidly turned over generating ADP-ribose in proximity to the long lived histones rich in lysine residues[16, 17].

The evaluation of a possible role of ADP-ribose in protein glycation and glycoxidation poses a major technical challenge since the enzyme catalyzed modification of proteins by ADP-ribose occurs at several different amino acid residues[10, 13], resulting in numerous different chemical linkages. To distinguish between glycation and enzymatic modification of proteins by ADP-ribose, previously this laboratory prepared model conjugates for ADP-ribose glycation and determined properties that distinguished glycation adducts chemically from enzymatic modifications[10]. This approach has been extended to study the products formed from ADP-ribose interactions with histones and to identify pathways for removal of these adducts.

3. RESULTS

3.1. Differentiation of Products Derived from ADP-Ribose Glycation and Enzymatic Modifications

All known enzymatic protein modifications involving ADP-ribose are stable at pH 9.0, but ketoamines formed from Schiff bases derived from nonenzymic interactions of protein amino groups and ADP-ribose are labile[10]. Studies of model glycation conjugates of ADP-ribose demonstrated that two primary products are released at pH 9.0, 5'-ADP and ADP-glycerate (G-ADP)[12]. The latter compound is most likely a unique glycation product of ADP-ribose and has potential utility for detection of specific protein glycation *in vivo*. The scheme in Fig. 1 shows how differences in chemical stability allow for the differentiation of nonenzymic and enzymic modifications to proteins by ADP-ribose using selective chemical release of the ADP-ribose adduct.

3.2. Formation of Ketoamine Adducts from Histone Glycation by ADP-Ribose

The potential of ADP-ribose as a protein glycating agent has been demonstrated using histones. This class of proteins was examined since they are likely intracellular targets

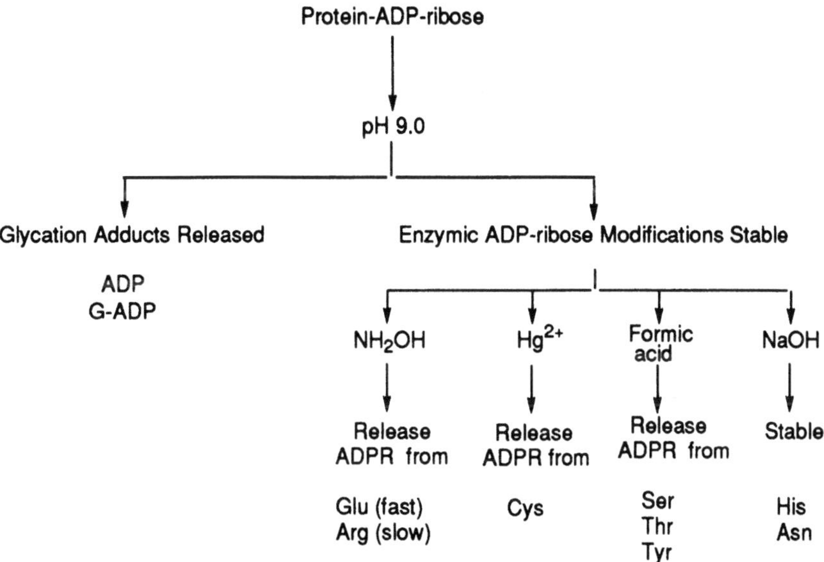

Figure 1. Differentiation of various classes of ADP-ribose:protein adducts by selective chemical release. Glycation adducts formed nonenzymically are distinguished from enzymic ADP-ribose modifications by incubation at pH 9.0. The primary products released are ADP and G-ADP. ADP-ribose is released by or remains stable to selective chemical treatments of enzymic modifications as shown in the diagram.

for ADP-ribose glycation due to their high content of lysine and their proximity to ADP-ribose polymer turnover. Earlier studies showed that incubation of histone H1 with ADP-ribose results in protein conjugates with the stability characteristics similar to ketoamine conjugates[11]. To examine glycation by ADP-ribose in more detail, histones H1, H2A, H2B, and H4 were studied using radiolabelled ADP-ribose. Analysis of the products by electrophoresis on acid-urea gels demonstrated that each of the histones was stably modified by ADP-ribose[12]. The amount of modification was concentration dependent for ADP-ribose. The ADP-ribose modification of histones was shown to involve formation of ketoamines, by studying the chemical stability of radiolabelled histone adducts compared to model glycation conjugates of ADP-ribose[10]. Like the model glycation conjugates, the histone conjugates were stable in formic acid and neutral hydroxylamine, but radiolabel was released at pH 9.0. The released material contained primarily 5'-ADP and G-ADP (see Fig. 1) as would be predicted for ketoamine adducts derived from ADP-ribose.

3.3. Degradation Products and the Formation of AGE

Studies of protein glycation by glucose have shown that ketoamine glycation adducts degrade to form AGE[6, 7]. These include CML and a heterogeneous group of complex modifications characterized by their high fluorescence and ability to cause protein-protein cross-links. The structures of CML and the well-characterized fluorescent product, pentosidine, are shown in Fig. 2. The portion of each compound derived from the sugar moiety is denoted by the dashed box. These compounds were used as reference standards in studies of the degradation products of glycation by ADP-ribose described in sections 3.4 and 3.5 below.

3.4. Formation of CML from Histone H1 Glycation by ADP-Ribose

The mechanism for the formation of G-ADP from a protein glycated by ADP-ribose[12] predicts that the release of G-ADP should result in the conversion of a protein lysine residue to a CML residue. This prediction has been tested by incubating histone H1 with ADP-ribose, followed by acid hydrolysis and amino acid analysis to examine for the presence of CML[12]. The analysis showed that most of the amino acid composition in the glycated sample was unchanged, except for lysine and arginine. Lysine content decreased proportionately to an increase in CML. Thus, it can be concluded that glycation of histone H1 by ADP-ribose leads to the formation of CML. The arginine decrease relates to the formation of fluorescent AGE described in section 3.4 below.

3.5. Formation of Fluorescent AGE with Histone Glycation by ADP-Ribose

Studies of protein glycation by glucose have shown that ketoamine glycation adducts degrade to release reactive dicarbonyl compounds such as 3-deoxyglucosone and undergo further reactions to generate a number of different fluorescent glycoxidation products[7, 18]. The study of Sell and Monnier[6] has shown that pentoses readily generate a number of different fluorescent AGE, including pentosidine. The formation of these fluorescent AGE begins with the initial glycation of a lysine amino group but further reactions involve both oxidation and reaction with protein arginine residues. If the lysine and arginine residues are present on different polypeptide chains, the formation of pentosidine results in protein-protein cross linking[6, 18]. To determine if the ketoamines formed from ADP-ribose on histone H1 undergo further reactions to generate fluorescent AGE products, experiments were conducted to identify which moieties of the ADP-ribose molecule are involved. Histone H1 was incubated with ADP-ribose radiolabeled with either ^{32}P in the adenosine proximal phosphate or ^{14}C in the ribose moiety containing the free aldehyde. Analysis of the amount and distribution of each radiolabel retained on H1 revealed that much of the ADP-ribose glycation is accompanied by further reactions that involve the loss of the radiolabeled phosphate, but retain the ^{14}C label in the ribose containing the free aldehyde. To determine if ADP-ribose glycation could lead to products that result in protein cross linking, histone H1 was incubated with ADP-ribose and subjected to analysis by SDS-PAGE. The H1 is almost completely converted to a dimer when the incubation is carried out at pH 9.0[12]. Dimerization also occurs at pH 7.0, but more slowly. Dimer forma-

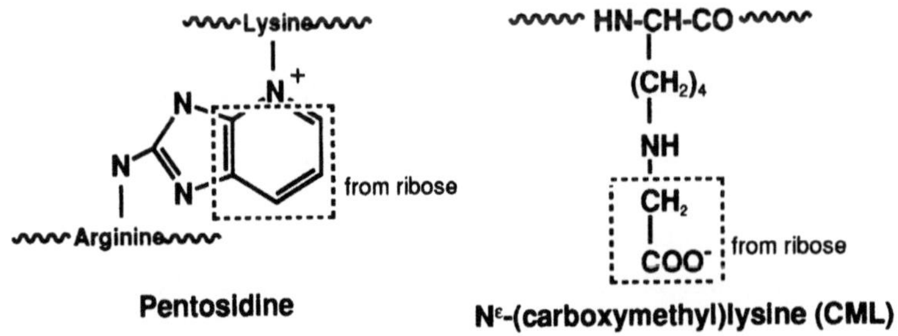

Figure 2. Structures of AGE products. The structures of pentosidine and Nᵉ-(carboxymethyl)lysine are shown where the dashed lines enclose the carbons derived from ribose.

tion is dependent upon the concentration of ADP-ribose. Further, ^{14}C in the ribose moiety containing the free aldehyde is incorporated into the H1 dimer. Aminoguanidine has been shown to serve as an effective inhibitor of protein glycoxidation[15] and this compound completely inhibits H1 dimer formation[12]. Thus ADP-ribose is effective in causing protein cross links by a mechanism involving glycation. To determine if the protein cross linking by ADP-ribose involves the formation of fluorescent AGE, ADP-ribose was incubated with H1 and fluorescence was monitored using the conditions of excitation and emission typical for pentosidine[6]. The data of Fig. 3, left panel, show that incubation of histone H1 with ADP-ribose results in the rapid formation of putative fluorescent AGE. Fluorescent products are readily detected after a lag of approximately 8 minutes and increase in a linear manner as a function of incubation time. The presence of the lag phase is consistent with a mechanism by which the formation of ketoamine glycation products precedes the formation of AGE. Fluorescent products of Fig. 3 were formed at pH 9.0 but also occur at pH 7.4, although the rate of formation is much slower. Incubation of either histone H1 or ADP-ribose alone does not result in any detectable fluorescence. When analyzed after a 20 minute incubation, formation of fluorescence is dependent upon the concentration of ADP-ribose in the incubation mixtures from 50 μM to 500 μM as seen in Fig. 3, middle panel. Further, the addition aminoguanidine inhibits the formation in a dose dependent manner (Fig. 3, right panel). No reaction between ADP-ribose and aminoguanidine is detectable, indicating that aminoguanidine is inhibiting oxidative steps that follow glycation, rather than limiting ADP-ribose availability for glycation. Further evidence that the mechanism involves reactive oxygen species derives from the finding that formation of the fluorescent products observed here is inhibited by reduced ascorbateFSchramm, D., and Cervantes-Laurean, D., unpublished observations.. A remarkable feature is that ADP-ribose is much more effective than either ribose or ribose 5-phosphate in the formation of fluorescent AGE. While incubation of ADP-ribose with histone H1 results in intense fluorescence in minutes, neither ribose or ribose 5-phosphate yields detectable fluorescence when incubated under the same conditions for several days.

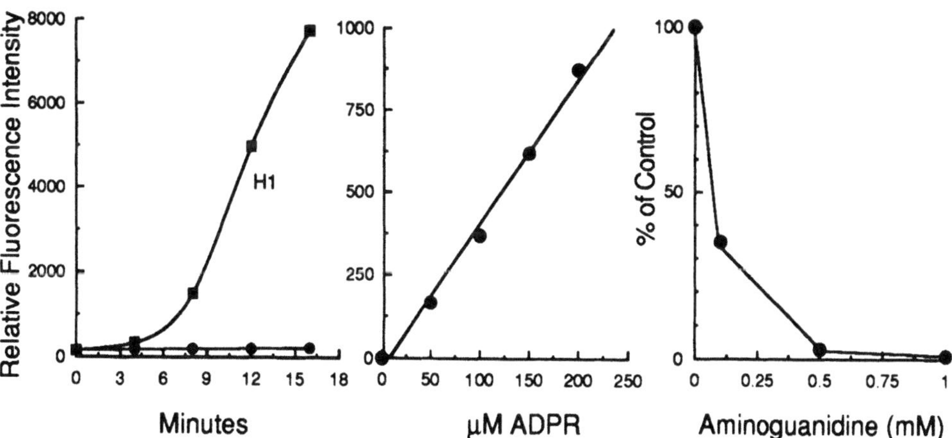

Figure 3. Formation of fluorescent products by incubation of histone H1 with ADP-ribose at pH 9.0. The squares in the left panel show the time course of formation of fluorescence following incubation of histone H1 with 500 μM ADP-ribose. The circles denote an incubation in which ADP-ribose was omitted. The middle panel shows the effect of varying the concentration of ADP-ribose on the formation of fluorescence in a 20 minute incubation and the right panel shows the effect of adding aminoguanidine to a 20 minute incubation containing 500 μM ADP-ribose and histone H1.

3.6. Formation of Fluorescent AGE from ADP-Ribose and Amino Acids

Since pentosidine can be formed by incubation of glucose or pentoses with lysine and arginine, experiments were conducted to determine if fluorescent AGE could be formed from ADP-ribose, lysine, and arginine. Using N-α-Boc groups to block the alpha amino groups of arginine and lysine to preclude their involvement in the formation of products, ADP-ribose was incubated with the amino acids alone and in combination. Incubation of ADP-ribose with N-α-Boc arginine alone results in a number of fluorescent compounds and incubation with N-α-Boc lysine alone yields a number of additional peaks[12]. Incubation of ADP-ribose with both N-α-Boc arginine and N-α-Boc lysine results in the formation of a fluorescent compound that co-migrates with a pentosidine standard and has excitation and emission spectra indistinguishable from pentosidine[12]. Thus, ADP-ribose also is capable of forming fluorescent AGE with free lysine and arginine.

3.7. Isolation of AGE from Histone H1:ADP-Ribose Glycation Products

To determine if pentosidine and/or related glycoxidation products are formed in proteins glycated by ADP-ribose, histone H1 was incubated with ADP-ribose and subjected to acid hydrolysis. Analyses showed that the glycated sample contains several fluorescent peaks, including one peak that co-elutes with pentosidine, showing that ADP-ribose can form AGE products with proteins[12].

4. DISCUSSION

The scheme shown in Fig. 4 summarizes what has been learned concerning the nonenzymic modification of protein amino groups by ADP-ribose. Concentrations of ADP-ribose of 50 to 500 µM result in readily detectable histone glycation involving ketoamine intermediates. These ketoamines are quite unstable and they degrade chemically by two primary pathways, releasing ADP and G-ADP from the protein. The formation of glycoxidation products, specifically histone cross links and other fluorescent AGE, likely relates to the degradation pathway that releases ADP[12]. The mechanism for the formation of AGE products, such as pentosidine, postulates the condensation of a ketoamine derived neutral five carbon fragment with a protein arginine residue[6]. Pathways for the degradation of ADP-ribose derived ketoamines that result in the presence of a residual phosphate would interfere with this condensation, but the release of the ADP moiety results in an uncharged 5 carbon fragment that should readily react with arginine and thus promote histone cross linking. The fluorescent products isolated from histone H1 are most likely derived from ADP-ribose since incubation of ADP-ribose only with lysine and arginine generates a product indistinguishable from pentosidine and these reactions are inhibited by the presence of the AGE inhibitor, aminoguanidine.

The second pathway of degradation of ADP-ribose derived ketoamines, which leads to the generation of G-ADP, predicts that protein glycation by ADP-ribose can result in the modification of protein lysine residues with carboxymethyl groups[12]. This prediction has been confirmed[12] and this pathway is of particular interest in that the release of G-ADP has potential utility for the detection of ADP-ribose specific glycation of proteins *in vivo*, since it likely represents a unique glycation product for this nucleotide.

The lability of the ketoamine glycation conjugates, the cross linking of histones, and the formation of fluorescent AGE are all readily detectable at pH values in the physiologi-

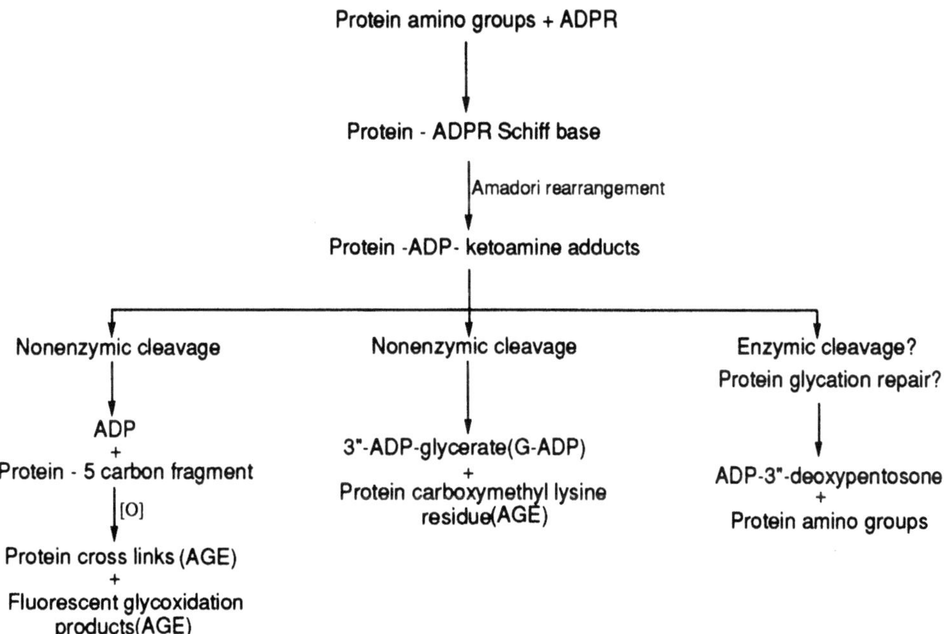

Figure 4. Summary of proposed degradation pathways for glycation adducts formed from ADP-ribose and protein amino groups.

cal range, indicating that these reactions are likely to occur in intact cells. While additional studies will be needed to determine if ADP-ribose causes histone glycation *in vivo*, a number of previous studies support this hypothesis. Hilz and co-workers[19] described ADP-ribose conjugates of histone H1 in hepatoma cells following DNA damage with chemical stability similar to the histone glycation conjugates described here, suggesting that these conjugates may represent histone glycation. Smulson and co-workers[20] reported a dimer of histone H1 and ADP-ribose polymers in HeLa cells. It is tempting to speculate that this dimer, similar to the one observed in the glycation product, is the result of histone glycoxidation initiated by ADP-ribose that was generated by polymer turnover. Fluorescent AGE have recently been detected in histones isolated from rats and were increased in streptozotocin-induced diabetic rats[21]. Streptozotocin is known to cause liver DNA damage and to stimulate poly(ADP-ribose) polymerase[13, 14]. This may have increased histone AGE due to ADP-ribose polymer turnover.

Figure 4 also suggests the possibility of a third pathway that may represent repair of protein glycation. Both bacteria[22] and fungi[23] contain glycation removal enzymes. Protein glycation by glucose leads to the release of 3-deoxyglucosone. The analogous product released following protein glycation by ADP-ribose would be ADP-3"-deoxypentosone. The proposed mechanism by which this product is formed is shown in Fig. 5. It is interesting that this compound has been reported to be the product of the enzyme ADP-ribosyl protein lyase[24], which is postulated to function in the turnover of ADP-ribose polymers by catalyzing the removal of the protein proximal ADP-ribose residue. From what has been learned here from the chemistry of ADP-ribose glycation adducts of histones and their degradation products it seems that ADP-ribosyl protein lyase should be further investigated as a possible glycation removal enzyme.

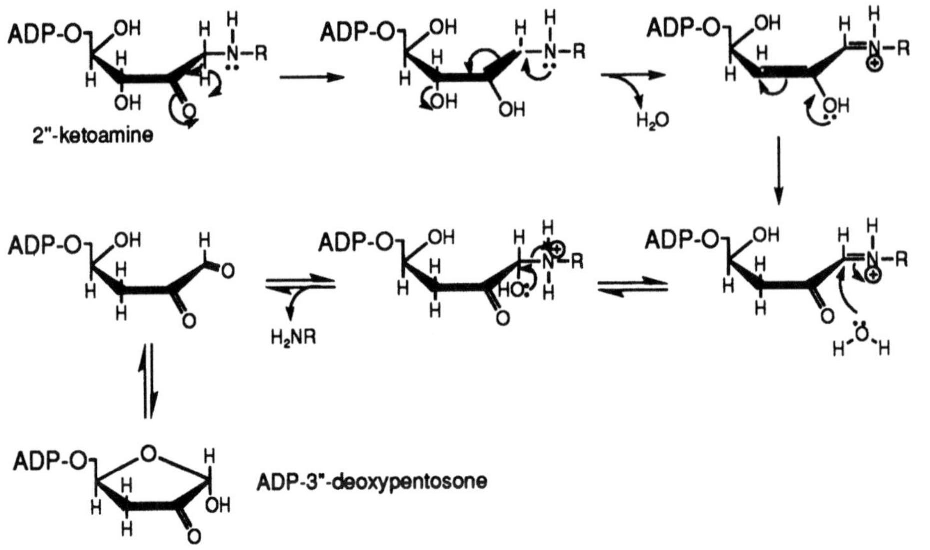

Figure 5. A mechanism proposed for the degradation of a ketoamine adduct formed from the interaction of ADP-ribose with a protein amino group.

It is clear from investigations to date that the chemistry of ADP-ribose and its interactions with protein amino groups are complex. The many ongoing studies of the role of ADP-ribose transfer reactions in biological functions should benefit from understanding the chemistry of both enzymic and nonenzymic protein modifications involving ADP-ribose. Further, identification of nonenzymic modifications of protein and/or DNA molecules as a result of ADP-ribose transfer, and their possible repair pathways, could have major implications in the pathophysiology of chronic diseases such as cancer and aging.

5. REFERENCES

1. Ahmed, M.U., S.R. Thorpe, & J.W. Baynes. 1986. Identification of N epsilon-carboxymethyllysine as a degradation product of fructoselysine in glycated protein. *J. Biol. Chem. 261*: 4889–4894.
2. Fu, M.X., K.J. Wells-Knecht, J.A. Blackledge, T.J. Lyons, S.R. Thorpe, & J.W. Baynes. 1994. Glycation, glycoxidation, and cross-linking of collagen by glucose. Kinetics, mechanisms, and inhibition of late stages of the Maillard reaction. *Diabetes* 43: 676–683.
3. Cerami, A., & M.J.C. Crabbe. 1986. Recent advances in ocular research. *Trends Pharmacol. Sci.* 7: 271–274.
4. Cerami, A., H. Vlassara, & M.J. Brownlee. 1986. Role of nonenzymatic glycosylation in atherogenesis. *Cell. Biochem. 30*: 111–120.
5. Monnier, V.M. 1990 Nonenzymatic glycosylation, the Maillard reaction and the aging process. *J. Gerontol.* 45: 8105–8111.
6. Sell, D.R., & V.M. Monnier. 1989. Structure elucidation of a senescence cross-link from human extracellular matrix. Implication of pentoses in the aging process. *J. Biol. Chem. 264*: 21597–21602.
7. Bucala, R., & A. Cerami. 1992. Advanced glycosylation: chemistry, biology, and implications for diabetes and aging. *Adv. Pharmacol. 23*: 1–34.
8. Dyrks, T., E. Dyrks, T. Hartmann, C.L. Masters, & K. Beyreuther. 1992. Amyloidogenicity of beta A4 and beta A4-bearing amyloid protein precursor fragments by metal-catalyzed oxidation. *J. Biol. Chem. 267*: 18210–18217.
9. Vlassara, H., R. Bucala, & L. Striker. 1994. Pathogenic effects of advanced glycosylation: biochemical, biologic, and clinical implications for diabetes and aging. *Lab. Invest. 70*: 138–151.

10. Cervantes-Laurean, D., D.E. Minter, E.L. Jacobson, & M.K. Jacobson. 1993. Protein glycation by ADP-ribose: studies of model conjugates. *Biochemistry 32*: 1528–1534.

11. Jacobson, E.L., D. Cervantes-Laurean, & M.K. Jacobson. 1994. Glycation of proteins by ADP-ribose. *Mol. Cell. Biochem. 138*: 207–212.

12. Cervantes-Laurean, D., E.L. Jacobson, & M.K. Jacobson. 1996. Glycation and glycoxidation of histones by ADP-ribose. *J. Biol. Chem. 271*: 10461–10469.

13. Williamson, K.C., & J. Moss. 1990. in ADP-ribosylating Toxins and G Proteins: Insights Into Signal Transduction. (Moss, J., & M. Vaughan, eds.) American Society of Microbiology, Washington, D.C., Vol. 25, 493–506.

14. Kim, H., E.L. Jacobson, & M.K. Jacobson. 1993 Synthesis and degradation of cyclic ADP-ribose by NAD glycohydrolases. *Science 261*: 1330–1333.

15. Lee, H.C. 1994. Cyclic ADP-ribose: a new member of a super family of signalling cyclic nucleotides. *Cell Signal. 6*: 591–600.

16. Jacobson, M.K., N. Aboul-Ela, D. Cervantes-Laurean, P.T. Loflin, E.L. Jacobson, & M.K. Jacobson. 1990. in ADP-ribosylating Toxins and G Proteins: Insights Into Signal Transduction. (Moss, J., & M. Vaughan, eds.) American Society of Microbiology, Washington, D.C., Vol. 25, 479–492.

17. Althaus, F.R., & C. Richter. 1987. ADP-ribosylation of Proteins. Enzymology and Biological Significance. Springer-Verlag, Heidelberg.

18. Baynes, J.W. Role of oxidative stress in development of complications in diabetes. 1991. *Diabetes 40*: 405–412.

19. Kreimeyer, A., K. Wielckens, P. Adamietz, & H. Hilz. 1984. DNA repair-associated ADP-ribosylation in vivo. Modification of histone H1 differs from that of the principal acceptor proteins. *J. Biol. Chem. 259*: 890–896.

20. Wong, M., Y. Kanai, M. Miwa, M. Bustin, & M. Smulson. 1983. Immunological evidence for the *in vivo* occurrence of a crosslinked complex of poly(ADP-ribosylated) histone H1. *Proc. Natl. Acad. Sci. U. S. A. 80*: 205–209.

21. Gugliucci, A., & M. Bendayan 1995. Histones from diabetic rats contain increased levels of advanced glycation end products. *Biochem. Biophys. Res. Commun. 212*: 56–62.

22. Gerhardinger, C., M.S. Marion, A. Rovner, M. Glomb, & V.M. Monnier. 1995. Novel degradation pathway of glycated amino acids into free fructosamine by a *Pseudomonas sp.* soil strain extract. *J. Biol. Chem. 270*: 218–224.

23. Horiuchi, T., & T. Kurokawa. 1991. Purification and properties of fructosylamine oxidase from *Aspergillus sp.* 1005. *Agri. Biol. Chem. 55*: 333–338.

24. Oka, J., K. Ueda, O. Hayaishi, H. Komura, & K. Nakanishi. 1984. ADP-ribosyl protein lyase purification, properties, and identification of the product. *J. Biol. Chem. 259*: 986–995.

50

INTRAMOLECULAR ADP-RIBOSE TRANSFER REACTIONS AND CALCIUM SIGNALLING[*]

Potential Role of 2'-Phospho-Cyclic ADP-Ribose in Oxidative Stress

Chinh Q. Vu,[1] Donna L. Coyle,[1] Hsin-Hsiung Tai,[1] Elaine L. Jacobson,[2,3] and Myron K. Jacobson[1,3]

[1]Division of Medicinal Chemistry and Pharmaceutics
College of Pharmacy
[2]Department of Clinical Sciences
[3]Lucille P. Markey Cancer Center
University of Kentucky
Lexington, Kentucky 40536

1. ABSTRACT

Intramolecular ADP-ribose transfer reactions result in the formation of cyclic ADP-ribose (cADPR) and 2'-phospho-cyclic ADP-ribose (P-cADPR) from NAD and NADP, respectively. The potent Ca^{2+} releasing activity of these cyclic nucleotides has led to the postulation that they function as second messengers of Ca^{2+} signalling. The synthesis and hydrolysis of cADPR and P-cADPR are catalyzed by NAD(P) glycohydrolases, but the metabolic signals that regulate their metabolism are poorly understood. To investigate the physiological roles of cADPR and P-cADPR, it is essential to have methods that allow the routine measurement of these nucleotides in cellular systems. As described here, a sensitive and selective radioimmunoassay (RIA) for cADPR has been adapted to search for the natural occurrence of P-cADPR in mammalian tissues. Perchloric acid extracts prepared from bovine tissues and purified by anion exchange chromatography were found to contain immunoreactive material which was identified as P-cADPR. P-cADPR may play an important role in oxidative stress as a link between NADP(H) metabolism and alteration of intracellular Ca^{2+} homeostasis.

* This work was supported in part by National Institutes of Health Grant CA43894.

ADP-Ribosylation in Animal Tissue, edited by Haag and Koch-Nolte
Plenum Press, New York, 1997

2. INTRODUCTION

The pyridine nucleotides, NAD(H)[†] and NADP(H), are ubiquitous coenzymes that participate in numerous reactions of energy metabolism. Both NAD and NADP have long been known to play fundamental roles as hydride acceptors in many cellular oxidation/reduction reactions central to biological chemistry. NAD is also a substrate for multiple classes of ADP-ribosyltransferases. These include poly(ADP-ribose) polymerase, which catalyzes the synthesis of ADP-ribose polymers involved in chromatin structural changes necessary for the recovery of cells from DNA damage,[1] and protein-mono-ADP-ribosyltransferases which are implicated in the regulation of adenylate cyclase,[2,3] muscle cell differentiation,[4,5] and membrane trafficking/secretion.[6] In addition, both NAD and NADP are substrates for NAD(P)ases, enzymes which have been known for over five decades, yet their physiological function is still an enigma. Recent studies have demonstrated the unique ability of many NAD(P)ases to catalyze both the synthesis of cADPR from NAD and the hydrolysis of cADPR to ADPR.[7,8] cADPR is a potent Ca^{2+} releasing agent which has been postulated as a second messenger although its physiological significance is still controversial.[9–11] NAD(P)ases also catalyze the synthesis and hydrolysis of P-cADPR, a newly discovered nucleotide that activates Ca^{2+} release by a mechanism apparently similar to cADPR but distinct from IP_3.[12] The synthesis of both cADPR and P-cADPR is achieved by the intramolecular transfer of the ADPR or P-ADPR to the adenine ring. Therefore, intramolecular ADP-ribose transfer reactions may link energy metabolism and cellular regulation by Ca^{2+} signalling.

3. RESULTS AND DISCUSSION

NAD(P)ases constitute one family of enzymes that catalyze ADP-ribose transfer reactions.[13] Many NAD(P)ases catalyze both the synthesis and hydrolysis of cADPR and P-cADPR from NAD and NADP, respectively (Fig. 1).[7,8,12] The synthetic pathway involves the release of nicotinamide followed by an intramolecular ADP-ribose transfer reaction in which the ADPR or P-ADPR moiety is cyclized to the N-1 position of the adenine ring, resulting in the formation of cyclic nucleotides with potent Ca^{2+} releasing activity.

Rapid changes in the cytosolic and nuclear levels of free Ca^{2+}, termed Ca^{2+} signalling, are involved in the regulation of diverse cellular events.[14] Cytosolic levels of free Ca^{2+} reflect a delicate balance between the flow of Ca^{2+} into the cytosolic and nuclear compartments from both extracellular and intracellular spaces through specific Ca^{2+} channels, and the removal of Ca^{2+} by active Ca^{2+} pumps and Ca^{2+} binding proteins. Ca^{2+} signalling is often initiated by the transient release of Ca^{2+} from internal stores by the IP_3-sensitive Ca^{2+} channels.[15,16] Another mechanism of intracellular Ca^{2+} release has been recently identified to be sensitive to cADPR, which appears to modulate the process of Ca^{2+}-induced Ca^{2+} release from ryanodine-sensitive Ca^{2+} channels.[9,17–19] Because cADPR is an endogenous metabolite of NAD, it represents a potential link between NAD(H) metabolism and regulation of Ca^{2+} homeostasis.

---

† The pyridine nucleotides are abbreviated as follows: NAD/NADP, oxidized forms; NADH/NADPH, reduced forms, NAD(H)/NADP(H), both oxidized and reduced forms; NAD(P), NAD or NADP. Other abbreviations used are: NAD(P)ases, NAD(P) glycohydrolases; cADPR, cyclic ADP-ribose, ADP-ribose or ADPR, adenosine diphosphoribose; P-cADPR, 2′-phospho-cyclic ADP-ribose; P-ADPR, 2′-phospho-ADP-ribose; IP_3, inositol trisphosphate; RIA, radioimmunoassay; HPLC, high pressure liquid chromatography.

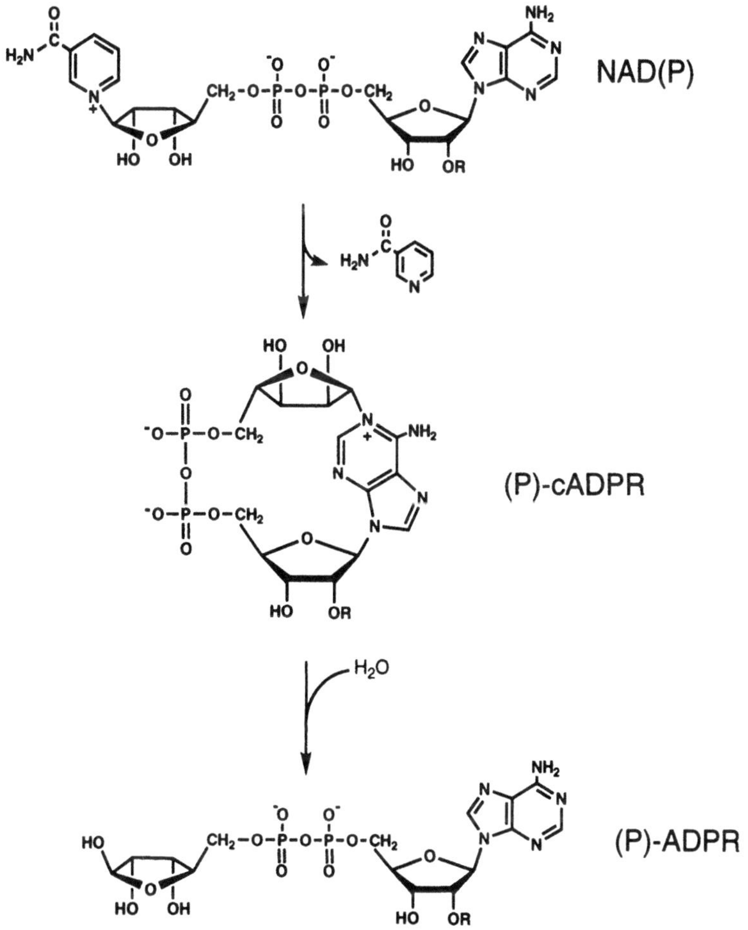

NAD(P)

(P)-cADPR

(P)-ADPR

Figure 1. Synthesis and hydrolysis of cADPR and P-cADPR by bifunctional NAD(P)ases. R = H in NAD, cADPR and ADPR; R= $(PO_3)^{2-}$ in NADP, P-cADPR, and P-ADPR.

Two lines of evidence from our laboratory have raised the possibility that Ca^{2+} signalling also may be linked to cellular NADP(H) metabolism.[12] First, an enzyme from canine spleen previously shown to catalyze both the synthesis of cADPR from NAD and the hydrolysis of cADPR to ADPR also utilized NADP and P-cADPR as substrates (Table I). Interestingly, the enzyme displayed a distinct kinetic preference for NADP over NAD as the specificity constant K_{cat}/K_m was approximately 13 times greater for NADP than for NAD. Second, the P-cADPR derived from NADP was as potent as cADPR in eliciting Ca^{2+} release from rat brain microsomes. P-cADPR appeared to act by a similar mechanism as cADPR based on similar kinetics of Ca^{2+} release, similar dose response curves, cross desensitization, and partial inhibition of release by procaine. Conversely, the P-cADPR-controlled mechanism was distinct from that of IP_3 as evidenced by the marked difference in Ca^{2+} release kinetics, lack of cross desensitization and the insensitivity of P-cADPR to heparin, a known antagonist of the IP_3 pathway. The results of this study suggested that P-cADPR may function as a component of Ca^{2+} signalling and as a link between NADP(H) metabolism and Ca^{2+} homeostasis.

Table 1. Kinetic parameters of canine spleen NAD(P)ase for different substrates

Substrate	K_m (μM)	V_{max} (nmole/min/μl)	K_{cat}/K_m (μM^{-1}/min^{-1})
NAD	9.9	1.41	0.14
NADP	1.6	2.96	1.85
cADPR	50	0.623	0.01
P-cADPR	137	0.064	0.0005

The kinetic parameters were calculated from Lineweaver-Burk plots. Each value represents the mean of at least three independent experiments.

An unanswered question extending from this work was whether P-cADPR is a naturally-occurring metabolite. To address this question, we adapted a method that was originally developed for the routine determination of cADPR in animal tissues or cultured cells.[20] This method utilizes antibodies raised against cADPR in a RIA. The RIA is performed in the presence of snake venom phosphodiesterase and alkaline phosphatase. A unique feature of P-cADPR and cADPR is their resistance to the action of snake venom phosphodiesterase.[7,12] However, P-cADPR is converted to cADPR by alkaline phos-

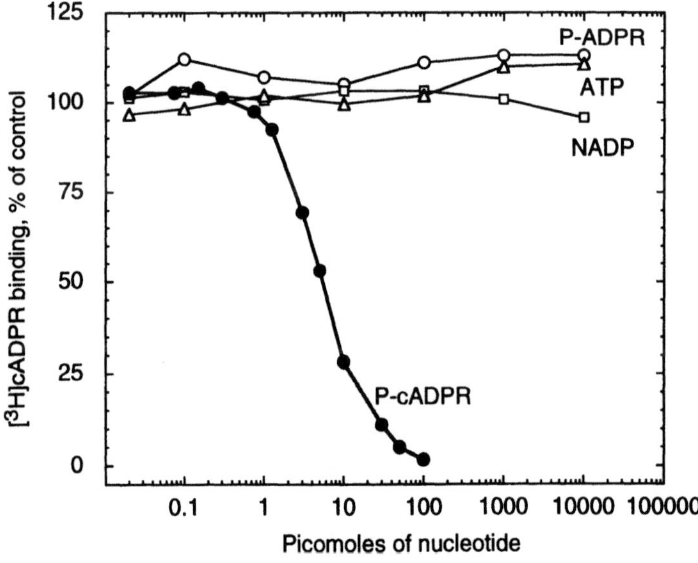

Figure 2. Competition of specific [³H]cADPR binding to cADPR antibodies by P-cADPR and some structurally related nucleotides. Rabbit antibodies were raised against cADPR as described previously.[20] Increasing concentrations of the nucleotides were pretreated with snake venom phosphodiesterase and bacterial alkaline phosphatase, and incubated with the antibodies in the presence of [³H]cADPR. The amount of [³H]cADPR bound to the antibodies was then determined by liquid scintillation counting.

phatase,[12] thus it gives a response identical to cADPR in the RIA. Fig. 2 shows the effect of P-cADPR and selected nucleotides on the binding of [3H]cADPR in the assay. The selectivity of the antibodies was demonstrated by the failure of many other structurally related nucleotides to compete for [3H]cADPR binding even at amounts 2000 times higher than P-cADPR.[21] Fig. 2 also illustrates the sensitivity of the assay where as little as 1 pmol of P-cADPR can be detected. To search for endogenous P-cADPR, perchloric acid extracts from bovine tissues were prepared and subjected to anion exchange HPLC under conditions in which P-cADPR was well separated from cADPR. Analysis of the fractions eluting at the position of P-cADPR by RIA consistently revealed the presence of immunoreactive material. To determine if the immunoreactive material represented P-cADPR, the extracts were subjected to additional fractionation by a second HPLC system and individual fractions were assayed. Fig. 3 shows the detection of immunoreactive material that eluted at the same position as P-cADPR. Further characterization studies presented elsewhere[21] were also performed to confirm the presence of endogenous P-cADPR.

The natural occurrence of P-cADPR in mammalian tissues, together with its Ca^{2+}-mobilizing activity suggest an important biological role for this nucleotide. It is conceivable that P-cADPR may serve as a link between NADP(H) metabolism and Ca^{2+} signalling and the possible significance of such a link is depicted in Fig. 4. While the NAD(H) pool is normally maintained in a highly oxidized state, the NADP(H) pool under normal physiological conditions is maintained in a highly reduced state to provide NADPH as a source of reducing equivalents for anabolic pathways.[22] The highly reduced state of the NADP(H) pool makes it a potentially sensitive indicator of metabolic conditions that cause oxidative stress. In many cases of oxidative stress, glutathione (G-SH) is oxidized to its dimeric form (G-S-S-G), and reduction of G-S-S-G is coupled to the oxidation of NADPH to NADP (Fig. 4). In the cytosolic compartment, NADP is converted to NADPH by the action of enzymes of the pentose phosphate pathway and by malic enzyme. If the

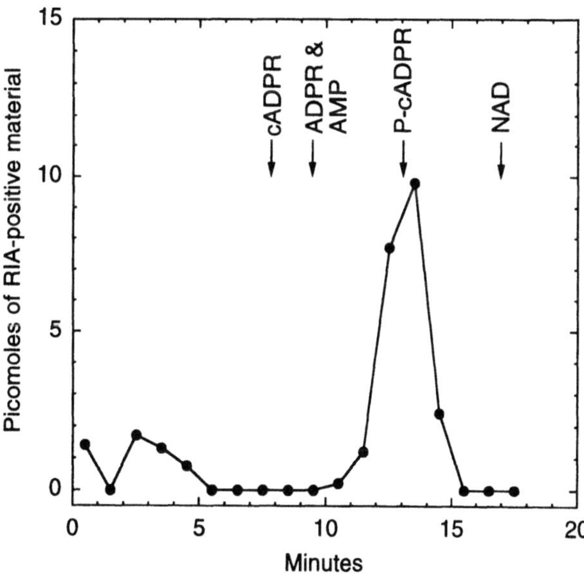

Figure 3. HPLC fractionation of purified bovine tissue extracts and analysis by RIA. Perchloric acid extracts from bovine tissues were prepared, purified, and fractionated by HPLC as described in reference 21. Each fraction was collected and subjected to the standard RIA as described previously.[20]

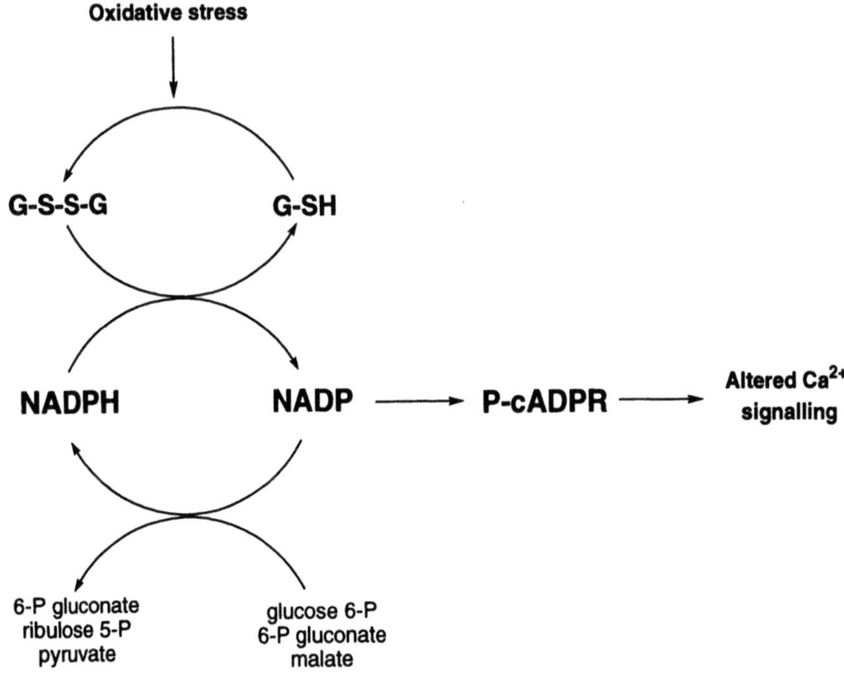

Figure 4. Potential role of P-cADPR in oxidative stress. G-SH, glutathione; G-S-S-G, dimeric form of oxidized glutathione.

amount of oxidation exceeds the capacity of these enzymes, the accumulation of NADP could provide substrate for the synthesis of P-cADPR. Studies over the past twenty years have linked oxidation of NADPH during oxidative stress to alteration of intracellular Ca^{2+} homeostasis,[23–28] yet neither the channels nor the regulatory events associated with this alteration have been identified. Several studies have attributed changes in Ca^{2+} homeostasis to the release of Ca^{2+} from both mitochondrial and extramitochondrial compartments.[25,26,28] Earlier studies from Richter and Kass[28] are particularly notable as they have closely linked agents that oxidize the NADP(H) pool to the activation of a mitochondrial NAD(P)ase followed by rapid Ca^{2+} release from mitochondria. Such phenomena may have important implications in oxidative stress-induced cell injury which has been implicated in a variety of neurological pathologies that include Parkinson's disease, Alzheimer's disease, and stroke.[29–31] The discovery of P-cADPR as a naturally occurring nucleotide raises the possibility that P-cADPR is the Ca^{2+}-mobilizing agent which links oxidative stress-induced NADPH oxidation to alteration of intracellular Ca^{2+} homeostasis.

4. REFERENCES

1. Althaus, F. R., & C. Richter. 1987. ADP-Ribosylation of Proteins: Enzymology and Biological Significance, Springer-Verlag, Berlin Heidelberg.
2. Williamson, K. C., & J. Moss. 1990. in ADP-ribosylating Toxins and G Proteins: Insights into Signal Transduction. (Moss, J. & M. Vaughan, eds.) American Society of Microbiology, Washington, D.C., Vol. 25, 479–492.
3. Quist, E. E., D. L. Coyle, R. Vasan, N. Satumtira, E. L. Jacobson, & M. K. Jacobson. 1994. Modification of cardiac membrane adenylate cyclase activity and Gsα by NAD and endogenous ADP-ribosyltransferase. *J. Mol. Cell. Cardiol. 26*: 251–260.

4. Peterson, J. E., J. S.-A Larew, & D. J Graves. 1990. Purification and partial characterization of arginine-specific ADP-ribosyltransferase from skeletal muscle microsomal membranes. *J. Biol. Chem. 265*: 17062–17069.

5. Zolkiewska, A., & J. Moss, 1993. Integrin $\alpha7$ as substrate for a glycosylphosphatidylinositol-anchored ADP-ribosyltransferase on the surface of skeletal muscle cells. *J. Biol. Chem. 268*: 25273–25276.

6. De Matteis, M. A., M. Di Girolamo, A. Colanzi, M. Pallas, G. Di Tullio, L. J. McDonald, J. Moss, G. Santini, S. Bannykh, D. Corda & A. Luini. 1994. Stimulation of endogenous ADP-ribosylation by brefeldin. *Proc. Natl. Acad, Sci, U. S. A.* 91: 1114–1118.

7. Kim, H., E. L. Jacobson, & M. K. Jacobson. 1993. Synthesis and hydrolysis of cyclic ADP-ribose by NAD glycohydrolases. *Science 261*: 1330–1333.

8. Howard, M., J. C. Grimaldi, J. F. Bazan, F. E. Lund, L. Santos-Argumedo, R. M. E. Parkhouse, T. F. Walseth, & H. C. Lee. 1993. Formation and hydrolysis of cyclic ADP-ribose catalyzed by lymphocyte antigen CD38. *Science 262*, 1056–1059.

9. Lee, H. C., A. Galione, & T. F. Walseth. 1994. Cyclic ADP-ribose: metabolism and calcium mobilizing function. *Vitam. Horm. 48*: 199–257.

10. Takasawa, S., K. Nata, H. Yonekura, & H. Okamoto. Cyclic ADP-ribose in insulin secretion from pancreatic β–cells. 1993. *Science 259*: 370–373.

11. Webb, D.-L., M. S. Islam, A. M. Efanow, G. Brown, G., M. Kö hler, O. Larsson, & P.-O. Berggren. 1996. Insulin exocytosis and glucose-mediated increase in cytoplasmic free Ca^{2+} concentration in the pancreatic β-cell are independent of cyclic ADP-ribose. *J. Biol. Chem. 32*: 19074–19074.

12. Vu, C. Q., P.-J Lu, C.-S. Chen, & M. K. Jacobson. 1996. 2′-Phospho-cyclic ADP-ribose, a calcium-mobilizing agent derived from NADP. *J. Biol. Chem. 271*: 4747–4754.

13. Price, S. R., & P. H. Pekala. 1987. in Pyridine Nucleotide Coenzymes: Chemical, Biochemical, and Medical Aspects. (Dolphin, D., R. Poulson & O. Avramovic, eds.) Wiley-Interscience, New York, Part B, 513–548.

14. Carafoli, E. 1987. Intracellular calcium homeostasis. *Ann. Rev. Biochem. 56*: 395–433.

15. Berridge, M. J. 1993. Inositol trisphosphate and calcium signalling. *Nature 361*: 315–325.

16. Putney, J. W., & G. J. Bird. 1993. The inositol phosphate-calcium signaling system in nonexcitable cells. *Endo. Rev. 14*: 610–631.

17. Galione, A., H. C. Lee, & W. B. Busa. 1991. Ca^{2+}-induced Ca^{2+} release in sea urchin egg homogenates: modulation by cyclic ADP-ribose. *Science 253*: 1143–1146.

18. Sitsapesan, R., S. J. McGarry, & A. J. William. 1995. Cyclic ADP-ribose, the ryanodine receptor and Ca^{2+} release. *Trends Pharmacol. Sci. 16*: 386–391.

19. Clementi, E., M. Riccio, C. Sciorati, G. Nistico, & J. Meldolesi. 1996. The type 2 ryanodine receptor of neurosecretory PC12 cells is activated by cyclic ADP-ribose. *J. Biol. Chem. 271*: 17739–17745.

20. Coyle, D. L., C. Q. Vu., X. Tong, H.-H. Tai, & M. K. Jacobson. 1996. Detection of cyclic ADP-ribose in animal tissues. Submitted for publication.

21. Vu, C. Q., D. L. Coyle, & M. K. Jacobson. 1996. Natural occurrence of 2′-phospho-cyclic ADP-ribose in mammalian tissues. Submitted for publication.

22. Veech, R.L. 1987. in Pyridine Nucleotide Coenzymes: Chemical, Biochemical, and Medical Aspects. (Dolphin, D., R. Poulson & O. Avramovic, eds.) Wiley-Interscience, New York, Part B, 80–104.

23. Lotscher, H. R., K. H. Winterhalter, E. Carafoli, & C. Richter. 1979. Hydroperoxides can modulate the redox state of pyridine nucleotides and the calcium balance in rat liver mitochondria. *Proc. Natl. Acad. Sci. 76*: 4340–4344.

24. Thor, H., M. T. Smith, P. Hartzell, G. Bellomo, S. A. Jewell, & S. Orrenius. 1982. The metabolism of menadione (2-methyl-1,4-naphthoquinone) by isolated hepatocytes. A study of the implications of oxidative stress in intact cells. *J. Biol. Chem. 257*: 12419–12425.

25. Bellomo, G., S. A. Jewell, H. Thor, & S. Orrenius. 1982. Regulation of intracellular calcium compartmentation: Studies with isolated hepatocytes and t-butyl hydroperoxide. *Proc. Natl. Acad. Sci. 79*: 1257–1259.

26. Bellomo, G., H. Thor, & S. Orrenius. 1984. Increase in cytosolic Ca^{2+} concentration during t-butyl hydroperoxide metabolism by isolated hepatocytes involves NADPH oxidation and mobilization of intracellular Ca^{2+} store. *FEBS Lett. 168*: 38–42.

27. Orrenius, S., M. J. Burkitt, G. E. N. Kass, J. M. Dypbukt, & P. Nicotera. 1992. Calcium ions and oxidative cell injury. *Ann. Neurol.32*: S33–42.

28. Richter, C., & G. E. N. Kass. 1991. Oxidative stress in mitochondria: Its relationships to cellular Ca^{2+} homeostasis, cell death, proliferation and differentiation. *Chem. Biol. Interact. 72*: 1–23.

29. Cohen, G. 1985. in Oxidative Stress in the Nervous System. (Sies, H., ed.) Academic Press, New York, 383–402.

30. Adams, J. D., & I. N. Odunze. 1991. Biochemical mechanisms of 1-methyl-4-phenyl-1,2,3,6-tetrahydropyridine toxicity: Could oxidative stress be involved in the brain? *Biochem. Pharmacol. 41*: 1099–1105.
31. Adams, J. D., B. Wang, L. K. Klaidman, C. P. LeBel, I. N. Odunze, & D. Shah. 1993. New Aspects of Brain Oxidative Stress induced by *tert*-butylhydroperoxide. *Free Radic. Biol. Med., 15*: 195–202.

51

HYDROPHOBIC PROPERTIES OF NAD GLYCOHYDROLASE FROM NEUROSPORA CRASSA CONIDIA AND INTERACTION WITH DIOXANE

Mario Pace,[1] Dario Agnellini,[2] Guido Lippoli,[2] and Robert L. Berger[2]

[1]Institute of Veterinary Physiology and Biochemistry
University of Milano, via G. Celoria
10-20133 Milano, Italy
[2]Lab. Biophysical Chemistry
NIH/NHLBI/IR, Bldg. 3 B1-03
9000 Rockville Pike
Bethesda, Maryland 20892

ABSTRACT

NAD glycohydrolase (NADase, EC 3.2.2.5) from Neurospora crassa conidia shows marked hydrophobic properties which are related to the self inhibition of the enzyme. Both aliphatic amines and carboxylic acids are able to inhibit noncompetitively the catalytic activity of the enzyme and the inhibition depends on the non-polar moiety of the substances. Also dioxane is an inhibitor of NAD glycohydrolase even though it apparently increases the specific activity of the enzyme. This effect can be explained by the fact that NADase is present as a dimer when the enzyme is concentrated or at high temperature, and dioxane binds the enzyme breaking the hydrophobic bonds in the dimeric enzyme and yielding the most active monomeric form which is only slightly inhibited by the organic solvent.

INTRODUCTION

NAD glycohydrolase (NADase, EC 3.2.2.5) is present in Neurospora crassa both in mycelium and in conidia, even though the characteristics of these preparations appear different. Everse and Kaplan [1] isolated the enzyme from the mycelium of the fungus grown in a zinc deficient medium and reported a few characteristics such as the content of sugars and the aminoacid composition. NADase from conidia was first purified by Menegus and Pace [2] using a competitive inhibitor, 4-methyl-NAD, bound to the agarose matrix through

ADP-Ribosylation in Animal Tissue, edited by Haag and Koch-Nolte
Plenum Press, New York, 1997

a hydrophilic spacer arm and later Pace et al. improved the procedure of purification by affinity chromatography using immobilized polyclonal antibodies [3].

NADase from conidia is a glycoprotein with a hydrophobic character [4] of 33,000 dalton containing about 20 % carbohydrates, whereas the enzyme from mycelium has 80 % carbohydrates and a molecular mass ranging between 31,000 and 35,500 dalton, and it is present as a dimer at the highest concentrations and temperatures [5]. Carboxylic acids are inhibitors but, while the bull semen enzyme is competitively inhibited by these compounds [6], NADase from Neurospora crassa conidia is inhibited non-competitively both by carboxylic acids and aliphatic amines and this effect should be due to the non polar moiety of these substances.

These characteristics answer for the self-inhibition of the activity which, however, seems different from that shown by NADases from bull semen and other sources [7–10]. On the other hand, dioxane apparently increases the specific activity of NADase from Neurospora crassa conidia and this paper reports the effect of substances with nonpolar moiety on the catalytic activity of NADase and the interaction of dioxane with the hydrophobic sites of the enzyme.

MATERIALS AND METHODS

Neurospora crassa (bd 1858/A strain) was grown in large Petri dishes on Vogel minimal medium [11]. NADase was purified according to the methods previously described[2,3]. Enzyme activity was determined with a titrimetric method [12] and, when a buffer was present, by high performance liquid chromatography [13,14]. All chemicals were of analytical grade.

Immobilization of NADase on Agarose Matrices

A) Direct Covalent Binding to the Gel. Five milliliters of Sepharose-4B (Pharmacia, Uppsala, Sweden) were suspended in 10 ml of 1 M trisodium phosphate and reacted for 12 min. with 200 mg of CNBr dissolved in 4 ml of dioxane. After washing at 2°C with water, 10 % dioxane, and 0.1 M phosphate buffer pH 7.0, the activated agarose was suspended in phosphate buffer containing 0.55 mg of NADase and allowed to react at 4°C for 3 h.

B) Hydrophobic Immobilization. n-Propylamine was coupled to Sepharose-4B activated with CNBr according to a method already described [4]. A column filled with 5 ml of propyl-Sepharose was washed with 2 M KCl and a crude extract of Neurospora crassa conidia (20 mg/ml in 0.17 M KCl) was continuously run at 25°C during 2 hours.

C) Covalent Binding Through a Hydrophilic Spacer Arm. A solution of purified NADase (0.48 mg of protein in 0.5 ml) was reacted overnight at 4°C with two gram of swollen Affi-Gel 10 (Bio-Rad Lab., Richmond, CA), according to the procedure described by the manufacturer.

The three NADase derivatives were exhaustively washed with water, 10 % dioxane (v/v), and 0.1 M phosphate buffer pH 7.0 and stored in the same buffer at 4°C.

Enzyme Activity and Determination of K_m's and K_i's

Michaelis constant of NADase immobilized onto agarose gels was determined at constant temperature by the pH-stat method, using repetitively the same amount of enzyme derivative, filtered and exhaustively washed after each assay. The assays were carried out with 100 mg of wet agarose derivative suspended in 2.0 ml of distilled water containing NAD^+ (1.8 x 10^{-4} M to 1.5 x 10^{-3} M) and the activity was determined titrating with 5 mM NaOH the hydrogen ions produced under a nitrogen atmosphere. Each assay was repeated in presence of dioxane 0.05 M and 0.5 M.

Michaelis constant of the enzyme in solution was measured at 10°C or 35°C using either 0.39 nM ("low concentration") or 1.21 μM ("high concentration") NADase in presence and absence of dioxane. Due to the strong buffer value of aliphatic acids and amines, the inhibitory effect of these substances was determined by HPLC according to the methods already described [13,14].

Interaction of NADase with Dioxane Assayed by Calorimetry

A Berger-Mudd stopped-flow microcalorimeter [15] was used for the study of the interaction of NADase with dioxane. In a mixing chamber, thermostatted at 10°C and 37°C, were rapidly injected 80 μl of NADase (0.1 mM) and 80 μl of a solution of dioxane (0.005-100 mM). Each experiment was repeated ten times and the average data, after subtraction of the blank, were taken as the heats of reaction of the enzyme with dioxane. The blank was the total heat obtained from the following mixtures:

1. water-water: for the calibration of the two microchambers
2. NADase-water: for the determination of the heat of dilution of the enzyme in water
3. dioxane-water: for the determination of the heat of dilution of dioxane in water.

Complete mixing occurred in 1 sec.

RESULTS

Inhibition by Amphipatic Compounds

Aliphatic amines and carboxylic acids are noncompetitive inhibitors of NADase from Neurospora crassa conidia, as demonstrated by double reciprocal plot with a series of compounds, some of which are reported in figure 1. The apparent inhibition constant of a number of aliphatic amines and carboxylic acids is shown in Table I. The effect of the nonpolar moiety of aliphatic compounds is pointed out by the relation between the enzyme activity and the number of methylene groups in the structure of amines and carboxylic acids. A linear relation was found by plotting the number of methylene groups into amines or carboxylic acids and the concentration of compound needed to obtain 50% inhibition (figure 2).

Effect of Dioxane

Despite its non polar structure (dielectric constant ε = 2.2), 1,4-dioxane apparently increases the specific activity of NADase, but this effect is due to the production of the more active monomeric form of the enzyme from the dimer which is usually present at high concentration and temperature [5]. Arrhenius plot of the data obtained with the enzyme in solution, within the range of temperature 5–37°C, showed a change in the activation energy at 25°C: between 5–25°C the activation energy resulted 4.1 kcal/mol and over

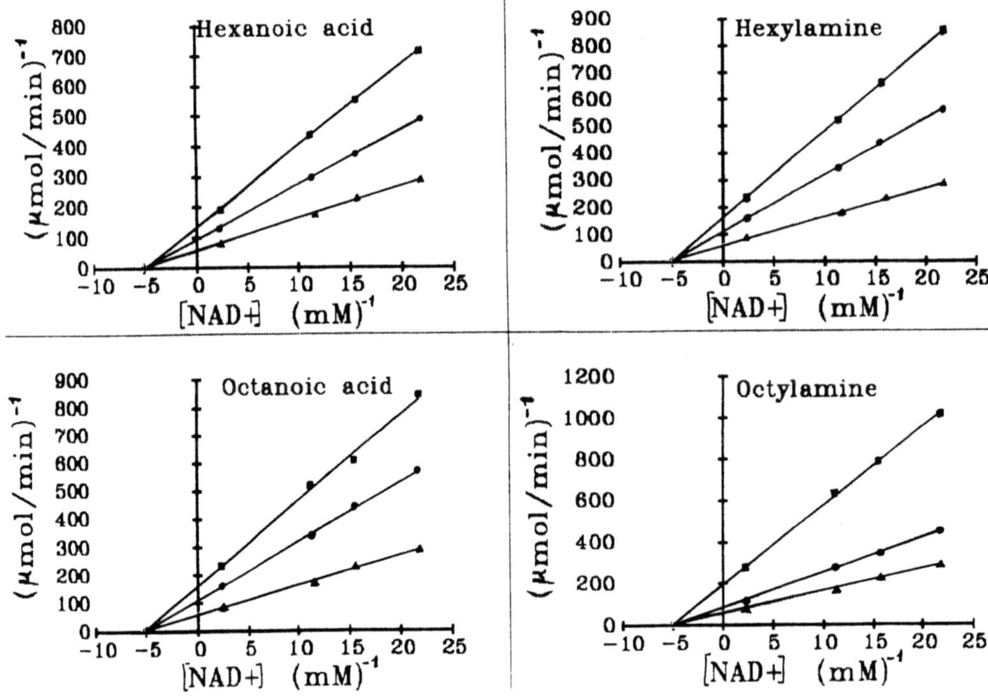

Figure 1. Double reciprocal plot of NADase activity in absence and in presence of some amines and carboxylic acids. The assays were made by the HPLC method using various amounts of inhibitor. Hexylamine, hexanoic acid and octanoic acid were 0.1 M and 0.2 M, octylamine was 0.01 M and 0.05 M.

25°C it raised to 14.23 kcal/mol. In presence of dioxane Arrhenius plot resulted linear within the whole range of temperature 5–37°C and the activation energy was 11.2 kcal/mol (see figure 3).

The maximum velocity and the Michaelis constant of NADase in presence or absence of dioxane were determined both at high concentration of protein and in diluted solution at different temperatures (10°C and 35°C). The same parameters were also calculated for the immobilized enzyme and the results are reported in figure 4.

Table 1. Inhibition constants of some aliphatic amines and carboxylic acids

Inhibitor	K_i (M)
Butanoic acid	3.56×10^{-1}
Butylamine	2.34×10^{-1}
Hexanoic acid	1.58×10^{-1}
Hexylamine	1.11×10^{-1}
Octanoic acid	5.6×10^{-2}
Octylamine	2.1×10^{-2}
Decanoic acid	1.5×10^{-2}
Decylamine	4.6×10^{-3}

The data were obtained by the HPLC assay because of the strong buffering power of the compounds used as inhibitors which hampered the titrimetric method.

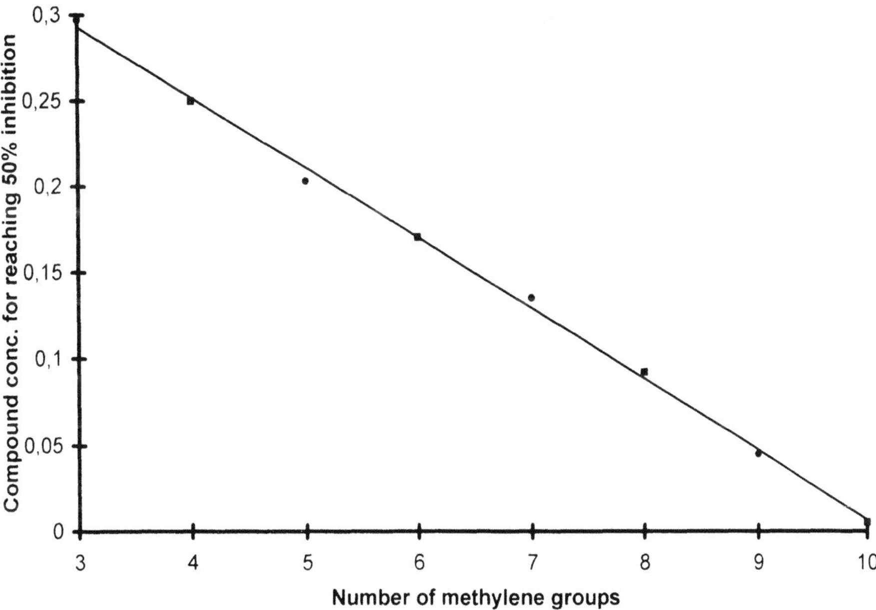

Figure 2. Relationship between the concentration of inhibitor required to obtain 50 % of the maximum velocity and the number of methylene groups in the molecule of amines (■) and carboxylic acids (●).

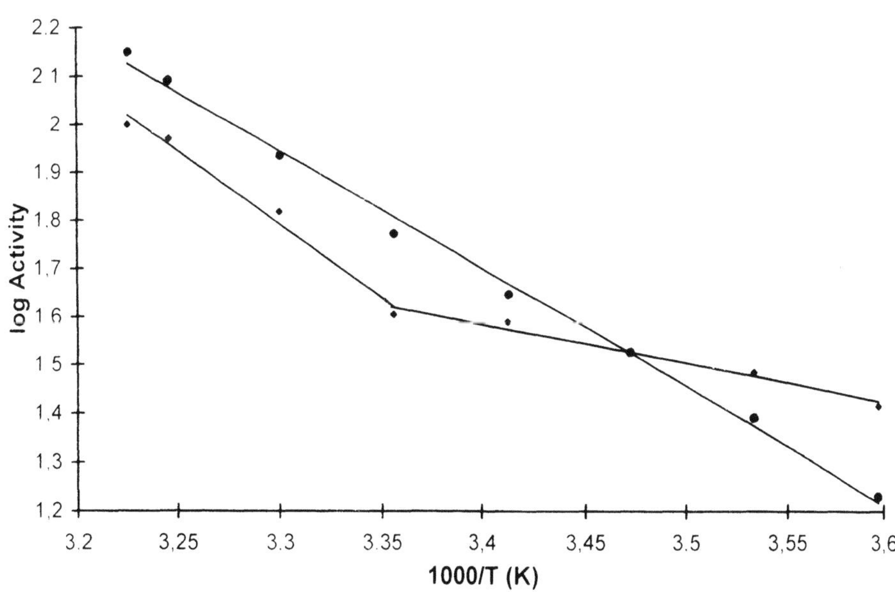

Figure 3. Arrhenius plot of the effect of temperature on the activity of soluble NADase in presence (●) and in absence (◆) of 0.5 M dioxane in the temperature range of 5–37°C.

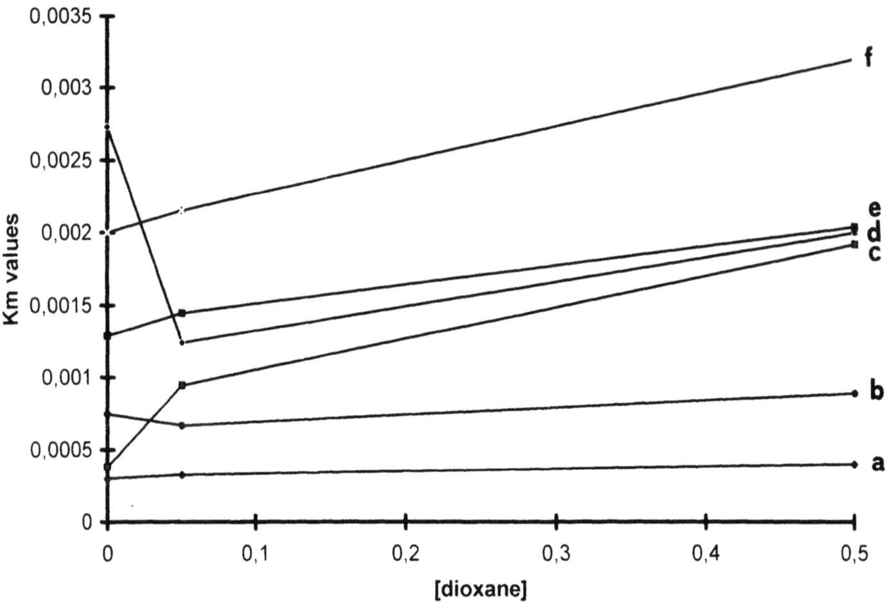

Figure 4. Effect of dioxane on Michaelis constant of NADase in different conditions: (a) in solution at 10°C, (b) in diluted solution at 35°C, (c) immobilized by hydrophobic interaction, (d) in concentrated solution at 35°C, (e) immobilized on Affi-Gel 10 and (f) immobilized by CNBr.

Interaction between NADase and Dioxane by Calorimetry

The values of the enthalpy for the interaction between NADase and dioxane, after correction for the blank, showed that the reaction is endothermic when carried out at 10°C. At 37°C the interpretation of the data was affected by the denaturation of the protein and by the monomer - dimer equilibrium. Therefore only the data obtained at 10°C, corresponding to an almost complete monomer, could be used for the determination of the dissociation constant of the complex enzyme-dioxane [20, 21]. Data obtained from a typical calorimetric experiment are shown in figure 5 and were developed as follows.

The equilibrium constant, $K_{eq} = [EL]/[E][L]$, can be expressed in terms of total amounts of enzyme ($[E]_o$) and ligand ($[L]_o$). The concentration of free enzyme ($[E]$) and unbound ligand ($[L]$) are respectively: $[E] = [E]_o - [EL]$ and $[L] = [L]_o - [EL]$ (where $[EL]$ is the concentration of the complex enzyme-ligand), therefore the equation of the equilibrium constant becomes:

$$K_{eq} = \frac{[EL]}{([E]_o - [EL])([L]_o - [EL])}$$

The molar fraction of the complex enzyme-ligand, that is the ratio between the concentration of bound enzyme and total enzyme, is: $R = [EL]/[E]_o$ and, by substituting in equation [1] the value of [EL] derived from the molar fraction, we obtain:

$$K_{eq} = \frac{R[E]_o}{([E]_o - R[E]_o)([L]_o - R[E]_o)}$$

which becomes:

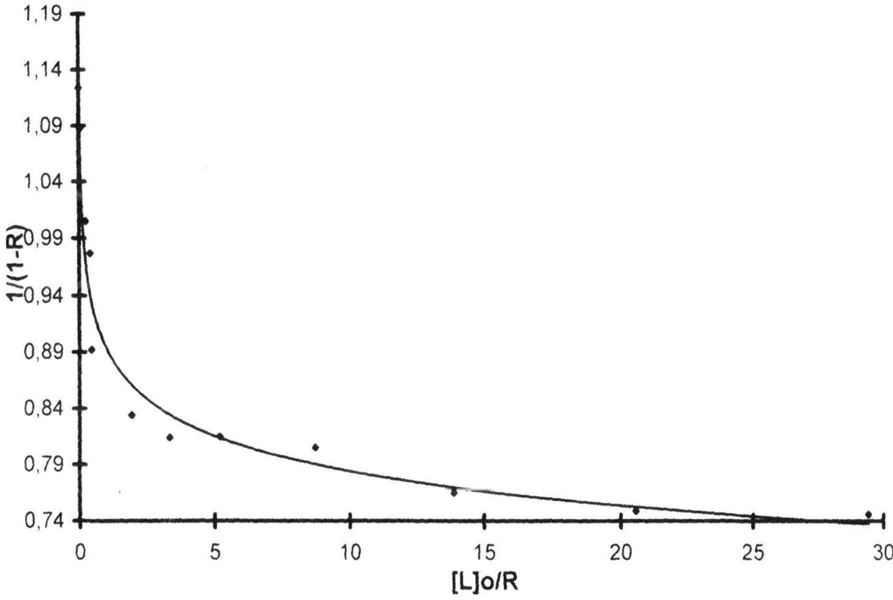

Figure 5. Binding studies of dioxane to NADase from Neurospora crassa conidia. The experiments were carried out at 10°C in a stopped-flow microcalorimeter using 80 µl of enzyme (0.1 mM) and 80 µl of dioxane at various concentrations.

$$\frac{1}{1-R} = K_{eq}\frac{[L]_0}{R} - K_{eq}[E]_0$$

The molar fraction (R) is proportional to the ratio of the calorimetric data of the enzyme partially saturated and the enzyme completely saturated, that is at very high concentration of ligand. Equation [3] represents a line when the values of $[L]_0/R$ are plotted against $1/(1-R)$. The value of the slope is the equilibrium constant (K_{eq}) and the intercept on the ordinate axis is $K_{eq}[E]_0$.

The curve shown in figure 5 indicates that there are at least two independent binding sites for dioxane in the NADase molecule with equilibrium constants of 0.013 mM^{-1} and 0.8 mM^{-1} respectively.

DISCUSSION

Unlike the enzyme from bull semen [6], NADase from Neurospora crassa conidia is inhibited noncompetitively by aliphatic compounds. The inhibition is due to the nonpolar moiety of these substances, as demonstrated by the similar effect of amines and carboxylic acids on the enzyme activity. In fact, by plotting the concentration of inhibitor which yields 50 % inhibition versus the number of methylene groups, a linear relation is found both for aliphatic amines and carboxylic acids. The two lines can be overlapped (see figure 2), indicating that the functional group, either acid or base, does not affect the enzyme activity.

As NADase from other sources, the enzyme from Neurospora crassa conidia shows a self inactivation, but this property seems to be a mechanism different from that already reported for the enzyme from bovine semen and erythrocytes, where it is a turnover-related process [7,10]. Since the Neurospora enzyme forms a dimer through hydrophobic bonds [5] and nonpolar substances inhibit the catalytic activity, self inhibition should be an effect of the interaction between two molecules of enzyme.

Membrane enzymes can modify their activation energy (E_a) without a change in reaction rate at the crystalline-to-liquid-crystalline transition temperature [18,19] and in this case not only E_a is subjected to a change but also the velocity constant is modified. Thus, non-linear Arrhenius plot was already reported for NADase bound to bovine erythrocyte ghosts [20], with a transition temperature of 28°C, indicating that an interaction with the lipid matrix of the erythrocyte membrane was present, but soluble calf-spleen NADase gave a linear Arrhenius plot when the activity was determined in the range of temperature 12–44°C [21]. On the contrary NADase from Neurospora crassa conidia shows a non-linear Arrhenius plot just when the enzyme is in solution and stripped of lipid membrane. Therefore the change in activation energy for NADase from Neurospora crassa conidia could be interpreted according to the theory of Wynn-Williams [22] with the difference that, instead of an interaction with a lipid membrane, here not present, this effect should be due to the interaction between the hydrophobic moieties of two protein molecules and to the prevalence of one enzyme form (the more active monomer or the self inhibited dimer) on the other [23]. Since it is known that hydrophobic interactions are endothermic, that is stronger when the temperature is raised, this accounts for the higher energy of activation over 25°C. Dioxane is able to produce the monomeric form of the enzyme [5], yielding a straight line in the Arrhenius plot, but at the same time it affects the energy of activation for the catalyzed reaction. The value of E_a obtained in presence of dioxane (11.2 kcal/mol) results therefore higher than that showed by the monomer (4.1 kcal/mol) but lower than the activation energy of the reaction catalyzed by the dimer (14.23 kcal/mol). Since the size of the nonpolar compound has great influence on the inhibition, as demonstrated by the inhibition constant of several amines and carboxylic acids, it is coherent that dioxane has a small effect compared to that of the nonpolar moiety of a second molecule of enzyme.

The inhibition by dioxane appears of mixed type, and more information on the effect of dioxane can be obtained from the values of Michaelis constant reported in figure 4. NADase shows little affinity for the substrate when the enzyme is in the dimeric form (i.e. high concentration of protein, 35°C). In presence of 0.05 M dioxane the K_m of the enzyme decreases, indicating that the monomer has higher affinity for the substrate than the dimer, but at higher concentration (0.5 M) dioxane has a more inhibitory effect. A similar behavior, but less striking, is shown at 35°C with a lower concentration of protein, when the monomer is prevalent on the dimer but not the only form of the enzyme. Finally at 10°C, a temperature that weakens the hydrophobic bonds in the protein, dioxane increases the Michaelis constant, showing only an inhibitory effect.

The negative effect of dioxane on the monomeric NADase is also pointed out by the increase of K_m when the enzyme is immobilized on agarose gels. After the procedures of binding and washing, the monomer is the only form of enzyme present in the derivative, therefore the increase of K_m in presence of dioxane indicates a decrease in the affinity of the monomeric NADase for the substrate. The polymeric support modifies the protein conformation, as it is demonstrated by the difference in K_m when NADase is covalently immobilized by direct bond or through a hydrophilic spacer arm. It is interesting to note that also the enzyme hydrophobically bound to propyl-Sepharose is slightly inhibited by dioxane, indicating the presence of at least a second hydrophobic site in the protein molecule that is confirmed by the results of the calorimetric data.

Therefore the results obtained can be interpreted as a double effect of dioxane on NADase, that is:

1. yield of the more active monomer from the self-inhibited dimer
2. direct inhibition of the monomer by the organic solvent.

REFERENCES

1. Everse, J. and Kaplan, N.O. 1968. Characteristics of microbial diphosphopyridine nucleotidases containing exceptionally large amounts of polysaccharides. *J. Biol. Chem.*, *243*: 6072–6074

2. Menegus, F. and Pace, M. 1981. Purification and some properties of NAD glycohydrolase from Neurospora crassa conidia. *Eur. J. Biochem. 113*: 485–490

3. Pace, M., Agnellini, D., Lippoli, G., Pietta, P.G., Mauri, P.L. and Cinquanta, S. 1991. Purification of NAD glycohydrolase from Neurospora crassa conidia by a polyclonal immunoadsorbent. *J. Chromatogr. (Biomed. Appl.): 539*: 517–523

4. Pace, M., Agnellini, D., Pietta, P.G., Cocilovo, A. and Bonizzi, L. 1984. Preparation of immobilized NAD glycohydrolase from Neurospora crassa conidia by hydrophobic interaction - Characteristics of the enzyme derivative. *Prep. Biochem. 14*: 349–362

5. Pace, M., Agnellini, D., Pietta, P.G., Mauri, P.L., Menegus, F. and Berger, R.L. 1988. NAD glycohydrolase from Neurospora crassa conidia - Association of the protein in solution *Plant Physiol. (Life Sci. Adv.), 7*: 115–118

6. Yuan, J.H. and Anderson, B.M. 1972. Bull semen NAD glycohydrolase - 4. Nonpolar interactions of inhibitors with the substrate binding site. *Arch. Biochem. Biophys. 149*: 419–424

7. Anderson, B.M, Yost, D.A. 1985. Study of self-inactivation of bovine seminal fluid NAD glycohydrolase. *Chem.Biol.Interact.*, *54*: 159–170

8. Green, S. and Dobriansky, A. 1971. pH-Dependent inactivation of nicotinamide-adenine dinucleotide glycohydrolase by its substrate, oxidized nicotinamide adenine dinucleotide. *Biochemistry, 10*: 2496–2500

9. Cayama, E., Apitz-Castro, R. and Cordes, E.H. 1973. Substrate-dependent, thiol dependent, inactivation of pig brain NAD glycohydrolase. *J. Biol. Chem.*, *248*: 6479–6483

10. Pekala, P.H., Yost, D.A. and Anderson, B.M. 1980. Self-inactivation of an erythrocyte NAD glycohydrolase. *Mol. Cell. Biochem. 31*: 49–56

11. Davis, R.H. and De Serres, F.J. 1970. in: *Methods in Enzymology* (Colowick, S. and Kaplan, N.O. eds), Vol. 17: pp. 79–143

12. Jacobsen, C.F., Léonis, J., Linderstrøm-Lang, K. and Ottesen, M. 1957. The pH-stat and its use in biochemistry, in *Methods of Biochemical Analysis* (Glick, D. ed.) Vol. 4, pp. 171–210, Wiley Interscience

13. Pietta, P.G., Pace, M. and Menegus, F. 1983. High performance liquid chromatography for assaying NAD glycohydrolase from Neurospora crassa conidia *Anal. Biochem., 131*: 533–537

14. Pace, M., Mauri, P.L., Pietta, P.G. and Agnellini, D. 1989. High-performance liquid chromatography determination of enzyme activities in the presence of small amounts of product *Anal. Bioch. 176*: 437–439

15. Mudd, C. and Berger, R.L. 1988. A complete controlled all Tantalum stopped flow microcalorimeter. *J. Biochem. Biophys. Meth. 17*: 161–191

16. Scatchard, G. 1949. The attraction of protein for small molecules and ions. *Ann. N.Y. Acad. Sci. 51*: 660–672

17. Edsall, J.T. and Gutfreund, H. 1983. *Biothermodynamics*, J. Wiley & Sons, Chichester pp. 159–165

18. Linden, C.D., Wright, K.L., McConnell, H.M. and Fox, C.F. 1973. Lateral phase separations in membrane lipids and the mechanism of sugar transport in Escherichia coli. *Proc. Natl. Acad. Sci. USA. 70*: 2271–2275

19. Silvius, J.R. and McElhaney R.N. 1981. Non-linear Arrhenius plots and the analysis of reaction and motional rates in biological membranes. *J. Theor. Biol. 88*: 135–152

20. Pekala, P.H and Anderson, B.M. 1978. Studies of bovine erythrocyte NAD glycohydrolase. *J. Biol. Chem. 253*: 7453–7459

21. Travo, P., Muller, H. and Schuber, F. 1979. Calf-spleen NAD glycohydrolase - Comparison of the catalytic properties of the membrane-bound and the hydrosoluble forms of the enzyme. *Eur. J. Biochem., 96*, 141–149

22. Wynn-Williams, A.T. 1976. An explanation of apparent sudden change in the activation energy of membrane enzymes. *Biochem. J. 157*: 279–281

23. Massey, V., Curti, B. and Ganther, H. 1966. A temperature-dependent conformational change in D-amino acid oxidase and its effect on catalysis. *J. Biol. Chem., 241*, 2347–2357

52

INVOLVEMENT OF BOVINE SPLEEN NAD$^+$ GLYCOHYDROLASE IN THE METABOLISM OF CYCLIC ADP-RIBOSE-MECHANISM OF THE CYCLIZATION REACTION

Hélène Muller-Steffner, Angélique Augustin, and Francis Schuber

Laboratoire de Chimie Bioorganique
URA CNRS 1386, Faculté de Pharmacie
74 route du Rhin, 67400
Illkirch, France

ABSTRACT

We have shown that highly purified bovine spleen NAD$^+$glycohydrolase (NADase), known so far to catalyze the hydrolysis of the nicotinamide-ribose bond of NAD(P)$^+$, was also able to convert NAD$^+$ into cyclic ADP-ribose (cADPR) and to hydrolyze cADPR into ADP-ribose. The kinetic parameters measured for the cyclic ADP-ribose hydrolase activity seem to exclude that cADPR is a kinetically competent reaction intermediate in the NADase catalyzed conversion of NAD$^+$ into ADP-ribose. The cyclase activity of bovine NADase was best evidenced by the transformation of NGD$^+$ into cyclic GDP-ribose which was the major reaction product whereas, in contrast, cADPR accounted for less than 2% of the products formed. For the formation of cADPR we propose a reaction mechanism that is based on the partitioning of an oxocarbenium reaction intermediate between an intramolecular attack by the N1-position of adenine and an intramolecular reaction by a water molecule. Accordingly, the difference in cyclization between NAD$^+$ and NGD$^+$ is accounted for by the difference in reactivity of the N1 and N7 positions of the purine ring in these dinucleotides.

INTRODUCTION

NAD$^+$glycohydrolases (NADases) are a class of enzymes (EC 3.2.2.5 & 3.2.2.6), widely distributed in eukaryotic and in some prokaryotic systems, that catalyze the hydrolysis of NAD$^+$ to adenosine diphosphate ribose (ADP-ribose) and nicotinamide [1]. The enzymes of mammalian origin, which are nearly all membrane-bound proteins, are also characterized by their ability to catalyze the cleavage of NADP$^+$ and transglycosida-

ADP-Ribosylation in Animal Tissue, edited by Haag and Koch-Nolte
Plenum Press, New York, 1997

tion reactions; this latter property has been much used for the preparation of pyridinium analogs of NAD(P)$^+$ [1,2]. Despite the fact that NAD$^+$glycohydrolases from different sources have been studied for many decades, both from catalytic and cellular perspectives, the physiological functions of these enzymes remain poorly understood. Importantly, although some NADase activity was found associated with intracellular compartments such as mitochondria [3,4], NAD$^+$glycohydrolases are overwhelmingly ecto-enzymes in cells that are rich in this enzyme [5,6].

Interest in NAD$^+$glycohydrolases was renewed with the recent discovery by Lee and co-workers of cyclic ADP-ribose (cADPR). This new metabolite of NAD$^+$, which was originally found in sea urchin eggs, is thought to be an endogenous regulator of the Ca^{2+}-induced Ca^{2+}-release process mediated by the ryanodine receptors [reviewed in 7]. In invertebrates cADPR is the exclusive reaction product obtained from NAD$^+$ by an ADP-ribosyl cyclase [8,9]. In mammalian tissues no equivalent enzyme could be detected; however, a high sequence homology was found between the cyclase from *Aplysia californica* and CD38, a human lymphocyte cell surface antigen, for which no biological activity was hitherto known [10]. CD38 revealed itself to be a multifunctional enzyme; i.e., besides catalyzing the hydrolytic cleavage of NAD$^+$ into ADP-ribose, it is also able to produce cADPR, albeit in small amounts (less than 2–3% of reaction products), and to hydrolyze cADPR into ADP-ribose [11]. Similar catalytic activities were established for a canine spleen enzyme [12] and BST-1, a GPI-anchored bone marrow stromal cell antigen [13]. The low yield of cADPR production by these mammalian systems was attributed to their multifunctionality: the cyclic metabolite does not accumulate because it is turned-over by the same enzyme that produces it [14]. In apparent agreement with this hypothesis, it was established by Graeff *et al.* [15] that NGD$^+$, an analogue of NAD$^+$, is converted in high yield by CD38 into cyclic GDP-ribose (cGDPR), a metabolite that was thought to be not hydrolyzable. Accordingly, since the net reaction catalyzed by the NAD$^+$glycohydrolases is that of an ADP-ribosyl cyclase coupled to a cADPR hydrolase (Figure 1), it was suggested that the NADases could in fact also be multifunctional enzymes whose cyclase activity had been overlooked [7].

Since the occurrence of CD38 was originally thought to be restricted to B and T lymphocytes [16,17] and erythrocytes [18], it was of interest to broaden the issue by investigating the possible contribution of the classical NAD(P)$^+$glycohydrolases to the metabolism of cADPR in mammalian systems. These enzymes have a much wider tissular and cellular distribution [1], than those attributed so far to CD38, *e.g.* they are also found associated with macrophages [5] including Kupffer cells in liver [19] and microglial cells [20]. We have been engaged for many years in the study of the molecular and cellular enzymology of bovine spleen NAD$^+$glycohydrolase and this knowledge might be helpful to better define the molecular mechanisms underlying the metabolism of cADPR. Thus, we have demonstrated that the bovine NADase catalyzes a dissociative mechanism; i.e., the nicotinamide-ribose bond of NAD$^+$ is cleaved to generate an enzyme-stabilized oxocarbenium ion-type intermediate that reacts in a non-rate limiting step with acceptors such as water (hydrolysis), methanol (methanolysis) or pyridines (transglycosidation) [21–25].

In the present work we show that highly purified calf spleen NAD(P)$^+$glycohydrolase [26] is a multifunctional enzyme. Besides its classical NADase function, it is also a cADPR hydrolase and a pyridine nucleotide cyclase that is able to convert NGD$^+$ into cGDPR in high yield and NAD$^+$ into small amounts of cADPR (less than 2% of the reaction products). Moreover, we provide evidence that the low net conversion of NAD$^+$ into cADPR by this enzyme is not due to a fast turnover of the cyclic metabolite, that prevents it from accumulating, but to a lesser reactivity of the adenine ring, compared to guanine, with the intermediary oxocarbenium ion. This mechanistic scheme might be general for mammalian cyclases.

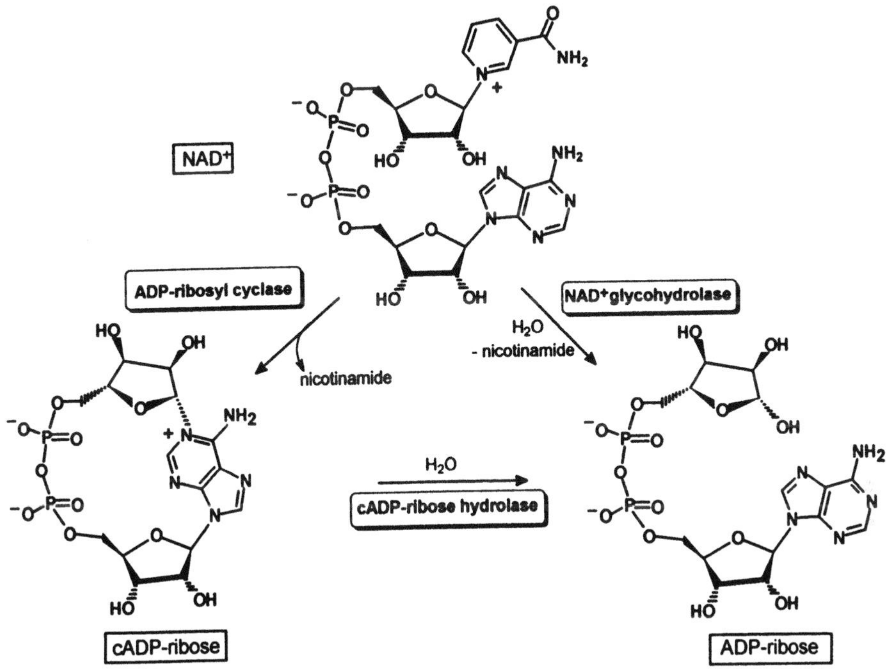

Figure 1. Reactions catalyzed by ADP-ribosyl cyclase, cyclic ADP-ribose hydrolase and by NAD⁺glycohydrolase.

RESULTS

Cyclic ADP-Ribose Hydrolase Activity of Bovine Spleen NAD⁺ Glycohydrolase

When incubated in the presence of bovine spleen NADase, cADPR was found to be transformed into ADP-ribose. This reaction could be easily followed by HPLC [27] and the reaction parameters determined under steady-state conditions (at 37°C and pH 7.4) yielded the following values: Km = 2.15 mM and Vmax = 210 μmol min^{-1} mg^{-1} protein. The high Km is in agreement with the low affinities that this enzyme generally displays for ligands lacking the pyridinium moiety. Under the same experimental conditions NAD⁺ gave: Km = 26 μM and Vmax 140 μmol min^{-1} mg^{-1} protein. This establishes that calf spleen NAD⁺glycohydrolase belongs to the family of cADP-ribose hydrolases. From a mechanistic view point, the important result is that the specificity constant of the enzyme (expressed in terms of Vmax/Km) is about 55-fold smaller for cADPR than for NAD⁺.

Pyridine Nucleotide Cyclase Activity of Bovine Spleen NAD⁺ Glycohydrolase

NGD⁺, which compared to NAD⁺ is converted in much higher yields into a cyclic derivative by CD38, is a convenient tool to study mammalian ADP-ribosyl cyclases [15]. The structure of cyclic GDP-ribose was recently reassessed and the cyclization involves the N7-position of the purine ring as opposed to the N1-position in the formation of cADPR [28,29]. The transformation of NGD⁺ into cyclic GDP-ribose, which can be conveniently followed by the spectral characteristics of the cyclic compound [15,29], was tested with calf spleen NAD$^+$glycohydrolase. When NGD$^+$ was incubated in the presence of the enzyme, an increase of absorbance at 300 nm was monitored as a function of time, which is indicative of the formation of cGDPR. The cyclization reaction was also established by measuring the increase of fluorescence at 410 nm; the progress curves reached a plateau and further addition of NAD$^+$glycohydrolase (up to 50 mU) did not produce a change in fluorescence indicating that under these experimental conditions the fluorescent reaction product is not turned over. Production of cyclic GDP-ribose was estimated by analyzing the reaction products by HPLC. The elution profiles obtained, which are similar to those described in literature with CD38 [15], showed that NAD$^+$glycohydrolase converted NGD$^+$ into cyclic GDP-ribose and GDP-ribose in a 2:1 ratio in favor of the cyclic compound (Table 1). The peak attributed to cyclic GDP-ribose was collected and authentified by its fluorescence spectra. From these experiments, we can conclude that bovine spleen NAD$^+$glycohydrolase is able to efficiently catalyze the conversion of NGD$^+$ into cGDPR and is therefore a member of the class of mammalian enzymes, that includes CD38 and BST-1, which catalyze the cyclization of pyridine dinucleotides.

Kinetic parameters of the transformation of NGD$^+$ by NAD$^+$glycohydrolase were calculated from the reaction rates obtained from the HPLC profiles; the following values were found: Km = 24 μM and Vm = 80 μmol min^{-1} mg^{-1} of protein. It appears therefore that NGD$^+$ is an excellent substrate for bovine NAD$^+$glycohydrolase, the only difference between this dinucleotide and NAD$^+$ lies in the products generated during enzymatic catalysis.

Because of the ability of NAD$^+$glycohydrolase to catalyze the conversion of NGD$^+$ into cyclic GDP-ribose, we have reassessed the formation of cADPR from NAD$^+$, which we had previously estimated to be very low [27]. Thus [^{14}C]NAD$^+$ was incubated in the presence of the enzyme and the reaction products were analyzed by HPLC using on-line UV and radioactivity detectors. The main product was ADP-ribose, but we could establish the formation, albeit in low amounts, of a product presenting an elution time identical, both by UV and radioactivity, to that of c[^{14}C]ADPR. The formation of this product was time-dependent (Fig. 2) and it amounted to about 1.5% (± 0.3, *n* = 6) of the products formed. In control experiments (boiled enzyme) no such peak appeared; moreover, the formation of cADPR became undetectable when the reaction was run in the presence of

Table 1. Transformation of NGD+ by NAD+glycohydrolase: competition between the reaction of cyclization and the reactions of methanolysis and hydrolysis. NGD$^+$ was incubated in the presence of bovine spleen NAD$^+$glycohydrolase (37°C, 10 mM potassium phosphate buffer, pH 7.4) and increasing concentrations of methanol. The proportions of reaction products were assessed by HPLC

Methanol (M)	$\dfrac{[\text{cGDP–ribose}]}{[\text{GDP–ribose}] + [\text{methly GDP–ribose}]}$
0	2.10
0.5	1.34
1	1.12
2	0.59
3	0.36

3M methanol (see below). Such a low percent for the formation of cADPR is very similar to those found for example for CD38 [11].

Methanolysis of NGD⁺ Catalyzed by Bovine Spleen NAD⁺ Glycohydrolase

The important difference in the formation of cADPR and cGDPR, by CD38 and other cyclases, is generally attributed to the "multifunctionality" of such enzymes and to the fast turnover of cADPR as opposed to the poor hydrolysis of cGDPR (see Introduction). Such an analysis was not borne out, at least for calf spleen NAD⁺glycohydrolase; i.e. because the specificity constant of the enzyme for NAD⁺ is 2 orders of magnitude higher than for cADPR (see above) one expects that the cyclic metabolite, once formed and released from the active site, would accumulate during the reaction. The fact that this is not observed rules out that cADPR is a kinetically competent reaction intermediate in the transformation of NAD⁺ into ADP-ribose by bovine spleen NAD⁺glycohydrolase. Therefore, since we have previously shown that the hydrolysis of pyridinium dinucleotides by this enzyme involves the formation, during the slow step of the reaction process, of a stabilized oxocarbenium ion reaction intermediate (see Introduction), we reasoned that the formation of the different reaction products is dependent on the reactivity of this intermediate with the acceptors. Thus, depending on the substrate used, i.e. NAD⁺ or NGD⁺, the intermediary oxocarbenium ion might react with water leading to A(G)DP-ribose, but also with N7 of the guanine ring yielding cGDPR and with the N1-position of the adenine to give cADPR, this latter process being unfavorable (Fig. 3).

To verify this hypothesis, which implies that these different acceptors compete for the same oxocarbenium ion, we decided to perform the reaction in presence of methanol. We have pre-viously demonstrated that the oxocarbenium intermediate formed during the NAD⁺glycohy-drolase-catalyzed transformation of NAD⁺ reacts about 50-fold faster with methanol than with water [24]. Methyl ADP-ribose is formed with retention of configuration and in the presence of high enough concentrations of methanol it becomes the preponderant reaction product. Thus, if a similar water/methanol partitioning ratio is found in the

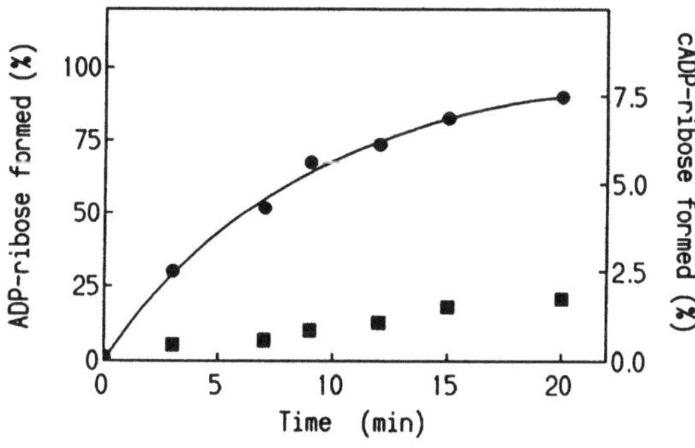

Figure 2. Formation of ADP-ribose and cyclic ADP-ribose from NAD⁺ catalyzed by bovine spleen NAD⁺glyco-hydrolase. [Adenine-U-¹⁴C]NAD⁺ was incubated (37°C, 10 mM potassium phosphate buffer, pH 7.4) in the presence of bovine NADase. Aliquots were removed at given time intervals and analyzed by HPLC. Quantitative estimates for the formation of ADP-ribose and cADPR (given as percent of total radioactivity) were obtained from radiochromatograms. Note the difference in scale for the formation of cADPR.

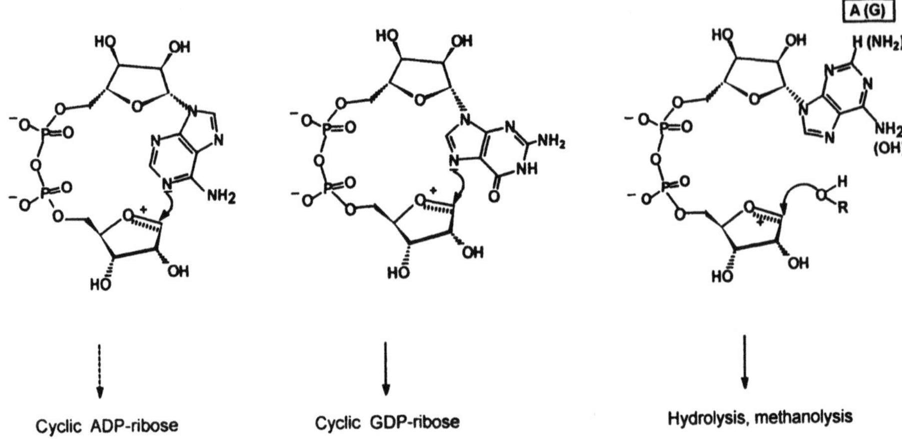

Figure 3. Partitioning of the oxocarbenium intermediate occurring in the reaction catalyzed by bovine spleen NAD⁺glycohydrolase. The enzyme-stabilized oxocarbenium intermediates generated from NA(G)D⁺ undergo intramolecular cyclization reactions involving the attack by N1 of adenine and N7 of guanine. The formation of respectively cADPR (minor pathway.) and cGDPR are in competition with the intermolecular reaction of the intermediate with water (formation of A(G)DPR) and methanol (formation of methyl A(G)DPR). The low amount of cADPR formed by the enzyme, compared to cGDPR, is proposed to be related to a lower rate of the intramolecular reaction which might be ascribed to a combination of a lower nucleophilicity of the N1-position of adenine, compared to N7 of guanine, and/or a less favorable orientation of the adenine ring for cyclization.

transformation of NGD⁺, one should observe the formation of methyl GDP-ribose at the expense of cGDPR.

Transformation of NGD⁺ catalyzed by calf spleen NAD⁺glycohydrolase in the presence of increasing concentrations of methanol was analyzed by HPLC (Fig. 4). In the presence of this acceptor a new reaction product was formed whose retention time was identical to one of the methyl GDR-ribose isomers obtained by spontaneous solvolysis of NGD⁺ in presence of 30% (v/v) methanol. Analysis of the data indicated that, similarly to NAD⁺, the partitioning ratio K ([methyl GDP-ribose] x [H₂O]/ [GDP-ribose] x [CH₃OH]) was 52.4 (± 4.6; $n = 6$) in favor of methanolysis. As indicated in the Table I, increasing concentrations of methanol led to increased formation of methyl GDP-ribose, at the expense of cGDPR, demonstrating that methanolysis as expected from our hypothesis competes with the cyclization process. It was also found that addition of methanol did not increase the turnover of NGD⁺, confirming that the reaction of the oxocarbenium ion with acceptors is a fast step in the catalytic process of NAD⁺glycohydrolase relative to the formation of this intermediate [21]. Importantly, in control experiments performed with the same concentration of enzyme and using a continuous fluorometric assay, NAD⁺glycohydrolase was found unable to transform cGDPR even in the presence of the highest concentrations of methanol used above.

DISCUSSION

We have shown that the « classical » bovine spleen NAD⁺glycohydrolase, which so far was only known to catalyze the hydrolysis of NAD(P)⁺ and transglycosidation reactions, is a « multifunctional » enzyme which, in analogy to CD38 and BST-1, is also endowed with ADP-ribosyl cyclase and cyclic ADP-ribose hydrolase activities. This finding

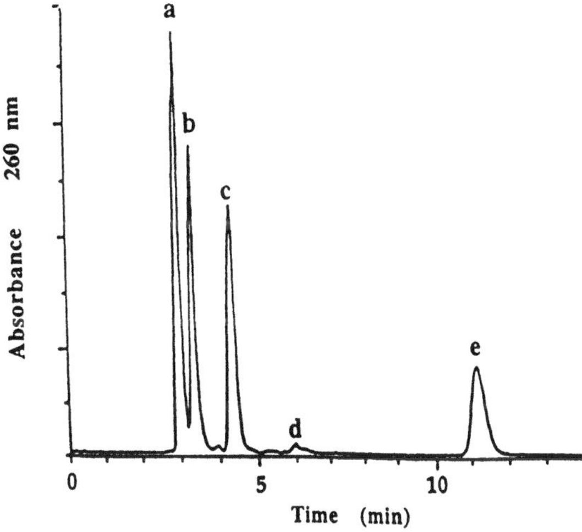

Figure 4. HPLC elution profile of the products obtained by incubation of NGD⁺ with bovine spleen NAD⁺glycohydrolase in the presence of methanol. NGD⁺ was incubated (37°C, 10 mM potassium phosphate buffer, pH 7.4) with the enzyme in the presence of 3 M methanol. HPLC analyses were obtained on a reverse-phase µBondapak C₁₈ (Waters) column; the compounds were eluted isocratically with a 10 mM ammonium phosphate buffer, pH 5.5, containing 1.2% (v/v) acetonitrile and detected by their UV absorbance at 260 nm. cGDP-ribose (a), GDP-ribose (b), methyl GDP-ribose (c), NGD⁺ (d), and nicotinamide (e).

raises the question of the classification of the different enzymes that are able to hydrolytically cleave the nicotinamide-ribose bond of NAD⁺. Mammalian NAD⁺glycohydrolases, which have been studied for many decades, form a very heterogeneous class of enzymes [1]; i.e. their catalytic and molecular properties may depend on their origin and in a same organism on their tissular localization. For example, NADases are known that differ in their ability to cleave NADP⁺ or perform transglycosidation reactions, in their mode of inhibition by isonicotinic acid hydrazide (isoniazid) or in their paracatalytic inactivation by NAD⁺ at basic pH. The ability of these enzymes to produce and hydrolyze cADPR represents a new criteria be added to that list. From our knowledge of these enzymes, we can predict that the NADases that are able to hydrolyze NADP⁺ and catalyze transglycosidation reactions should also be multifunctional, i.e. participate to the metabolism of cADPR. Finally, NADases vary also widely by their molecular properties such as molecular weights or their mode of association with membranes (transmembrane domain or GPI-anchored). Bovine spleen NAD⁺glycohydrolase shares with CD38 some structural features, e.g. it is an ecto-enzyme [6] and an amphipathic glycoprotein that consists of two domains, a hydrophilic domain that contains the active site and a hydrophobic domain that anchors the enzyme to the membrane [30]. The structural homology of these enzymes remains to be established, and we hope to be able to clarify that point in the next future.

The fact that bovine spleen NAD⁺glycohydrolase, in analogy with CD38, forms very little cADPR from NAD⁺ (less than 2% of the reaction products), is an important issue from a physiological perspective but it raises also the question of the molecular mechanism of the cyclization process. As discussed above, the poor performance of the mammalian ADP-ribosyl cyclases was generally attributed to the multifunctionality of these enzymes; i.e., it is believed that cADPR does not accumulate because it is quickly hydrolyzed by the very same enzyme that produces it [14]. The kinetic parameters determined for the hydrolysis of cADPR by bovine NADase predict, however, that if the cyclic metabolite is a compulsory reaction intermediate in the transformation of NAD^+ into ADP-ribose (i.e. a kinetically competent intermediate) and is released from the active site, it should accumulate in the medium. This was not observed indicating that cADPR is, like ADP-ribose, a reaction product. This very limited capacity of bovine spleen NAD^+glycohydrolase to generate cADPR contrasts with the transformation of NGD^+ that yields predominantly cGDPR. We believe that this is a direct consequence of the reaction mechanism of the enzyme. Thus, we have conclusively demonstrated with the experiment of NGD^+ methanolysis that NADase-catalyzed cyclization reaction is in competition with the hydrolysis/methanolysis reactions. It follows that the intramolecular cyclizations which involve the N1-position of adenine and the N7-position of guanine in respectively NAD^+ and NGD^+, are in competition with an intermolecular nucleophilic attack of a common intermediary oxocarbenium ion by a water molecule that yields A(G)DP-ribose (Fig. 3). The question is raised on the origin of such a marked difference in the extent of cADPR and cGDPR formation. It is well known that in adenine the most nucleophilic position is N1 whereas in guanine it is the N7-position [31,32]; it is therefore probably not a coincidence that the cyclic compounds that are formed are precisely the ones that involve these nucleophilic centers in the purine rings. Moreover, the fact that 1,N^6-etheno NAD^+, whose N1 is masked, is also cyclized at position N7 [28,29] indicates that in the active site of these enzymes, the purine rings do not have a fixed binding pattern. It appears therefore that the differences in the formation in cADPR and cGDPR in this intramolecular cyclization reaction might reflect the differences in nucleophilicity of the two nitrogens in the active site of the enzyme; indeed, it is well known that in many reactions, involving the alkylation of bases, N7 of the guanine ring is more nucleophilic than N1 of adenine [32 - 34]. Interestingly, a high sensitivity of NAD^+glycohydrolases to the nucleophilicity of attacking pyridines, yielding by transglycosidation pyridinium analogues of NAD^+, has been noted before [23,35]. The much favored formation of cGDPR might also indicate the occurrence of additional factors such as, e.g. a better positioning, within the active site, of the guanine with respect to the reaction with the oxocarbenium ion. Interestingly, during the enzyme-catalyzed cyclization step the attacking purine rings must adopt opposite configurations, i.e. *syn* and *anti* respectively for cADPR and cGDPR, which are the less favored in solution.

That NAD^+glycohydrolase is an efficient GDP-ribosyl cyclase and hydrolyzes cADPR is also an important indication with regard to the conformation of the substrates when bound to its active site. In contrast to toxins [36] and to oxidoreductases having a "Rossmann fold" [37], where NAD^+ is bond in an extended conformation (i.e. the adenine and nicotinamide moieties are apart), in the present case the substrates must adopt a more "closed" conformation, i.e. a nicotinamide-adenine stacked conformation, reminiscent of the form that is found in solution [38,39]. Such a conformation might even be more compact on binding to the active site inducing some strain, such as unfavorable non-bonded interactions in the NMN^+ moiety, which could be released when going to the transition state and thus explain some of the catalysis in the N-glycosidic bond destabilization.

In conclusion, bovine spleen NAD^+glycohydrolase is a multifunctional enzyme that, similarly to CD38, is involved in the metabolism of cADPR. The low proportion of cADPR formed from NAD^+ might be nevertheless high enough to be of physiological significance in Ca^{2+}-signaling [40]. Importantly the mechanism we propose for the formation of the cyclic metabolite, which is based on the partitioning of the oxocarbenium reaction intermediate between an intramolecular attack by the N1 position of adenine and an intramolecular reaction by a water molecule (Fig. 5), predicts that small changes in the conformation of the active site of the enzyme might have dramatic consequences on the extent of

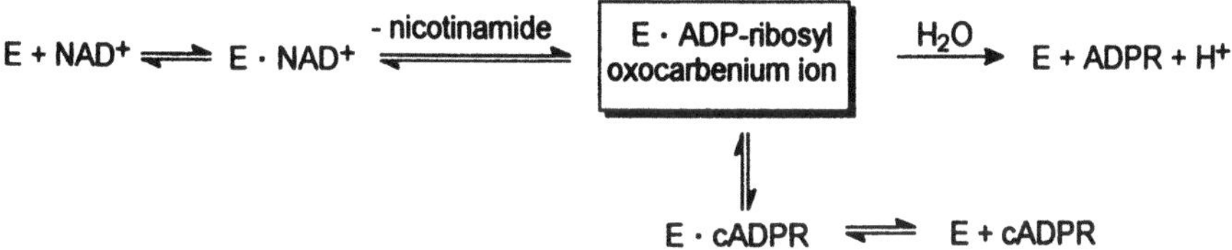

Figure 5. Minimal mechanistic scheme for the transformation of NAD⁺ catalyzed by bovine spleen NAD⁺glycohydrolase. The rate limiting step(s) involve the transformation of NAD⁺ into a stabilized E.ADP-ribosyl oxocarbenium ion complex. This reaction intermediate reacts intramolecularly with N1 of adenine to yield cADPR and intermolecularly with water. It should be noted that within the active site of the enzyme, the ADP-ribosyl oxocarbenium ion might not be a single species with regards to the configuration of the adenine-ribose bond. Thus, when generated from NAD⁺, its methanol/water partitioning ratio is 50 (which might be an average value for the different conformers) whereas when generated from cADPR this ratio drops to 30 [27], indicating that an oxocarbenium ion generated from a species where the adenine has exclusively a *syn* configuration is less stabilized by the active site.

cyclization. Thus a non-bonded interaction ("cross-talk") with a neighboring protein (receptor ?) of this ecto-enzyme might change the topology of its active site and favor the cyclization reaction.

REFERENCES

1. Price, S.R., and Pekala, P.H. (1987) Pyridine nucleotide-linked glycohydrolases. In *Pyridine Nucleotide Coenzymes: Chemical, Biochemical, and Medical Aspects* (Dolphin, D., Poulson, R., and Avramovic, O., Eds), Vol 2B, pp. 513–548, John Wiley.

2. Anderson, B.M. (1982) Analogs of pyridine nucleotide coenzymes. *In The Pyridine Nucleotide Coenzymes* (Everse, J. Anderson, B. and You, K.S., eds.), pp. 91–133, Academic Press, New York.

3. Moser, B., Winterhalter, K.H. and Richter, C. (1983) Purification and properties of a mitochondrial NAD⁺ Glycohydrolase. *Arch. Biochem. Biophys. 224*: 358–364.

4. Masmoudi, A. and Mandel, P. (1987) ADP-Ribosyl transferase and NAD⁺glycohydrolase activities in rat liver mitochondria. *Biochemistry 26*: 1965–1969.

5. Artman, M. and Seeley, R.J. (1979) Nicotinamide adenine dinucleotide splitting enzyme: A plasma membrane protein of murine macrophages. *Arch. Biochem. Biophys. 195*: 121–127.

6. Muller, H.M., Muller, C.D. and Schuber, F. (1983) NAD⁺glycohydrolase, an ecto-enzyme of calf spleen cells. *Biochem. J. 212*: 459–464.

7. Lee, H.C. (1994) Cyclic ADP-ribose: A calcium mobilizing metabolite of NAD⁺. *Mol. Cell. Biochem. 138*: 229–235.

8. Lee, H.C., and Aarhus, R. (1991) ADP-ribosyl cyclase: an enzyme that cyclizes NAD⁺ into a calcium-mobilizing metabolite. *Cell Regul. 2*: 203–209.

9. Hellmich, M.R., and Strumwasser, F. (1991) Purification and characterization of a molluscan egg-specific NADase, a second-messenger enzyme. *Cell Regul. 2*: 193–202.

10. States, D.J., Walseth, T.F., and Lee, H.C. (1992) Similarities in amino acid sequences of Aplysia ADP-ribosyl cyclase and human lymphocyte antigen CD38. *Trends Biochem. Sci. 17*: 495.

11. Howard, M., Grimaldi, J.F., Bazan, J.F., Lund, F.E., Santos-Argumedo, L., Parkhouse, R.M.E., Walseth, T.F., and Lee, H.C. (1993) Formation and hydrolysis of cyclic ADP-Ribose catalyzed by lymphocyte antigen CD38. *Science 262*: 1056–1059.

12. Kim, H., Jacobson, E.L., and Jacobson, M.K. (1993) Synthesis and degradation of cyclic ADP-Ribose by NADglycohydrolases. *Science 261*: 1330–1333.

13. Hirata, Y., Kimura, N., Sato, K., Ohsugi, Y., Takasawa, S., Okamoto, H., Ishikawa, T., Ishihara, K. and Hirano, T. (1994) ADP-ribosyl cyclase activity of a novel bone marrow stromal cell surface molecule, BST-1. *FEBS Lett. 356*: 244–248.

14. Lee, H.C., Graeff, R., and Walseth, T.F. (1995) Cyclic ADP-ribose and its metabolic enzymes. *Biochimie 77*: 354–355.

15. Graeff, R.M., Walseth, T.F., Fryxell, K., Branton, W.D., and Lee, H.C. (1994) Enzymatic synthesis and characterizations of cyclic GDP-ribose. *J. Biol. Chem. 269*: 30260–30267.

16. Malavasi, F., Funaro, A., Roggero, S., Horenstein, A., Calosso, L., and Mehta, K. (1994) Human CD38: a glycoprotein in search of a function. *Immunol. Today 15*: 95–97.

17. Lund, F., Solvason, N., Grimaldi, J.C., Parkhouse, R.M.E., and Howard, M. (1995) Murine CD38: an immunoregulatory ectoenzyme. *Immunol. Today 16*: 469–473.

18. Zocchi, E., Franco, L., Guida;, Benatti, U., Bargellesi, A., Malavasi, F., Lee, H.C., and De Flora, A. (1993) A single protein immunologically identified as CD38 displays NAD$^+$ glycohydrolase, ADP-ribosyl cyclase and cyclic ADP-ribose hydrolase activities at the outer surface of human erythrocytes. *Biochem. Biophys. Res. Commun. 196*: 1459–1465.

19. Amar-Costesec, A., Prado-Figueroa, M., Beaufay, H., Nagelkerke, J.F., and Van Berkel, T.J.C. (1985) Analytical study of microsomes and isolated subcellular membranes from rat liver. IX. Nicotinamide adenine dinucleotide glycohydrolase: A plasma membrane enzyme prominently found in Kupffer cells. *J. Cell Biol. 100*: 189–197.

20. Bocchini, V., Rebel, G., Massarelli, R., Schuber, F. and Muller, C.D. (1988) Latex beads phagocytosis and ecto-NAD$^+$glycohydrolase activity of in vitro cultivated rat brain microglia cells. *Int. J. Devl. Neuroscience 6*: 525–534.

21. Schuber, F., Travo, P., and Pascal, M. (1976) Calf spleen Nicotinamide-adenine dinucleotide glycohydrolase kinetic mechanism. *Eur. J. Biochem. 69*: 593–602.

22. Schuber, F., Travo, P., and Pascal, M. (1979) On the mechanism of action of calf spleen NAD$^+$ glycohydrolase. *Bioorg. Chem. 8*: 83–90.

23. Tarnus, C., and Schuber, F. (1987) Application of linear free-energy relationships to the mechanistic probing of nonenzymatic and NAD$^+$-glycohydrolase-catalyzed hydrolysis of pyridine dinucleotides. *Bioorg. Chem. 15:* 31–42.

24. Tarnus, C., Muller, H.M., and Schuber, F. (1988) Chemical evidence in favor of a stabilized oxocarbonium-ion intermediate in the NAD$^+$glycohydrolase-catalyzed reactions. *Bioorg. Chem. 16*: 38–51.

25. Handlon, A.L., Xu, C., Muller-Steffner, H.M., Schuber, F., and Oppenheimer, N.J. (1994) 2′-Ribose substituent effects on the chemical and enzymatic hydrolysis of NAD$^+$. *J. Am. Chem. Soc. 116*: 12087–12088.

26. Muller-Steffner, H., Schenherr-Gusse, I., Tarnus, C., and Schuber, F. (1993) Calf spleen NAD$^+$glycohydrolase: Solubilization, purification, and properties of the intact form of the enzyme. *Arch. Biochem. Biophys. 304*: 154–162.

27. Muller-Steffner, H., Muzard, M., Oppenheimer, N., and Schuber, F. (1994) Mechanistic implications of cyclic ADP-Ribose hydrolysis and methanolysis catalyzed by calf spleen NAD$^+$glycohydrolase. *Biochem. Biophys. Res. Commun. 204*: 1279–1285.

28. Zhang, F.-J., and Sih, C.J. (1995) Novel enzymatic cyclisations of pyridine nucleotide analogs: cyclic-GDP-Ribose and cyclic-HDP-Ribose. *Tetrahedron Lett. 36*: 9289–9292.

29. Graeff, R.M., Walseth, T.F., Hill, H.K., and Lee, H.C. (1996) Fluorescent analogs of cyclic ADP-Ribose: synthesis, spectral characterization, and use. *Biochemistry 35*: 379–386.

30. Schuber, F., Muller, H. and Schenherr, I. (1980) Amphipathic properties of calf spleen NAD glycohydrolase. *FEBS Lett. 1980, 109*: 247–251.

31. Jones, J.W., and Robins, R.K. (1963) Purine nucleosides. III Methylation studies of certain naturally occurring purine nucleosides. *J. Am. Chem. Soc. 85*: 193–201.

32. Singer, B. (1975) The chemical effects of nucleic acid alkylation and their relation to mutagenesis and carcinogenesis. *In Progress in Nucleic Acid Research and Molecular Biology* (Cohen, W.E., Ed.), Vol. 15, pp. 219–332, Academic Press, New York.

33. Brown, D.M. (1974) Chemical Reactions of Polynucleotides and Nucleic Acids. In *Basic Principles in Nucleic Acid Chemistry* (Ts'o, P.O.P., Ed.) Vol II, pp 1–82, Academic Press, New York.

34. Beranek, D.T., Weis, C.C., and Swenson, H. (1980) A comprehensive quantitative analysis of methylated and ethylated DNA using high pressure liquid chromatography. *Carcinogenesis 1*: 595–606.

35. Yost, D.A., and Anderson, B.M. (1983) Adenosine diphosphoribose transfer reactions catalyzed by Bungarus fasciatus venom NADglycohydrolase. *J. Biol. Chem. 258*: 3075–3080.

36. Bell, C.E., and Eisenberg, D., (1996) Crystal structure of diphteria toxin bound to nicotinamide adenine dinucleotide. *Biochemistry 35*: 1137–1149.

37. Rossmann, M.G., Liljas, A., Brändén, C.-I., and Banaszak, L.J. (1975) Evolutionary and structural relationships among dehydrogenases. *Enzymes 11*: 61–102.

38. Reisbig, R.R., and Woody, R.W. (1978) Characterization of a long-wavelength feature in the absorption and circular dichroism spectra of beta-nicotinamide adenine dinucleotide. Evidence for a charge transfer transition. *Biochemistry 17*: 1974–1984.

39. Oppenheimer, N.J. (1982) Chemistry and solution conformation of the pyridine coenzymes. In *The Pyridine Nucleotide Coenzymes* (Everse, J., Anderson, B. and You, K.-S. eds.) pp. 51–89, Academic Press, New York.

40. Guse, A.H., da Silva, C.P., Emmrich, F., Ashamu, G.A., Potter, B.V.L., and Mayr, G.W. (1995) Characterization of cyclic adenosine diphosphate-ribose-induced Ca^{2+} release in T lymphocyte cell lines. *J. Immunol. 155*: 3353–3359.

53

ADP-RIBOSYL CYCLASE AND CD38

Multi-Functional Enzymes in Ca^{+2} Signaling

Hon Cheung Lee, Richard M. Graeff, and Timothy F. Walseth

Department of Physiology and Pharmacology
University of Minnesota
Minneapolis, Minnesota 55455

ABSTRACT

Mobilization of internal Ca^{+2} is an important signaling mechanism in cells. In addition to the inositol trisphosphate pathway, cyclic ADP-ribose (cADPR) and nicotinic acid adenine dinucleotide (NAADP) have been shown to mobilize Ca^{+2} via independent mechanisms. Although the structures of cADPR and NAADP are totally distinct, both nucleotides can be synthesized by ADP-ribosyl cyclase or CD38, a lymphocyte antigen. Both enzymes cyclize NAD to cADPR. In the presence of nicotinic acid the two enzymes catalyze a base exchange reaction resulting in the synthesis of NAADP from NADP. The switch between these two modes of catalysis is regulated by pH. Furthermore, both enzymes can also cyclize nicotinamide guanine dinucleotide (NGD) to produce a fluorescent product, cyclic GDP-ribose (cGDPR), which has a site of cyclization different from cADPR. A model is proposed to account for the multi-functionality of these enzymes. In order to be able to verify the model, a soluble ADP-ribosyl cyclase has been crystallized and X-ray diffraction shows that it is a dimer. Solution of the crystal structure of the cyclase should provide valuable insight into the structural features necessary for its multiple catalytic functions.

BACKGROUND

ADP-ribosyl cyclase catalyzes the cyclization of NAD to cADPR, a novel cyclic nucleotide that functions as a modulator or messenger in the mobilization of intracellular Ca^{+2} stores (reviewed in 1). The cyclase was first discovered in sea urchin eggs (2) and later shown to be present also in mammalian tissues (3). It is now known to be ubiquitous, from bacteria (4) to human (5). In most of the tissues tested, the cyclase is associated with membranes. A soluble form is present in abundance in *Aplysia* ovotestis, from which it

ADP-Ribosylation in Animal Tissue, edited by Haag and Koch-Nolte
Plenum Press, New York, 1997

has been purified and sequenced (6,7). A search on the GenBank database reveals that the *Aplysia* cyclase is homologous with a lymphocyte antigen, CD38 (8). Subsequent work shows that CD38 not only catalyzes the synthesis of cADPR but also its hydrolysis (9–11). CD38 thus belongs to a family of bifunctional enzymes involved in the metabolism of cADPR. The first member of this family is a membrane bound enzyme previously purified from spleen (12). BST-1, a surface molecule on bone marrow stromal cell also has similar catalytic properties (13). In addition to catalyzing the synthesis and hydrolysis of cADPR, both the *Aplysia* cyclase and CD38 also catalyze the exchange of the nicotinamide base in NADP with nicotinic acid, leading to the synthesis of NAADP (14). Not only is the Ca^{+2} release mechanism activated by NAADP totally distinct from that activated by cADPR, the Ca^{+2} stores mobilized by these two metabolites are also different (15). The cyclase and CD38 are thus multi-functional enzymes that can synthesize two different Ca^{+2} messengers under appropriate conditions, suggesting their crucial role in Ca^{+2} signaling. In this paper, we review our work concerning this novel multi-functionality in catalysis.

RESULTS AND DISCUSSION

The Cyclization Reaction

The existence of an enzyme that cyclizes NAD was discovered during an investigation of the Ca^{+2} mobilizing action of NAD. The Ca^{+2} release activity of NAD showed a prominent delay, which was shown to be due to the conversion of NAD to an active metabolite (2). Purification and structural determination indicated the metabolite is a novel cyclic nucleotide and it was named cADPR (16). The cyclic structure has now been confirmed by X-ray crystallography, which also established definitively that the site of cyclization between the terminal ribose and N1 of the adenine ring (17). As indicated in Figure 1, both the N1-ribosyl and the N9-ribosyl linkages are in the β-conformation (17). Any model of catalysis proposed for the synthesizing enzyme must be consistent with these structural features of the product.

Because of the novelty of cADPR, the synthesizing enzyme was named ADP-ribosyl cyclase (6) to distinguish it from NAD-glycohydrolase (NADase) which converts NAD to ADP-ribose. Indeed, the classical NADase from *Neurospora* neither synthesized nor hydrolyzed cADPR (18). In addition to NAD, both the *Aplysia* cyclase and CD38 also cyclized various analogs of NAD, including the guanine analog, NGD (18). The resulting product was also a cyclic molecule since it could be hydrolyzed by heat to GDP-ribose through the addition of a molecule of water. Mass spectrometry measurements showed that the product has the same mass as would be expected for cGDPR (18). Unlike cADPR, cGDPR is a fluorescent molecule. This novel property allows for the continuous measurement of the cyclization reaction by simply monitoring the appearance of the fluorescent cGDPR product. There is no need to separate the substrate from the products since neither the substrate, NGD, nor the hydrolysis product, GDP-ribose, is fluorescent, (18).

Based on an analogy with cADPR, it was originally proposed that the site of cyclization in cGDPR is also at the N1 position of the guanine ring (18). Results from chemical synthesis suggested that the linkage site may be at the N7 position instead (19). A systematic study of the various N1- and N7-substituted guanine analogs showed that only the N7-substitution compounds are fluorescent. The pH dependence of the fluorescence and absorbance of cGDPR was also similar to the 7-substituted compounds, such as 7-methyl

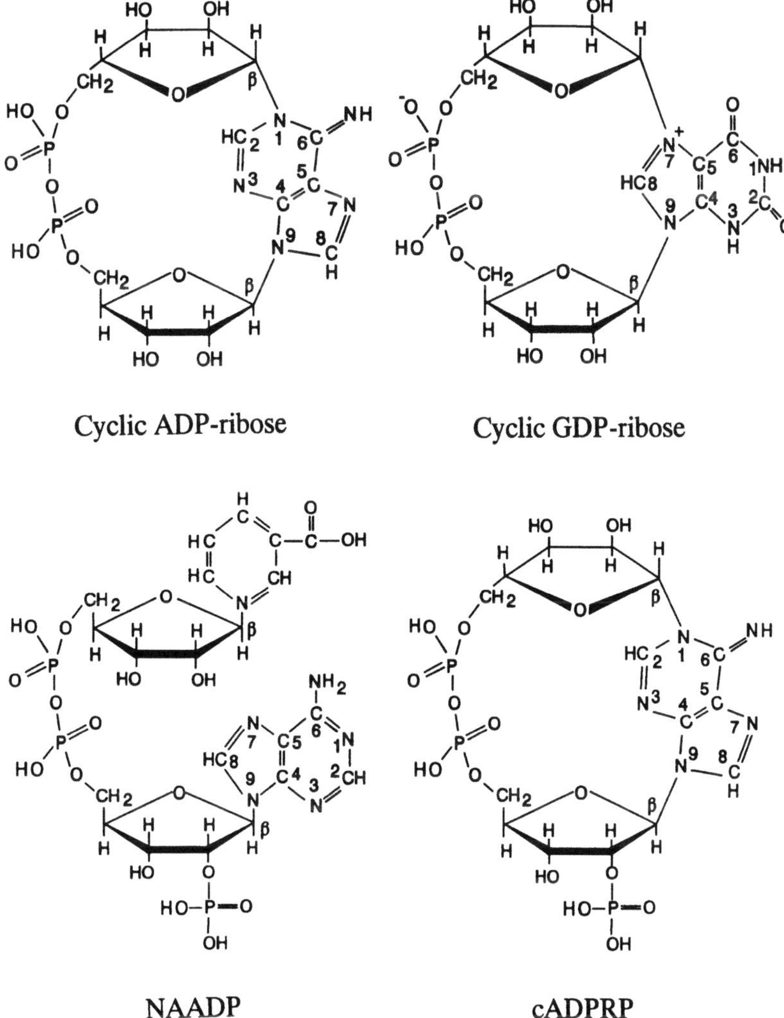

Figure 1. Chemical structures of the products of ADP-ribosyl cyclase and CD38.

GMP, but distinct from those with substitution at the N1-position (20). These results indicated the site of cyclization in cGDPR is at the N7-position, and as can be seen from Figure 1, this mechanism requires that the guanine base be in the *anti*-orientation, which is opposite to the *syn*-orientation of the adenine in cADPR. Since both products are produced by the same enzymes, these results indicate that the active site of the enzymes must allow relatively free rotation of the base ring around the N9-ribosyl linkage.

The Hydrolysis Reaction

The cyclic linkage, the N1-ribosyl bond, in cADPR is very stable even under acidic conditions. The half-time of hydrolysis at 37 °C and pH 2.0 is about 24 hours (21). At neutral pH, the rate of hydrolysis is about half of that in acidic pH. It is thus expected specific enzymes must be present in cells that catalyze the hydrolysis of cADPR and thus terminate its Ca^{+2} signaling function. This appears to be the case, as tissue extracts from marine invertebrate to human have been shown to be able to catalyze the hydrolysis of the N1-ribosyl linkage of cADPR to produce ADP-ribose (21). The hydrolytic enzymes have not been purified and their identities are unknown. The finding that bifunctional enzymes such as CD38 can hydrolyze cADPR suggests that the hydrolytic activity detected in tissue extracts may be catalyzed by similar enzymes.

Because of the hydrolytic activity of CD38, it has been very difficult to functionally distinguish it from NADase. The cADPR synthesized by the enzyme can be hydrolyzed to ADP-ribose. The overall reaction is the conversion of NAD to ADP-ribose, identical to that catalyzed by NADase. This confusion has been clarified by the development of a fluorimetric assay using NGD. The fluorescent product, cGDPR, is highly resistant to hydrolysis and accumulates once synthesized by CD38 (18). The rate of hydrolysis of cGDPR catalyzed by CD38 is about 35 times less than that of cADPR (22). NADases, such as that from *Neurospora*, on the other hand, do not cyclize NGD, even though they effectively hydrolyze it to GDP-ribose, a non-fluorescent product (18). By using the alternative substrate, NGD, and measuring the increase in cGDPR fluorescence, the two classes of enzymes can easily be distinguished.

The hydrolytic function of CD38 can, nevertheless, be monitored by using another fluorescent analog of cADPR, cyclic inosine dinucleotide phosphate ribose (cIDPR), which, unlike cGDPR, is susceptible to hydrolysis by CD38 (20,22). This assay provides a more convenient way of obtaining kinetic constants of the hydrolytic reaction (20,22). The fact that CD38 can readily hydrolyze cADPR and its analog, cIDPR, as well as synthesize them from NAD and NHD, respectively, indicates that its active site must be able to bind not only the substrates (NAD and NHD) but also the products, cADPR and cIDPR. From these observations, it is clear that CD38 is a very novel enzyme.

Site directed mutagenesis studies show that the hydrolytic activity of CD38 depends critically on two specific cysteine residues (23). Mutations of cysteine 119 to lysine and/or cysteine 201 to glutamate eliminate the hydrolytic activity of CD38 but preserve the cyclase activity. Since these cysteine residues are likely to be involved in the formation of disulfide bridges, the conformation of CD38 may be important for its hydrolytic activity of the molecule.

The Base Exchange Reaction

The observation that initiated our investigation of the base exchange reaction catalyzed by ADP-ribosyl cyclase and CD38 was that NADP had Ca^{+2} mobilizing activity. (2). Addition of NADP to sea urchin egg microsomes activates Ca^{+2} release without a delay. The release cannot be blocked by 8-amino-cADPR or heparin, specific antagonists respectively of the cADPR- and inositol trisphosphate-mediated mechanisms, indicating it is an independent pathway that is hitherto unknown (2,15). A subsequent investigation showed that the active component is not NADP itself but a minor contaminant in the commercial samples, which was identified as NAADP (15). It was shown previously that spleen NADase catalyzes the exchange of the nicotinamide base with nicotinic acid to produce NAADP and this reaction has recently been confirmed (24,25). We have shown that both ADP-ribosyl cyclase and CD38 can use NADP as a substrate and catalyze the same base exchange reaction (14). In addition to nicotinic acid, the base exchange reaction also requires acidic pH. At neutral and alkaline pH, the cyclase catalyzes, instead, the cyclization of NADP to produce cyclic ADP-ribose phosphate (cADPRP, ref. #14). The structural identification of this phosphorylated form of cADPR is based on the observation that the treatment with alkaline phosphatase removes the 2'-phosphate and converts it from a metabolite with no Ca^{+2} release activity in sea urchin egg homogenates to the Ca^{+2} messenger, cADPR (14). Although cADPRP does not release Ca^{+2} from sea urchin egg microsomes, it can do so with brain microsomes (26, 27). The characteristics of the release by cADPRP indicate it is activating the same mechanism as cADPR (27).

The base exchange reaction catalyzed by CD38 shows the same acidic pH requirement as the cyclase. At neutral and alkaline pH, CD38 catalyzes the hydrolysis of NADP to ADP-ribose-phosphate (14). It is not known whether this hydrolysis reaction represents the combination of synthesis of cADPRP and its subsequent hydroysis to ADP-ribose phosphate. This circumstance would be similar to that observed with NAD as substrate. Indeed, both the cyclase and CD38 can catalyze the base exchange reaction using NAD as substrate to produce nicotinic acid adenine dinucleotide (14). Therefore, the action of the two enzymes on NAD or NADP appears to be quite similar. This indicates that the active site of the two enzymes must not be recognizing the 2'-position of the ribose and can bind and act on NAD regardless of whether its 2'-position is phosphorylated or not. If this is the case, one would expect the equivalence of NAD and NADP as substrates may well carry over to their respective guanine or hypoxanthine analogs. Thus, it would be of interest to determine if both enzymes can cyclize the guanine analog of NADP, NGDP, to produce a fluorescent guanine analog of cADPRP, cGDPR-phosphate. We have shown recently that the cyclase can produce fluorescent analogs from NHD or NHDP (unpublished observations).

A Model for the Multiplicity of the Catalysis by ADP-Ribosyl Cyclase and CD38

The model depicted in Figure 2 proposes, firstly, that the active sites of both enzymes recognize and bind NAD in a folded conformation resembling the product, cADPR. It is generally believed that the hydrophobic stacking between the nicotinamide group and the adenine ring enable NAD to assume such a conformation. In this manner, the N1-position of the adenine can be brought close to the anomeric carbon of the terminal ribose so that the cyclization reaction can occur. The fact that CD38 can also effectively hydrolyze cADPR as well as synthesize it from NAD is consistent with the active site recognizing a conformation common to both substrates. On the other hand, neither the cyclase nor CD38 can cyclize ADP-ribose (6,28). Without the nicotinamide group, it is unlikely that ADP-ribose can assume the necessary folded configuration recognizable by the active sites. Indeed, even at a concentration as high as 10 mM, ADP-ribose has no effect on the cyclase reaction (28). Both enzymes are lenient with respect to the purine base and either NAD or NADP can serve as substrates for all three reactions; cyclization, hydrolysis and the base-exchange reaction (14), indicating the 2'-position of the adeninyl ribose is not important for substrate recognition by the enzymes. Taken together, these results indicate the portions of NAD recognized by the active sites are primarily the pyrophosphate linkage and the terminal ribose.

The next step is likely to be the catalytic attack on the anomeric carbon of the ribose by the enzyme, resulting in the formation of an ADP-ribosyl intermediate and the release of the nicotinamide group. This type of intermediate has previously been proposed as part of the catalysis of NADases (12), in which the anomeric carbon of the ADP-ribosyl intermediate is in an activated state. Since NAD, as discussed above, is bound in a folded conformation, the N1 of the adenine ring is in close proximity of the activated carbon. Hydrophilic attack by the N1 on the activated ribosyl carbon would result in cyclization. The consecutive reactions at the anomeric carbon generate a double inversion, preserving

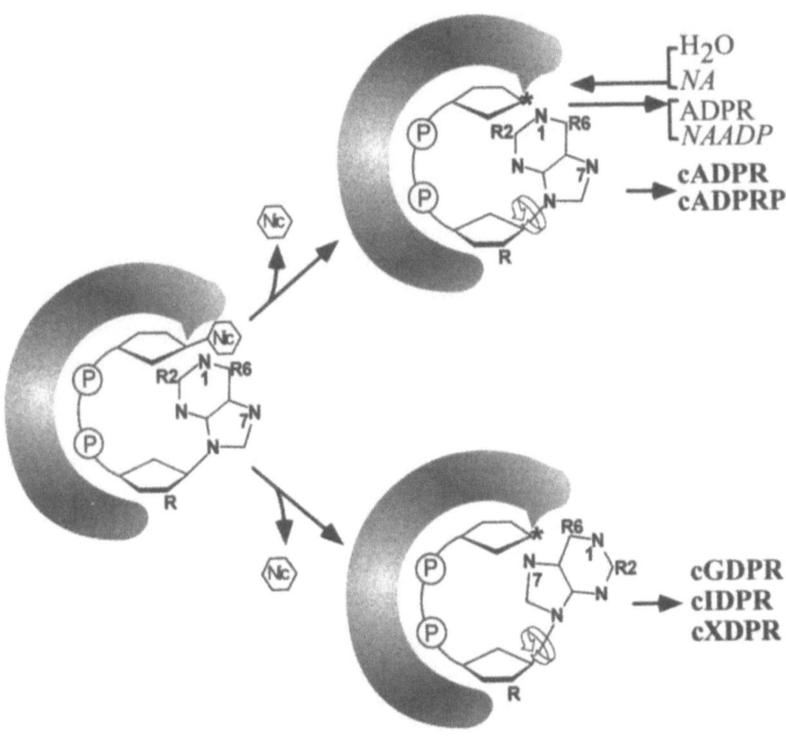

Figure 2. A model for the multiple catalytic pathways of ADP-ribosyl cyclase and CD38. The model depicts the active site of the enzymes recognizing the pyrophosphate and the terminal ribose of the various substrates in the folded conformation. The catalysis involves the formation of an activated intermediate (*) and the release of the nicotinamide group (Nic). The purine-ring is depicted as being free to rotate around the N9-ribosyl bond. The intermediate is susceptible to attack by N1 or N7 of the purine ring resulting in cyclization, hydrolysis, or base-exchange by nicotinic acid (NA). The group R can be -OH (NAD) or P_i (NADP). The group R6 can be NH_2 (adenine) or O (guanine, inosine or xanthine). The group R2 can be H (adenine or inosine), NH_2 (guanine) or O (xanthine). The details of the model are described in the text.

the N1-ribosyl bond in the β-configuration as in NAD. The subsequent release of the cADPR formed from the active site could be the consequence of the cyclization linkage compacting the molecule and thereby reducing its affinity for the active site.

In the case of cGDPR, it is the N7 of the guanine that is linked to the terminal ribose. In order to bring the N7 into position for attack, the guanine ring needs to rotate around the N9-ribosyl bond from the *syn-* to the *anti-*orientation. It can be envisioned that the base of the bound nucleotide is rotating freely around the N9-ribosyl. Because the reactivity of the N1 in the guanine is apparently reduced, cyclization results, instead, when the ring rotates to the *anti-*orientation, bringing the N7 close to the activated carbon of the ribosyl intermediate. In addition to cGDPR, cyclization at N7 is also observed in the inosine (cIDPR) and xanthosine (cXDPR) analogs of cADPR (20). The common feature of the three is the presence of a keto at the C6-position of the base, which may cause reduced reactivity at the N1-position.

The hydrolytic function of CD38 is also compatible with the model, if it is proposed that the ribosyl enzyme intermediate in CD38 is more accessible to water. Attack by water before the cyclization at N1 would result in the formation of ADP-ribose. Thus the main reaction catalyzed by CD38 is the conversion of NAD to ADP-ribose, and only a small amount of cADPR is formed (9,10). The fact that CD38 can also hydrolyze cADPR to ADP-ribose indicates that the active site of CD38 has much higher affinity for cADPR than that of the cyclase. The high affinity of CD38 for cADPR allows cADPR to bind and be converted to the ribosyl enzyme intermediate. Subsequent attack by water can lead to formation of ADP-ribose. The high affinity of the active site for cADPR would mean that the cADPR formed from NAD also remains on the site longer, which can increase the chance of its reacting with water to form ADP-ribose instead of being released as free cADPR. The differences between the active site of CD38 and the cyclase are therefore, according to the model, higher affinity for cADPR and more accessibility to water.

The active site of CD38 apparently has much less affinity for cGDPR. Once formed, it is mainly released as free cGDPR. Because of the cyclization at N7 and the *anti*-orientation of the guanine ring, it is expected that cGDPR is a more compact molecule than cADPR. This structural difference could reduce the fit between cGDPR and the active site, and contribute to its lower affinity. The N7-ribosyl bond in cGDPR is more stable than the N1-ribosyl bond in cADPR since the rate of chemical hydrolysis of cGDPR induced by heat (85 °C) is 2.8 times slower than cADPR (18). The intrinsic stability and the low affinity for the active site all contribute to it being released by CD38 as free cGDPR instead of being hydrolyzed to GDP-ribose. Indeed, the rate of hydrolysis of cGDPR by CD38 has been measured to be 36 times slower than cADPR (22).

Assuming the active site is accessible to nicotinic acid, the same model can also account for the base-exchange reaction. The attack by nicotinic acid on the anomeric carbon of the ribosyl intermediate would result in the formation of NAAD if NAD is the substrate or NAADP if NADP is the substrate. Both reactions have been observed (14). The acidic pH requirement for the base-exchange reaction (14) could be due to conformational changes of the enzymes. On the other hand, perhaps the simplest explanation is that, at low pH, the negative charges on both the enzyme and nicotinic acid are reduced, allowing easier access of the nicotinic acid to the active site. This explanation does not require specific pH-dependent conformations and is consistent with the fact that the base-exchange reaction catalyzed by two different enzymes, CD38 and the cyclase, show a very similar pH dependency (14).

The Crystal Structure of ADP-Ribosyl Cyclase

Some features of the model described above can be tested. For example, the binding affinity of cADPR and cGDPR can be measured with radioactive probes. The structural aspects of the active site, on the other hand, can best be examined by X-ray crystallography. A very efficient method has been developed to purify *Aplysia* ADP-ribosyl cyclase (6,22). Using a specific type of cation exchange chromatography, about 40 mg of the cyclase can be purified from 59 g of *Aplysia* ovotestis in a single step (6,22). Three different forms of diffraction quality crystals of the enzyme have been obtained (29). Form I is hexagonal-shaped crystals of 0.4–0.6 mm in size. Form II is rod shaped crystals of 0.6–0.8 mm in length. Form III is prismatic plate-like crystals of 1.0 mm in length. In each of the three crystal forms the cyclase crystallizes as a dimer. Since three different crystallization conditions all result in dimer, it is likely that the cyclase has an intrinsic tendency to form a dimer. If the dimer proves to be the functional unit, it could have important implications for our understanding of the multi-functionality of the enzyme. The solution of the crystal structure promises to provide valuable insight about the structural features necessary for its catalytic functions.

ACKNOWLEDGMENTS

The work is supported by NIH grants HD17484 and HD32040 (to H.C.L) and DA08131 (to T.F.W.).

REFERENCES

1. Lee, H.C. 1996. Modulator and messenger functions of cyclic ADP-ribose in calcium signaling. *Recent Prog. Hormone Res. 52*: 357–391.
2. Clapper, D.L., Walseth, T.F., Dargie, P.J. and H.C. Lee. 1987. Pyridine nucleotide metabolites stimulate calcium release from sea urchin egg microsomes desensitized to inositol trisphosphate. *J. Biol. Chem. 262*: 9561–9568.
3. Rusinko, N. and H.C. Lee. 1989. Widespread occurrence in animal tissues of an enzyme catalyzing the conversion of NAD into a cyclic metabolite with intracellular Ca^{+2} mobilizing activity. *J. Biol. Chem. 264*:11725–11731.
4. Karasawa, T., Takasawa, S., Yamakawa, K., Yonekura, H., Okamoto, H. and S. Nakamura. 1995. NAD^+-glycohydrolase from streptococcus pyogenes shows cyclic ADP-ribose forming activity. *FEMS Microbiol. Lett. 130*: 201–204.
5. Lee, H.C., Zocchi, E., Guida, L., Franco, L., Benatti, U., and A. De Flora. 1993. Production and hydrolysis of cyclic ADP-ribose at the outer surface of human erythrocytes. *Biochem. Biophys. Res. Commun. 191*: 639–645.
6. Lee, H.C. and R. Aarhus. 1991. ADP-ribosyl cyclase: an enzyme that cyclizes NAD^+ into a calcium-mobilizing metabolite. *Cell Regul 2*: 203–209.
7. Glick, D.L., Hellmich, M.R., Beushausen, S., Tempst, P., Bayley H., F. Strumwasser. 1991. Primary structure of a molluscan egg-specific NADase, a second-messenger enzyme. *Cell Regul 2*: 211–218.
8. States, D.J., Walseth, T.F. and H.C. Lee. 1992. Similarities in amino acid sequences of *Aplysia* ADP-ribosyl cyclase and human lymphocyte antigen CD38. *Trends Biochem. Sci. 17*: 495.
9. Howard, M., Grimaldi, J.C., Bazan, J.F., Lund, F.E., Santos-Argumedo, L., Parkhouse, R.M.E, Walseth, T.F, and H.C. Lee. 1993. Formation and hydrolysis of cyclic ADP-ribose catalyzed by lymphocyte antigen CD38. *Science 262*: 1056–1059.
10. Takasawa, S., Tohgo, A., Noguchi, N., Koguma, T., Nata, K., Sugimoto, T., Yonekura, H., and H. Okamoto. 1993. Synthesis and hydrolysis of cyclic ADP-ribose by human leukocyte antigen CD38 and inhibition of the hydrolysis by ATP. *J. Biol. Chem. 268*: 26052–26054.
11. Zocchi, E., Franco, L., Guida, L., Benatti, U., Bargellesi, A., Malavasi, F., Lee, H.C., A. De Flora. 1993. A single protein immunologically identified as CD38 displays NAD^+ glycohydrolase, ADP-ribosyl cyclase and cyclic ADP-ribose hydrolase activities at the outer surface of human erythrocytes. *Biochem. Biophys. Res. Commun. 196*: 1459–1465.
12. Kim, H., Jacobson, E.L., and M.K. Jacobson. 1993. Synthesis and degradation of cyclic ADP ribose by NAD glycohydrolases. *Science 261*: 1330–1333.
13. Hirata, Y., Kimura, N., Sato, K., Ohsugi, Y., Takasawa, S., Okamoto, H., Ishikawa, J., Kaisho, T., Ishihara, K. and T. Hirano. 1994. ADP-ribosyl cyclase activity of a novel bone marrow stromal cell surface molecule, BST-1. *FEBS Lett. 356*: 244–248.
14. Aarhus, R., Graeff, R.M., Dickey, D.M., Walseth, T.F. and H.C. Lee. 1995. ADP-ribosyl cyclase and CD38 catalyze the synthesis of a calcium mobilizing metabolite from NADP. *J. Biol. Chem. 270*: 30327–30334.
15. Lee, H.C. and R. Aarhus. 1995. A derivative of NADP mobilizes calcium stores insensitive to inositol trisphosphate and cyclic ADP-ribose. *J. Biol. Chem. 270*: 2152–2157.
16. Lee, H.C., Walseth, T.F., Bratt, G.T., Hayes, R.N., and D.L. Clapper. 1989. Structural determination of a cyclic metabolite of NAD^+ with intracellular Ca^{+2} mobilizing activity. *J. Biol. Chem. 264*: 1608–1615.
17. Lee, H.C., Aarhus, R. and D. Levitt. 1994. The crystal structure of cyclic ADP-ribose. *Nature Struct. Biol. 1*: 143–144.
18. Graeff, R., Walseth, T.F., Fryxell, K., Branton, D. and H. C. Lee. 1994. Enzymatic synthesis of cyclic GDP-ribose. A procedure for distinguishing enzymes with ADP-ribosyl cyclase activity. *J. Biol. Chem. 269*: 30260–30267.
19. Yamada, S., Gu, Q.M. and C.J. Sih. 1994. Cyclic ADP-ribose via stereoselective cyclization of β-NAD^+. *J. Am. Chem. Soc. 116*: 10787–10788.

20. Graeff, R.M., Walseth, T.F., Hill, H.K. and H.C. Lee. 1996. Fluorescent analogs of cyclic ADP-ribose. Synthesis, spectral characterizations and use. *Biochem. 35*: 379–386.

21. Lee, H. C. and R. Aarhus. 1993. Wide distribution of an enzyme that catalyzes the hydrolysis of cyclic ADP-ribose. *Biochim. Biophys. Acta 1164*: 68–74.

22. Lee, H.C., R.M. Graeff, C.B. Munshi, T.F. Walseth and Aarhus, R. 1996. Large scale purification of *Aplysia* ADP-ribosyl cyclase and measurement of its activity by a fluorimetric assay. Methods Enzymol. (in press).

23. Tohgo, A., Takasawa, S., Noguchi, N., Koguma, T., Nata, K., Sugimoto, T., Furuya, Y., Yonekura, H. and H. Okomoto. 1994. Essential cysteine residues for cyclic ADP-ribose synthesis and hydrolysis by CD38. *J. Biol. Chem. 269*: 28555–28557.

24. Bernofsky, C. 1980. Nicotinic acid adenine dinucleotide phsophate (NAADP⁺). *Methods Enzymol. 66*: 105–112.

25. Chini, E.N., Beers, K.W., and T.P. Dousa. 1995. Nicotinate adenine dinucleotide phosphate (NAADP) triggers a specific calcium release system in sea urchin eggs. *J. Biol. Chem. 270*: 3216–3223.

26. Zhang, F.J., Gu, Q.M., Jing, P., and C.J. Sih. 1995. Enzymatic cyclization of nicotinamide adenine dinucleotide phosphate (NADP). *Bioorg. Med. Chem. 5*:2267–2272.

27. Vu, C.Q., Lu, P.-J., Chen, C.-S., and M.K. Jacobson. 1996. 2′-phospho-cyclic ADP-ribose, a calcium-mobilizing agent derived from NADP. *J. Biol. Chem. 271*: 4747–4754.

28. Inageda, K., Takahashi, K., Tokita, K., Nishina, H., Kanaho, Y., Kukimoto, I., Kontani, K., Hoshino, S. and T. Katada. Enzyme properties of *Aplysia* ADP-ribosyl cyclase: Comparison with NAD glycohydrolase of CD38 antigen. *J. Biochem. 117*: 125–131.

29. Prasad, G.S., Levitt, D.G., Lee, H.C., C.D. Stout. 1996. Preliminary crystallographic studies of ADP-ribosyl cyclase from *Aplysia california. Proteins 24*: 138–140.

54

SIGNAL TRANSDUCTION VIA THE CD38/NAD$^+$ GLYCOHYDROLASE

Kenji Kontani, Iwao Kukimoto, Yasunari Kanda, Shin-ichi Inoue,
Hiroyuki Kishimoto, Shin-ichi Hoshino, Hiroshi Nishina,
Katsunobu Takahashi, Osamu Hazeki, and Toshiaki Katada

Department of Physiological Chemistry
Faculty of Pharmaceutical Sciences
University of Tokyo
Tokyo 113, Japan

ABSTRACT

The human cell surface CD38 molecule is a 46-kDa type-II transmembrane glyco-protein with a short N-terminal cytoplamic domain and a long Cys-rich C-terminal ex-tracellular one. We previously demonstrated that an ecto-form NAD$^+$ glycohydrolase (NADase) activity induced by all-*trans* retinoic acid in HL-60 cells is due to the extracel-lular domain of CD38. In the present study, we investigated a possible signal transduction mediated through CD38 in the retinoic acid-differentiated HL-60 cells with anti-CD38 monoclonal antibodies (mAbs). The addition of selected anti-CD38 mAbs to the cells in-duced rapid tyrosine phosphorylation of the cellular proteins with the molecular weights of 120,000, 87,000 and 77,000; the phosphorylated 120-kDa protein was identified as the c-*cbl* proto-oncogene product, p120$^{c\text{-}cbl}$. Furthermore, the phosphorylated p120$^{c\text{-}cbl}$ associ-ated with the 85-kDa subunit of phosphatidylinositol 3-kinase. To determine the relation-ship between the amino acid sequence responsible for the NADase activity and epitopes recognized by the stimulatory mAbs, we produced its carboxy-terminal deletion mutants in COS-7 cells. The mutants with less than 15 amino acids deleted from the carboxyl ter-minus of the 300-amino acid wild-type molecule still maintained NADase activity, but those with more than 27 amino acids deleted did not. Introduction of site-directed muta-tion of a cysteine residue (Cys^{275}), located in the 273–285 sequence, completely abolished the NADase activity. These CD38 mutants were also used for an epitope mapping of anti-CD38 mAbs. All the epitopes recognized by the mAbs inducing the tyrosine phosphoryla-tion were mapped on the same Cys^{275}-containing sequence of 273–285. Thus, the discrete carboxy-terminal sequence not only plays a key role in its ecto-NADase activity, but also contains the epitopes of the agonistic anti-CD38 mAbs for the transmembrane signaling. We also found that the agonistic mAbs markedly potentiate superoxide generation induced by the stimulation of G protein-coupled chemotactic receptors. Our results suggested that

ADP-Ribosylation in Animal Tissue, edited by Haag and Koch-Nolte
Plenum Press, New York, 1997

the stimulation of CD38 might generate an accessory signal(s) to enhance the G protein-mediated signaling, probably though the protein-tyrosine phosphorylation.

BACKGROUND

The human cell surface CD38 molecule is a 46-kDa type-II transmembrane glycoprotein with a short amino-terminal cytoplasmic domain and a long carboxy-terminal extracellular one (1). The analysis of the expression of CD38 is widely used as a phenotypic marker of differentiation or activation of human T and B lymphocytes (2). We previously demonstrated that NAD^+ glycohydrolase (NADase) activity induced by all-*trans* retinoic acid (RA) in human leukemic HL-60 cells is due to the extracellular domain of CD38 (3) and that the same molecule is able to bind hyaluronate (4). Human, mouse and rat CD38s displayed an amino acid sequence similar to *Aplysia* ADP-ribosyl cyclase, an enzyme catalyzing the formation of cyclic ADP-ribose from NAD^+ (5). The biological relevance of this cyclic nucleotide relies on the hypothesis that it is a new mediator or modulator of Ca^{2+} release from intracellular stores insensitive to inositol 1, 4, 5-trisphosphate (see Ref. 6 for review). Indeed, CD38 catalyzes not only the hydrolysis of NAD^+, but also the formation and hydrolysis of cyclic ADP-ribose (7, 8), though ADP-ribosyl cyclase activity of CD38 is quite low compared to its NADase activity (9). We recently developed a specific radioimmunoassay for the measurement of cellular cyclic ADP-ribose: the results indicate a close correlation between intracellular concentrations of cyclic ADP-ribose and surface expression of CD38 (10). However, the physiological meaning(s) of the accumulation of cyclic ADP-ribose in CD38-producing cells remains obscure.

CD38 is considered as an important regulatory molecule in the immune system, as inferred from the observation that CD38 ligation by agonistic mAbs is followed by elicitation of a variety of cell functions, e.g. cell proliferation, lymphopoiesis, apoptosis, adhesion and cytokine production (11–16). The CD38 ligation by the mAbs is believed to mimic some of the events triggered by the binding with a natural ligand. In this regard, a cell surface protein with the M_r of 120,000, which is predominantly produced in endothelial cells, has recently been identified as one of the CD38-binding proteins (17). However, the molecular mechanisms underlying CD38-mediated signal transduction are still awaiting to be defined.

In the present study, we show that selected anti-CD38 mAbs induce tyrosine phosphorylation of several cellular proteins (18). To determine the relationship between the amino acid sequence responsible for the NADase activity and epitopes recognized on CD38 by the agonistic mAbs, we produced deletion and site-direction mutants of CD38. The results obtained here clearly indicate that a limited sequence near the carboxyl terminus of CD38 plays a key role in the enzyme activity and that all the epitopes recognized by the agonistic mAbs are localized on the same carboxy-terminal sequence of CD38. Interestingly, the agonistic anti-CD38 mAbs markedly potentiated superoxide generation induced by the stimulation of G protein-coupled chemotactic receptors.

RESULTS

Stimulation of Protein-Tyrosine Phosphorylation by Anti-CD38 mAbs in RA-Treated HL-60 Cells

We first investigated the possibility that tyrosine phosphorylation of cellular proteins might be induced upon stimulation of CD38 with its mAbs. HL-60 cells that had been cultured with RA to produce CD38 were stimulated with an anti-CD38 mAb, T16, and then a lysate was prepared from the cells. Tyrosine-phosphorylated proteins in the cell lysate, after being immunoprecipitated with an anti-phosphotyrosine (PY) mAb, PY20, were separated by SDS-PAGE and visualized by means of immunoblotting with an anti-PY polyclonal antibody (pAb). As shown in Fig. 1A, the addition of T16 to the RA-differentiated cells stimulated tyrosine phosphorylation of cellular proteins with the M_r of 120,000, 87,000 and 77,000 (*lane 3*). For simplicity, the major tyrosine-phosphorylated proteins are henceforth referred to as p120, p87 and p77, respectively. Another anti-CD38 mAb, HB-7, also stimulated protein tyrosine phosphorylation, the pattern being the same as that observed for T16 (*lane 4*). The specificity of the action of HB-7 was confirmed by its inability to stimulate the tyrosine phosphorylation in undifferentiated (*Un*) or dibutyryl cAMP (*Bt₂-cAMP*)-differentiated HL-60 cells (Fig. 1B), in which CD38 is not produced (3). These results clearly indicated that selected anti-CD38 mAbs stimulate the tyrosine phosphorylation of cellular proteins through CD38 produced on the surface of HL-60 cells.

We next measured a tyrosine kinase activity in the lysate of HL-60 cells which had been stimulated with HB-7. The cell lysate was immunoprecipitated with PY20, and the *in vitro* tyrosine kinase assay was carried out with a synthetic peptide as the substrate. The anti-CD38 mAb HB-7 rapidly stimulated the tyrosine kinase activity in the PY20-immunoprecipitated fraction; the maximum stimulation was observed at about 2 min. The tyrosine kinase activity in the fraction prepared from undifferentiated HL-60 cells was, however, not affected by the addition of HB-7. These results suggested that stimulation of CD38 activates a cellular tyrosine kinase(s) which may be responsible for the tyrosine phosphorylation of p120, p87 and p77.

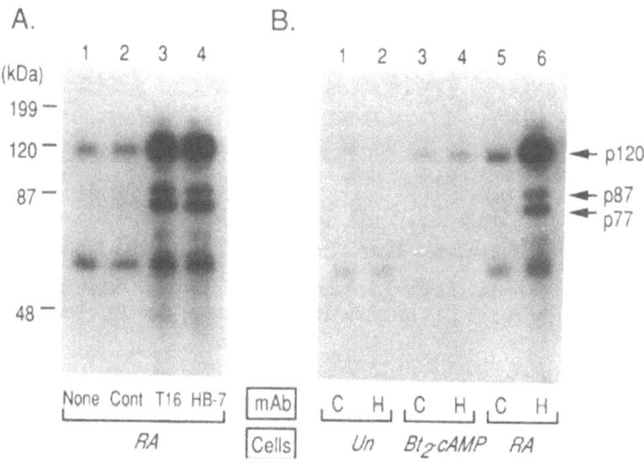

Figure 1. Tyrosine phosphorylation of cellular proteins after stimulation of RA-treated HL-60 cells with anti-CD38 mAbs. *Panel A*, HL-60 cells which had been cultured with 1 µM RA for 48 h were incubated at 37°C for 2 min without (*lane 1*) or with 7.5 µg/ml of the control IgG1 (*lane 2*), or the anti-CD38 mAb, T16 (*lane 3*) or HB-7 (*lane 4*). Tyrosine-phosphorylated proteins in the cell lysate, after being immunoprecipitated with PY20, were separated by SDS-PAGE and then visualized with anti-PY polyclonal Ab. The molecular-weight markers used were obtained from Bio-Rad and are indicated in *kDa*. *Panel B*, HL-60 cells which had been cultured for 48 h without (*lanes 1* and *2*; *Un*) or with 0.5 mg/ml of dibutyryl cAMP (*lanes 3* and *4*; *Bt₂-cAMP*) or 1 µM RA (*lanes 5* and *6*; *RA*) were stimulated at 37°C for 2 min with the control IgG1 (*C*), or the anti-CD38 mAb (*H*), and then tyrosine-phosphorylated proteins were visualized as described above.

Identification of Tyrosine-Phosphorylated Proteins Induced by Anti-CD38 mAbs

Recent reports revealed that tyrosine phosphorylation of the c-*cbl* proto-oncogene product, p120$^{c\text{-}cbl}$, may play a critical role in signal transduction mediated by T cell receptor, B cell receptor, Fc receptor for IgG, and EGF receptor (19–22). These observations led us to examine whether p120$^{c\text{-}cbl}$ is involved in transmembrane signaling mediated by CD38. To investigate this possibility, RA-treated HL-60 cells were stimulated with HB-7 or the control IgG1, and the cell lysates, after being immunoprecipitated with anti-p120$^{c\text{-}cbl}$ pAb, were immunoblotted with anti-PY pAb. The addition of HB-7 to the cells stimulated the tyrosine phosphorylation of p120$^{c\text{-}cbl}$ (Fig. 2A, *lane 2*), though the immunoprecipitated amounts of p120$^{c\text{-}cbl}$ were the same for the control IgG1- and HB-7-treated HL-60 cells (*lanes 3 and 4*).

The relationship between this p120$^{c\text{-}cbl}$ and p120 identified as the major tyrosine-phosphorylated protein in the same RA-differentiated HL-60 cells was further investigated as follows. A cell lysate obtained from HB-7-stimulated HL-60 cells was first treated with the anti-p120$^{c\text{-}cbl}$ antibody to deplete p120$^{c\text{-}cbl}$ from the lysate. The treated lysate, after being immunoprecipitated with PY20, was separated by SDS-PAGE and then subjected to immunoblotting with the anti-p120$^{c\text{-}cbl}$ pAb or anti-PY pAb (Fig. 2B). As expected, the depletion of p120$^{c\text{-}cbl}$ from the cell lysate was almost completely achieved (*lanes 1 and 2*). When the cell lysate was analyzed by immunoblotting with the anti-PY pAb, a marked decrease in the immunoreactivity of p120 to the pAb was observed without significant changes in any other tyrosine-phosphorylated proteins (*lanes 3 and 4*). These results indicated that the major fraction of the tyrosine-phosphorylated p120 observed on the stimula-

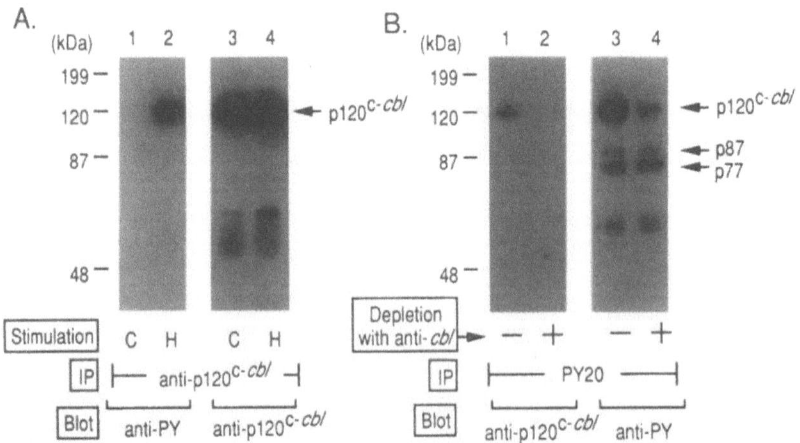

Figure 2. Tyrosine phosphorylation of the c-cbl proto-oncogene product upon stimulation of CD38. *Panel A*, a cell lysate was prepared from RA-treated HL-60 cells which had been incubated with the control IgG1 (*C*) or anti-CD38 mAb HB-7 (*H*) at 37°C for 2 min, and then p120$^{c\text{-}cbl}$ in the cell lysate was immunoprecipitated with its pAb. The precipitated proteins, after being separated by SDS-PAGE, were subjected to immunoblot analysis with the anti-PY pAb (*lanes 1 and 2*) or anti-p120$^{c\text{-}cbl}$ pAb (*lanes 3 and 4*). The position of p120$^{c\text{-}cbl}$ is indiated by the *arrow*. *Panel B*, a cell lysate prepared from HB-7-stimulated HL-60 cells was first subjected to three rounds of immunoprecipitation with the control IgG1 (*lanes 1 and 3*) or anti-p120$^{c\text{-}cbl}$ pAb (*lanes 2 and 4*). The lysates, after being immunoprecipitated with PY20, were separated by SDS-PAGE and then analyzed by immunoblotting with the anti-p120$^{c\text{-}cbl}$ pAb (*lanes 1 and 2*) or anti-PY pAb (*lanes 3 and 4*).

tion of CD38 represents the phosphorylated $p120^{c\text{-}chl}$ in the HL-60 cells. Furthermore, our preliminary results suggested that the phosphorylated $p120^{c\text{-}chl}$ associated with the 85-kDa subunit of phosphatidylinositol 3-kinase and that $p72^{syk}$ was also tyrosine-phosphorylated upon the stimulation of CD38.

Production of CD38 Mutants in COS-7 Cells and Identification of the Catalytic Domain on CD38/NADase

We next investigated the amino acid sequence required for NADase activity of CD38 by construction of carboxy-terminal deletion mutants (Fig. 3). The mutant proteins, together with the full-length human CD38, were synthesized in COS-7 cells by means of transient expression of their cDNAs. A chimeric protein of *Aplysia* ADP-ribosyl cyclase, with the amino terminus anchored to the transmembrane region of CD38 (*CD-AC*), was also prepared as a comparison. Rabbit pAbs raised against a recombinant CD38 fused with maltose-bing protein and ADP-ribosyl cyclase purified from *Aplysia* ovotestis (9) were used to detect the mutant proteins produced in COS-7 cell membranes; the deletion mutants of *D298*, *D294*, *D285* and *D273*, together with the wild-type CD38 and *CD-AC*, were expressed in the membranes to similar extents.

NAD⁺ was very rapidly hydrolyzed to ADP-ribose and nicotinamide upon incubation with the membranes containing the wild-type CD38; there was no formation of cyclic ADP-ribose. In contrast, there was time-dependent accumulation of cyclic ADP-ribose without the formation of ADP-ribose, when the membranes containing the *Aplysia* cyclase-type chimera (*CD-AC*) were incubated with NAD⁺. These enzymatic characteristics were essentially the same as those observed for the preparation of HL-60 cell membranes

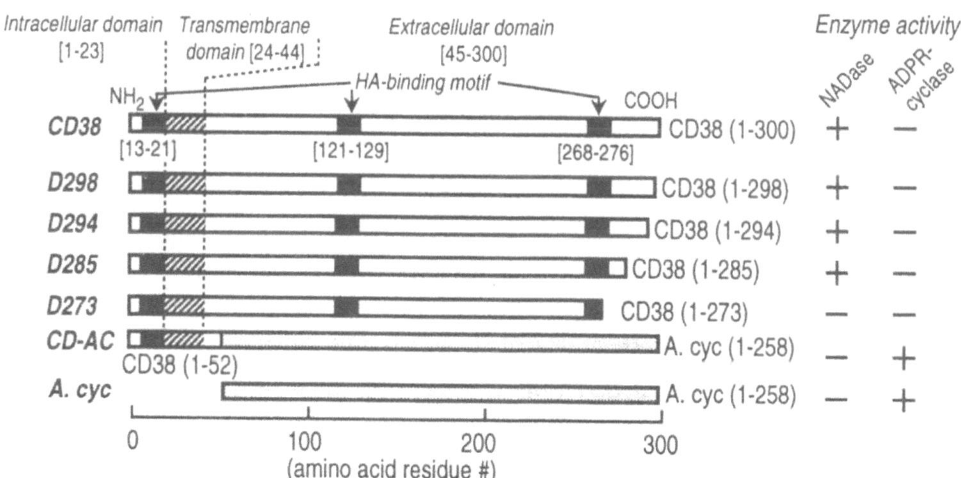

Figure 3. Schematic overview of the carboxy-terminal deletion mutants of CD38 produced in COS-7 cells. The sequences of human CD38 (CD38) and *Aplysia* ADP-ribosyl cyclase (A. cyc) are indicated by open and dotted columns, respectively. The hatched columns and closed boxes show the transmembrane domain of CD38 and a putative hyaluronate (HA)-binding [B(X_7)B] motif, in which B is either R or K, and X_7 contains no acidic residues and at least one basic amino acid, respectively. Human CD38 and the mature form of *Aplysia* ADP-ribosyl cyclase consist of 300 and 258 amino acid residues, respectively. The numbers in parentheses are the numbers of amino acid residues from their amino termini.

and the purified enzyme from *Aplysia* ovotestis (9). Membranes containing the CD38 mutant proteins were also subjected to assaying NADase activity. There were no marked differences in the enzyme activity among the deletion mutants of *D298*, *D294* and *D285*, and their specific activities almost overlapping that of the wild-type CD38. NADase activity of CD38 was, however, completely abolished upon the deletion of more than 27 amino acids from its carboxyl terminus. Thus, the amino acid sequence of 273–285 appears to be essential for the NADase activity of CD38.

Cysteine Residue Located Near Carboxyl Terminus Is Responsible for the Enzymatic Activities of CD38/NADase and Aplysia ADP-Ribosyl Cyclase

Ten cysteine residues in human CD38 are full identical to those found in *Aplysia* ADP-ribosyl cyclase (23). Among these cysteine residues, Cys^{275} is present in the carboxy-terminal sequence of 273–285. Therefore, we substituted this cysteine residue with serine and produced a CD38 mutant (*C275S*) in the COS-7 cell system. The *C275S* mutant did not display any detectable NADase activity, though the product of the *C275S* mutant was apparent as expressed by the cell membranes. Considering the conservation of this cysteine residue in *Aplysia* cyclase, the mutation of Cys^{254} (which corresponds to Cys^{275} in CD38) may have similar effects on the ADP-ribosyl cyclase activity. To validate this hypothesis, Cys^{254} in *Aplysia* cyclase was substituted with serine (*C254S*). As expected, the *C254S* mutant protein did not exhibit detectable ADP-ribosyl cyclase activity. These results indicate that the conserved cysteine residues located near the carboxyl terminus of either CD38 or *Aplysia* cyclase play important roles in their enzyme activities.

The Epitopes of the Agonistic MAbs Are Mapped on the Same Carboxy-Terminal Domain on CD38/NADase

The binding of mAbs to CD38 has been shown to trigger a wide variety of cell functions including cell proliferation, protection from apoptosis, inhibition of cell adhesion and cytokine production, as described previously. The analysis included three agonistic mAbs, namely HB-7 and T16 (previously shown to stimulate tyrosine phosphorylation of cellular proteins in RA-treated HL-60 cells) and IB4 (reported to induce protein-tyrosine phosphorylation in human B cell lines, ref. 24). The epitopes of the stimulatory anti-CD38 mAbs were mapped by means of immunoblot analysis using CD38 mutants as targets.

In addition to the wild-type CD38, its mutants with less than 15 carboxy-terminal amino acids deleted (*D298*, *D294*, and *D285*) were clearly recognized by any of the three mAbs. However, none of the mAbs reacted with the CD38 mutant with more than 27 amino acids deleted (*D273*). The analysis also included as negative control the OKT10, an anti-CD38 mAb, of which addition did not induce any detectable tyrosine phosphorylation. The functional feature was confirmed at a molecular level: indeed, the wild-type and *D298* molecules were, but *D294*, *D285*, and *D273* were not, recognized by OKT10 in Western blot. Thus, the OKT10 epitope appeared to be mapped on the sequence of 294–298. The *C275S* mutant of CD38 was also used for the epitope mapping of the individual mAbs. As expected, OKT10 reacted with the *C275S* mutant protein, while the agonistic T16, HB-7, and IB4 mAbs failed to react with the protein. The conclusion is that the epitopes of mAbs inducing tyrosine phosphorylation are identical or closely located in the 273–285 CD38 sequence containing Cys^{275}.

Potentiation by Anti-CD38 mAb of Chemotactic Receptor-Mediated Superoxide Generation in Human Myeloid Cell Lines

CD38 was also induced in human monocytic leukemia THP-1 cells after their treatment with Bt_2-cAMP (25). The Bt_2-cAMP-treated THP-1 cells generated superoxide in response to fMLP. Interestingly, the addition of the agonistic anti-CD38 mAb HB-7 markedly enhanced the fMLP-induced superoxide generation, though it alone had small effect on superoxide generation. Fig. 4A shows the effect of HB-7 on the various concentrations of fMLP in the Bt_2-cAMP-treated THP-1 cells. The anti-CD38 mAb potentiated the fMLP-induced superoxide generation without changing the EC_{50} of fMLP. HB-7 enhanced superoxide generation induced by another chemoattractant, C5a (data not shown), of which receptors are also coupled to heterotrimeric G proteins. On the contrary, HB-7 failed to enhance superoxide generation stimulated by phorbol 12-myristate acetate (PMA), which bypassed the receptors/G proteins to stimulate the cells (Fig. 4B). Thus, stimulation of CD38 appeared to specifically potentiate superoxide generation mediated by the stimulation of G protein-coupled receptors.

Potentiation of Superoxide Generation by Anti-CD38 mAb as Associated with Protein-Tyrosine Phosphorylation

We investigated the relationship between the potentiating action of anti-CD38 mAb and tyrosine phosphorylation with the different concentrations of HB-7. The half-maximum effect of HB-7 on the fMLP-induced superoxide generation was observed at the concentrations of 2–3 µg/ml. HB-7-induced tyrosine phosphorylation was also dependent on its concentration, and the concentration-dependency for the overall phosphorylation pattern appeared to be similar to that observed for the potentiation of fMLP-induced superoxide generation. Possible involvement of the tyrosine phosphorylation in the potentiation of superoxide generation was further investigated with two types of anti-CD38 mAbs. As mentioned above, HB-7 induced both protein-tyrosine phosphorylation and potentiation of

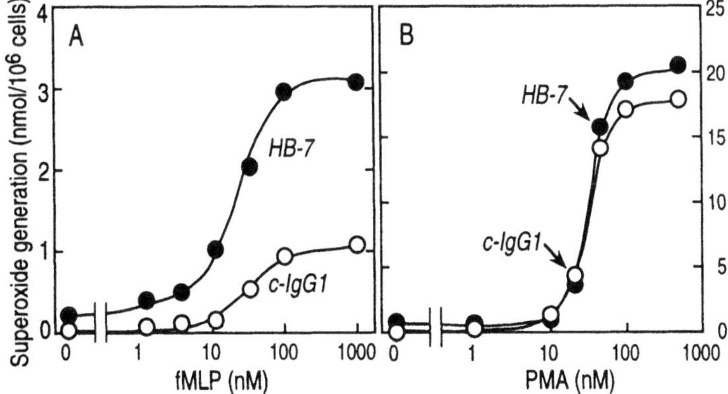

Figure 4. Potentiation by the agonistic anti-CD38 mAb HB-7 specifically observed for the receptor-mediated superoxide generation. THP-1 cells, which had been treated with 0.5 mM Bt_2-cAMP for 3 days, were incubated at 37°C for 10 min with the indicated concentrations of fMLP (*panel A*) or PMA (*panel B*), in Hank's buffer containing 5 µg/ml of the control IgG1 (m) or HB-7 (l). Superoxide generation was then measured and is expressed as nmol of cytochrome *c* reduction per 10^6 cells.

superoxide generation in the Bt_2-cAMP-treated THP-1 cells. On the contrary, no potentiation of the fMLP-induced superoxide generation was observed with another anti-CD38 mAb, RFT-10, which did not stimulate tyrosine phosphorylation. These results suggest that tyrosine phosphorylation of cellular proteins may be responsible for CD38-mediated potentiation of the fMLP-induced superoxide generation.

DISCUSSION

In the present study, we found that a Cys^{275}-containing sequence (the amino acid residues of 273–285) near the carboxyl terminus of CD38 not only plays a key role in its ecto-NADase activity, but also contains the epitopes of anti-CD38 mAbs inducing protein-tyrosine phosphorylation (see Fig. 5). This supports an idea that some of the biological functions mediated by the agonistic anti-CD38 mAbs may be related to its NADase activity, though there is no report showing that the intrinsic enzyme activity (hydrolase or cyclase) of CD38 is modified by the presence of anti-CD38 mAbs. Anti-CD38 mAb-induced tyrosine phosphorylation of cellular proteins or enzymes has also been reported in mice (26, 27) and human (24) B cells, in addition to HL-60 (18) and THP-1 cells (25). Among them, phosphatidylinositol 3-kinase has been identified as one of the tyrosine-phosphorylated proteins (24), which may be directly involved in the fMLP receptor-mediated signal transduction pathway leading to superoxide generation (28). Moreover, $p120^{c-cbl}$, which was also identified as a CD38-induced tyrosine-phosphorylated protein in the present study (18), has been shown to bind the SH2 and SH3 domains of many important signaling molecules including tyrosine kinases, adaptor proteins and phosphatidylinositol 3-kinase. Thus, the potentiation of fMLP-induced superoxide generation by the agonistic anti-CD38 mAbs might be mediated by the protein-tyrosine phosphorylation. Further study is currently under investigation in our laboratory to identify the accessory molecule(s) responsible for the enhancement of fMLP-receptor/G protein-mediated signaling.

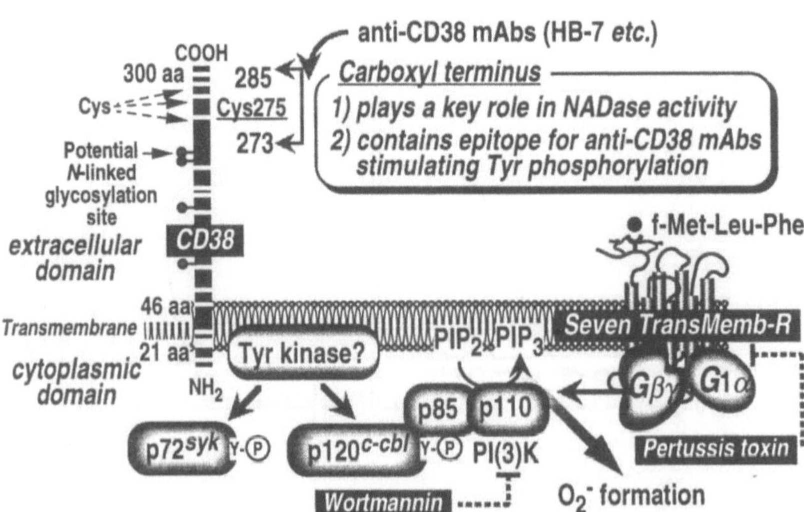

Figure 5. Molecular characterization of and cellular signal transduction via CD38/NADase. See text for explanation.

ACKNOWLEDGMENTS

We would like to thank Dr. Fabio Malavasi, Department of Genetics, Biology and Medical Chemistry, University of Turin, for providing us some of the agonistic anti-CD38 mAbs. This work was supported in part by research grants from the Scientific Research Fund of the Ministry of Education, Science, Sports, and Culture of Japan.

REFERENCES

1. Jackson, D. G. & Bell, J. I. 1990. Isolation of a cDNA encoding the human CD38 (T10) molecule, a cell surface glycoprotein with an unusual discontinuous pattern of expression during lymphocyte differentiation. *J. Immunol. 144*: 2811–2815.

2. Malavasi, F., Funaro, A., Roggero, S., Horenstein, A., Calosso, L. & Mehta, K. 1994. Human CD38: a glycoprotein in search of a function. *Immunol. Today 15*: 95–97.

3. Kontani, K., Nishina, H., Ohoka, Y., Takahashi, K. & Katada, T. 1993. NAD glycohydrolase specifically induced by retinoic acid in human leukemic HL-60 cells. Identification of the NAD glycohydrolase as leukocyte cell surface antigen CD38. *J. Biol. Chem. 268*: 16895–16898.

4. Nishina, H., Inageda, K., Takahashi, K., Hoshino, S., Ikeda, K. & Katada, T. 1994. Cell surface antigen CD38 identified as ecto-enzyme of NAD glycohydrolase has hyaluronate-binding activity. *Biochem. Biophys. Res. Commun. 203*: 1318–1323.

5. Lee, H. C. & Aarhus, R. 1991. ADP-ribosyl cyclase: an enzyme that cyclizes NAD+ into a calcium-mobilizing metabolite. *Cell Regul. 2*: 203–209.

6. Lee, H. C., Galione, A. & Walseth, T. F. 1994. Cyclic ADP-ribose: metabolism and calcium mobilizing function. [Review]. *Vitam. Horm. 48*: 199–257.

7. Howard, M., Grimaldi, J. C., Bazan, J. F., Lund, F. E., Santos, A. L., Parkhouse, R. M., Walseth, T. F. & Lee, H. C. 1993. Formation and hydrolysis of cyclic ADP-ribose catalyzed by lymphocyte antigen CD38. *Science 262*: 1056–1059.

8. Takasawa, S., Tohgo, A., Noguchi, N., Koguma, T., Nata, K., Sugimoto, T., Yonekura, H. & Okamoto, H. 1993. Synthesis and hydrolysis of cyclic ADP-ribose by human leukocyte antigen CD38 and inhibition of the hydrolysis by ATP. *J. Biol. Chem. 268*: 26052–26054.

9. Inageda, K., Takahashi, K., Tokita, K., Nishina, H., Kanaho, Y., Kukimoto, I., Kontani, K., Hoshino, S. & Katada, T. 1995. Enzyme properties of Aplysia ADP-ribosyl cyclase: comparison with NAD glycohydrolase of CD38 antigen. *J. Biochem.117*: 125–131.

10. Takahashi, K., Kukimoto, I., Tokita, K., Inageda, K., Inoue, S., Kontani, K., Hoshino, S., Nishina, H., Kanaho, Y. & Katada, T. 1995. Accumulation of cyclic ADP-ribose measured by a specific radioimmunoassay in differentiated human leukemic HL-60 cells with all-trans-retinoic acid. *FEBS Lett. 37*: 204–208.

11. Funaro, A., Spagnoli, G. C., Ausiello, C. M., Alessio, M., Roggero, S., Delia, D., Zaccolo, M. & Malavasi, F. 1990. Involvement of the multilineage CD38 molecule in a unique pathway of cell activation and proliferation. *J. Immunol. 145*: 2390–2396.

12. Santos, A. L., Teixeira, C., Preece, G., Kirkham, P. A. & Parkhouse, R. M. 1993. A B lymphocyte surface molecule mediating activation and protection from apoptosis via calcium channels. *J. Immunol. 151*: 3119–3130.

13. Zupo, S., Rugari, E., Dono, M., Taborelli, G., Malavasi, F. & Ferrarini, M. 1994. CD38 signaling by agonistic monoclonal antibody prevents apoptosis of human germinal center B cells. *Eur. J. Immunol. 24*: 1218–1222.

14. Dianzani, U., Funaro, A., DiFranco, D., Garbarino, G., Bragardo, M., Redoglia, V., Buonfiglio, D., De, M. L., Pileri, A. & Malavasi, F. 1994. Interaction between endothelium and CD4+CD45RA+ lymphocytes. Role of the human CD38 molecule. *J. Immunol. 153*: 952–959.

15. Kumagai, M., Coustan, S. E., Murray, D. J., Silvennoinen, O., Murti, K. G., Evans, W. E., Malavasi, F. & Campana, D. 1995. Ligation of CD38 suppresses human B lymphopoiesis. *J. Exp. Med. 181*: 1101–1110.

16. Ausiello, C. M., Urbani, F., Sala, A. L., Funaro, A. & Malavasi, F. 1995. CD38 ligation induces discrete cytokine mRNA expression in human cultured lymphocytes. *Eur. J. Immunol. 25*: 1477–1480.

17. Deaglio, S., Dianzani, U., Horenstein, A. L., Fernandez, J. E., van Kooten, C., Bragardo, M., Funaro, A., Garbarino, G., Virgilio, F. D., Banchereu, J. & Malavasi, F. 1996. Human CD38 ligand: A 120-KDa protein predominantly expressed on endothelial cells. *J. Immunol. 156*: 727–734.

18. Kontani, K., Kukimoto, I., Nishina, H., Hoshino, S., Hazeki, O., Kanaho, Y. & Katada, T. 1996. Tyrosine phosphorylation of the c-*cbl* proto-oncogene product mediated by cell surface antigen CD38 in HL-60 cells. *J. Biol. Chem. 271*: 1534–1537.

19. Donovan, J. A., Wange, R. L., Langdon, W. Y. & Samelson, L. E. 1994. The protein product of the c-cbl protooncogene is the 120-kDa tyrosine-phosphorylated protein in Jurkat cells activated via the T cell antigen receptor. *J. Biol. Chem. 269*: 22921–22924.

20. Cory, G. O., Lovering, R. C., Hinshelwood, S., MacCarthy, M. L., Levinsky, R. J. & Kinnon, C. 1995. The protein product of the c-cbl protooncogene is phosphorylated after B cell receptor stimulation and binds the SH3 domain of Bruton's tyrosine kinase. *J. Exp. Med. 182*: 611–615.

21. Marcilla, A., Rivero, L. O., Agarwal, A. & Robbins, K. C. 1995. Identification of the major tyrosine kinase substrate in signaling complexes formed after engagement of Fc gamma receptors. *J. Biol. Chem. 270*: 9115–9120.

22. Galisteo, M. L., Dikic, I., Batzer, A. G., Langdon, W. Y. & Schlessinger, J. 1995. Tyrosine phosphorylation of the c-cbl proto-oncogene protein product and association with epidermal growth factor (EGF) receptor upon EGF stimulation. *J. Biol. Chem. 270*: 20242–20245.

23. States, D. J., Walseth, T. F. & Lee, H. C. 1992. Similarities in amino acid sequences of Aplysia ADP-ribosyl cyclase and human lymphocyte antigen CD38. *Trends Biochem. Sci. 17*: 495.

24. Silvennoinen, O., Nishigaki, H., Kitanaka, A., Kumagai, M., Ito, C., Malavasi, F., Lin, Q., Conley, M. E. & Campana, D. 1996. CD38 signal transduction in human B-cell precursors: rapid induction of tyrosine phosphorylation, activation of *syk* tyrosine kinase, and phosphorylation of phospholipase C-γ and phosphatydilinositol 3-kinase. *J. Immunol. 156*: 100–107.

25. Kontani, K., Kukimoto, I., Kanda, Y., Inoue, S., Kishimoto, H., Hoshino, S. & Katada, T. 1996. Induction of CD38/NADase and its monoclonal antibody-induced tyrosine phosphorylation in human leukemia cell lines. *Biochem. Biophys. Res. Commun.* (in press).

26. Kirkham, P. A., Santos, A. L., Harnett, M. M. & Parkhouse, R. M. 1994. Murine B-cell activation via CD38 and protein tyrosine phosphorylation. *Immunology 83*: 513–516.

27. Kikuchi, Y., Yasue, T., Miyake, K., Kimoto, M. & Takatsu, K. 1995. CD38 ligation induces tyrosine phosphorylation of Bruton tyrosine kinase and enhanced expression of interleukin 5-receptor α chain: Synergistic effects with interleukin 5. *Proc. Natl. Acad. Sci. U.S.A. 92*: 11814–11818.

28. Okada, T., Sakuma, L., Fukui, Y., Hazeki, O. & Ui, M. 1994. Blockage of chemotactic peptide-induced stimulation of neutrophils by wortmannin as a result of selective inhibition of phosphatidylinositol 3-kinase. *J. Biol. Chem. 269*: 3563–3567.

55

Ca^{2+}-SIGNALLING IN HUMAN T-LYMPHOCYTES

Potential Roles for Cyclic ADP-Ribose and 2′-Phospho-Cyclic ADP-Ribose

Andreas H. Guse,[1] Cristina P. da Silva,[1] Barry V. L. Potter,[2] and Georg W. Mayr[1]

[1]Universität Hamburg
Institut für Physiologische Chemie
Abt. für Enzymchemie, Grindelallee 117
D-20146 Hamburg, Germany
[2]University of Bath
School of Pharmacy and Pharmacology
Claverton Down, Bath BA2 7AY
United Kingdom

1. ABSTRACT

Intracellular Ca^{2+}-signals belong to the major events transducing extracellular signals into living cells. The discovery of (i) a caffeine-sensitive intracellular Ca^{2+}-pool in Jurkat T-lymphocytes [1] and (ii) cyclic adenosine diphosphoribose (cADPR) as an agent that mobilizes Ca^{2+} from a caffeine- and ryanodine sensitive Ca^{2+}-store in sea urchin egg homogenates [2] prompted us to investigate the potential role of this compound in T-lymphocyte Ca^{2+}-signalling. cADPR, as well as its 2′-phosphorylated derivative, 2′-phospho-cADPR (2′-P-cADPR), released Ca^{2+} in a dose-dependent, specific manner from intracellular, non-endoplasmic reticular stores of permeabilized Jurkat and HPB.ALL T cells. In addition, attempts were made to prove the presence of endogenous cADPR and 2′-P-cADPR by HPLC. Several HPLC protocols, including microbore-HPLC were tested resulting in the detection of endogenous cADPR by sequential separation on strong-anion exchange HPLC and reverse-phase ion-pair HPLC.

ADP-Ribosylation in Animal Tissue, edited by Haag and Koch-Nolte
Plenum Press, New York, 1997

2. BACKGROUND

Ca^{2+}-release from intracellular stores in many cases is one of the major events transducing extracellular signals into living cells. Beside the well characterized system of D-myo-inositol 1,4,5-trisphosphate (Ins(1,4,5)P3)-induced Ca^{2+}-release from the endoplasmic reticulum [3], recently a cyclic metabolite of NAD+, cyclic ADP-ribose (cADPR) has been described to release Ca^{2+} from internal stores of living cells [for reviews see 4,5]. Interestingly, the effect of cADPR was not mediated by Ins(1,4,5)P3-receptors since it was sensitive to ruthenium red, but not to heparin, indicating the presence of two different intracellular Ca^{2+}-release systems either sensitive to cADPR or to Ins(1,4,5)P3 [2]. Although most of the experimental results have been obtained in sea urchin eggs, there is growing evidence for a role of cADPR acting on the ryanodine receptor also in mammalian cells, e.g. pancreatic acinar cells [6,7], skeletal and cardiac muscle cells [8,9], brain microsomes [10] and GH4C1 pituitary cells [11].

2. RESULTS AND DISCUSSION

2.1. cADPR- and 2′-Phospho-cADPR-Induced Ca^{2+}-Release

We have recently demonstrated cADPR-induced Ca^{2+}-release in permeabilized Jurkat and HPB.ALL T-lymphocytes (Fig.1, ref.12). It is characterized (i) by a relatively high half-maximal effective concentration (2.25 mM), (ii) by being independent of the Ins(1,4,5)P3-mediated Ca^{2+}-release, (iii) by its sensitivity to the inhibitors 8-NH2-cADPR, 8-Br-cADPR and ruthenium red, and (iv) by the fact that the cADPR-sensitive Ca^{2+}-pool was not thapsigargin-sensitive, but overlapped with the caffeine-sensitive Ca^{2+}-pools of Jurkat T cells [12]. Also mouse T-lymphoma cells have been shown to contain a cADPR-sensitive, ryanodine receptor-containing intracellular compartment, which is not part of the ER but seems to represent a class of small vesicles being localized close to the plasma membrane [13].

Very recently, we found that also the 2′-phosphorylated derivative of cADPR, 2′-P-cADPR released Ca^{2+} from the same intracellular Ca^{2+}-pool of permeabilized Jurkat T cells as compared to cADPR (Fig.1; Guse et al., manuscript submitted). This finding is in accordance with recent reports from Zhang et al. [14] and Vu et al. [15] demonstrating Ca^{2+}-release from rat brain microsomes by 2′-P-cADPR.

These data, in combination with a recent report demonstrating expression of ryanodine receptors in Jurkat T cells [16], indicate a role for cADPR in the regulation of intracellular Ca^{2+}-signalling in T-lymphocytes (Fig.2). Whether a direct, receptor-mediated regulation of the intracellular concentration of cADPR takes place in T cells is not yet clear (see also below). However, the Ca^{2+}-release activity of cADPR may be regulated indirectly by inorganic phosphate. Evidence for the latter has been obtained recently by demonstrating that inorganic phosphate (i) could modulate cADPR-induced Ca^{2+}-release (Fig.2), and (ii) was elevated in response to stimulation of the T cell receptor/CD3-complex (Fig.2; Guse et al., submitted).

2.2. Analysis of Endogenous cADPR by HPLC

The current view of a second messenger implies that there is regulation of its intracellular concentration by extracellular agonists. Therefore, we have used a 2-step HPLC

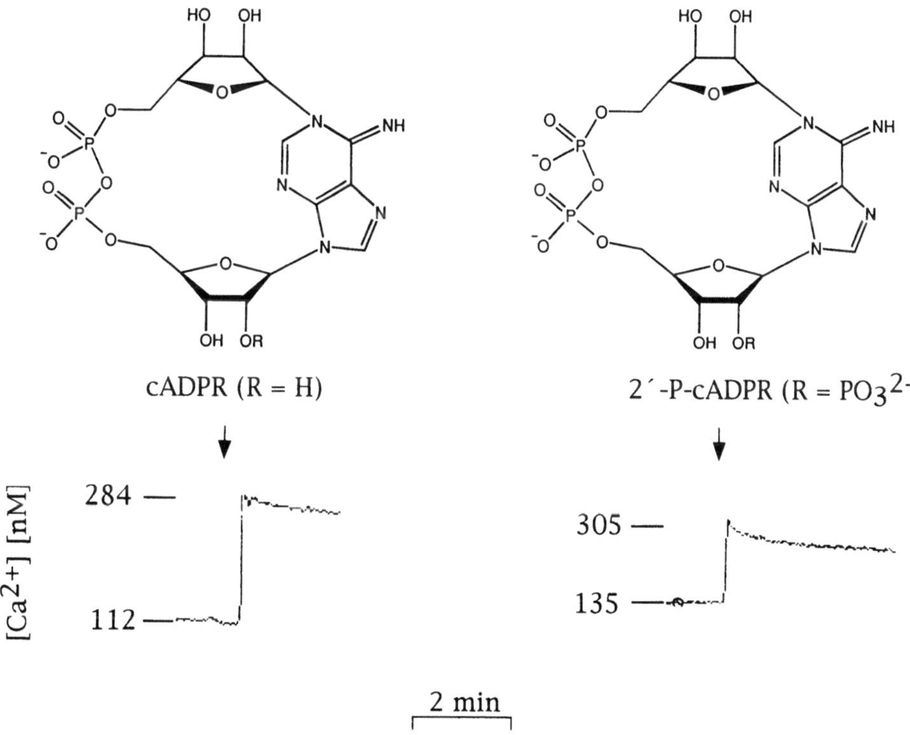

Figure 1. Structures of cADPR and 2′ -P-cASPR and their Ca²⁺- release activity in permeabilized. Jurkat T-lymphocytes Jurkat T-lymphocytes were permeabilized and [Ca²⁺] was measured as detailed in [12]. After charging the Ca²⁺- pools by addition of ATP (1mM) and an ATP-regenerating system, cADPR or 2′-P-cADPR-induced Ca²⁺-release are shown (from 3 to 6 independent experiments).

system consisting of anion-exchange- and reversed-phase column steps to measure intracellular concentrations of cADPR [12]. Using this HPLC system, the endogenous levels of cADPR in unstimulated Jurkat and HPB.ALL T cells were found to be in the range of 3 to 4.5 nmol/108 cells. A second, single step microbore-HPLC system preceeded by solid phase extraction confirmed our results (Fig.3; ref.12). Interestingly, up to now significant changes in the level of endogenous cADPR in response to stimulation of a cell surface receptor, e.g. the T cell receptor/CD3-complex, have not been observed [12]. However, due to the potential errors which may arise even when different HPLC systems are used for analysis, the only definite conclusion at the moment is that cADPR was present in the cell extracts. Indeed, initial experiments using a reversed-phase column with higher selectivity in the second HPLC-step indicate that the peak co-eluting with cADPR as described in [12] may be composed of more than one compound. Therefore, the intracellular levels of cADPR may be lower as previously published [12]. Further work to develop more reliable mass analysis techniques of cADPR therefore is necessary.

3. CONCLUSION

Our results suggest that cADPR and perhaps also 2′-P-cADPR are involved in the regulation of T-lymphocyte Ca²⁺-signalling. However, the following unresolved aspects are be subject of further investigations in our laboratory: (i) exact quantification of

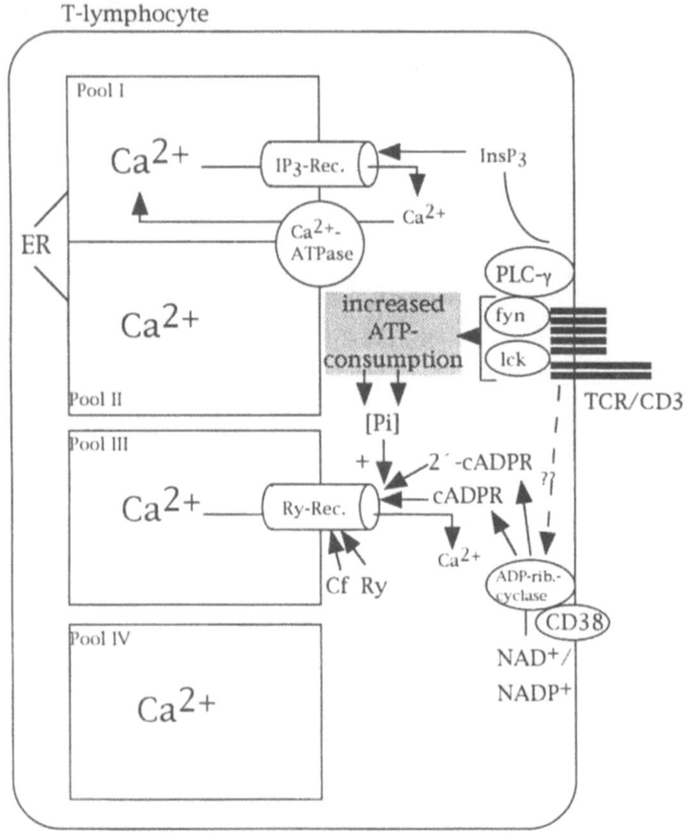

Figure 2. Model of T-lymphocyte Ca^{2+}-pool organization: potential sites of action of Ca^{2+}-mobilizing second messengers. Jurkat T cells contain at least 4 different Ca^{2+}-pools [1]. Pool I is sensitive to Ind(1, 4, 5,) P_3 due to its specific receptor (IP_3-Rec.). Pools I and II are sensitive to thapsigargin and are therefore thought to be localized in the endoplasmic reticulum (ER). Pool III, originally detected as a caffeine (Cf)-sensitive Ca^{2+}-store [1], is also sensitive to ryanodine (Ry) - therefore it appears to contain a ryanodine receptor (Ry-Rec.) - and to cADPR and 2´ -P-cADPR, which are suspected to be synthesized by a so far unknown ADP-ribosyl cyclase or the ectoenzyme CD38 from NAD^+ and $NADP^+$. Whether one of these enzymes or both can be controlled by the T cell receptor/CD3-complex is (TCR/CD3) currently unknown. The activation of the T cell recptor/CD3-complex generally leads to an increased ATP (and NTP) consumption resulting in a marked increase in inorganic phosphate (Pi) which in turn can modulate cADPR-induced Ca^{2+}-release in permeabilized T cells. Other abbreviations: PLC-, phospholipase C-˜; fyn, p59^{fyn}; lck, p56^{lck}.

cADPR by different methods in unstimulated vs. receptor-stimulated cells, and (ii) analysis of the enzymes and binding proteins involved in cADPR-metabolism and function.

ACKNOWLEDGMENTS

We are grateful to Karin Weber and Karin Müller for excellent technical assistance. This work was supported in part by grants from the Deutsche Forschungsgemeinschaft (Gu 360/2–1 to AHG and GWM), MRC and the British-German ARC Programme (to

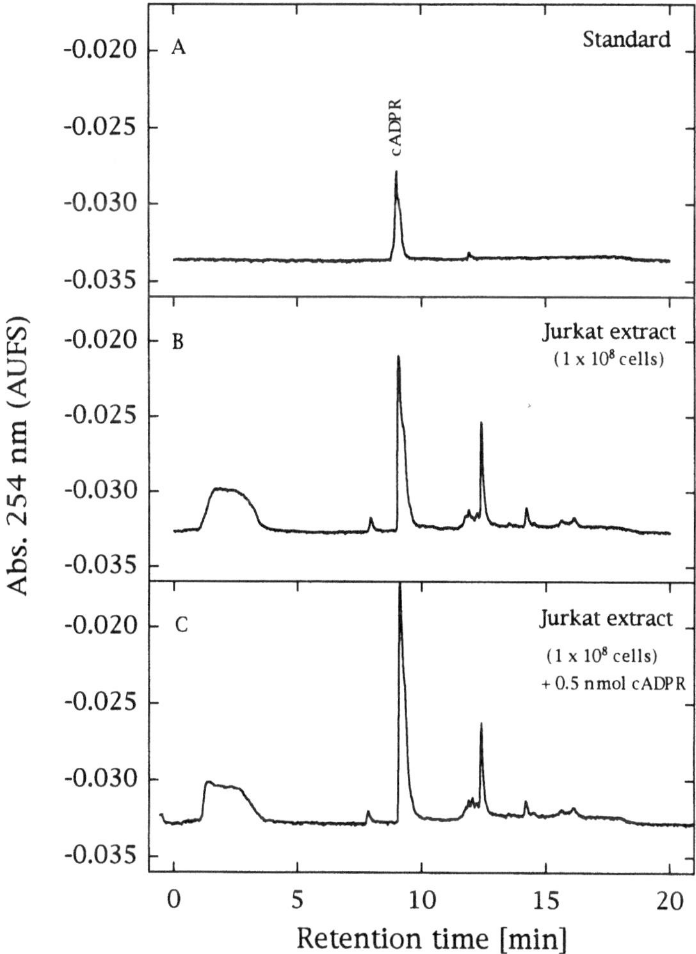

Figure 3. HPLC analysis of endogenous cADPR by microbore-HPLC. Jurkat T cells (1 x 10^8) were extracted by a perchloric acid procedure as described [12]. The extract was pre-purified by solid-phase extraction on small (1 ml volume) Q-Sepharose FF columns by stepwise elution with trifluoroacetic acid (TFA). The pool containing cADPR was lyophilized, redissolved in Tris-buffer (1 mM, pH 8.0) and analyzed using a SMART System (Pharmacia, Biotech, Freiburg, Germany) equipped with a MiniQ PC 3.2/3 column (Pharmacia). A complex gradient from 1 mM Tris-HCl, pH 8.0, to 150 mM TFA in 15 min at 400 ml/min was used. Panels show chromatograms for (A) standard cADPR, (B) an extract from Jurkat cells, and (C) the same extract as in (B) spiked with 0.5 nmol standard cADPR.

BVLP), and JNICT (Portugal, Praxis XXI-BPD/4228/94 to CPS). BVLP is a Lister Institute Research Professor.

4. REFERENCES

1. Guse, A. H., Roth, E., & Emmrich, F. 1993. Intracellular Ca^{2+}-pools in Jurkat T-lymphocytes. *Biochem. J.* *291*: 447–451.
2. Galione, A., H. C. Lee, & W. A. Busa. 1991. Ca^{2+}-induced Ca^{2+} release in sea urchin egg homogenates: modulation by cyclic ADP-ribose. *Science 253*: 1143–1146.
3. Berridge, M. J. 1993. Inositol trisphosphate and calcium signalling. *Nature 361*: 315–325.

4. Galione, A. & A. White. 1994. Ca^{2+}-release induced by cyclic ADP-ribose. *Trends Cell Biol. 4*: 431–436.

5. Lee, H. C. 1994. Cyclic ADP-Ribose: a new member of a superfamily of signalling cyclic nucleotides. *Cell. Signalling 6*: 591–600.

6. Thorn, P., O. Gerasimenko, & O. H. Petersen. 1994. Cyclic ADP-ribose regulation of ryanodine receptors involved in agonist evoked cytosolic Ca^{2+} oscillations in pancreatic acinar cells. *EMBO J. 13:* 2038–2043.

7. Gerasimenko, O. V., J. V. Gerasimenko, A. V. Tepikin, & O. H. Petersen. 1995. ATP-dependent accumulation and inositol trisphosphate- or cyclic ADP-ribose-mediated release of Ca^{2+} from the nuclear envelope. *Cell 80*: 439–444.

8. Morrissette, J., G. Heisermann, J. Cleary, A. Ruoho, & R. Coronado. 1993. Cyclic ADP-ribose induced Ca^{2+}-release in rabbit skeletal muscle sarcoplasmic reticulum. *FEBS Lett. 330*: 270–274.

9. Meszaros, L. G., J. Bak, & A. Chu. 1993. Cyclic ADP-ribose as an endogenous regulator of the non-skeletal type ryanodine receptor Ca^{2+}-channel. *Nature 364*: 76–79.

10. White, A. M., S. P. Watson, & A. Galione. 1993. Cyclic ADP-ribose-induced Ca^{2+}-release from rat brain microsomes. *FEBS Lett. 318*: 259–263.

11. Koshiyama, H., H. C. Lee, & A. H. Tashjian Jr. 1991. Novel mechanism of intracellular calcium release in pituitary cells. *J. Biol. Chem. 266*: 16985–16988.

12. Guse, A.H., da Silva, C.P., Emmrich, F., Ashamu, G.A., Potter, B.V.L., & Mayr, G.W. 1995. Characterization of Cyclic adenosine diphosphate-ribose-induced Ca^{2+}-release in T-lymphocyte cell lines. *J. Immunol. 155*: 3353–3359.

13. Bourguignon, L.Y.W., Chu, A., Jin, H. & Brandt, N.R. 1995. Ryanodine receptor-ankyrin interaction regulates internal Ca^{2+}-release in mouse T-lymphoma cells. *J. Biol. Chem. 270*: 17917–17922.

14. Zhang, F.-J., Gu, Q.-M., Jing, P., & Sih, C.J. 1995. Enzymatic cyclization of nicotinamide adenine dinucleotide phosphate. *Bioorg. Med. Chem. Lett. 5*: 2267–2272.

15. Vu, C.Q., Lu, P.-J., Chen, C.-S., & Jacobson, M.K. 1996. 2′-Phospho-cyclic ADP-ribose, a calcium-mobilizing agent derived from NADP.*J. Biol. Chem. 271*: 4747–4754.

16. Hakamata, Y., Nishimura, S., Nakai, J., Nakashima, Y., Kita, T., & Imoto, K. 1994. Involvement of the brain type ryanodine receptor in T-cell proliferation. *FEBS Lett.* 352: 206–210

56

METABOLISM AND ACTIONS OF ADP-RIBOSES IN CORONARY ARTERIAL SMOOTH MUSCLE

Pinlan Li, Ai-Ping Zou, and William B. Campbell

Departments of Pharmacology & Toxicology and Physiology
Medical College of Wisconsin
8701 Watertown Plank Road
Milwaukee, Wisconsin 53226

ABSTRACT

The present study examined the metabolism of ADP-ribose (ADPR) and cyclic ADP-ribose (cADPR) in small bovine coronary arterial homogenates and characterized the effects of these nucleotides on the activity of potassium (K^+) channels in coronary smooth muscle cells. ADPR and cADPR were produced from NAD^+ (1mM) by homogenates from small bovine coronary arteries. The conversion rate was 2.81 ± 0.19 nmol/min/100 µg protein for ADPR and 1.37 ± 0.03 nmol/min/100 µg protein for cADPR. In patch clamp experiments, ADPR produced a concentration-dependent increase in the activity of a calcium activated K (K_{Ca}) channel in inside-out membrane patches of coronary arterial smooth muscle cells at concentrations of 0.1, 1 and 10 µM. The open state probability (NPo) of K_{Ca} channel was maximally increased 5-fold at a concentration of 10 µM. cADPR reduced the activity of K_{Ca} channel at concentrations of 1 and 10 µM. The NPo was decreased by 45% and 75%, respectively. The results indicate that there is an enzymatic pathway in the coronary arterial smooth muscle to produce ADPR and cADPR. These nucleotides may play a role in the control of coronary vascular tone by altering the activity of the K_{Ca} channel in vascular smooth muscle cells.

BACKGROUND

cADPR and ADPR have been detected in a variety of tissues including the heart, liver, spleen, brain and red blood cells, lymphocytes, pituitary cells and cultured renal epithelial cells (1–3). Recent studies indicated that cADPR causes the mobilization of intracellular calcium by a mechanism completely independent of IP_3 (4, 5), and cADPR-mediated intracellular calcium-mobilization participates in the regulation of the

ADP-Ribosylation in Animal Tissue, edited by Haag and Koch-Nolte
Plenum Press, New York, 1997

secretion of insulin, the fertilization of eggs and the effect of nitric oxide in non-muscle tissue (3–5). However, there is no evidence indicating the metabolism and action of cADPR and ADPR in the vasculature. The purpose of the present study is to characterize the production of ADPR and cADPR in coronary arteries and to determine the role of these nucleotides in the gating of K^+ channels in coronary arterial smooth muscle cells.

EXPERIMENTAL PROCEDURES

Small coronary arteries of bovine heart were microdissected, pooled and stored in ice-cold phosphate buffer saline. The dissected arteries were homogenized with a glass homogenator in ice-cold HEPES buffer, and homogenates were prepared as described previously (2). To determine the production of ADPR and cADPR, the homogenates (50–100 µg) were incubated for 10 min with 1–3 mM β-NAD^+ at 37° C and the reaction products were separated and analyzed by HPLC using a Supelcosil LC-18 column (4.6x150 mm i.d.).

Coronary arterial smooth muscle (VSM) cells were dissociated and single channel K^+ currents were recorded using the patch-clamp technique as described previously (6). The effects of ADPR and cADPR on K^+ channel activity were examined in the inside-out patch mode.

RESULTS

A representative reverse-phase HPLC chromatogram depicting the profiles of NAD^+ and its metabolites is presented in figure 1A. The homogenates prepared from bovine coronary arteries produced cADPR and ADPR when incubated with 1 mM NAD^+ for 10 minutes. The products with retention times of 3.2, 4.1 and 14.2 min coeluted with synthetic cADPR, nicotinamide (Ni) and ADPR standards, respectively. NAD^+ has a retention time of 5.4 min (figure 1B). The production rate of ADPR in the homogenates was 2.81 ± 0.19 nmol/min/100 µg protein and that of cADP-R was 1.37 ± 0.03 nmol/min/100 µg protein. Using purified ADPRcyclase and NADase, cADPR or ADPR production was confirmed using HPLC analysis (Fig 1C and 1D). The conversion rates of ADPR and cADPR were 580.4 ± 22 nmol/min/100 µg protein and 25,897 ± 218 nmol/min/ 100 µg protein in the presence of 1 mM NAD^+.

Figure 2A is a representative tracing depicting unitary K^+ currents at a membrane potential of +40 mmHg in an inside-out patch from coronary arterial smooth muscle cells. The unitary K^+ current has a mean slope conductance of 256.3 ± 5 pS and the biophysical properties are consistent with a calcium-activated large conductance K^+ channel (K_{Ca} channel). cADPR reduced the activity of the K_{Ca} channel when applied to the internal surface of inside-out excised patches (Fig 2B). The open probability (NPo) of K_{Ca} channel was decreased by 75% at a concentration of 100 nM cADPR (Fig. 2D). The amplitude of the K_{Ca} channel was unaltered by addition of cADPR (Fig. 2B). In contrast with the effect of cADPR to decrease the activity of the K_{Ca} channel, ADPR increased the activity of the K_{Ca} channel when added to bath solution in the inside-out patch mode (Fig. 2C). The NPo of the K_{Ca} channel was increased 5 fold at an ADPR concentration of 100 nM (Fig. 2D). ADPR had no effect on the single current amplitude of the K_{Ca} channel (Fig. 2C).

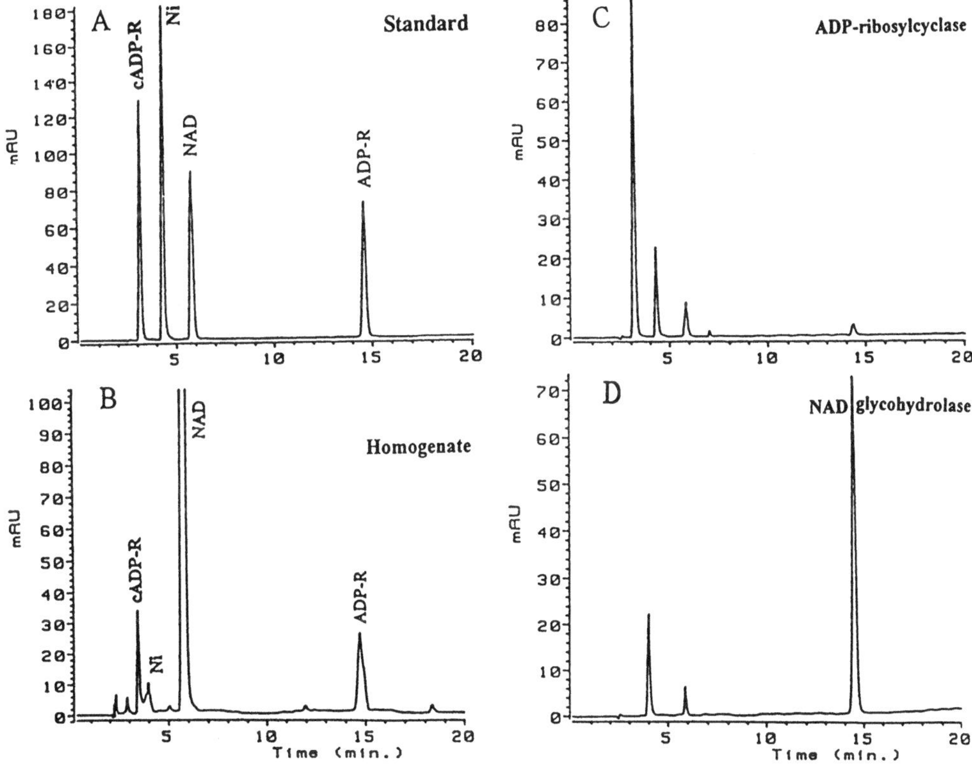

Figure 1. HPLC chromatograms of NAD$^+$ and its metabolites (A). Coronary arterial homogenates were incubated with NAD$^+$ (100 μM). The products of this reaction has retention times of 3.1 and 14.2 minutes and coeluted with synthetic cADPR and ADPR standards (B). Purified ADPRcyclase or NADase was also incubated with NAD$^+$ and the products coeluted with synthetic cADPR and ADPR (C and D), respectively.

DISCUSSION

In the present study, we have found that the homogenates from small bovine coronary arteries produced cADPR and ADPR when incubated with NAD$^+$. The conversion rate for ADPR was higher than that for cADPR. This provided the first evidence indicating cADPR and ADPR synthesis by the vasculature. Recent studies have indicated that NAD$^+$ can be cyclized to cADPR via ADPRcyclase and hydrolysed to ADPR via NADase, and cADPR can also be hydrolysed to form ADPR (5,7). Using purified ADPRcyclase and NADase, we confirmed that the NAD$^+$ metabolites produced from coronary arterial homogenates are same as that produced by corresponding purified enzymes.

The role of cADPR and ADPR in the control of vascular tone remains unknown. Recent studies indicated that cADPR binds to the ryanodine receptor and mobilizes intracellular calcium (5, 8). So far there is no evidence suggesting that this cADPR-mediated calcium mobilization participates in the regulation of vascular tone. In this regard, NO has been reported to increase the production of cADPR and induce intracellular calcium mobolization in non-vascular tissue (5, 9). However, this cADPR-mediated calcium mobilization could not mediate the vasodilatory effect of NO, since rise in intracellular Ca^{++} causes vasoconstriction. It seems that cADPR has a tissue-specific effect.

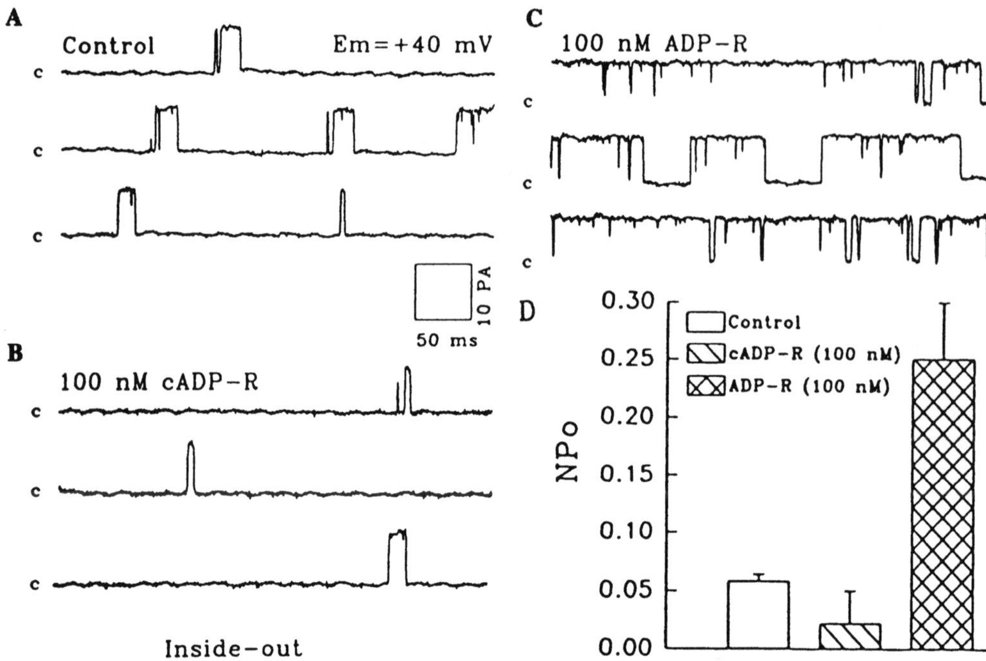

Figure 2. Effect of cADPR and ADPR on the activity of the K_{Ca} channel in excised inside-out membrane patches of smooth muscle cells isolated from small bovine coronary arteries. * indicates significant difference from control.

The present study demonstrated that cADPR and ADPR altered the activity of K^+ channels in coronary arterial smooth muscle. ADPR markedly activated K_{Ca} channels and cADPR inhibited this channel. The activation of K_{Ca} channels may lead to the hyperpolarization of vascular smooth muscle, a reduction in intracellular Ca^{++} and vasodilation. It is possible that ADPR and cADPR participate in the control of vascular tone by gating the K_{Ca} channels in vascular smooth muscle cells. It remains to be further studied whether ADPR and cADPR mediate the effect of any vasoactive agonists which serve as K^+ channel activator such as endothelium-derived vasodilators, NO, epoxyeicosatrienoic acids or prostaglandin I_2.

In summary, our results suggest that ADPR and cADPR may serve as intracellular signalling molecules that regulate the K_{Ca} channel and control vascular tone in coronary microcirculation.

This work was supported by grants from the National Heart, Lung and Blood Institute (HL-51055) and American Heart Association, Wisconsin Affiliate (95-GB-52).

REFERENCES

1. Beers, K., E. N. Chini, H. C. Lee, T. P. Dousa. 1995. Metabolism of cyclic ADP-ribose in opossum kidney renal epithelial cells. Am. J. Physiol. 268(cell Physiol. 37): C741-C746.
2. Lee, H. C. and R. Aarhus. 1993. Wide distribution of an enzyme that catalyzes the hydrolysis of cyclic ADP-ribose. Biochim. Biophys. Acta 1164:68–74.
3. Takesawa, S., K. Nata, H. Yonekura, H. Okamoto. 1993. Cyclic ADP-ribose in insulin secretion from pancreatic b cells. Science 259:370–373.

4. Lee, H. C., R. Aarhus, and T.F. Walseth. 1993. Calcium mobilization by dual receptors during fetilization of sea urchin eggs. Science Wash. DC261:352–355.

5. Lee, H. C. 1994. A signaling pathway involving cyclic ADP-ribose, cGMP, and nitric oxide.NIPS 9: 134–137.

6. Hamill, O. P., M. R. Neher, B. Sakmann. 1981. Improved patch clamp techniques for high-resolution current recording from cell and cell-free membrane patches. Pflugers Archiv. 391:85–100.

7. Kim, H., E. L. Jacobson, M. K. Jacobson. 1993. Synthesis and degradation of cyclic ADP-ribose by NAD glycohydrolases. Science 261:1330–1333.

8. Galion, A.1993. Cyclic ADP-ribose: a new way to control calcium. Science 259:325–326.

9. Galione, A., H. White, N. Willmott, M. Turner, B.V.L. Potter, and S.P.Watson. 1993. cGMP mobilizes intracellular Ca2+ in sea urchin eggs by stimulating cyclic ADP-ribose synthesis. Nature Lond. 365: 456–459.

57

BOVINE LIVER MITOCHONDRIAL NAD$^+$ GLYCOHYDROLASE

Relationship to ADP-Ribosylation and Calcium Fluxes

Mathias Ziegler, Dierk Jorcke, Andrés Herrero-Yraola, and Manfred Schweiger

Freie Univ. Berlin
Institut f. Biochemie
Thielallee 63, 14195 Berlin
Germany

ABSTRACT

Mitochondrial NAD$^+$ glycohydrolase (NADase) has been proposed to be required for (nonenzymatic) ADP-ribosylation and subsequent activation of a Ca^{2+} release pathway. In our studies it has been found that several agents including nicotinamide, dithiothreitol, and EDTA exert no or little effect on ADP-ribosylation in isolated bovine liver mitochondria, while strongly inhibiting the NADase. The NADase did, however, catalyze the formation of cyclic purine nucleoside diphosphoriboses (similar to cyclic ADP-ribose) from NAD$^+$ analogs. It appears possible, therefore, that this enzyme may be involved in the regulation of mitochondrial Ca^{2+} fluxes by forming a potent Ca^{2+}-mobilizing agent, rather than by providing the substrate for non-enzymatic ADP-ribosylation.

INTRODUCTION

Increased degradation of intramitochondrial pyridine nucleotides has been reported to be a consequence of "oxidative stress". Stimulation of both ADP-ribosylation of mitochondrial proteins and Ca^{2+} release have been observed following incubation of the organelles with prooxidants[1]. A hypothesis has been presented suggesting that peroxidation reactions may activate NAD$^+$ glycohydrolase, resulting in the accumulation of ADP-ribose within mitochondria. Free ADP-ribose would then, in a nonenzymatic, but specific reaction, modify protein(s) causing the activation of a mitochondrial calcium release pathway (reviewed in 1). The physiological significance of the nonenzymatic mitochondrial ADP-ribosylation has been called into question by observations demonstrating that nicoti-

ADP-Ribosylation in Animal Tissue, edited by Haag and Koch-Nolte
Plenum Press, New York, 1997

namide and 3-aminobenzamide reduced ADP-ribosylation only to a limited extent at concentrations sufficient to nearly completely block the NADase[2]. Our studies have been aimed at clarifying the possible role of the NADase in mitochondrial ADP-ribosylation and its potential involvement in the regulation of mitochondrial calcium fluxes.

MATERIALS AND METHODS

Bovine liver mitochondrial NADase was partially purified following solubilization by Triton X-100 as described[3]. NADase activity was measured fluorimetrically using the substrate analog 1,N^6-etheno NAD$^+$ (ε–NAD$^+$)[4]. ADP-ribosylation of mitochondrial proteins was assessed by incubating isolated mitochondria in the presence of 25 μM [*adenylate*-^{32}P]-NAD$^+$. Bound label was detected autoradiographically following SDS-PAGE. For some experiments [*adenine*- ^{14}C]-NAD$^+$ or [*nicotinamide*- ^{14}C]-NAD$^+$ were used instead of [*adenylate*-^{32}P]-NAD$^+$.

RESULTS AND DISCUSSION

Bovine liver mitochondrial NADase was purified previously[5]. The final preparation contained a single protein band on SDS-PAGE with an apparent molecular mass of about 32,000. Enzyme activity was correlated with a protein band directly within the gel by a fluorescent assay[3]. Nicotinamide inhibited the enzyme, the K_I being about 150 μM. At 5 mM nicotinamide no residual activity could be detected. Disulfide bond(s) appear to be of importance for enzyme activity (or stability), because reducing agents such as dithiothreitol (DTT) caused substantial loss of activity. Incubation of the mitochondrial NADase in the presence of 10 mM DTT led to inactivation of the enzyme within a few minutes. Addition of EDTA substantially inhibited the enzyme (see Jorcke et al., this volume).

The availibility of strong inhibitors of the NADase, e.g. nicotinamide, EDTA, and DTT, enabled the study of mitochondrial ADP-ribosylation in the absence of NADase activity. Table 1 summarizes the influence of these NADase inhibitors on ADP-ribosylation. Neither EDTA nor DTT inhibited ADP-ribosylation. Nicotinamide reduced the extent of mitochondrial ADP-ribosylation. It is concluded from these results that NADase activity is not required for ADP-ribosylation of bovine liver mitochondrial proteins using NAD$^+$ as substrate. Additional evidence in support of this conclusion has been provided by the observation that a twenty-fold molar excess of free ADP-ribose over [*adenylate*-^{32}P]-NAD$^+$ did not substantially reduce the amount of incorporated label (Tab. 1). Moreover, 1 mM

Table 1. Influence of various effectors on mitochondrial ADP-ribosylation. ADP-ribosylation assays were carried out using 25 μM [^{32}P]-NAD$^+$ as substrate. ↑, stimulation; ↓, inhibition; ↓↓, no label detected

Effector	Effect on mitochondrial ADP-ribosylation
5 mM nicotinamide	↓
2 mM EDTA	none
10 mM DTT	none
0.5 mM ADP-ribose	↓
2 mM MgCl2	↓↓↓
1 mM ATP	↑

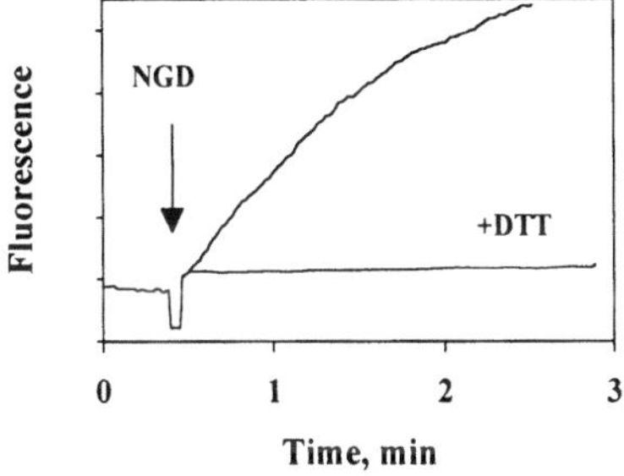

Figure 1. Synthesis of cyclic GDP-ribose by mitochondrial NADase. Partially purified NADase was incubated with 1 mM NGD⁺. Fluorescence was followed at 410 nm, the excitation wavelength was set to 310 nm. For the lower trace (+DTT), the enzyme was preincubated with 10 mM dithiothreitol for 10 min.

ATP appeared to stimulate ADP-ribosylation, while NADase activity was inhibited by more than 50%. Taken together these results strongly suggest the existence of a specific monoADP-ribosyl transferase in bovine liver mitochondria. ADP-ribosylation in these organelles would appear to occur primarily by way of an enzymatic mechanism from NAD⁺ rather than nonenzymatically utilizing free ADP-ribose generated by NADase.

Experiments using [¹⁴C]-labeled NAD⁺ demonstrated that the adenine, but not the nicotinamide moiety was present in the modified proteins ruling out a direct modification by NAD⁺. This control is important, since covalent modification of glyceraldehyde-3-phosphate dehydrogenase by NAD⁺, and not ADP-ribose, has been demonstrated[6]. It was also observed that addition of 2 mM Mg^{2+} drastically reduced the extent of ADP-ribosylation (Tab. 1), whereas NADase activity was unaffected[3]. Further experiments indicated that this effect was due to accelerated removal of the label from the modified proteins (not shown).

As the NADase was not required for ADP-ribosylation, it was attempted to define another possible function for this protein. It is demonstrated in Fig. 1 that, similarly to some other NAD⁺ glycohydrolases, the mitochondrial NADase catalyzed the formation of a fluorescent product from nicotinamide guanine dinucleotide (NGD⁺). It has been shown that this product is cyclic GDP-ribose[7], because noncyclic derivatives are not fluorescent. Similarly to the NADase activity, no GDP-ribosyl cyclase activity was detected when the enzyme was preincubated with 10 mM DTT (Fig. 1, lower trace). The ability to synthesize a cyclic metabolite from an NAD⁺ analog strongly suggests that the mitochondrial NADase may serve as an ADP-ribosyl cyclase[7,8]. Therefore, this enzyme may participate in the regulation of mitochondrial calcium fluxes by producing cyclic ADP-ribose, a strong intracellular calcium-mobilizing agent[9], rather than by providing the substrate for nonenzymatic ADP-ribosylation. Our results also indicate that mitochondrial ADP-ribosylation may be involved in processes unrelated to the regulation of calcium fluxes.

REFERENCES

1. Richter, C., & and G.E.N. Kass. 1991. Oxidative stress in mitochondria: its relationship to cellular Calcium homeostasis, cell death, proliferation, and differentiation. *Chem.-Biol. Interactions* **77**, 1–23
2. Masmoudi, A., & P. Mandel. 1987. ADP-ribosyl transferase and NAD glycohydrolase activities in rat liver mitochondria. *Biochemistry* **26**, 1965–1969

3. Ziegler, M., D. Jorcke, J. Zhang, R. Schneider, H. Klocker, B. Auer, & M. Schweiger. 1996. Characterization of detergent-solubilized beef liver mitochondrial NAD glycohydrolase and its truncated hydrosoluble form. *Biochemistry,* **35,** 5207–5212

4. Barrio, J. R., J. A. Secrist III, & N. J. Leonard. 1972. A fluorescent analog of nicotinamide adenine dinucleotide. *Proc. Natl. Acad. Sci. U.S.A.,* **69,** 2039–2042

5. Zhang, J., M. Ziegler, R. Schneider, H. Klocker, B. Auer, & M. Schweiger. 1995. Identification and purification of bovine liver mitochondrial NAD glycohydrolase. *FEBS Lett.* **377,** 530–534

6. McDonald, L. J., & J. Moss. 1993. Stimulation by nitric oxide of an NAD linkage to glyceraldehyde-3-phosphate dehydrogenase. *Proc. Natl. Acad. Sci. U.S.A.* **90,** 6238- 6241

7. Graeff, R. M., T. F. Walseth, K. Fryxell, W. D. Branton, & H. C. Lee. 1994. Enzymatic synthesis and characterizations of cyclic GDP-ribose. A procedure for distinguishing enzymes with ADP-ribosyl cyclase activity. *J. Biol. Chem.* **269,** 30260–30267

8. Graeff, R. M., T. F. Walseth, H. K. Hill, & H. C. Lee. 1996. Fluorescent analogs of cyclic ADP-ribose: synthesis, spectral characterization, and use. *Biochemistry* **35,** 379–386

9. Lee, H. C. 1994. Cyclic ADP-ribose: A new member of a super family of signalling cyclic nucleotides. *Cell. Signalling* **6,** 591–600

58

CHARACTERIZATION OF HYDROSOLUBLE AND DETERGENT-SOLUBILIZED FORMS OF MITOCHONDRIAL NAD$^+$ GLYCOHYDROLASE FROM BOVINE LIVER

Dierk Jorcke, Mathias Ziegler, and Manfred Schweiger

Freie Universität Berlin
Institut f. Biochemie
Thielallee 63, 14195 Berlin
Germany

ABSTRACT

Treatment of isolated bovine liver mitochondria with either detergents or a crude pancreatic lipase, steapsin, resulted in solubilization of NAD$^+$ glycohydrolase (NADase) activity. The two forms of this enzyme can be visualized directly in SDS-polyacrylamide gels (PAGs) by a fluorescence assay utilizing 1,N^6-etheno-NAD$^+$ (ε-NAD$^+$) as substrate. Only a slight difference of about 2,000 in the apparent molecular masses was detected. Values of 28,000 and 30,000 for the steapsin- and detergent-solubilized enzyme, respectively, were estimated. The catalytic properties as well as the dependence on temperature, pH, and ionic strength were found to be similar for both forms of the enzyme. One important difference regarding their sensitivity against bivalent metal ions was observed. While the detergent-solubilized NADase was activated in the presence of, for example, Zn^{++}, and inhibited by EDTA, the truncated enzyme seemed to be unaffected under these conditions.

INTRODUCTION

NAD$^+$ glycohydrolases represent one of three classes of enzymes which catalyze the transfer of the ADP-ribose moiety from NAD$^+$ to nucleophilic acceptors. In this reaction NADases utilize water as the acceptor molecule, as opposed to the other two classes, including the poly(ADP-ribosyl)polymerase and the mono(ADP-ribosyl)transferases, which modify target proteins. The NADases described so far from mammalian origin are primarily membrane-bound and in most cases associated with the microsomal fraction or the plasma membrane [1]. Moreover, NADase activity has been detected in mitochondria from

ADP-Ribosylation in Animal Tissue, edited by Haag and Koch-Nolte
Plenum Press, New York, 1997

rat and bovine tissues [2–6]. Evidence has been presented indicating that mitochondrial NADase may play a key role in a non-enzymatic ADP-ribosylation of proteins, possibly involved in a Ca^{++} release pathway, by generating free ADP-ribose [2,4]. On the other hand, ADP-ribosylation in mitochondria may occur under conditions causing strong inhibition of the NADase [3,7].

Some NADases are able to catalyze the formation of cyclic ADP-ribose (cADPR) from NAD^+. cADPR was shown to be a potent Ca^{++}-mobilizing agent acting on internal calcium stores [8,9].

In a recent study, the bovine liver mitochondrial NADase could only be sufficiently purified by including a denaturation step [5]. To further characterize the enzyme it was highly desirable to obtain a more soluble preparation. Our approach to solubilize the mitochondrial NADase with steapsin, similarly to a microsomal NADase from calf spleen [10], yielded a hydrosoluble, fully active form of the enzyme with almost identical catalytic properties compared to the detergent-solubilized one.

MATERIALS AND METHODS

Solubilization of mitochondrial NADase with either detergent or steapsin was performed as described [6]. Purification of both solubilized forms was accomplished by four chromatographic steps, employing the following resins : hydroxyapatite, CM-Trisacryl-Cellulose, DEAE-52-Cellulose, and Cibacron Blue F3GA-Sepharose. Enzyme activity was measured continuously in a spectrofluorimeter (Perkin Elmer LS50B) or was detected directly in the gel matrix after SDS-PAGE utilizing the substrate analog ε-NAD^+ [11].

RESULTS AND DISCUSSION

In a previous report bovine liver mitochondrial NADase was purified to apparent homogeneity and was shown to be tightly membrane-bound [5]. Treatment of isolated mitochondria with steapsin, a crude pancreatic lipase, yielded a fully active, water-soluble enzyme with a somewhat reduced apparent molecular mass as compared to the detergent-solubilized one.

In order to compare and characterize both forms of the enzyme we partially purified the detergent- and steapsin-solubilized NADases by several chromatographic steps.

Incubation of SDS-PAGs with the substrate analog ε-NAD^+ following protein separation by gel electrophoresis permitted to visualize NADase activity fluorimetrically under UV-light (Fig. 1). The detergent-solubilized NADase (Fig. 1, lane 2) as well as the steapsin-solubilized form (Fig. 1, lane 1) were detected as single homogenous bands containing enzyme activity. Marking the position of the fluorescent bands and subsequent staining of the gel with Coomassie Blue revealed apparent molecular masses of 30,000 and 28,000 for the detergent- and steapsin-solubilized enzyme, respectively, by comparison with standard proteins.

Analysis of the reaction products by thin layer chromatography [12], led us to the conclusion that only a single activity utilizing NAD^+, a NAD^+ glycohydrolase, was present in both preparations [6].

Table 1 summarizes the enzymatic properties of the two preparations. The enzyme assay made use of the about tenfold fluorescence enhancement following cleavage of the substrate analog $1,N^6$-etheno-NAD^+. We determined the binding affinities towards ε-

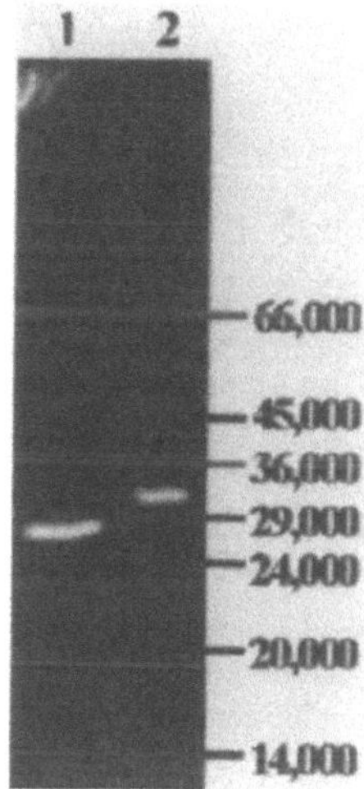

Figure 1. Direct detection of mitochondrial NADase activity after SDS-PAGE. After renaturation following electrophoresis the gel was incubated with the substrate analog ε-NAD$^+$ (150 μM). Lane 1, steapsin-solubilized NADase; lane 2, detergent-solubilized NADase. The numbers on the right indicate the positions of molecular weight standards.

NAD$^+$. Apparent K_M values of 10 and 9 μM for the detergent- and steapsin-solubilized enzyme, respectively, were calculated. Both forms were efficiently inhibited by nicotinamide (Fig. 2). At a nicotinamide concentration of 3 mM no remaining enzyme activity could be detected. Dixon plot analysis (not shown) of data from experiments as shown in Figure 2 with different substrate concentrations revealed K_I values of 128 and 150 μM for the steapsin- and detergent-solubilized forms, respectively. In addition both preparations

Table 1. Properties of mitochondrial NAD$^+$ glycohydrolase solubilized by detergent or steapsin

Solubilization by	Detergent	Steapsin
Apparent Mol. Mass	30,000	28,000
K_M (ε-NAD$^+$)	10 μM	9 μM
K_I (nicotinamide)	150 μM	128 μM
pH-Optimum	8.3	8.5
Temperature Optimum	39 °C	36 °C
__Remaining Activity at:__		
NaCl, 1 M	20 %	12 %
DTT, 10 mM	none	none
EDTA, 2 mM	5 %	97 %

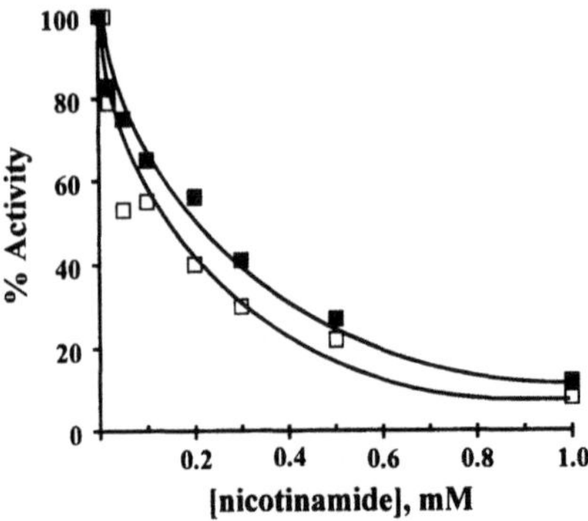

Figure 2. Inhibition of mitochondrial NADase by nicotinamide. Activities of the steapsin-solubilized NADase (■) and the detergent-solubilized NADase (□) were determined at 5 µM ε-NAD⁺ following a 5 min preincubation of the enzyme with the indicated concentration of nicotinamide.

exhibited almost the same characteristics with regard to changes in temperature, pH, and ionic strength (Table 1).

Reducing agents such as β-mercaptoethanol or dithiothreitol (DTT) led to considerable loss of activity. Preincubation of both NADases with 10 mM DTT for 5 minutes appeared to inactivate enzyme activity. Furthermore, we were unable to visualize enzyme activity directly in SDS-PAGs, if the sample buffer contained β-mercaptoethanol. Therefore, it appears that an internal disulfide bridge stabilizes the enzyme during electrophoresis or may be of importance for the catalytic activity. A marked difference between both solubilized forms was observed relating to their sensitivity towards bivalent metal ions. The detergent-treated enzyme was strongly activated by Zn^{++}, but not by up to 2.5 mM Mg^{++}, and, in addition, was inhibited in the presence of 2 mM EDTA. The steapsin-solubilized enzyme seemed to be unaffected under these conditions.

CONCLUSIONS

Treatment of mitochondria with steapsin renders the membrane-bound NADase more water-soluble, presumably by cleaving the portion attaching it to the membrane. The mechanism of the solubilization has still remained unknown. However, the observation of a single band containing activity in the fluorescence staining (Fig. 1) suggests a specific cleavage site. It was surprising that the loss of such a small part (approximately 2,000 in the molecular mass) converted the enzyme from a hydrophobic to a water-soluble protein.

It is important that the activity of the truncated form was independent of the presence of bivalent metal ions, whereas the detergent-solubilized enzyme was activated. A possible explanation for this phenomenon is that the putative membrane anchor confers metal sensitivity which may be of importance for the regulation of the enzyme.

Nicotinamide, EDTA, and DTT were potent inhibitors of the NADase activity (Table 1). With regard to the mitochondrial ADP-ribosylation it was found that under these conditions the extent of ADP-ribosylation was only slightly diminished, suggesting that the NADase may not be required for this modification.

Furthermore, there are first evidences that the mitochondrial NADase may serve as cADPR generating enzyme [7]. Therefore, further investigations are required to define the role of the mitochondrial NADase. The availability of a water-soluble protein will enable such studies under conditions that avoid complications related to the otherwise necessary use of detergents.

REFERENCES

1. Price, S. R., & P. H. Pekala. 1987. Pyridine nucleotide-linked glycohydrolases. (In: Dolphin, D., O. Avramovic, R. Poulson ; eds.; *Pyridine nucleotide coenzymes: chemical, biochemical, and medical aspects*). *Whiley-Interscience, New-York*, pp. 513–548
2. Hilz, H., R. Koch, W. Fanick, K. Klapproth, & P. Adamietz. 1984. Nonenzymic ADP-ribosylation of specific mitochondrial polypeptides. *Proc. Natl. Acad. Sci. USA 81*, 3929–3933
3. Masmoudi, A., & P. Mandel. 1987. ADP-ribosyl transferase and NAD glycohydrolase activities in rat liver mitochondria. *Biochemistry 26*, 1965–1969
4. Richter, C., & G. E. N. Kass. 1991. Oxidative stress in mitochondria: its relationship to cellular Ca^{2+} homeostasis, cell death, proliferation, and differentiation. *Chem.-Biol. Interactions 77*, 1–23
5. Zhang, J., M. Ziegler, R. Schneider, H. Klocker, B. Auer, & M. Schweiger. 1995. Identification and purification of a bovine liver mitochondrial NAD⁺-glycohydrolase. *FEBS Lett. 377*, 530–534
6. Ziegler, M., D. Jorcke, J. Zhang, R. Schneider, H. Klocker, B. Auer, & M. Schweiger. 1996. Characterization of detergent-solubilized beef liver mitochondrial NAD⁺-glycohydrolase and its truncated hydrosoluble form. *Biochemistry 35*, 5207–5212.
7. Ziegler, M., D. Jorcke, A. Herrero-Yraola, & M. Schweiger. 1996. Bovine liver mitochondrial NAD⁺ glycohydrolase: relationship to ADP-ribosylation and calcium fluxes, *this volume*.
8. Lee, H. C. 1994. Cyclic ADP-ribose: a new member of a super family of signalling cyclic nucleotides. *Cell. Signalling 6*, 591–600
9. Jacobson, M. K., J-C. Amé, W. Lin, D. L. Coyle, & E. L. Jacobson. 1995. Cyclic ADP-ribose: a new component of calcium signaling. *Receptor 5*, 43–49
10. Schuber, F., & P. Travo. 1976. Calf-spleen nicotinamide-adenine dinucleotide glycohydrolase: solubilization purification and properties of the enzyme. *Eur. J. Biochem. 65*, 247–255
11. Barrio, J. R., J. A., III, Secrist, and N. J. Leonard. 1972. A fluorescent analog of nicotinamide adenine dinucleotide. *Proc. Natl. Acad. Sci. USA. 69*, 2039–2042
12. Lötscher, H-R., K. H. Winterhalter, E. Carafoli, & C. Richter. 1980. Hydroperoxide-induced loss of pyridine nucleotides and release of calcium from rat liver mitochondria. *J. Biol.Chem. 255*, 9325–9330

59

CONTEMPLATIONS ON THE EVOLUTION OF PRO- AND EUKARYOTIC MONO(ADP-RIBOSYL)TRANSFERASES IN THE CONTEXT OF THE IMMUNE SYSTEM

Jonathan C. Howard

Institute of Genetics
University of Cologne
Zülpicher Str. 47
D-50674 Köln
Germany

The mammalian cell-surface mono-ADP-ribosyltransferase, known as RT6 in the rat in which it was first uncovered (1–3), presents us with a common problem in modern biology, namely the definition of function in a molecule found by biochemistry (in which I include serology) and not by genetics. First detected as a so-called "polymorphic marker" by classical serological investigations, it immediately joined the queue, hugely extended a little later by the advent of monoclonal antibody technology, of functionally undefined cell surface molecules whose primary interest lay at that time in their utility either as cell subset or genetic markers. It was always important to ask functional questions about markers, rather than accept that they were basically flags established for the convenience of immunologists, but the tools to do so have only gradually become available. Most of these tools have now been applied to RT6, and so far, with great respect to all who have been involved in the enterprise, the functional identity of RT6 remains enigmatic. It would have been much easier had RT6 originally turned up as a functional defect: in this sense genetics is a better approach to analyse the complexity of biological material than biochemistry. However even genetics fails when the function of interest is specified by redundant genes, unless it happens that they are very close together and disappear together in a single deletion. Since the modern analytic is to do the genetics last and one gene at a time, mouse RT6 is about to be knocked out, but we must be prepared for an initial disappointment if neither of the two mouse genes Rt6–1 and Rt6–2 alone carries the functional burden.

I want to focus first on one aspect of the RT6 problem which may provide at least a new way of looking at it, if not a solution, and this is the aspect of polymorphism. In a sense, this is Nature doing the genetics for us and its existence is full of functional implications which deserve analysis. What should we make of the existence of two distinct alleles of RT6 in the rat? If the distinction had been a couple of silent base changes or a single conservative substitution, it would be easy to dismiss it as simply mutational drift seen in process. After all, proteins and nucleic acids change with time in a conservative way, and there is no way for such changes to occur except via transient polymorphism of greater or lesser duration, but essentially no functional meaning. But the RT6 alleles differ by 12 amino acids, many of them highly non-conservative, with a further 6 silent base changes (4). This pattern of variation has all the hallmarks of selection at the protein level, some, perhaps most of the amino acid substitutions having functional meaning and being subject to selection, others perhaps including the two found in the post-translationally cleaved C-terminal peptide, as well as the silent exchanges, representing mutational drift captured in one allele or the other.

It is relatively unusual to find direct evidence for natural selection in protein structure. Most of the proteins we are familiar with were invented and perfected many hundreds, even thousands of millions of years ago, and evolution since then has consisted of maintaining the structure in the face of mutational drift. Polymorphism such as that shown by RT6 in the rat looks very different. The presence of two such distinct alleles suggests a dynamic situation in which both alleles, appropriately modified by past selection, function in a selectively complex situation in which both enjoy an advantage, either via heterozygous advantage, or through a selective pressure which continually changes direction. The multiple substitutions which distinguish the two alleles speak of ample opportunities in evolutionary time for one or the other allele to gain supremacy, yet these opportunities have not apparently been taken, and both alleles persist. The polymorphism in rat RT6 must be relatively recent since no large amount of silent substitutions or conservative changes has accumulated between the two alleles, and since a somewhat different situation, to which I shall soon turn, obtains in the mouse and man. Thus it is likely that the differential structure of the two RT6 alleles of the rat reflect present or at least recent selection on this protein.

An odd fact, to which my attention was first drawn by Geoff Butcher, is that the rat seems to collect strikingly complex polymorphisms in genes involved in pathogen resistance. RT6 is one of three such, the other two being the κ light chain of immunoglobulin (5, 6) and the TAP2 chain of the MHC-linked peptide transporter (7, 8). In all these cases the character of the polymorphism looks functionally significant, although only in the case of TAP2 has a clear functional difference and probably selective explanation also been offered (8).

RT6 is a mammalian mono-ADP-ribosylating enzyme, and a member of a family with increasingly well-understood structure-function relationships extending back to the prokaryotic world (9, 10) and Bazan (this meeting). The active site shows the extraordinary conservation over a thousand million years of evolution that is now familiar from other great families such as the ATP-binding cassette transporters. With such stability over such a long period one may predict that the recent modifications of structure implied by the RT6 polymorphism will not having anything directly to do with the enzymatic activity. And indeed it is apparent on close inspection of the available data that, with the possible exception of the Thr-Lys exchange at residue 158, the polymorphic residues, highly non-conservative as they are, are in general distant from the active site as defined by evolution and by modelling (although irritatingly one has to do the analysis oneself because no publication yet aligns the polymorphic residues to the structurally conserved motifs and putative active site). If recent evolution has been accumulating substitutions in RT6 alleles, one might guess that it relates to the specificity for substrate for ADP-ribosylation. Indeed it is difficult to see what else it can be. That indeed RT6.1 and RT6.2 may have different substrate specificities is already indicated by the fact that RT6.2 alone shows active auto-modification while both alleles are active NAD-glycohydrolases (11), although this observation may merely reflect the fact that the arginine targets for auto-modification are better distributed in RT6.2. As well-emphasized elsewhere, the identity of the natural substrates of mono-ADP-ribosylation by RT6 and its homologs is a key issue in trying to understand their physiological role (10).

RT6 is remarkably dynamic in recent periods of evolution. In the rat we have already discussed the striking dimorphism. In as close a relative as the mouse, a member of the same subfamily, the situation has already changed again, with two closely-linked active genes, Rt6–1 and Rt6–2, neither yet described as polymorphic, and both substantially modified from, and equidistant from the two rat alleles (12). And yet again this structural divergence is reflected in a modification of functional activity. Unlike either of the rat alleles and Rt6–2, Rt6–1 surprisingly has no NAD-glycohydrolase activity, while both protein products, unlike either of the rat products, are capable of ADP-ribosylating a (presumably unphysiological) exogenous substrate (10). As if further to emphasize the lability of the RT6 function, the RT6 gene of humans appears to be degraded (13), and if the RT6 function is to be expressed by human lymphocytes it must be through a different, albeit related, molecule.

If RT6 is being thrown so violently around during relatively recent periods of mammalian evolution, does this mean that its function is dispensable, or alternatively that it is exposed to especially violent forces of selection? I have already partially answered this question in the context of the rat alleles, which carry all the hallmarks of selection rather than drift, and there is no reason to change that view for the mouse. But the loss of the gene altogether, as in man, may suggest that here things really have changed, and the selective forces which pushed RT6 evolution in the murine rodents have finally spent themselves during the evolution of the human lineage. My guess is quite other. The situation perhaps resembles that in the non-classical class I genes of the MHC, where one may search in vain in man for the true homolog of H-2M3, a non-classical mouse class I gene structurally adapted for (14) and functionally active in (15) the presentation of short exogenous peptides carrying an N-terminal f-Met, normally presumably derived from pathogenic bacteria (16). It is impossible to believe that such a function is irrelevant to man. More plausibly, it is served by a different and distantly related molecule of the antigen presenting system, perhaps a member of the CD1 family. Likewise, human CD1b can present glycolipid epitopes of bacterial origin to T cells (17) while mice lack a close CD1b homolog. It is conceivable that this function may be served in mice by H-2M3 (16, 18). Thus in these interesting cases there is no reason to believe that the selective pressure goes away, rather that a different gene product picks up the responsibility of dealing with it. Perhaps the most striking of all such transferred functions in the man-mouse relationship is the case of the NK cell receptor for MHC class I molecules, where, so far as is presently known, one gene family (Ly-49) provides the specificity in mouse and an essentially unrelated one (KIR) provides the specificity in man (reviewed in (19, 20)). Whether, as some believe, this situation is due for a revision with the discovery of the Ly-49 family in man and the KIR family in mouse remains to be seen, but at present it is still arguable that NK receptor function has been allocated to different protein families during relatively recent evolution. It may be surprising to imagine the NK receptor under severe pressure from natural selection, but a recent analysis of the relationship between the human cytomegalovirus and its host has made such a scenario highly plasuible. HCMV plays some remarkable tricks to undermine the classical immune system. The US11 protein manages to persuade recently synthesized class I molecules to return, probably backwards down the translocon, to the cytoplasm where they are rapidly degraded by the proteasome (21). Another still uncharacterized protein further limits class I expression at the cell surface (22). Eliminating class I expression from the infected cell seems to be a good thing because viral protein synthesis is thereby rendered invisible to the host immune system. But life as a virus is not so simple in the presence of the mammalian immune system, because NK cells are specifically charged with destroying cells with unnaturally low expression of peptide-loaded class I MHC molecules (23, 24). An extraordinary viral response to this problem has now been attributed to HCMV. It has been known for many years that HCMV codes for a class I-like protein, UL18 (25), which binds host β2-microglobulin. UL18 has now been shown to bind peptides apparently similar in character to those bound by a human class I molecule and it is proposed that HCMV synthesises peptide-loaded UL18 as a decoy designed to persuade NK cells falsely that all is well inside (26). I do not propose to analyse the implications of this model any further, and it may well be wrong (how, for example, does UL18 avoid being an antigen in its own right?), but it suffices to indicate that when we are dealing with the battle to the death between fast-evolving pathogens and their slowly evolving, especially warm-blooded, hosts, there is esentially nothing in point of the bizarre, the baroque and the byzantine which is not already in place.

There are still three other arguments or suggestions which each helps in its way to support the thesis that RT6, although it is a cell surface mono-ADP-ribosyltransferase, is critically involved in pathogen resistance. Most trivial of these, perhaps, is the simple fact that RT6 is expressed in the immune system only. Maybe this is no more interesting than saying that PKCβ is critically involved in pathogen resistance because it is expressed only in the immune system, and when knocked out has only an immunological phenotype (27). The function of kinases in immune cell regulation may be indispensable for immune resistance, and their deployment characteristic of the differentiation state of lymphocytes, but they are not directly part of the host parasite-interface and they do not evolve especially fast or strangely. This is, however, not an entirely trivial comparison because it is valid to compare the function of protein modification by ADP-ribosylation with protein modification by phosphorylation, and many of the same considerations such as functional modification and reversibility may well apply. Other mono-ADP-ribosyltransferases may function in other tissues (28), just as other PKCs do, but the fact that RT6 functions on immune cells is certainly in favour of it having a function limited to the immune system. And the fact that it sticks out rather than in, however inconvenient it may be for the provision of NADH, is hard to explain unless the substrate is extracellular as well, where most pathogens live most of the time.

A second argument for relating RT6 expression to the host-pathogen relationship, and again a highly circumstantial one, is the presence of a well-organized IRF-1 site in the promoter at -55/-43, as well as a (somewhat distant) NF-kB site at -716/-707 (29). The combination of IRF-1 and NF-kB sites, associated with synergistic induction by IFN-γ and TNF-α is highly suggestive of a gene involved in some way or other in the host-pathogen relationship. No direct studies of the induction of RT6 by these cytokines have been reported, and such an analysis would obviously be of great interest.

A third argument, albeit circumstantial, which I personally find peculiarly convincing in support of the thesis that RT6 is a direct agent of pathogen resistance, is that it is massively over-expressed at what is categorically the main pathogen barrier, namely the gut (30). Here it is expressed on intra-epithelial lymphocytes in a most interesting and unexpected way, namely as a context or microenvironmental marker showing essentially indistinguishable high expression on $\alpha\beta$ TCR+ or TCR-, on CD4+, CD8+, or double positive or double negative populations (31). Even in *nu/nu* rats whose other peripheral populations are RT6 negative, all IEL are strongly RT6 positive, independently of the expression even of CD3 (31). These results strongly suggest that high RT6 expression is induced in the gut microenvironment, and that all lymphocytes of the T cell lineage, including $\alpha\beta$ and $\gamma\delta$ cells and NK cells, are equipped with the ability to respond to the RT6-inductive stimulus in this site. If one was looking for a protein target or a function for mono-ADP-ribosylation by RT6, surely after these results one would start in the gut. Is it detoxifying for microbial toxins in the gut environment? If so, why have it on lymphocytes rather than gut epithelial cells? Is it intended to mark or modify the surfaces of gut bacteria that trespass into the epithelial lining, and if so, what are the consequences for the bacteria? If bacterial products or surfaces are ADP-ribosylated, is there a receptor for ADP-ribosylated proteins or particles somewhere? It is never too early to speculate. Why do we speculate about such a direct interaction between RT6 and pathogens and their products? After all, surface proteins involved in lymphocyte homing to the gut are also induced in the gut microenvironment but it is not proposed that they contribute directly to the host-pathogen interface. The reason is that, like PKCβ, adhesion molecules and their ligands do not evolve in an interesting way, and this is the key to the function of RT6.

ACKNOWLEDGMENTS

I am extremely grateful to Drs Fritz Koch-Nolte, Friedrich Haag and Prof. Dr. Heinz-Günther Thiele for sharing their interest in RT6 with me, and for many years of friendship.

REFERENCES

1. Miller, C., and C.W. DeWitt. 1973. Cellular and humoral responses to major and minor histocompatibility antigens. *Transplant. Proc.* 5:303.
2. Lubaroff, D.M. 1973. An alloantigenic marker on rat thymus-derived and marrow-derived cells. *Transplant. Proc.* 5:115.
3. Howard, J.C., and D.W. Scott. 1974. The identification of sera distinguishing marrow-derived and thymus-derived lymphocytes in the rat thoracic duct. *Immunology* 27:903–922.
4. Haag, F., F. Koch, and H.-G. Thiele. 1990. Polymorphism between rat T-cell alloantigens RT6.1 and RT6.2 is based on multiple amino acid substitutions. *Transplant Proc.* 22:2541–2542.
5. Sheppard, H.W., and G.A. Gutman. 1981. Allelic forms of rat k chain genes: evidence for strong selection at the level of nucleotide sequence. *Proc. Natl. Acad. Sci. USA* 78:7064–7068.
6. Frank, M.B., R.M. Besta, P.R. Baverstock, and G.A. Gutman. 1987. The structure and evolution of immunoglobulin kappa chain constant region genes in the genus *rattus*. *M_olecular Imunology* 24:953–961.
7. Powis, S.J., E.V. Deverson, W.J. Coadwell, A. Ciruela, N.S. Huskisson, H. Smith, G.W. Butcher, and J.C. Howard. 1992. Effect of polymorphism of an MHC-linked transporter on the peptides assembled in a class I molecule. *Nature* 357:211–215.
8. Powis, S.J., L.L. Young, E. Joly, P.J. Barker, L. Richardson, C.J. Thorpe, P.J. Travers, R.P. Brandt, C.J. Melief, J.C. Howard, and G.W. Butcher. 1996. The rat *cim* effect: TAP allele dependent changes in a class I MHC anchor motif and evidence against C-terminal trimming of peptides in the ER. *Immunity* 4:159–165.
9. Domenighini, M., C. Magagnoli, M. Pizza, and R. Rappuoli. 1994. Common features of the NAD-binding and catalytic site of ADP-ribosylating toxins. *Mol. Microbiol.* 14:41–50.
10. Koch-Nolte, F., D. Petersen, S. Balasubramanian, F. Haag, D. Kahlke, T. Willer, R. Kastelein, F. Bazan, and H.-G. Thiele. 1996. Mouse T cell membarne proteins Rt6–1 and Rt6–2 are arginine/protein mono(ADP-Pribosyl)transferases and share secondary structure motifs with ADP-ribosylating bacterial toxins. *J. Biol. Chem.* 271:7686–7693.

11. Haag, F., V. Andresen, S. Karsten, F. Koch-Nolte, and H.-G. Thiele. 1995. Both allelic forms of the rat T cell differentiation marker RT6 display nicotinamide adenine dinucleotide (NAD) -glycohydrolase activity, yet only RT6.2 is capable of automodification upon incubation with NAD. *Eur. J. Immunol.* in press.

12. Hollmann, C., F. Haag, M. Schlott, A. Damaske, H. Bertuleit, M. Matthes, M. Kühl, H.-G. Thiele, and F. Koch-Nolte. 1996. Molecular characterization of mouse T-cell ecto-ADP-ribosyltransferase RT6. Cloning of a second functional gene and identification of the RT6 gene products. *Mol. Immunol.* 33:807–817.

13. Haag, F., F. Koch-Nolte, M. Kühl, S. Lorenzen, and H.-G. Thiele. 1994. Premature stop codons inactivate the RT6 genes of the human and chimpanzee species. *J. Mol. Biol.* 243:537.

14. Wang, C.R., A.R. Castano, P.A. Peterson, C. Slaughter, K.F. Lindahl, and J. Deisenhofer. 1995. Nonclassical binding of formylated peptide in crystal structure of the MHC class Ib molecule H2-M3. *Cell* 82:655–64.

15. Lindahl, K.F., V.M. Dabhi, R. Hovik, G.P. Smith, and C.R. Wang. 1995. Presentation of N-formylated peptides by H2-M3. *Biochem Soc Trans* 23:669–74.

16. Lenz, L.L., B. Dere, and M.J. Bevan. 1996. Identification of an H-2-M3-restricted Listeria epitope: implications for antigen presentation by M3. *Immunity* 5:63–72.

17. Sieling, P.A., D. Chatterjee, S.A. Porcelli, T.I. Prigozy, R.J. Mazzaccaro, T. Soriano, B.R. Bloom, M.B. Brenner, M. Kronenberg, P.J. Brennan, and et al. 1995. CD1-restricted T cell recognition of microbial lipoglycan antigens. *Science* 269:227–30.

18. Nataraj, C., M.L. Brown, R.M. Poston, S.M. Shawar, R.R. Rich, K.F. Lindahl, and R.J. Kurlander. 1996. H2-M3(wt)-restricted, listeria monocytogenes-specific CD8 T cells recognize a novel, hydrophobic, protease-resistant, periodate-sensitive antigen. *International Immunology* 8:367–378.

19. Gumperz, J.E., and P. Parham. 1995. The enigma of the natural killer cell. *Nature* 378:245–248.

20. Lanier, L. 1996. NK cell receptors and MHC class I interactions. *Curr. Opinion in Immunol.* in press.

21. Wiertz, E.J.H., T.R. Jones, L. Sun, M. Bogyo, H.J. Geuze, and H.L. Ploegh. 1996. The human cytomegalovirus US11 gene product dislocates MHC class I heavy chains from the Endoplasmic Reticulum to the cytosol. *Cell* 84:769–779.

22. Koszinowski, U. 1996. Emptying Pandora's box. *Trends in Microbiol.* 4:338–339.

23. Ljunggren, H.G., and K. Karre. 1990. In search of the 'missing self': MHC molecules and NK cell recognition. *Immunol Today* 11:237–44.

24. Correa, I., and D.H. Raulet. 1995. Binding of diverse peptides to MHC class I molecules inhibits target cell lysis by activated natural killer cells. *Immunity* 2:61–71.

25. Beck, S., and B.G. Barrell. 1988. Human cytomegalovirus encodes a glycoprotein homologous to MHC class-I antigens. *Nature* 331:269–72.

26. Fahnestock, M.L., J.L. Johnson, R.M. Feldman, J.M. Neveu, W.S. Lane, and P.J. Bjorkman. 1995. The MHC class I homolog encoded by human cytomegalovirus binds endogenous peptides. *Immunity* 3:583–90.

27. Leitges, M., C. Schmedt, R. Guinamard, J. Davoust, S. Schaal, S. Stabel, and A. Tarakhovsky. 1996. Immunodeficiency in protein-kinase C-beta-deficient mice. *Science* 273:788–791.

28. Koch-Nolte, F., F. Haag, R. Braren, M. Kühl, J. Hoovers, S. Balasubramanian, F. Bazan, and H.-G. Thiele. 1996. Two novel members of an emerging mammalian gene family related to ADP-ribosylating bacterial toxins.

29. Haag, F., G. Kuhlenbäumer, F. Koch-Nolte, E. Wingender, and H.-G. Thiele. 1996. Structure of the gene encoding the rat T cell ecto-ADP-ribosyltransferase RT6.2.

30. Fangmann, J., R. Schwinzer, and K. Wonigeit. 1991. Unusual phenotype of intestinal intraepithelial lymphocytes in the rat: predominance of T cell receptor alpha/beta+/CD2- cells and high expression of the RT6 alloantigen. *Eur J Immunol* 21:753–60.

31. Fangmann, J., R. Schwinzer, H.J. Hedrich, and K. Wonigeit. 1993. Demonstration of RT6 expression on a CD3- population of intestinal intraepithelial lymphocytes of athymic nude rats. *Transplant Proc* 25:2789–90.

APPENDIX

THE VERTEBRATE GENE FAMILY OF MONO(ADP-RIBOSYL)TRANSFERASES

Proposal for a Unified Nomenclature

Friedrich Haag and Friedrich Koch-Nolte

Department of Immunology
University Hospital
D-20246 Hamburg, Germany

A panel of experts including the principal investigators that have reported the cloning of vertebrate mono(ADP-ribosyl)transferase genes[*] met at the workshop in Hamburg to draft a proposal for a unified nomenclature for members of this emerging gene family.

Consensus was reached that the gene family symbol " *ART* " for mono(*ADP-ribo-syl)transferase* as proposed by the nomenclature committees of the human and mouse genome projects is appropriate. Following established conventions "*Art*" (i.e. lower case letters for second and third position) should be used for mouse genes and "*ART*" (i.e. all capital letters) for other species. Italics should be used when referring to genes or gene transcripts, normal case when referring to proteins.

To date six apparent members of the gene family have been cloned from the mouse and three from the chicken (Refs 1–16, see also Chapters 13–19, 39, this volume). Most of the mouse genes have also been identified in the human. It was agreed that these genes should be numbered sequentially as indicated in Table I. Figure 1 shows the aligned amino acid sequences of representative gene products. Although the chicken enzymes have the highest degree of sequence identity to the mammalian muscle enzyme, it is unlikely that they constitute the direct species homologues of the muscle transferase. Thus, they have been assigned different numbers.

---

[*] Fernando Bazan, Palo Alto; Gunther Dennert, Los Angeles; Georges Guellaen, Paris; Friedrich Haag, Hamburg; Friedrich Koch-Nolte, Hamburg; Edward Leiter, Bar Harbor; Joel Moss, Bethesda; Sydney Shall, Brighton Falmer; Makoto Shimoyama, Izumo; Heinz-Günter Thiele, Hamburg; Mikako Tsuchiya, Izumo; Anna Zolkiewska, Bethesda

ADP-Ribosylation in Animal Tissue, edited by Haag and Koch-Nolte
Plenum Press, New York, 1997

Figure 1. Alignment of the deduced amino acid sequences of representative ART gene family members. Alignment was performed with the DNA-star Software on a Macintosh personal computer. Not included are sequences for N- and C-terminal signal peptides as these do not show any significant sequence identities. Sequences are as deposited in the public database (see also Table I): ART1: S74683, Art2a: X52991, Art2b: X87612, ART3: U47054, ART4: X95826, Art5: Y08028, ART6A: D31864, ART6B: D31865, ART7: X82397.

Distinct genes, which have arisen by an evidently recent gene duplication event (i.e., a gene that is restricted to a particular locus in a specific mammalian lineage) are given the same number followed by a letter on the same line, e.g., mouse *Art2a* and *Art2b*, chicken *ART6A* and *ART6B*. Alleles are indicated by superscript letters, e.g. mouse *Art2a[a]*, *Art2a[b]*, and *Art2a[c]*. The letter "P" is appended to the name of non-expressed pseudogenes (e.g., human *ART2P*).

New family members should be numbered chronologically in the order that their primary structures are determined and deposited in the public database.

When established other names for gene family members are used in the future, at least one crossreference to the new *ART*-gene family number shown in Table I should be given (e.g. *RT6 = ART2*).

Joel Moss at the NIH, Bethesda, agreed to serve the community as a score keeper for the numbering of novel *ART* genes.

Fernando Bazan at the DNAX Institute, Palo Alto, offered to set up a world-wide-web site for the community to facilitate communication and the exchange of news and reagents. The site will provide access to the sequences of published gene-family members. The sequence of a presumptive novel gene family member can be compared to the pub-

Table 1. Members of the emerging ART-gene family

gene name	other names	accession number (species)	year & source of database deposition	published references	chapters in this volume
ART1	RABNAART	M98764 (rabbit)	93 Bethesda	1-5	13, 15, 19, 39
		S74683 (human)	94 Bethesda		
	YAC1	U31510 (mouse)	95 Bethesda		
		X95842 (mouse)	96 Los Angeles		
	MART	X95825 (mouse)	96 Hamburg		
		C03716 (human EST)	96 Tokyo		
		W08722 (mouse EST)	96 IMAGE		
ART2	RT6	M85193 (rat, allele B)	89 Hamburg	6-10	13, 15, 19-23
		M31138 (rat, allele A)	90 Hamburg		
Art2a		X52991 (mouse, gene 1)	90 Hamburg		
		X65050 (human pseudogene)	92 Hamburg		
Art2b		X87612 (mouse, gene 2)	95 Hamburg		
		Y08030 (rabbit)	96 Hamburg		
ART3		R35364 (human EST)	95 IMAGE	11-13	14, 19
	HTMART	U47054 (human)	96 Paris		
	TART1	X95827 (human)	96 Hamburg		
		Y08027 (mouse)	96 Hamburg		
ART4	LART	T70606 (human EST)	95 IMAGE	13	19
		X95826 (human)	96 Hamburg		
		Y08029 (rabbit)	96 Hamburg		
		Y08300 (mouse)	96 Hamburg		
ART5	YAC2	U60881 (mouse)	96 Bethesda	14	15, 19
	TART2	Y08028 (mouse)	96 Hamburg		
		W12489 (mouse EST)	96 IMAGE		
ART6A	CHAT1	D31864 (chicken)	94 Izumo	15	16
ART6B	CHAT2	D31865 (chicken)	94 Izumo		
ART7	CEAT	X82397 (chicken)	95 Brighton	16	17

lished sequences and the results of these analyses can be deposited at the site until the sequence itself is released.

The panel will meet again at appropriate occasions to update and amend the family nomenclature.

REFERENCES

1. Zolkiewska, A., M. S. Nightingale & J. Moss. 1992. Molecular characterization of NAD:arginine ADP-ribosyltransferase from rabbit skeletal muscle. *Proc. Natl. Acad. Sci. USA 89*: 11352–11356.
2. Okazaki, I. J., A. Zolkiewska, M. S. Nightingale & J. Moss. 1994. Immunological and structural conservation of mammalian skeletal muscle glycosylphosphatidylinositol-linked ADP-ribosyltransferases. *Biochemistry 33*: 12828–12836.
3. Okazaki, I. J., H.-J. Kim & J. Moss. 1996. Molecular characterization of a glycosylphosphatidylinositol-linked ADP-ribosyltransferase from lymphocytes. *Blood 88*: 915–921.
4. Koch-Nolte, F., M. Kühl, F. Haag, M. Cetkovich-Cvrlje, E. H. Leiter & H. G. Thiele. 1996. Assignment of the human and mouse genes for muscle ecto-mono(ADP-ribosyl)transferase to a conserved linkage group on human Chromosme 11p15 and mouse Chromosome 7. *Genomics 36*: 215–216.
5. Tanaka, T., A. Ogiwara, I. Uchiyama, T. Takagi, Y. Yazaki & Y. Nakamura. 1996. Construction of a Normalized Directionally Cloned cDNA Library from Adult Heart and Analysis of 3040 Clones by Partial Sequencing. *Genomics 35*: 231–235.
6. Koch, F., F. Haag, A. Kashan & H. G. Thiele. 1990. Primary structure of rat RT6.2, a nonglycosylated phosphatidylinositol-linked surface marker of postthymic T cells. *Proc. Natl. Acad. Sci. USA 87*: 964–967.
7. Haag, F., F. Koch & H. G. Thiele. 1990. Nucleotide and deduced amino acid sequence of the rat T-cell alloantigen RT6.1. *Nucleic Acids Res. 18*: 1047.
8. Koch, F., F. Haag & H. G. Thiele. 1990. Nucleotide and deduced amino acid sequence for the mouse homologue of the rat T-cell differentiation marker RT6. *Nucleic Acids Res. 18*: 3636.

9. Haag, F., F. Koch-Nolte, M. Kühl, S. Lorenzen & H. G. Thiele. 1994. Premature stop codons inactivate the *RT6* genes of the human and chimpanzee species. *J. Mol. Biol. 243*: 537–546.

10. Hollmann, C., F. Haag, M. Schlott, A. Damaske, H. Bertuleit, M. Matthes, M. Kühl, H. G. Thiele & F. Koch-Nolte. 1996. Molecular characterization of mouse T-cell ecto-ADP-ribosyltransferase Rt6: cloning of a second functional gene and identification of the Rt6 gene products. *Mol. Immunol. 33*: 807–817.

11. Pawlak, A., C. Toussaint, I. Levy, F. Bulle, M. Poyard, R. Barouki & G. Guellaen. 1995. Characterization of a large population of mRNAs from human testis. *Genomics 26*: 151–8.

12. Lévy, I., Y. Q. Wu, N. Roeckel, F. Bulle, A. Pawlak, S. Siegrist, M. G. Mattéi & G. Guellaen. 1996. Human testis specifically expresses a homologue of the rodent T lymphocytes RT6 mRNA. *FEBS Lett. 382*: 276–280.

13. Koch-Nolte, F., F. Haag, R. Braren, M. Kühl, J. Hoovers, S. Balasubramanian, F. Bazan & H. G. Thiele. 1996. Two novel human members of an emerging mammalian gene family related to mono-ADP-ribosylating bacterial toxins. *Genomics* in press.

14. Okazaki, I. J., H—J.Kim & J. Moss. 1996. Cloning and characterization of a novel membrane-associated lymphocyte NAD:arginine ADP-ribosyltransferase. *J. Biol. Chem. 271*: 22052–22057.

15. Tsuchiya, M., N. Hara, K. Yamada, H. Osago & M. Shimoyama. 1994. Cloning and expression of cDNA for arginine-specific ADP-ribosyltransferase from chicken bone marrow cells. *J. Biol. Chem. 269*: 27451–27457.

16. Davis, T. & S. Shall. 1995. Sequence of a chicken erythroblast mono(ADP-ribosyl)transferase-encoding gene and its upstream region. *Gene 164*: 371–372.

INDEX

NAD-glycohydrolases, *see also* DRAG
  of bovine liver mitochondria, 444, 448
  of bovine spleen, 399–402
  of canine spleen, 384
  cysteine dependent, 276
  of Neurospora crassa condia, 389–392
  of rat liver, 345
  relation to mono(ADP-ribosyl)transferases and
      ADP-ribosyl cyclases, 8, 192, 401
Natural killer cells, 235–238, 258
Neuromodulin: *see* B50/GAP43
Neurons, effect of mADPRT inhibitors on, 4
Neurospora crassa NAD glycohydrolase,
  dimer/monomer conversion, 391
  immobilization of, 390–393
NGD, Nicotinamide guanine dinucleotide as substrate
      of ADP-ribosyl cyclases and NADase,
      400–403, 412, 414, 445
Nicotinamide
  as inhibitor of ADP-ribosylation, 205, 251
  as inhibitor of NAD-glccohydrolysis, 450
  effect on phosphorylation, 251, 252
Nicotinamide adenine dinucleotide: *see* NAD
Nicotinamide guanine dinucleotide: *see* NGD
Nitric oxide, 203–206, 261, 439
Nitric oxide synthetase
  as mediator of macrophage cytotoxicity, 203
  inhibition of induction by ADP-ribosylation inhibi-
    tors, 206, 251
Nitroprusside, 159
NK cells, 235–238, 258
Northern blot analysis
  of ADP-ribosylarginine hydrolase, 30
  of lymphatic tissue mono(ADP-ribosyl)transferase,
    164
  of mono(ADP-ribosyl)transferases, 6
  of muscle mono(ADP-ribosyl)transferase, 299
  of RT6, 272
  of testis mono(ADP-ribosyl)transferase, 125
  of Yac-1 mono(ADP-ribosyl)transferase, 132
Novobiocin: *see* Inhibitors
Nucleotide phosphodiesterase. PDE, 31, 199, 243, 325
  association with Golgi complex, 344, 345
  processing of ADP-ribosylated integrin by, 302
  requirement for divalent cations, 345
  requirement for guanine nucleotides, 345
Nude rats, 234, 235
NZW mice, deletion of Rt6–2 gene in, 272–274

Oxidative stress, 385, 386, 443

p32 of CHO cells, mono ADP-ribosylation of, 85–86
p33
  of chicken heterophils, 130
  of human macrophages, mono ADP-ribosylation of,
    250–251
p40 of mouse cyotoxic T cells, mono ADP-ribosyla-
    tion of, 5, 27, 130, 131, 198, 200

$p56^{lck}$, 27, 130, 196–198, 200, 367, 434
$p59^{fyn}$, 198, 367
$p72^{syk}$, 425
Palmitoylation of detergent resistant domains, 359
PARP: *see* Poly ADP-ribose polymerase
PDE: *see* Nucleotide phosphodiesterase
Pentosidine, 373, 374
  cross-linking of proteins by, 374, 375
  structure of, 374
Peripheral nervous system, endogenous ADP-ribosyla-
    tion in, 289
Pertussis toxin, 4, 31, 35, 40, 50, 83–86, 87–90, 187, 275
Phage mono(ADP-ribosyl)transferases
  of Corynebacterium diphtheriae, 36
  of E. coli, 3, 71–75; *see also* Bacteriophage T4
    transferase
Phosphatidyl-inositol phosphate kinase, 53, 280, 350
Phospho-cyclic ADP-ribose, P-cADPR, 382–386,
    432–434
Phosphodiesterase: *see* Nucleotide, phosphodiesterase
Phosphoinositide-3-kinase, 53, 351, 425, 428
Phospholipase A_2, 351
Phospholipase C, 94, 326
Phospholipase D, 53, 350
  activation by ARF, 318
Phosphorylation, effect of nicotinamide on, 251, 252
Phycomyces blakesleeanus mono(ADP-ribosyl)trans-
    ferase, 156–161
  effects of $MgCl_2$ on activity, 159
  effects of inhibitors on activity, 160
  purification of, 157
Phylogenetic tree: *see* Evolutionary tree
Plant mono(ADP-ribosyl)transferases, 9
PMNs, Polymorphonuclear neutrophil leukocytes,
    241–244
Poly ADP-ribose polymerase, PARP, 344
  alignment of catalytic core with mono(ADP-ribo-
    syl)transferases, 102
  cloning of, 8
  crystal structure of, 8
  knock out mice, 19
  role in DNA excision repair, 18
Polymorphonuclear neutrophil leukocytes, PMNs,
    241–244
Potassium channels, regulation by cADPR and ADPR,
    437–440
Proliferative response
  of mouse cytotoxic T-cells, 131, 173, 194
  of rat T-cells, 172, 173, 230
Promoter
  of chicken erythroblast mADPRT, 149
  of RT6, 110, 238, 265–267, 456
  T4 bacteriophage early, 73
Protein kinase C, 350, 456
Pseudomonas aeruginosa exotoxin A, 4, 35, 40, 187
Pseudomonas aeruginosa exotoxin S, 187
PT: *see* Pertussis toxin
PTK1 cell line, 339